Panagiotis Chountas, Ilias Petrounias and Janusz Kacprzyk (Eds.)

Intelligent Techniques and Tools for Novel System Architectures

Studies in Computational Intelligence, Volume 109

Editor-in-chief
Prof. Janusz Kacprzyk
Systems Research Institute
Polish Academy of Sciences
ul. Newelska 6
01-447 Warsaw
Poland
E-mail: kacprzyk@ibspan.waw.pl

Dr. Panagiotis Chountas
Harrow School of Computer Science
The University of Westminster
Watford Road
Northwick Park
London HA1 3TP
UK
chountp@wmin.ac.uk

Dr. Ilias Petrounias
School of Informatics
The University of Manchester
Oxford Road
Manchester M13 9PL
UK
Ilias.Petrounias@manchester.ac.uk

Prof. Janusz Kacprzyk
Systems Research Institute
Polish Academy of Sciences
Ul. Newelska 6
01-447 Warsaw
Poland
kacprzyk@ibspan.waw.pl

ISBN 978-3-540-77621-5   e-ISBN 978-3-540-77623-9

Studies in Computational Intelligence ISSN 1860-949X

Library of Congress Control Number: 2008920251

Cover design: Deblik, Berlin, Germany

Printed on acid-free paper

9 8 7 6 5 4 3 2 1

springer.com

Panagiotis Chountas
Ilias Petrounias
Janusz Kacprzyk
(Eds.)

Intelligent Techniques and Tools for Novel System Architectures

With 192 Figures and 89 Tables

 Springer

Foreword

The purpose of this volume is to foster and present new directions and solutions in broadly perceived intelligent systems. The emphasis is on constructive approaches that can be of utmost important for a further progress and implementability.

The volume is focused around a crucial prerequisite for developing and implementing intelligent systems, namely to computationally represent and manipulate knowledge (both theory and information), augmented by an ability to operationally deal with large-scale knowledge bases, complex forms of situation assessment, sophisticated value-based modes of reasoning, and autonomic and autonomous system behaviours.

These challenges exceed the capabilities and performance capacity of current open standards, approaches to knowledge representation, management and system architectures. The intention of the editors and contributors of this volume is to present tools and techniques that can help in filling this gap.

New system architectures must be devised in response to the needs of exhibiting intelligent behaviour, cooperate with users and other systems in problem solving, discovery, access, retrieval and manipulation of a wide variety of "data" and knowledge, and reason under uncertainty in the context of a knowledge-based economy and society.

This volume provides a source wherein academics, researchers, and practitioners may derive high-quality, original and state-of-the-art papers describing theoretical aspects, systems architectures, analysis and design tools and techniques, and implementation experiences in intelligent systems where information and knowledge management should be mainly characterised as a *net-centric infrastructure riding on the fifth wave of "distributed intelligence."*

An urgent need for editing such a volume has occurred as a result of vivid discussions and presentations at the "IEEE-IS' 2006 – The 2006 Third International IEEE Conference on Intelligent Systems" held in London, UK, at the University of Westminster in the beginning of September, 2006. They have triggered our editorial efforts to collect many valuable inspiring works written by both conference participants and other experts in this new and challenging field.

LONDON   *P. Chountas*
2007   *I. Petrounias*
                                                         J. Kacprzyk

Contents

Part V Knowledge Discovery and Management

Part VI Intuitonistic Fuzzy Sets and Systems

Part I

Intelligent-Enterprises and Service Orchestration

Applying Data Mining Algorithms to Calculate the Quality of Service of Workflow Processes

Jorge Cardoso

Department of Mathematics and Engineering, 9000-390 Funchal, Portugal
jcardoso@uma.pt

Summary. Organizations have been aware of the importance of Quality of Service (QoS) for competitiveness for some time. It has been widely recognized that workflow systems are a suitable solution for managing the QoS of processes and workflows. The correct management of the QoS of workflows allows for organizations to increase customer satisfaction, reduce internal costs, and increase added value services. In this chapter we show a novel method, composed of several phases, describing how organizations can apply data mining algorithms to predict the QoS for their running workflow instances. Our method has been validated using experimentation by applying different data mining algorithms to predict the QoS of workflow.

1 Introduction

The increasingly global economy requires advanced information systems. Business Process Management Systems (BPMS) provide a fundamental infrastructure to define and manage several types of business processes. BPMS, such as Workflow Management Systems (WfMS), have become a serious competitive factor for many organizations that are increasingly faced with the challenge of managing e-business applications, workflows, Web services, and Web processes. WfMS allow organizations to streamline and automate business processes and re-engineer their structure; in addition, they increase efficiency and reduce costs.

One important requirement for BMPS and WfMS is the ability to manage the Quality of Service (QoS) of processes and workflows [1]. The design and composition of processes cannot be undertaken while ignoring the importance of QoS measurements. Appropriate control of quality leads to the creation of quality products and services; these, in turn, fulfill customer expectations and achieve customer satisfaction. It is not sufficient to just describe the logical or operational functionality of activities and workflows. Rather, design of workflows must include QoS specifications, such as response time, reliability, cost, and so forth.

J. Cardoso: *Applying Data Mining Algorithms to Calculate the Quality of Service of Workflow Processes*, Studies in Computational Intelligence (SCI) **109**, 3–18 (2008)
www.springerlink.com   © Springer-Verlag Berlin Heidelberg 2008

One important activity, under the umbrella of QoS management, is the prediction of the QoS of workflows. Several approaches can be identified to predict the QoS of workflows before they are invoked or during their execution, including statistical algorithms [1], simulation [2], and data mining based methods [3, 4].

The latter approach, which uses data mining methods to predict the QoS of workflows, has received significant attention and has been associated with a recent new area coined as Business Process Intelligence (BPI). In this paper, we investigate the enhancements that can be made to previous work on BPI and business process quality to develop more accurate prediction methods.

The methods presented in [3, 4] can be extended and refined to provide a more flexible approach to predict the QoS of workflows. Namely, we intend to identify the following limitations that we will be addressing in this paper with practical solutions and empirical testing:

1. In contrast to [4], we carry out QoS prediction based on path mining and by creating a QoS activity model for each workflow activity. This combination increases the accuracy of workflow QoS prediction.
2. In [4], time prediction is limited since workflow instances can only be classified to "have" or "not to have" a certain behavior. In practice, it means that it is only possible to determine that a workflow instance will have, for example, the "last more than 15 days" behavior or will not have that behavior. This is insufficient since it does not give an actual estimate for the time a workflow will need for its execution. Our method is able to deduce that a workflow wi will probably take 5 days and 35 min to be completed with a prediction accuracy of 78%.
3. In [4], the prediction of the QoS of a workflow is done using decision trees. We will show that MultiBoost Naïve Bayes outperforms the use of decision trees to predict the QoS of a workflow.

This chapter is structured as follows: In Sect. 2, we present our method of carrying out QoS mining based on path mining, QoS activity models, and workflow QoS estimation. Section 3 describes the set of experiments that we have carried out to validate the QoS mining method we propose. Section 4 presents the related work in this area. Finally, Sect. 5 presents our conclusions.

2 Motivation

Nowadays, a considerable number of organizations are adopting workflow management systems to support their business processes. The current systems available manage the execution of workflow instances without any quality of service management on important parameters such as delivery deadlines, reliability, and cost of service.

Let us assume that a workflow is started to deliver a particular service to a customer. It would be helpful for the organization supplying the service to be able to predict how long the workflow instance will take to be completed or the cost associated with its execution. Since workflows are non-deterministic and concurrent, the time it takes for a workflow to be completed and its cost depends not only on which activities are invoked during the execution of the workflow instance, but also depends on the time/cost of its activities. Predicting the QoS that a workflow instance will exhibit at runtime is a challenge because a workflow schema w can be used to generated n instances, and several instances w_i $(i \leq n)$ can invoke a different subset of activities from w. Therefore, even if the time and cost associated with the execution of activities were static, the QoS of the execution of a workflow would vary depending on the activities invoked at runtime.

For organizations, being able to predict the QoS of workflows has several advantages. For example, it is possible to monitor and predict the QoS of workflows at any time. Workflows must be rigorously and constantly monitored throughout their life cycles to assure compliance both with initial QoS requirements and targeted objectives. If a workflow management system identifies that a running workflow will not meet initial QoS requirements, then adaptation strategies [5] need to be triggered to change the structure of a workflow instance. By changing the structure of a workflow we can reduce its cost or execution time.

3 QoS Mining

In this section we focus on describing a new method that can be used by organizations to apply data mining algorithms to historical data and predict QoS for their running workflow instances. The method presented in this paper constitutes a major and significant difference from the method described in [4]. The method is composed of three distinct phases (Fig. 1) that will be explained in the following sections.

In the first phase, the workflow log is analyzed and data mining algorithms are applied to predict the path that will be followed by workflow instances at

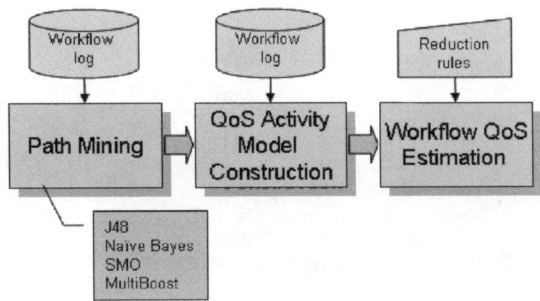

Fig. 1. Phases of workflow QoS mining

runtime. This is called path mining. Path mining identifies which activities will most likely be executed in the context of a workflow instance. Once we know the path, we also know the activities that will be invoked at runtime. For each activity we construct a QoS activity model based on historical data which describes the runtime behavior (duration and cost) of an activity. In the last phase, we compute the QoS of the overall workflow based on the path predicted and from the QoS activity models using a set of reduction rules.

3.1 Path Mining

As we have stated previously, the QoS of a workflow is directly dependent on which activities are invoked during its execution. Different sets of activities can be invoked at runtime because workflows are non-deterministic. Path mining [6,7] uses data mining algorithms to predict which path will be followed when executing a workflow instance.

Definition. *(Path): A path P is a continuous mapping P: [a, b] → C°, where P(a) is the initial point, P(b) is the final point, and C° denotes the space of continuous functions. A path on a workflow is a sequence $\{t_1, t_2, \ldots, t_n\}$ such that $\{t_1, t_2\}, \{t_2, t_3\}, \ldots, \{t_{n-1}, t_n\}$ are transitions of the workflow and the t_i are distinct. Each t_i is connected to a workflow activity.*

A path is composed of a set of activities invoked and executed at runtime by a workflow. For example, when path mining is applied to the simple workflow illustrated in Fig. 2, the workflow management system can predict the probability of paths A, B, and C being followed at runtime. Paths A and B have each six activities, while path C has only four activities. In Fig. 2, the symbol ⊕ represented non-determinism (i.e., a xor-split or xor-join).

To perform path mining, current workflow logs need to be extended to store information indicating the values and the type of the input parameters

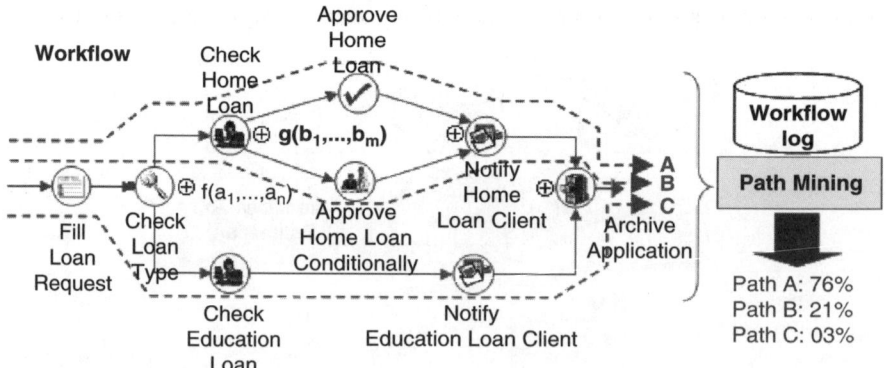

Fig. 2. Path mining

Table 1. Extended workflow log

	Workflow log extension	
...	Parameter/Value	Path
...	int SSN = 7774443333; string loan-type = "car-loan" ...	...
...	string name = jf@uma.pt; ...	{FillLoanRequest, CheckLoanType, CheckCarLoan, ApproveCarLoan, NotifyCarLoanClient, ArchiveApplication}
...	...	...

passed to activities and the output parameters received from activities. The values of inputs/outputs are generated at runtime during the execution of workflow instances. Table 1 shows an extended workflow log which accommodates input/output values of activity parameters that have been generated at runtime. Each 'Parameter/Value' entry as a type, a parameter name, and a value (for example, string loan-type= "car-loan").

Additionally, the log needs to include path information: a path describing the activities that have been executed during the enactment of a process. This information can easily be stored in the log. From the implementation perspective it is space efficient to store in the log only the relative path, relative to the previous activity, not the full path. Table 1 shows the full path approach because it is easier to understand how paths are stored in the log.

During this phase, and compared to [3,4], we only need to add information on paths to the log. Once enough data is gathered in the workflow log, we can apply data mining methods to predict the path followed by a process instance at runtime based on instance parameters. In Sect. 4.2, we will show how the extended workflow log can be transformed to a set of data mining instances. Each data mining instance will constitute the input to machine learning algorithm.

3.2 QoS Activity Model Construction

After carrying out path mining, we know which activities a workflow instance will be invoking in the near future. For each activity that will potentially be invoked we build what we call a QoS activity model. The model includes information about the activity behavior at runtime, such as its cost and the time the activity will take to execute [1].

Each QoS activity model can be constructed by carrying out activity profiling. This technique is similar to the one used to construct operational profiles. Operational profiles have been proposed by Musa [8,9] to accurately predict future the reliability of applications. The idea is to test the activity based on specific inputs. In an operational profile, the input space is partitioned into domains, and each input is associated with a probability of being selected during operational use. The probability is employed in the input domain to guide input generation. The density function built from the probabilities is called the operational profile of the activity. At runtime, activities have a probability associated with each input. Musa [9] described a detailed procedure for developing a practical operational profile for testing purposes. In our case, we are interested in predicting, not the reliability, but the cost and time associated with the execution of workflow activities.

During the graphical design of a workflow, the business analyst and domain expert construct a QoS activity model for each activity using activity profiles and empirical knowledge about activities. The construction of a QoS model for activities is made at design time and re-computed at runtime, when activities are executed. Since the initial QoS estimates may not remain valid over time, the QoS of activities is periodically re-computed, based on the data of previous instance executions stored in the workflow log.

The re-computation of QoS activity metrics is based on data coming from designer specifications (i.e. the initial QoS activity model) and from the workflow log. Depending on the workflow data available, four scenarios can occur (Table 2) (a) For a specific activity a and a particular dimension Dim (i.e., time or cost), the average is calculated based only on information introduced by the designer (Designer Average$_{Dim}(a)$); (b) the average of an activity a dimension is calculated based on all its executions independently of the workflow that executed it (MultiWorkflow Average$_{Dim}(a)$); (c) the average of the dimension Dim is calculated based on all the times activity a was executed in any instance from workflow w (Workflow Average$_{Dim}(t, w)$); and (d) the average of the dimension of all the times activity t was executed in instance i of workflow w (Instance Average$_{Dim}(t, w, i)$).

Let us assume that we have an instance i of workflow w running and that we desire to predict the QoS of activity $a \in w$. The following rules are used to choose which formula to apply when predicting QoS. If activity a has never

Table 2. QoS dimensions computed at runtime

(a)	$QoS_{Dim}(a) =$	Designer Average$_{Dim}(a)$
(b)	$QoS_{Dim}'(a) =$	$wi_1{}^*$ Designer Average$_{Dim}(a) +$
		$wi_2{}^*$ MultiWorkflow Average$_{Dim}(a)$
(c)	$QoS_{Dim}(a, w) =$	$wi_1{}^*$ Designer Average$_{Dim}(a) +$
		$wi_2{}^*$ MultiWorkflow Average$_{Dim}(a) +$
		$wi_3{}^*$ Workflow Average$_{Dim}(a, w)$
(d)	$QoS_{Dim}(a, w, i) =$	$wi_1{}^*$ Designer Average$_{Dim}(a) +$
		$wi_2{}^*$ MultiWorkflow Average$_{Dim}(a) +$
		$wi_3{}^*$ Workflow Average$_{Dim}(a, w) +$
		$wi_4{}^*$ Instance Workflow Average$_{Dim}(a, w, i)$

been executed before, then formula (a) is chosen to predict activity QoS, since there is no other data available in the workflow log. If activity a has been executed previously, but in the context of workflow w_n, and $w \: ! = w_n$, then formula (b) is chosen. In this case we can assume that the execution of a in workflow w_n will give a good indication of its behavior in workflow w. If activity a has been previously executed in the context of workflow w, but not from instance i, then formula (c) is chosen. Finally, if activity a has been previously executed in the context of workflow w, and instance i, meaning that a loop has been executed, then formula (d) is used.

The workflow management system uses the formulae from Table 2 to predict the QoS of activities. The weights wi_k are manually set. They reflect the degree of correlation between the workflow under analysis and other workflows for which a set of common activities is shared. At this end of this second phase, we already know the activities of a workflow instance that will most likely be executed at runtime, and for each activity we have a model of its QoS, i.e. we know the time and cost associated with the invocation of the activity.

3.3 Workflow QoS Estimation

Once we know the path, i.e. the set of activities which will be executed by a workflow instance, and we have a QoS activity model for each activity, we have all the elements required to predict the QoS associated with the execution of a workflow instance.

To compute the estimated QoS of a process in execution, we use a variation of the Stochastic Workflow Reduction (SWR) algorithm [1]. The variation of the SWR algorithm that we use does not include probabilistic information about transitions. The SWR is an algorithm for computing aggregate QoS properties step-by-step. At each step a reduction rule is applied to shrink the process. At each step the time and cost of the activities involved is computed. This is continued until only one activity is left in the process. When this state is reached, the remaining activity contains the QoS metrics corresponding to the workflow under analysis. For the reader interested in the behavior of the SWR algorithm we refer to [1].

For example, if the path predicted in the first phase of our QoS mining method includes a parallel system, as show in Fig. 3, the parallel system reduction rule is applied to a part of the original workflow (Fig. 3a) and a new section of the workflow is created (Fig. 3b).

A system of parallel activities $t_1, t_2, \ldots, t_n$, an *and* split activity t_a, and an *and* join activity t_b can be reduced to a sequence of three activities t_a, t_{1n}, and t_b. In this reduction, the incoming transitions of t_a and the outgoing transition of activities t_b remain the same. The only outgoing transitions from activity t_a and the only incoming transitions from activity t_b are the ones shown in the figure below.

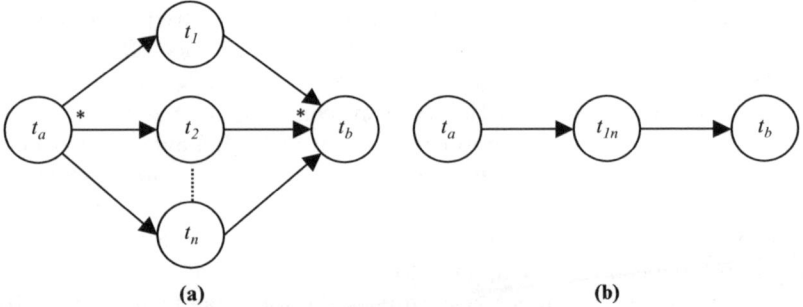

Fig. 3. Parallel system reduction

The QoS of the new workflow is computed using the following formulae (the QoS of tasks t_a and t_b remain unchanged):

$$\text{Time}(t_{1n}) = \text{Max}_{i \in \{1..n\}} \{\text{Time}(t_i)\} \text{ and}$$

$$\text{Cost}(t_{1n}) = \sum_{1 \leq i \leq .n} \text{Cost}(t_i)$$

Reduction rules exist for sequential, parallel, conditional, loop, and network systems [1]. These systems or pattern are fundamental since a study on fifteen major workflow management systems [10] showed that most systems support the reduction rules presented. Nevertheless, additional reduction rules can be developed to cope with the characteristics and features of specific workflow systems.

Our approach to workflow QoS estimation – which uses a variation of the SWR algorithm – addresses the third point that we raised in the introduction and shows that the prediction of workflow QoS can be used to obtain actual metrics (e.g. the workflow instance w will take 3 days and 8 h to execute) and not only information that indicates if an instance takes "more" than D days or "less" than D days to execute.

4 Experiments

In this section, we describe the data set that has been used to carry out workflow QoS mining, how to apply different data mining algorithms and how to select the best ones among them, and finally we discuss the results obtained. While we describe the experiments carried out using the loan process application (see Fig. 4), we have replicated our experiments using a university administration process. The conclusions that we have obtained are very similar to the one presented in this section.

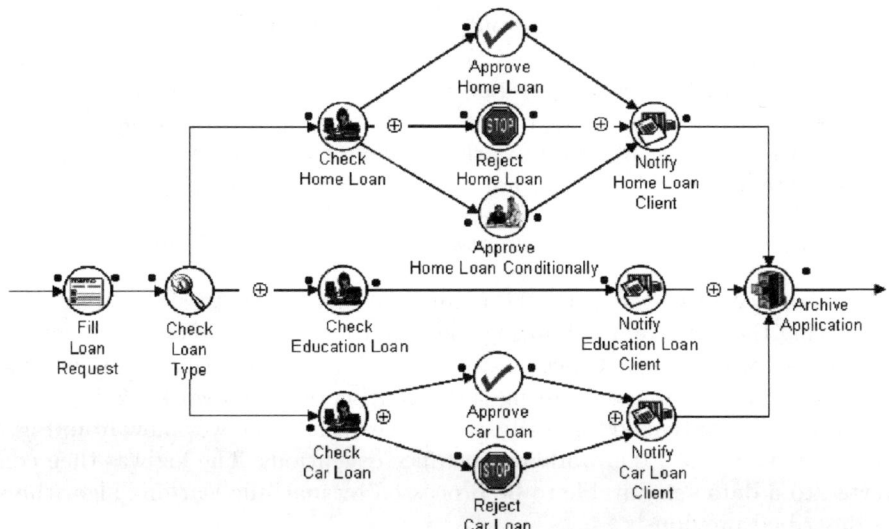

Fig. 4. The loan process

4.1 Workflow Scenario

A major bank has realized that to be competitive and efficient it must adopt a new and modern information system infrastructure. Therefore, a first step was taken in that direction with the adoption of a workflow management system to support its processes. One of the services supplied by the bank is the loan process depicted in Fig. 4. While the process is simple to understand, a complete explanation of the process can be found in [6].

4.2 Path Mining

To carry out path mining we need to log information about the execution of workflow instances. But before storing workflow instances data we need to extended our workflow management log system, as explained in Sect. 3.1, to store information indicating the values of the input parameters passed to activities and the output parameters received from activities (see [6, 7] for an overview of the information typically stored in the workflow log). The information also includes the path that has been followed during the execution of workflow instances.

To apply data mining algorithms to carry out path mining, the data present in the workflow log need to be converted to a suitable format to be processed by data mining algorithms. Therefore, we extract data from the workflow log to construct data mining instances. Each instance will constitute an input to machine learning and is characterized by a set of six attributes:

income, loan_type, loan_amount, loan_years, Name, SSN

The attributes are input and output parameters from the workflow activities. The attributes income, loan_amount, loan_years and SSN are numeric, whereas the attributes loan_type and name are nominal. Each instance is also associated with a class (named *[path]*) indicating the path that has been followed during the execution of a workflow when the parameters were assigned specific values. Therefore, the final structure of a data mining instance is:

$$income, loan\_type, loan\_amount, loan\_years, Name, SSN, [path]$$

In our scenario, the path class can take one of six possible alternatives indicating the path followed during the execution of a workflow when activity parameters were assigned specific values (see Fig. 4 to identify the six possible paths that can be followed during the execution of a loan workflow instance).

Having our extended log ready, we have executed the workflow from Fig. 4 and logged a set of 1,000 workflow instance executions. The log was then converted to a data set suitable to be processed by machine learning algorithms, as described previously.

We have carried out path mining to our data set using four distinct data mining algorithms: J48 [11], Naïve Bayes (NB), SMO [12], and Multi-Boost [13]. J48 was selected as a good representative of a symbolic method, Naïve Bayes as a representative of a probabilistic method, and the SMO algorithm as representative of a method that has been successfully applied in the domain of text-mining. Multiboost is expected to improve performance of single classifiers with the introduction of meta-level classification.

Since when we carry out path mining to a workflow not all the activity input/ouput parameters may be available (some activities may not have been invoked by the workflow management system when path mining is started), we have conducted experiments with a variable number of parameters (in our scenario, the parameters under analysis are: income, loan_type, loan_amount, loan_years, name, and SSN) ranging from 0 to 6. We have conducted 64 experiments (2^6); analyzing a total of 64000 records containing data from workflow instance executions.

Accuracy of Path Mining

The first set of experiments was conducted using J48, Naïve Bayes, and SMO methods with and without the Multiboost (MB) method. We obtained a large number of results that are graphically illustrated in Fig. 5. The chart indicates for each of the 64 experiments carried out, the accuracy of path mining.

The chart indicates, for example, that in experiment no 12, when we use two parameters to predict the path that will be followed by a workflow instance from Fig. 4, we achieve a prediction accuracy of 87.13% using the J48 algorithm. Due to space limitation, the chart in Fig. 4 does not indicate which parameters or the number of parameters that have been utilized in each experiment.

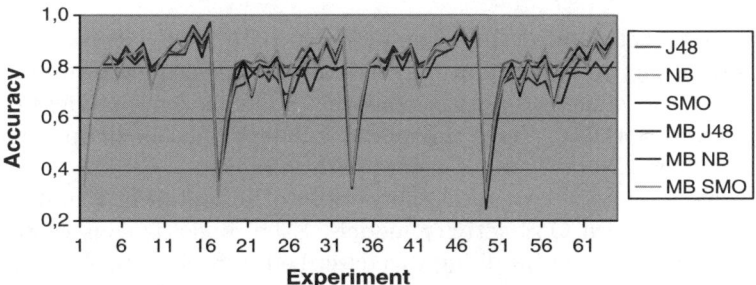

Fig. 5. Accuracy analysis of path mining

Table 3. Summary results of accuracy analysis of path mining

	J48	NB	SMO
Avg acc.	75.43%	78.84%	77.79%
Min acc.	24.55%	30.84%	29.04%
Max acc.	93.41%	96.41%	93.11%
	MB J48	MB NB	MB SMO
Avg acc.	79.74%	81.11%	78.28%
Min acc.	24.55%	30.84%	29.04%
Max acc.	94.61%	97.31%	96.11%

For reasons of simplicity and as a summary, we computed the average, the minimum, and the maximum accuracy for each method for all the experiments carried out. The results are shown in Table 3.

On average the Naïve Bayes approach performs better than all other single methods when compared to each other. When the number of parameters is increased, the accuracy of Naïve Bayes improves. It can be seen that all the methods produced more accurate results when a more appropriate set of parameters was proposed. The worst results were produced by the J48 and SMO algorithms. It is safe to assume that these algorithms overfitted and were not able to find a generalized concept. That is probably a result of the nature of the dataset that contains parameters and that introduced noise. These results address the third point that was raised in the introduction and show that path prediction using MultiBoost Naïve Bayes outperforms the use of decision trees.

Next we added the meta-level of the multiboost algorithm and repeated the experiments. As expected, the multiboost approach made more accurate prognoses. All the classifiers produced the highest accuracy in Experiment 16, since this experiment includes the four most informative parameters (i.e. income, loan_type, loan_amount, and loan_years). In order to evaluate which parameters are the most informative, we have used information gain.

4.3 QoS Activity Model Construction

Once we have determined the most probable path that will be followed by
a workflow at runtime, we know which activities a workflow instance will be
invoking. At this stage, we need to construct a QoS activity model from each
activity of the workflow. Since this phase is independent of the previous one,
in practice it can be carried out before path mining.

Since we have 14 activities in the workflow illustrated in Fig. 4, we need
to construct fourteen QoS activity models. Each model is constructed using
a profiling methodology (profiling was described in Sect. 3.2). When carrying
out activity profiling we determine the time an activity will take to be executed
(i.e. Activity Response Time (ART)) and its cost (i.e. Activity cost (AC)).
Table 4 illustrates the QoS activity model constructed for the Check Home
Loan activity in Fig. 4 using profiling.

This static QoS activity model was constructed using activity profiling.
When a sufficient number of workflows have been executed and the log has a
considerable amount of data, we re-compute the static QoS activity at runtime, originating a dynamic QoS activity model. The re-computation is done
based on the functions presented in Table 2. Due to space limitations we do
not show the dynamic QoS activity model. It has exactly the same structure
as the model presented in Table 4, but with more accurate values since they
reflect the execution of activities in the context of several possible workflows.

4.4 Workflow QoS Estimation

As we have already mentioned, to compute the estimated QoS of a workflow
in execution, we use a variation of the Stochastic Workflow Reduction (SWR)
algorithm. The SWR aggregates the QoS activity models of each activity step-
by-step. At each step a reduction rule is applied to transform and shrink the
process and the time and cost of the activities involved is computed. This
is continued until only one activity is left in the process. When this state is
reached, the remaining activity contains the QoS metrics corresponding to
the workflow under analysis. A graphical simulation of applying the SWR
algorithm to our workflow scenario is illustrated in Fig. 6.

The initial workflow (a) is transformed to originate workflow (b) by applying the conditional reduction rule to two conditional structures identified in
the figure with a box (dashed line). Workflow (b) is further reduced by applying the sequential reduction rule to three sequential structures also identified

Table 4. QoS activity model for the Check Home Loan activity

	Static QoS model		
	Min value	Avg value	Max value
Time (min)	123	154	189
Cost (euros)	4.80	5.15	5.70

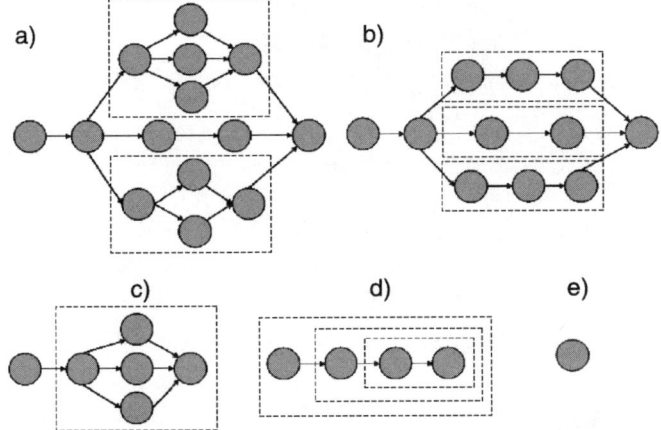

Fig. 6. SWR algorithm applied to our workflow example

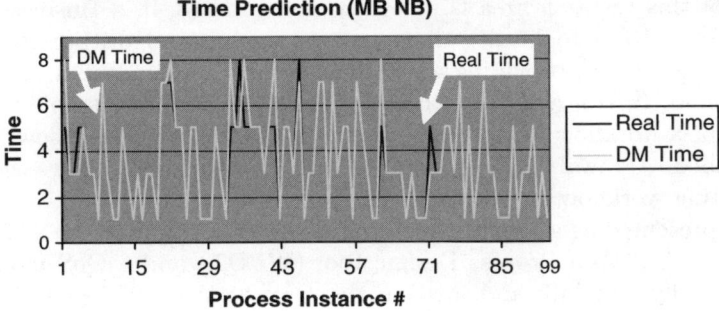

Fig. 7. QoS prediction for time

with a box (dashed line). The resulting workflow, workflow (c), is transformed several times to obtain workflow (d) and, finally, workflow (e). The final workflow (e) is composed of only one activity. Since at each transformation step SWR algorithm aggregates the QoS activity models involved in the transformation, the remaining activity contains the QoS metrics corresponding to the initial workflow under analysis.

4.5 QoS Experimental Results

Our experiments have been conducted in the following way. We have selected 100 random workflow instances from our log. For each instance, we have computed the real QoS (time and cost) associated with the instance. We have also computed the predicted QoS using our method. The results of QoS prediction for the loan process are illustrated in Fig. 7.

The results clearly show that the QoS (Fig. 8) mining method yields estimations that are very close to the real QoS of the running processes.

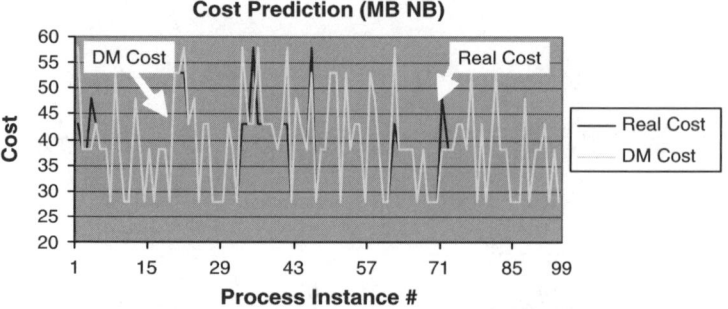

Fig. 8. QoS prediction for cost

5 Related Work

Process and workflow mining is addressed in several papers and a detailed survey of this research area is provided in [14]. In [3, 4], a Business Process Intelligence (BPI) tool suite that uses data mining algorithms to support process execution by providing several features, such as analysis and prediction is presented. In [15] and [16] a machine learning component able to acquire and adapt a workflow model from observations of enacted workflow instances is described. Agrawal et al. [17] propose an algorithm that allows the user to use existing workflow execution logs to automatically model a given business process presented as a graph. Chandrasekaran et al. [2] describe a simulation coupled with a Web Process Design Tool (WPDT) and a QoS model [1] to automatically simulate and analyze the QoS of Web processes. While the research on QoS for BMPS is limited, the research on time management, which is under the umbrella of QoS process, has been more active and productive. Eder et al. [18] and Pozewaunig et al. [19] present an extension of CMP and PERT frameworks by annotating workflow graphs with time, in order to check the validity of time constraints at process build-time.

6 Conclusions

The importance of QoS (Quality of Service) management for organizations and for workflow systems has already been much recognized by academia and industry. The design and execution of workflows cannot be undertaken while ignoring the importance of QoS measurements since they directly impact the success of organizations. In this paper we have shown a novel method that allows us to achieve high levels of accuracy when predicting the QoS of workflows. Our first conclusion indicates that workflow QoS mining should not be applied as a one-step methodology to workflow logs. Instead, if we use a methodology that includes path mining, QoS activity models, and workflow

QoS estimation, we can obtain very good prediction accuracy. Our second conclusion indicates that the MultiBoost (MB) Naïve Bayes approach is the data mining algorithm that yields the best workflow QoS prediction results.

References

1. Cardoso, J. et al., *Modeling Quality of Service for workflows and Web Service Processes.* Web Semantics: Science, Services and Agents on the World Wide Web Journal, 2004. **1**(3): pp. 281–308
2. Chandrasekaran, S. et al., *Service Technologies and Their Synergy with Simulation.* in *Proceedings of the 2002 Winter Simulation Conference (WSC'02).* 2002. San Diego, California. pp. 606–615
3. Grigori, D. et al., *Business Process Intelligence.* Computers in Industry, 2004. **53**: pp. 321–343
4. Grigori, D. et al., *Improving Business Process Quality through Exception Understanding, Prediction, and Prevention.* in *27th VLDB Conference.* 2001. Roma, Italy
5. Cardoso, J. and A. Sheth. *Adaptation and Workflow Management Systems.* in *International Conference WWW/Internet 2005.* 2005. Lisbon, Portugal. pp. 356–364
6. Cardoso, J., *Path Mining in Web processes Using Profiles*, in *Encyclopedia of Data Warehousing and Mining*, J. Wang, Editor. 2005, Idea Group Inc. pp. 896–901
7. Cardoso, J. and M. Lenic. *Web Process and Workflow Path mining Using the multimethod approach.* Journal of Business Intelligence and Data Mining (IJBIDM). submitted
8. Musa, J.D., *Operational Profiles in Software-Reliability Engineering.* IEEE Software, 1993. **10**(2): pp. 14–32
9. Musa, J.D., *Software reliability engineering: more reliable software, faster development and testing.* 1999, McGraw-Hill, New York
10. van der Aalst, W.M.P., et al., *Workflow patterns homepage.* 2002, http://tmitwww.tm.tue.nl/research/patterns
11. Weka, *Weka.* 2004
12. Platt, J., *Fast Training of Support Vector Machines Using Sequential Minimal Optimization*, in *Advances in Kernel Methods – Support Vector Learning*, B. Scholkopf, C.J.C. Burges, and A.J. Smola, Editors. 1999, MIT, Cambridge, MA. pp. 185–208
13. Webb, I.G., *MultiBoosting: A Technique for Combining Boosting and Wagging.* Machine Learning, 2000. **40**(2): pp. 159–196
14. van der Aalst, W.M.P. et al., *Workflow Mining: A Survey of Issues and Approaches.* Data and Knowledge Engineering (Elsevier), 2003. **47**(2): pp. 237–267
15. Herbst, J. and D. Karagiannis. *Integrating Machine Learning and Workflow Management to Support Acquisition and Adaption of Workflow Models.* in *Ninth International Workshop on Database and Expert Systems Applications.* 1998. pp. 745–752
16. Weijters, T. and W.M.P. van der Aalst. *Process Mining: Discovering Workflow Models from Event-Based Data.* in *13th Belgium-Netherlands Conference on Artificial Intelligence (BNAIC 2001).* 2001. Amsterdam, The Netherlands. pp. 283–290

17. Agrawal, R., D. Gunopulos, and F. Leymann. *Mining Process Models from Workflow Logs.* in *Sixth International Conference on Extending Database Technology.* 1998. Springer, Valencia, Spain. pp. 469–483

18. Eder, J. et al., *Time Management in Workflow Systems.* in *BIS'99 3rd International Conference on Business Information Systems.* 1999. Springer Verlag, Poznan, Poland. pp. 265–280

19. Pozewaunig, H., J. Eder, and W. Liebhart. *ePERT: Extending PERT for Workflow Management systems.* in *First European Symposium in Advances in Databases and Information Systems (ADBIS).* 1997. St. Petersburg, Russia. pp. 217–224

Utilisation Organisational Concepts and Temporal Constraints for Workflow Optimisation

D.N. Wang and I. Petrounias

School of Informatics, University of Manchester, UK
dorothy.wang@postgrad.manchester.ac.uk,
ilias.petrounias@manchester.ac.uk

Summary. Workflow systems have been recognised as a way of modelling business processes. The issue of workflow optimisation has received a lot of attention, but the issue of temporal constraints in this area has received significantly less. Issues that come from the enterprise, such as actors performing tasks, resources that these tasks utilise, etc. have not been taken into account. This chapter proposes a combination of utilisation of enterprise modelling issues and temporal constraints in order to produce a set of rules that aid workflow optimisation and therefore, business process improvement.

1 Introduction

Business processes are the key elements to achieving competitive advantage. Organisational effectiveness is depending on them. To meet new business challenges and opportunities, improving existing business processes is an important issue for organisations. A Business Process is the execution of a series of tasks leading to the achievement of business results, such as creation of a product or service. Workflows have been considered as a means to model business processes. Time and cost constraints are measurements for business process performance. The execution time of a single business task can be improved, but, the overall performance of the business process is hard to optimise. This is further complicated by the following factors:

- There are different types of workflow instances and if any task changes in a workflow, this may or may not effect other tasks, depending upon the before mentioned types.
- The execution time of each task can be fixed, not fixed or even indefinite.
- An actor is somebody (or something) that will perform business tasks. The actor's workload and availability are hard to compute. The actor may participate in multiple workflow tasks, have different availability schedules and the business task may not be executed straight away.

D.N. Wang and I. Petrounias: *Utilisation Organisational Concepts and Temporal Constraints for Workflow Optimisation*, Studies in Computational Intelligence (SCI) **109**, 19–42 (2008)
www.springerlink.com   © Springer-Verlag Berlin Heidelberg 2008

Thus, it is necessary to consider these factors, and also the interrelationships between tasks also need to be observed.

This chapter is proposing a new approach to the overall improvement of business processes that addresses the limitations of existing workflow solutions. It attempts to answer the following questions: How do we find what can be improved? When can a task and whole process be improved? The first question is answered by looking at each task within a workflow and examining the concepts related to them with an enterprise model. The second question is answered by a set of general rules proposed by this study and they address the cases in which processes can be improved and tasks executed in parallel. These questions have not been explicitly addressed in previous studies. The rest of the chapter is organised as follows. Section 2 discusses existing work in business process improvement. Section 3 reviews the enterprise modelling concepts. Section 4 identifies the possible workflow routings by using Allen's temporal interval inferences. Section 5 describes the approach used to examine the concepts of tasks and processes within an enterprise model. Section 6 describes a set of possible cases in which processes can be improved and tasks executed in parallel. Section 7 describes a case study by applying these rules. Section 8 summarises the proposed approach and suggests further work.

2 An Overview of Existing Work

Business process improvement involves optimising the process in workflow specification. Previous studies are based on two categories: workflow optimisation and modelling temporal constraints for workflow systems.

Workflow optimisation has received a lot of attention in the area of workflow scheduling, elimination of handoffs and job shop scheduling. [1] proposed a new methodology designed to optimally consolidate tasks in order to reduce the overall cycle time. This methodology takes into account the following parameters: precedence of information flow, loss of specification, alignment of decision rights, reduction in handoffs and technology support costs. Consequently, the organisation could achieve better results due to the elimination of handoffs. Baggio et al. [2] suggest a new approach: 'the Guess and Solve Technique'. The approach applies scheduling techniques to workflows by mapping a workflow situation into a job-shop scheduling problem. As a result, it minimises the number of late jobs in workflow systems. Dong et al. [3] present a framework for optimising the physical distribution of workflow schemes. The approach focuses on compile-time analysis of workflow schemas and mapping of parallel workflows into flowcharts. The total running time for processing a workflow instance and maximum throughput have been considered in this approach.

Modelling temporal constraints and time management for workflow systems recently started to be addressed. Little has been done on the time management of process modelling and avoiding deadline violations. Event calculus axioms, timed workflow graphs and project management tools have been purposed to represent the time structure [4–6]. [4] presents a technique for modelling, checking and enforcing temporal constraints by using the Critical Path Method (CPM) in workflow processes containing conditionally executed tasks. This ensures that the workflow execution avoids violating temporal constraints. Two enactment schedules: 'free schedules' and 'restricted due-time schedules' are purposed in [7]. In a free schedule, an agent may use any amount of time between a minimum and a maximum time to finish the task; in a restricted due-time one, an agent can only use up to the declared maximum time. [7] also proposed to enhance the capabilities of workflow systems to specify quantitative temporal constraints on the duration of activities and their synchronisation requirements. [5] introduced a new concept for time management in workflow systems consisting of calculating internal deadlines for all tasks within a workflow, checking time constraints and monitoring time at run-time. PERT diagrams are used to support the calculation of internal deadlines. Previous approaches in optimising workflow systems haven't taken enough consideration of process complexity, the interrelationships between tasks and temporal constraints. To the authors' knowledge, no previous approach considers the use of an enterprise model to optimise workflow systems. We propose such an approach to improve the process, looking at the concepts within each process, the interrelationships among tasks, and the management of tasks cross-functionally. In the rest of the chapter, we discuss how processes can be improved by using the enterprise modelling technique.

3 Enterprise Modelling

The use of Enterprise Modelling [8] in different applications shows that the main issue of success is not only the Enterprise Model itself, but also the management of business processes and requirements engineering [9]. An enterprise model describes the current and future state of an organization and provides a way to describe different aspects of that organisation by using a set of interrelated models, e.g. Goals Model, Business Rules Model, Concepts Model, Business Process Model, Actors and Resources Model and Technical Components and Requirements Model. We want to use the Enterprise Model to examine the concepts related to workflow processes and the tasks they consist of, and identify the possible cases in which processes can be improved. A workflow models a business process and contains a collection of tasks, and their order of execution follows the workflow routing. The enterprise model is used in order to identify the interrelationships between tasks, and to examine the concepts related to each task within the workflow. Allen's temporal interval inference rules are applied to the workflow patterns, and the possible workflow routings are identified.

4 Identifying Possible Workflow Routings

Workflow specification addresses business requirements. It can be addressed from a number of different perspectives [10, 11]. The control-flow perspective describes tasks and their execution routings. The data perspective defines business and processing data on the control-flow perspective. The resource perspective addresses the roles part within the workflow. The operational perspective describes the elementary actions executed by activities. The control-flow perspective provides a big picture of workflow execution orders, addressing what we believe identify the workflow specification's effectiveness. These workflow execution orders need to be addressed in order to support business requirements from simple to complex. [12] describes possible workflow routing constructs from basic to complex to meet business requirements.

A time interval is an ordered pair of points with the first point less than the second. In these workflow routings, [12] provides an insight into the relations of different intervals. [13] describes a temporal representation that takes the notion of a temporal interval as primitive and provides an inference algebra to combine two different measures of the relation of two points. [13] also describes the possible relations between unknown intervals. In the workflow routings, described by [12], some relations between tasks, e.g. sequence routing, are already provided. We use the possible relations between the parallel activities that can be identified by applying Allen's 13 possible temporal interval inferences [13] (see Fig. 1) to existing workflow routings. In addition, three types of workflow patterns are identified: sequential routing, single task triggering multiple tasks routing, multiple tasks triggering single task routing.

- Sequential Routing: Sequence, Exclusive choice, Simple merge, Arbitrary cycles, Cancellation patterns. In a sequential routing (Fig. 2), task C is always executed after task A. Both exclusive choice pattern and simple merge pattern can be considered as sequential routing: task C (B) always meets or will be after the previously executed task. In Multiple Merge, Synchronizing Merge and Discriminator patterns, if only one task is chosen, this workflow's flow can be considered as a sequential routing [Routing 1].

Relation	Symbol	Symbol for Inverse	Pictoral Example
X *before* Y	<	>	XXX YYY
X *equal* Y	=	=	XXX YYY
X *meets* Y	m	mi	XXXYYY
X *overlaps* Y	o	oi	XXX YYY
X *during* Y	d	di	XXX YYYYYY
X *starts* Y	s	si	XXX YYYYY
X *finishes* Y	f	fi	XXX YYYYY

Fig. 1. The 13 possible relationships

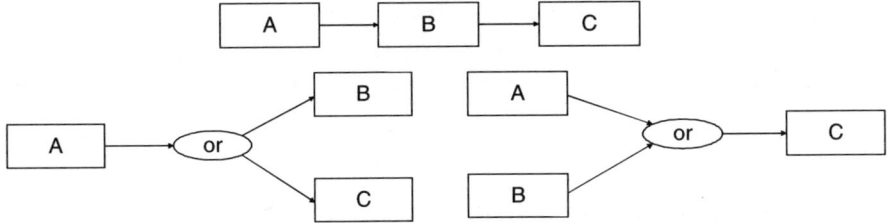

Fig. 2. (a) Sequence (b) Or-Split (c) Or-Join

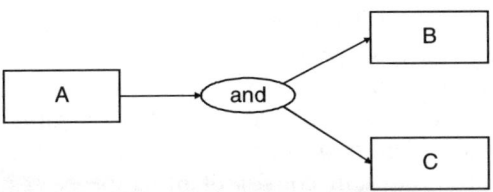

Fig. 3. And-Split

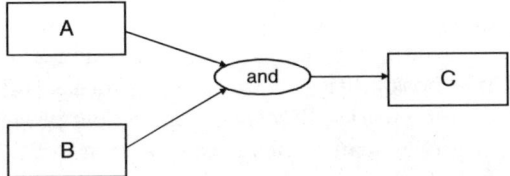

Fig. 4. And-Join

- Single task triggers multiple tasks: parallel split, multiple choice (see Fig. 3)

$$A\ldots\ldots(<, m) \to B \Leftrightarrow B\ldots\ldots(>, mi) \to A$$
$$A\ldots\ldots(<, m) \to C$$

Using Allen's temporal interval, $B\ldots\ldots(<, >, o, oi, m, mi, d, di, s, si, =, f, fi) \to C$
If the relation between A and B, and A and C are already given, A may meet or be before B and C. Then, by using Allen's temporal interval inference, the relation between its output B and C could be any of those 13 intervals above [Routing 2].

- Multiple tasks trigger single task: Synchronization (see Fig. 4)

$$B\ldots\ldots(<, m) \to C \Leftrightarrow C\ldots\ldots(>, mi) \to B$$
$$A\ldots\ldots(<, m) \to C$$

Using Allen's temporal intervals, $A\ldots\ldots(<, >, o, oi, m, mi, d, di, s, si, =, f, fi) \to B$

If the relations between A and C, and B and C are already given, C may meet or be after the execution of A and B and by using Allen's temporal interval inference, the relation between its inputs A and B could be any of the 13 intervals above [Routing 3].

5 Examining Concepts Related to the Processes Within an Enterprise Model

The enterprise model is used for modelling the organisation and examining concepts related to business processes. A high level enterprise metamodel is defined with the following concepts: actor, resource, product, goal and duration (Fig. 5). These will be examined within the three types of routings identified above. One should note the 'recursive' link on the concept 'process'. This means that processes can consist of other processes. At a lower level of decomposition processes will be reduced to tasks (making up an overall process), which can also, using this metamodel, consist of subtasks.

- Actor: Actors are the people who perform the process. An actor can be a single person or a group, who plays more than one roles.
  An actor has three possibilities to work on a process [14]:
  – Direct work: Actor works directly on the whole process.
  – Delegation: A process can be delegated by an actor; this can be done by delegating the whole process or dividing the process into sub-process and eventually tasks to other actors (this is shown by the 'recursive' link to process in Fig. 5).
  – Sub-processes: An actor can initialise another workflow model to fulfill the task/process (again Fig. 5).
  These cases are analysed with the existing workflow routings:
  – Direct Work: For the sequential execution (see Fig. 6), if these tasks are being performed by the same actor, task B can be executed after task A finishes, and task C can be executed after task B finishes. Even

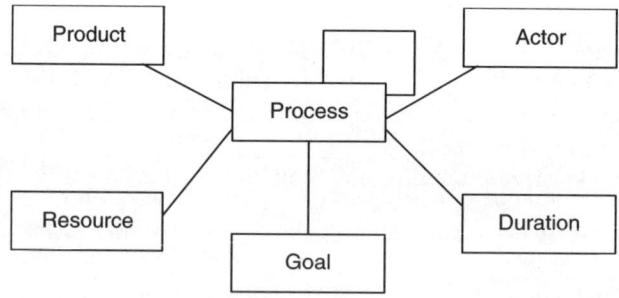

Fig. 5. Enterprise model

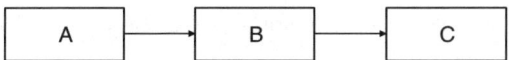

Fig. 6. Sequential execution

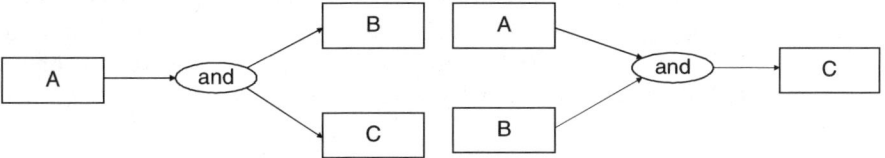

Fig. 7. (a) And-Split (b) And-Join

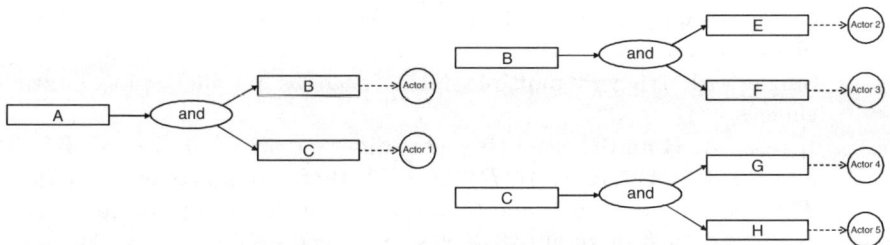

Fig. 8. (a) And-Split with same actors (b) (c) And-Split with different actors

if tasks A, B and C are being performed by different actors, these tasks are still executed sequentially.

Task B and task C can be executed in parallel only if they are being performed by different actors (see Fig. 7).

- Delegation: If an actor delegates a task, this task can be divided into sub-tasks to other actors. If these sub-tasks are being performed by different actors, then, task B and task C can be executed parallel. And the sub-tasks of process B and C can be executed in parallel if they are being performed by different actors (see Fig. 8).

- Resource: There are two types of resources: shared and private. Each shared resource can be accessed by different tasks within a workflow or from different workflows. A private resource can only be accessed by one task. For each shared resource, we can use a locking mechanism to control the concurrent access of it in two different modes: shared and mode. In the shared mode, a task acquires a shared lock on a resource if this resource can be shared simultaneously by other tasks and the access does not change the state of the resource-read only access. On the other hand, in the exclusive mode, the task acquires an exclusive lock on the resource and the access changes the state of the resource-read and write access [15]. These workflow routings are considered with three different resource sharing cases, and thus, the possible relations between tasks are defined.

- Read only access: Shared resources can be accessed by different tasks simultaneously.
- Read and write access: Shared resources can only be accessed by a single task at one time.
- Private resources can only be accessed by a specified task at one time.

1. Single task triggers single thread of tasks, such as sequential routing. Task B is always executed after task A completes, and task C always executed after task B completes.

$$A\ldots(<,=) \to B$$
$$B\ldots(<,=) \to C$$

If tasks A, B and C need to access the same shared resource R1, there are no resource conflicts. If tasks A, B and C need to access different private resources R1, R2 and R3, there are no resource conflicts.

2. Single task triggers multiple tasks, such as parallel split, multiple choices.

If tasks A, B and C need to access different private resources R1, R2 and R3, $R1(A) \cap R2(B) \cap R3(C) = \varnothing$, there are no resource conflicts. The relation between B and C can be any of those 13 possible intervals $B\ldots\ldots(<,>,o,oi,m,mi,d,di,s,si,=,f,fi) \to C$. If tasks B and C are resource dependent, $R2(B) \cap R3(C) \neq \varnothing$, and both tasks acquire exclusive access (read and write accesses), then task B cannot execute simultaneously with task C, in order to avoid the resource conflict, the possible relations can only be $B\ldots..(<,>,m,mi) \to C$. If tasks B and C are resource dependent, $R2(B) \cap R3(C) \neq \varnothing$, and both tasks acquire access (read only access), then task B can be executed simultaneously with task C; there is no resource conflict. The possible relations can be $B\ldots\ldots(<,>,o,oi,m,mi,d,di,s,si,=,f,fi) \to C$.

3. Multiple tasks trigger single task, such as synchronisation.

If tasks A, B and C need to access different private resources R1, R2 and R3, $R1(A) \cap R2(B) \cap R3(C) = \varnothing$, these are no resource conflicts. The relation between A and B and be any of those 13 possible intervals, $A\ldots\ldots(<,>,o,oi,m,mi,d,di,s,si,=,f,fi) \to B$. If tasks A and B are resource dependent, $R1(A) \cap R2(B) \neq \varnothing$, and both tasks acquire exclusive access (read and write accesses), then task A cannot execute simultaneously with task B, in order to avoid the resource conflict. The possible relations can only be $A\ldots..(<,>,m,mi) \to B$. If tasks A and B are resource dependent, $R1(A) \cap R2(B) \neq \varnothing$, and both tasks acquire access (read only access), then task A can be executed simultaneously with task B; there is no resource conflict. The possible relations can be $A\ldots\ldots(<,>,o,oi,m,mi,d,di,s,si,=,f,fi) \to B$.

- *Goal.* The goal is the objective of the task and is not affected by the relationship with other tasks.
- *Product.* If the output of one task is not the input of another one, then these two tasks can be executed parallel.

If the output of one task is the input of another one, these two tasks can only be executed sequentially.

- *Time.* The time of a task can be a time point or time interval [13]. Time point is a precise point in time, e.g. "12 o'clock". Time interval is a time period, which could be fixed, fuzzy or indefinite [16].
  - Fixed duration has exact beginning and end, for example, my semester started on the 15th of January and finished on the 28th of March.
  - Fuzzy duration, the duration is known (3–5 days) and it has an earliest and latest start time and an earliest and latest finish time.
  - Indefinite duration, the end of the interval cannot be determined or estimated. By examining Allen's interval algebra (the 13 basic relations), the rule is: If the finish time of one task is after the start time of another one, then these tasks can be executed in parallel. A.....$(d, di, s, si, f, fi, o, oi, =) \rightarrow B$.
- *Parallelism conditions.* The parallelism heuristic is a way of optimising the workflow [17]. We believe if the tasks can be executed in parallel, the throughput time may be reduced. From the three workflow routings identified above, the possible relations between two tasks A and B could be any of these 13 possible relations A...$(<, >, o, oi, m, mi, d, di, s, si, =, f, fi) \rightarrow B$ that can be divided into two categories:
  - Parallel execution relations A.....$(o, oi, d, di, s, si, f, fi =) \rightarrow B$
  - Sequential execution relations A.....$(<, >, m, mi) \rightarrow B$

As mentioned above, a process can be quite complex. It may consist of different actors performing different tasks; it may also need to access different resources etc. An enterprise model can be used to model the organisation and examine the five concepts related to processes/tasks. We use a reverse reasoning method to address the conditions in which tasks can be executed in parallel, and those in which tasks can be only executed sequentially. (See Tables 1 and 2: X = different, $\sqrt{}$ = same)

Table 1. Parallel execution relations

	Actor	Resource		Private	Goal	Product	Time
		Shared					
		Read-only	Read and write				
O	x	$\sqrt{}$	X	$\sqrt{}$ or x	$\sqrt{}$	x	$\sqrt{}$
Oi	x	$\sqrt{}$	X	$\sqrt{}$ or x	$\sqrt{}$	x	$\sqrt{}$
D	x	$\sqrt{}$	X	$\sqrt{}$ or x	$\sqrt{}$	x	$\sqrt{}$
Di	x	$\sqrt{}$	X	$\sqrt{}$ or x	$\sqrt{}$	x	$\sqrt{}$
=	x	$\sqrt{}$	X	$\sqrt{}$ or x	$\sqrt{}$	x	$\sqrt{}$
S	x	$\sqrt{}$	X	$\sqrt{}$ or x	$\sqrt{}$	x	$\sqrt{}$
Si	x	$\sqrt{}$	X	$\sqrt{}$ or x	$\sqrt{}$	x	$\sqrt{}$
F	x	$\sqrt{}$	X	$\sqrt{}$ or x	$\sqrt{}$	x	$\sqrt{}$
fi	x	$\sqrt{}$	X	$\sqrt{}$ or x	$\sqrt{}$	x	$\sqrt{}$

Table 2. Sequential executions relations

Actor	Resource			Goal	Product	Time
	Shared		Private			
	Read-only	Read and write				
<	√	√	√	√ or x	√	√	x
>	√	√	√	√ or x	√	√	x
M	√	√	√	√ or x	√	√	x
mi	√	√	√	√ or x	√	√	x

From the above, to execute tasks in parallel, five conditions have to be satisfied:

1. These tasks need to be performed by different actors.
2. These tasks can only acquire read-only access to the shared resource or acquire access to different private resources.
3. These tasks can address the same or different goals.
4. The product of the task cannot be the input of other tasks.
5. The task finishing time is after the start time of other tasks.

6 General Rules of Process Improvement

By examining the concepts of the process/task with an enterprise model, a set of rules is derived in which processes can be improved and tasks can be executed in parallel:

1. If two or more tasks are being performed by different actors, then these tasks can be executed in parallel.
2. If two or more tasks are being performed by the same actor, and the sub-tasks of these have different actors, then these tasks can be executed in parallel.
3. In a composite relationship, if sub-tasks are being performed by different actors, then these sub-tasks can be executed in parallel.
4. If two or more tasks need to access different private resources, then these tasks can be executed in parallel.
5. If two or more tasks acquire read only access to the same shared resources, then these tasks can be executed in parallel.
6. If two or more tasks acquire read and write access to different shared resources, then these tasks can be executed in parallel.
7. In a composite relationship, if sub-tasks acquire access to different private resources, then these sub-tasks can be executed in parallel.
8. In a composite relationship, if sub-tasks acquire read only access to the same shared resources, then these sub-tasks can be executed in parallel.

9. In a composite relationship, if sub-tasks acquire read and write access to different shared resources, then these sub-tasks can be executed in parallel.
10. If the output of one task is not the input of another task, then these tasks can be executed in parallel.
11. If the output of a sub-task is not the input of another sub-task, then these sub-tasks can be executed in parallel.
12. If the finish time of one task is after the start time of another one, then these tasks can be executed in parallel.

7 Case Study on Electricity Utility System Improvement Process

To illustrate the improvement rules identified above, an electricity installation process is used, which is based on the Electricity Supply Industry Case Study [18]. It is the process of receiving customer applications and providing electricity. In it there are four actors: customer, customer service department (service administration), studies department (service provision) and construction department. Each task has a unique number so that it can be identified and it has assigned an appropriate time constraint expressed in time units, i.e. days (d). We assume that in some cases, tasks have a definite duration, e.g. the duration of submitting an application is 1 day. In other cases, tasks have an associated time-interval, e.g. the duration of investigating a site is between 1 and 3 days. This is due to the existence of different workflow instances: different sites require different time to investigate, i.e. a local site takes 1 day to investigate and a site in another city may take more than 1 day. Other tasks may never be completed, e.g. customers may never notify the customer service department with their decision. These have an infinite interval; deadlines are assigned, i.e. $\infty = 14$ days. In this case study, we assume customers accept the offer. Figure 9 shows the logical view of existing task executions in this process. In order to illustrate the process improvement procedures, the workflow model of existing execution process is divided into eight execution patterns for analysis (Fig. 10).

Step 1. Pattern 1 follows sequential routing [Routing 1]. In order to optimise the sequential tasks, the parallel execution rules identified in the previous section are used to examine these tasks (see Table 3). The improved tasks are shown in Fig. 11.

Step 2. Pattern 2 follows sequential routing [Routing 1]. In order to optimise the sequential tasks, the parallel execution rules identified in the previous section are used to examine these tasks (see Table 4). The improved tasks are shown in Fig. 12.

Step 3. Pattern 3 is mapped into an OR-Split construct, which follows sequential routing [Routing 1]. Task T15 triggers either task T16 or task T20, which is dependent on the condition, the execution routing cannot be changed (Fig. 13).

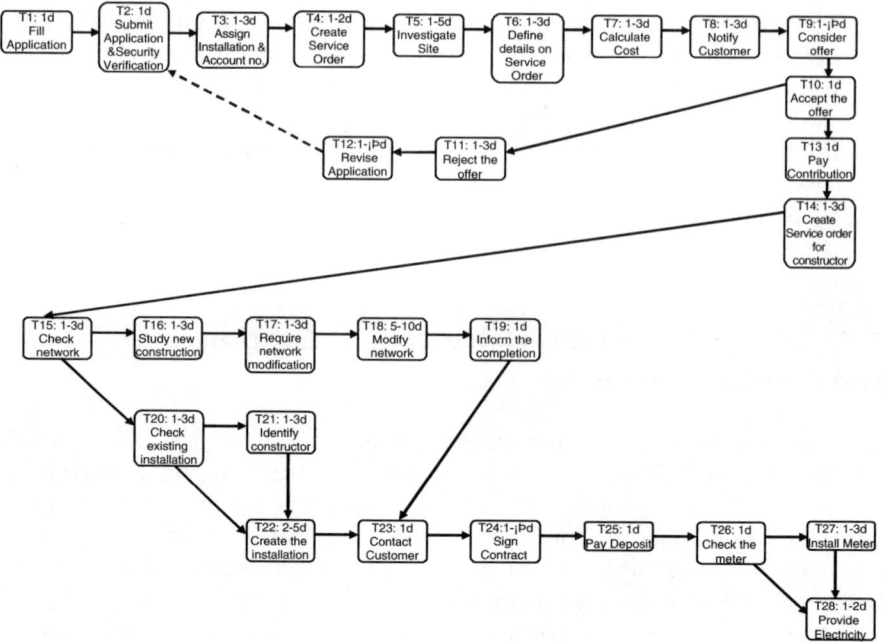

Fig. 9. Existing process of electricity installation

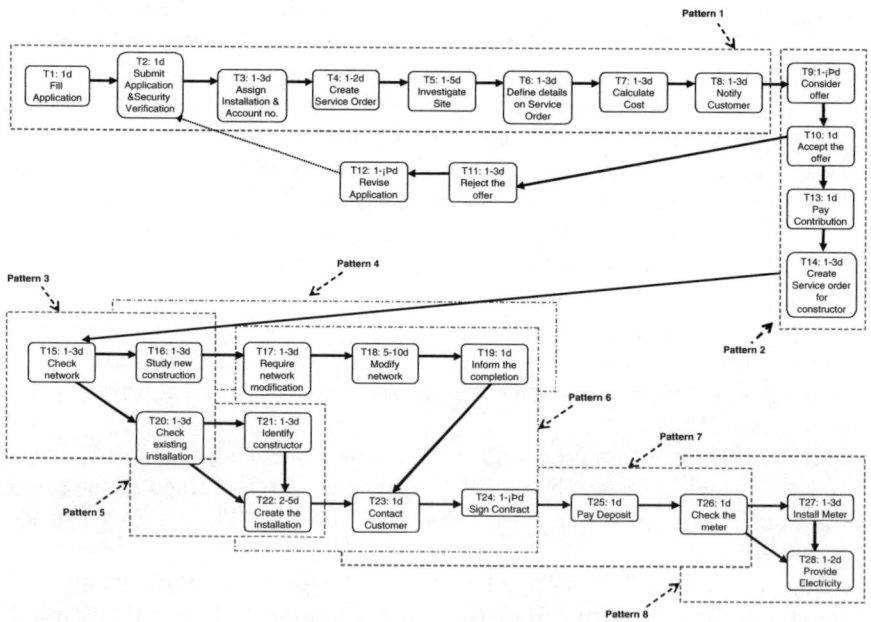

Fig. 10. Existing process of electricity installation execution patterns

Table 3. Pattern 1 analysis

Pattern name			Pattern 1		
Execution routing			Sequential routing		
Enterprise model	Actor	Resource	Product	Goal	Time
T1	Customer	Application form	Application form	Apply electricity installation	T1......(<, m) → T2
T2	Customer	Application form and security info	Application form and security info	Apply electricity installation	T2......(<, m) → T3
T3	Customer service Dept: *Staff 1*	Installation account	Installation account	Assign installation and account no.	T3......(<, m) → T4
T4	Customer service dept: *Staff 2*	Service order	Service order	Create service order	T4......(<, m) → T5
Task T5	Studies dept: *Staff 3*	Installation account	Installation details	Investigate site	T5......(<, m) → T6
T6	Studies dept: *Staff 4*	Service order	Service order details	Define details on service order	T6......(<, m) → T7
T7	Customer service dept: *Staff 5*	Cost	Service cost	Calculate cost	T7......(<, m) → T8
T8	Customer service dept: *Staff 6*	Customer	Customer knows the service cost and details	Notify customer	T8......(<, m) → T9

Interpretation: In this case study, we assume that apart from the customer who is a single individual, we may have different actors who are responsible for different activities within the same department. For example, in the customer service dept, one person deals with the task assign installation and account number, while another person deals with create service order. In the existing process, tasks T3 and T4 are assigned to different actors to perform; Task T3 and T4 require read and write access to different resources, customer account and service order; Task T4 is not dependant on the output of task T3, which satisfies the parallel execution rules [Rule 1,4,10]. Therefore, task T3 and task T4 can be executed in parallel. In the existing process, tasks T5 and T6 are assigned to different actors to perform; Task T5 and T6 require read and write access to different resources: customer site and service order; Task T6 is not dependant on the output of task T5, which satisfies the parallel execution rules [Rule 1, 4, 10]. Therefore, task T5 and task T6 can be executed in parallel. Since task T3 and task T5 require read-write access to the same shared resource i.e. customer installation account; task T4 and task T6 require read-write access to the same shared resource i.e. service order. Task T5 can be only executed after the completion of T3, and task T6 can be only executed after the completion of T4. Task T7 depends on the output of T5 and T6, customer service dept can only calculate the cost after investigate the site and define the service order details, task T7 is executed after the completion of task T5 and task T6, which follows the execution rule [Rule 10]. Task T8 depends on the output of T7, customer service dept can only notify the customer after calculating the installation cost. Thus, task T9 is executed after the completion of task T8, these two tasks can not be executed in parallel [Rule 10].

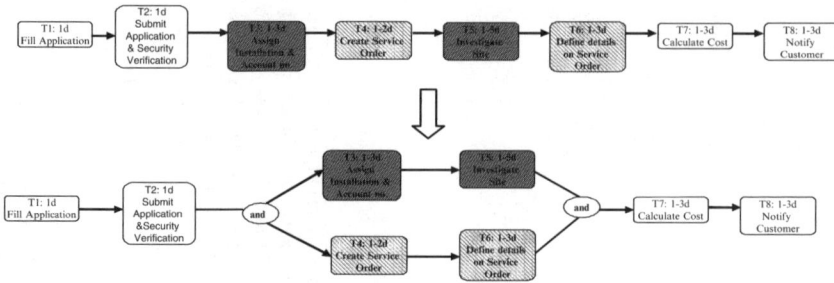

Fig. 11. Pattern 1 optimisation

Table 4. Pattern 2 analysis

Pattern name			Pattern 2		
Execution routing			Sequential routing		
Enterprise model	Actor	Resource	Product	Goal	Time
T9	Customer	Service order	Make decision	Consider offer	T9......(<, m) → T10
T10	Customer	Service order	Accept offer	Accept the offer	T10......(<, m) → T13
Task T13	Customer	Payment	Payment received	Pay contribution	T13......(<, m) → T14
T14	Customer service dept	Service order	Service order for installation	Create service order for installation construction	T14......(<, m) → T15

Interpretation: In the existing process, customer responsibles to execute task T9, T10 and task T13; Since customer can and accept the offer after the consideration, and he/she only need to pay contribution after accept the offer, Thus, task T9, T10 and task T13 need to executed in sequential order. Customer Service Dept is the actor, who is responsible to execute task T14; it requires access to different resource from task T13 and is not dependant on the output of task T13. Therefore, these two tasks can be executed in parallel, parallel execution rules [Rule 1, 4, 10] are applied.

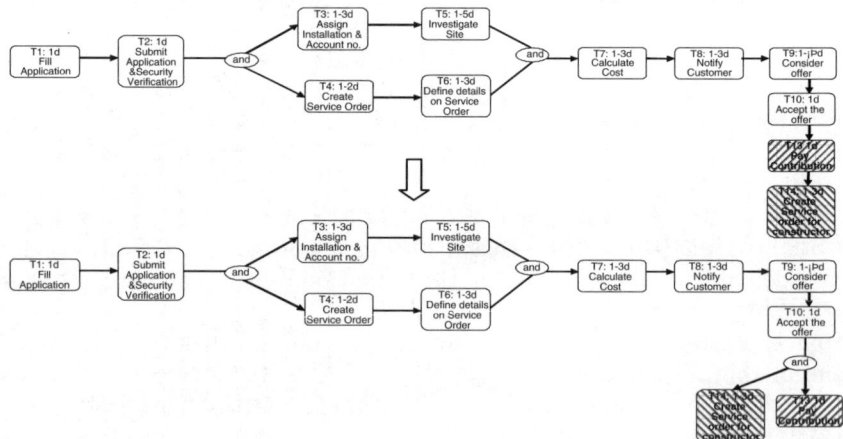

Fig. 12. Pattern 2 optimisation

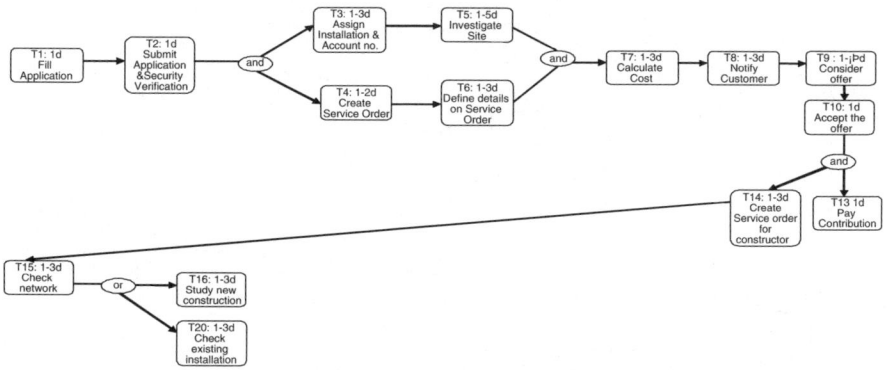

Fig. 13. Pattern 3 analysis

Table 5. Pattern 4 analysis (a)

Pattern name			Pattern 4		
Execution routing			Sequential Routing		
Enterprise model	Actor	Resource	Product	Goal	Time
Task					
T16	Studies dept	Service order	New construction details	Study new construction	T16 . . . . . . (<, m) → T17
T17	Studies dept	Service order	Network modification requirement	Require net-work modifi-cation	T17 . . . . . . (<, m) → T18
T18	Construction Dept	Service order	Network modification	Modify net-work	T18 . . . . . . (<, m) → T19
T19	Construction Dept	Service order	Inform the completion	Inform the completion	

Interpretation: Tasks T16, T17, T18 and T20 are require the same shared access, the input of task T17 is depend on the output of task T16, the input of task T18 is depend on the output of task T17, the input of T19 is depend on the output of T18, which don't satisfy with parallel executions rules. These tasks can only be executed in sequential order.

Step 4. Pattern 4 follows sequential routing [Routing 1]. In order to optimise the sequential tasks, the parallel execution rules identified in the previous section are used to examine these tasks (see Table 5). The improved tasks are shown in Fig. 14.

Step 5. Pattern 5 is mapped into an OR-Split construct, which follows sequential routing [Routing 1]. Task T20 triggers either task T21 or task T22, which is dependent on the condition, the execution routing can not be changed (Fig. 15).

Step 6. Pattern 6 is mapped into an OR-Join construct, which follows sequential routing [Routing 1]. Task T23 is triggered by either task T19 or task T22, which is depend on the condition, the execution routing can not be changed (Fig. 16).

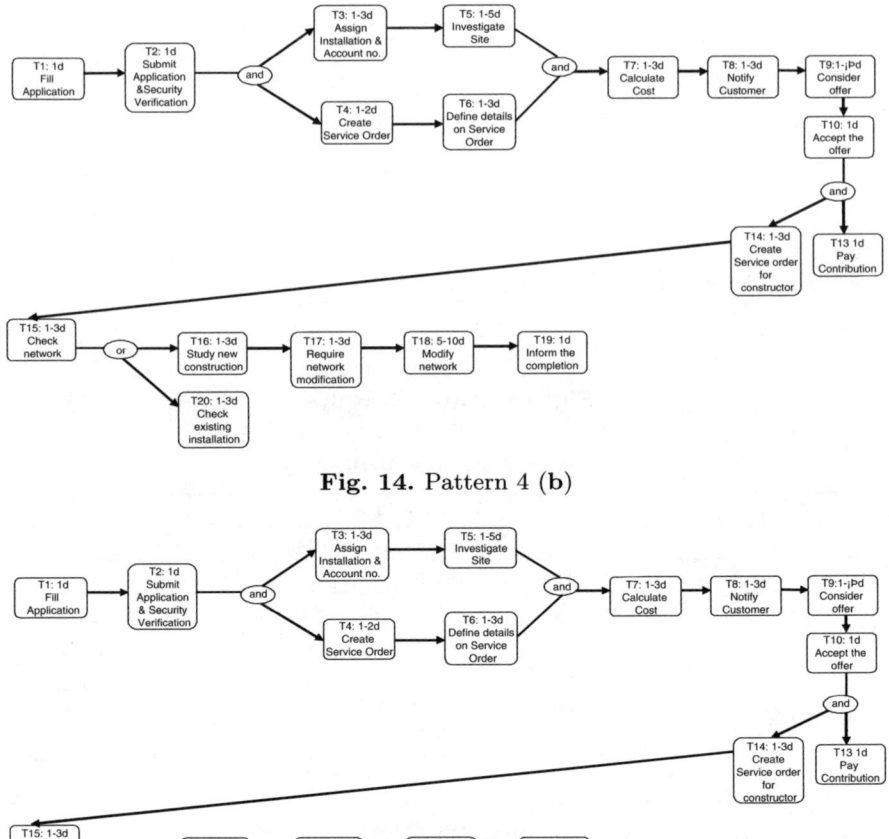

Fig. 14. Pattern 4 (b)

Fig. 15. Pattern 5 analysis

Step 7. Pattern 7 follows sequential routing [Routing 1]. In order to opti-mise the sequential tasks, the parallel execution rules identified in the previous section are used to examine these tasks (see Table 6). The improved tasks are shown in Fig. 17.

Step 8. Pattern 8 is mapped into an OR-Split construct, which follows sequential routing [Routing 1]. Task T26 triggers either task T27 or task T28, which is dependent on the condition. The execution routing can not be changed (Fig. 18).

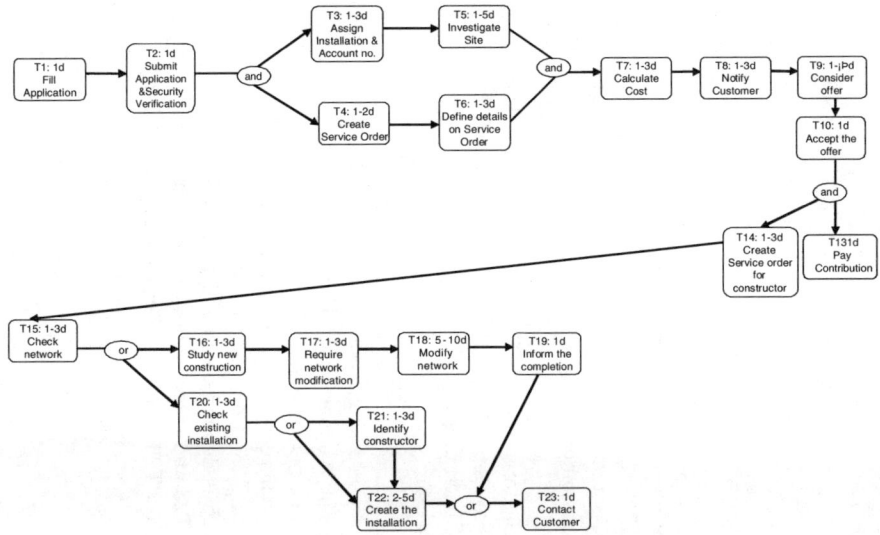

Fig. 16. Pattern 6 analysis

7.1 Execution Time Comparisons

Existing Process

The six possible task execution paths and the throughput time are shown below (see Tables 7 and 8).

Due to the conditional execution of activities, the following time information can be associated with the end event of activity: E^{BS}, E^{WS}, E^{BF}, E^{WF}, L^{BS}, L^{WS}, L^{BF}, and L^{WF} [19]:

- E^{BF} The earliest point the process can finish when the shortest condition and alternative path are chosen.
- E^{BS} The earliest point the process can finish when the longest condition and alternative path are chosen.
- E^{WF} The earliest point when the longest condition and the shortest alternative path are chosen.
- E^{WS} The earliest point when the longest condition and the longest alternative path are chosen.
- L^{BF} The latest point the process can finish when the shortest condition and alternative path are chosen.
- L^{BS} The latest point the process can finish when the longest condition and alternative path are chosen.
- L^{WS} The latest point when the longest condition and the longest alternative path are chosen.
- L^{WF} The latest point when the longest condition and the shortest alternative path are chosen.

Table 6. Pattern 7 analysis

Pattern name		Pattern 7			
Execution routing		Sequential Routing			
Enterprise model	Actor	Resource	Product	Goal	Time
T23	Customer service dept – Staff 7	Construction details	Network installation/completion	Contact customer	T23......(<, m) → T24
Task T24	Customer	Contract	Contract signed	Sign contract	T24......(<, m) → T25
T25	Customer	Deposit	Deposit paid	Pay deposit	T25......(<, m) → T26
T26	Customer service dept – Staff 8	Meter	Meter checked	Check the meter	

Interpretation: In this case study, we assume that apart from the customer who is a single individual, each one of the actors can be seen as a department. We assume that within the same department, we may have different staffs who are responsible for different activities. For example, in the customer service dept, one person deals with contact Customer, while another person check the meter. Task T23 and task T26 are assigned to different actors. Task T24 and task T25 are performed by customer, who needs to sign contract and pay deposit. Task T26 is depend on task T25, customer need to pay deposit before customer service dept to check the meter. Task T25 is depend on task T24, customer need to sign contract first before pay the deposit. Task T24 is not depend on task T23 and there is no resource conflict with task T23 and task T24. Therefore task T23 can be executed in parallel with Task T25 and T26, Parallel execution rules [Rule 1, 4, 10] are applied.

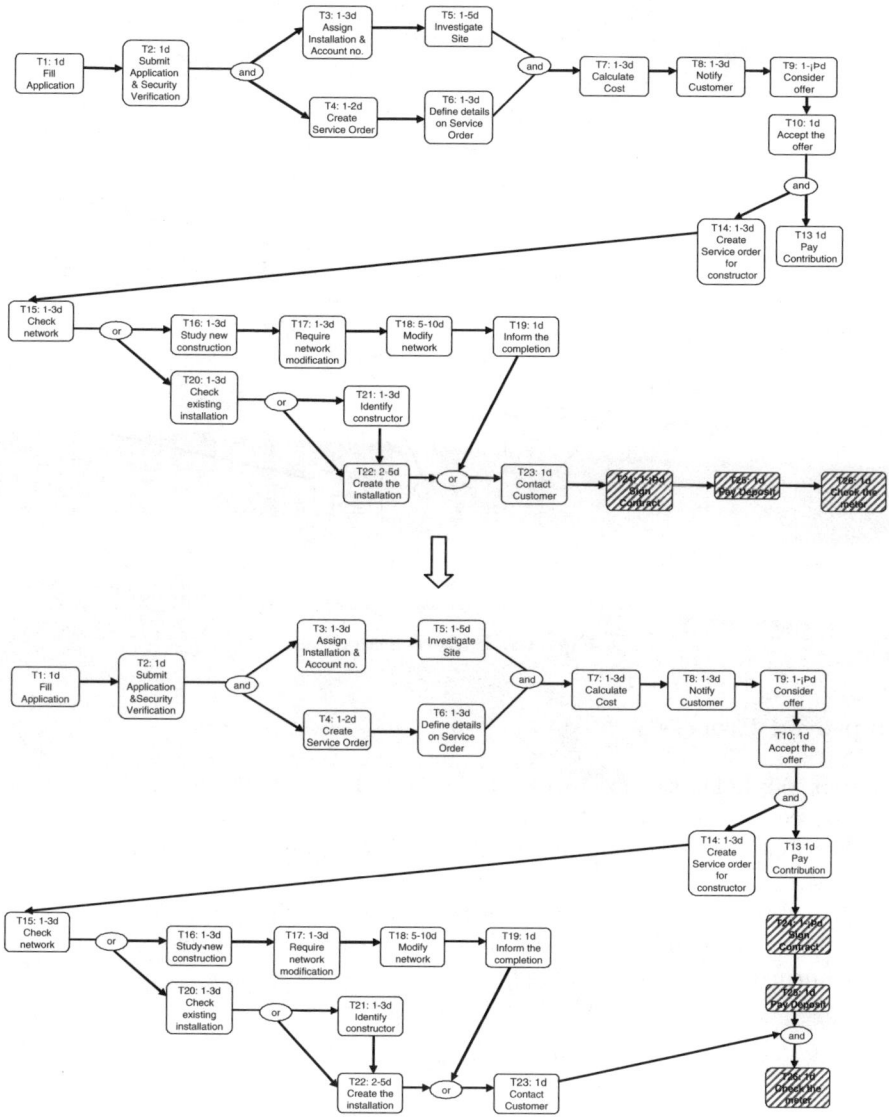

Fig. 17. Pattern 7 optimisation

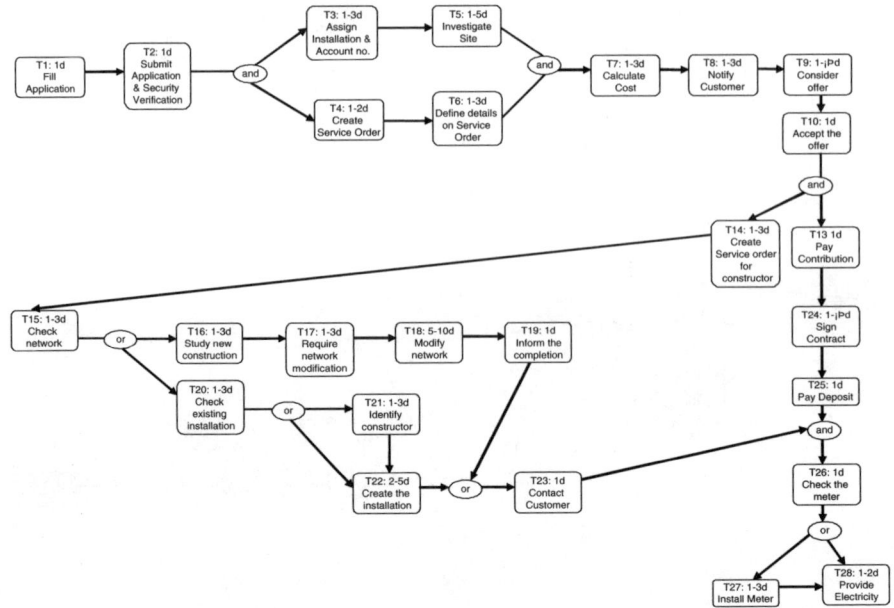

Fig. 18. Pattern 8 analysis

Improved Process

The six possible task execution paths and the throughput time are shown below (see Tables 9 and 10).

The following formulae represent how throughput time is improved in the process.

$f(n)$ represents the duration of task, n is the number of the task. If we don't know the duration variables of each task, the existing execution time (in sequence) can be represented as,

$$\text{Total time}_{\text{sequentialexecution}} = \sum_{k=1}^{n} f(k)$$

If two or more tasks are executed in parallel, the throughput time of these tasks is $\max_{k=1}^{n}(f(k))$

$$\text{Time saving} = \sum_{k=1}^{n} f(k) - \max_{k=1}^{n}(f(k))$$

For x parallel tasks, if all tasks have same fixed minimum duration D, (the time unit in this case study is 1 day). Time saving = Dx − D

Table 7. Six possible task execution paths in existing process

Path 1: T1 → T2 → T3 → T4 → T5 → T6 → T7 → T8 → T9 → T10 → T13 → T14 → T15 → T16 → T17 → T18 → T19 → T23 → T24 → T25 → T26 → T27 → T28

Path 2: T1 → T2 → T3 → T4 → T5 → T6 → T7 → T8 → T9 → T10 → T13 → T14 → T15 → T16 → T17 → T18 → T19 → T23 → T24 → T25 → T26 → T28

Path 3: T1 → T2 → T3 → T4 → T5 → T6 → T7 → T8 → T9 → T10 → T13 → T14 → T15 → T20 → T21 → T22 → T23 → T24 → T25 → T26 → T27 → T28

Path 4: T1 → T2 → T3 → T4 → T5 → T6 → T7 → T8 → T9 → T10 → T13 → T14 → T15 → T20 → T21 → T22 → T23 → T24 → T25 → T26 → T28

Path 5: T1 → T2 → T3 → T4 → T5 → T6 → T7 → T8 → T9 → T10 → T13 → T14 → T15 → T20 → T22 → T23 → T24 → T25 → T26 → T27 → T28

Path 6: T1 → T2 → T3 → T4 → T5 → T6 → T7 → T8 → T9 → T10 → T13 → T14 → T15 → T20 → T22 → T23 → T24 → T25 → T26 → T28

Table 8. Existing process execution time calculation

	Path 1		Path 2		Path 3		Path 4	Path 5	Path 6	
		E^{WS}		E^{WF}		E^{BS}				E^{BF}
		L^{WS}		L^{WF}		L^{BS}				L^{BF}
Best case	27		26		23		22	22	21	
Worst case	82		79		76		73	73	70	

Table 9. Six possible task execution paths improved process

Path 1: T1 → T2 → (T3 → T5||T4 → T6) → T7 → T8 → T9 → T10 → (T13 → T24 → T25||T14 → T15 → T16 → T17 →
T18 → T19 → T23) → T26 → T27 → T28

Path 2: T1 → T2 → (T3 → T5||T4 → T6) → T7 → T8 → T9 → T10 → (T13 → T24 → T25||T14 → T15 → T16 → T17 →
T18 → T19 → T23) → T26 → T28

Path 3: T1 → T2 → (T3 → T5||T4 → T6) → T7 → T8 → T9 → T10 → (T13 → T24 → T25||T14 → T15 → T20 → T21 →
T22 → T23) → T26 → T27 → T28

Path 4: T1 → T2 → (T3 → T5||T4 → T6) → T7 → T8 → T9 → T10 → (T13 → T24 → T25||T14 → T15 → T20 → T21 →
T22 → T23) → T26 → T28

Path 5: T1 → T2 → (T3 → T5||T4 → T6) → T7 → T8 → T9 → T10 → (T13 → T24 → T25||T14 → T15 → T20 → T22 →
T23) → T26 → T27 → T28

Path 6: T1 → T2 → (T3 → T5||T4 → T6) → T7 → T8 → T9 → T10 → (T13 → T24 → T25||T14 → T15 → T20 → T22 →
T23) → T26 → T28

Table 10. Improved process execution time calculation

	Improved process calculation (days)										
	Path 1		Path 2		Path 3		Path 4	Path 5		Path 6	
Best case	22	E^{WS}	21	E^{WF}	18	E^{BS}	17	17	17	E^{BF}	
Worst case	61	L^{WS}	58	L^{WF}	55	L^{BS}	52	53	50	L^{BF}	

8 Conclusions

This chapter has demonstrated how the enterprise model is used to model an organisation, examined the concepts related to processes/tasks and identified possible cases in which processes can be improved and tasks executed in parallel. We have illustrated the modelling of different workflow routings within an enterprise model, and proposed a list of general rules for parallel task execution. As parallel execution is a more efficient method in workflow systems, these general rules can be implemented in a system, in order to find the processes that can be executed in parallel. A case study on the Electricity Supply Process is used to demonstrate how these execution rules are applied. We showed how tasks are examined and applied by the enterprise model and demonstrated the general rules in each step, calculating how the throughput time is improved. In this chapter, we have established a framework to optimise a single workflow. Future work concerns the optimisation of muti-workflow systems where tasks may belong to different workflows and may be performed by the same or different actors. We are intending to extend our parallel execution rules to accommodate multi-workflow optimisation in order to further prove their validity.

References

1. Dewan, R., Seidmann, A., Walter, Z. (1998) *Workflow Optimization Through Task Redesign in Business Information Processes*, Proceedings of Hawaii International Conference on System Science
2. Baggio, G., Wainer, J., Ellis, C. (2004) *Applying Scheduling Techniques to Minimize the Number of Late Jobs in Workflow Systems*, Proceedings of the 2004 ACM Symposium on Applied Computing – Nicosia, Cyprus, pp. 1396–1403
3. Dong, Z., Hull, R., Kumar, B., Su, J., Zhou, G. (2000) *A Framework for Optimizing Distributed Workflow Executions*, Proceedings of 7th International Workshop on Database Programming Languages: Research Issues in Structured and Semi structured Database Programming, pp. 152–167
4. Eder, J., Gruber, W., Panagos, E. (2000) *Temporal Modeling of Workflows with Conditional Execution Paths*, Proceedings of the 11th International Conference on Database and Expert Systems Applications, pp. 243–253
5. Pozewaunig, H., Eder, J., Liebhart, W. (1997) *ePERT: Extending PERT for Workflow Management Systems*, 1st East European Symposium on Advances in Database and Information Systems (ADBIS'97)

6. Çiçekli, N.K. (1999) *A Temporal Reasoning Approach to Model Workflow Activities*, LNCS 1649, Pinter, R. and Tsur, S. (eds.), NGITS'99, Israel
7. Bettini, C., Wang, X.S., Jajodia, S. (2002) *Temporal Reasoning in Workflow Systems*, Distributed and Parallel Databases, 11(3), pp. 269–306
8. Persson, A., Stirna, J., (2001) *EKD User Guide*, Royal Institute of Technology (KTH), Sweden
9. Kirikova, M., Bubenko, J.A. (1994) *Enterprise Modelling: Improving the Quality of Requirements Specifications*, Information Research Seminar, IRIS 17, Finland
10. Van der Aalst, W.M.P. and van Hee, K.M. (2002) *Workow Management: Models, Methods, and Systems*, MIT, Cambridge, MA
11. Jablonski, S. and Bussler, C. (1996) *Workflow Management: Modeling Concepts, Architecture, and Implementation*, International Thomson Computer Press, London
12. Van der Aalst, W.M.P., ter Hofstede, A.H.M., Kiepuszewski, B., Barros, A.P. (2003) *Workflow Patterns*. Distributed and Parallel Databases, 14(3), pp. 5–51
13. Allen, J.F. (1983) *Maintaining Knowledge about Temporal Intervals*, Communications of the ACM, 26(11), pp. 832–843
14. Faustmann, G. (2000) *Configuration for Adaptation-a Human-Centred Approach to Flexible Workflow Enactment*, Computer Supported Cooperative Work: The Journal of Collaborative Computing, 9(3–4), pp. 413–434
15. Li, H., Yang, Y., (2004) *Dynamic Checking of Temporal Constraints for Concurrent workflows*, Sixth Asia-Pacific Web Conference (APWeb2004), pp. 804–813
16. Visser, U., Hubner, S. (2003) *Temporal Representation and Reasoning for the Semantic Web*, Technical Report, TZI-Bericht Nr.28
17. Reijers, H.A. (2003) *Design and Control of Workflow Processes: Business Process Management for the service industry*, Springer, Berlin Heidelberg New York
18. Loucopoulos, P. (2006) *A Series of Lectures on Enterprise System Modelling: Designing for change*, The University of Manchester, Manchester
19. Eder, J., Panagos, E. (2000) *Managing Time in Workflow Systems*. Fischer, L. (ed.), Workflow Handbook 2001, Future Strategies INC. In Association with Workflow Management Coalition 2000, ISBN 0-0703509-0-2, pp. 109–132

Extending the Resource-Constrained Project Scheduling Problem for Disruption Management

Jürgen Kuster and Dietmar Jannach

Institute for Applied Informatics, University Klagenfurt, A-9020 Austria
jkuster@ifit.uni-klu.ac.at, dietmar@ifit.uni-klu.ac.at

Summary. This chapter describes how the Resource-Constrained Project Scheduling Problem (RCPSP) can be used as a basis for comprehensive disruption management, concerned with both rescheduling as well as potential process variations. It is illustrated how the RCPSP can be extended by the possibility to represent alternative activities and how the respective constructs can be used to describe various forms of typical interventions. Moreover, an approach for schedule optimization and the resolution of the generalized problem is presented, based on the combination of well-established methodologies and specific evolutionary operators. In an illustrative example it is finally shown how the proposed framework can be applied for the development of real-time decision support systems in the domain of airport ground process management.[1]

1 Introduction

Uncertainty is an intrinsic and pervasive aspect of the real world [1]. Whenever it unfolds, deviations from a predetermined plan are likely: A so-called disruption occurs. Disruption management (DM, see [1,2]) is concerned with the resolution of respective problems and the continuous optimization of the relationship between real and planned processes: As predetermined plans and schedules are typically optimized according to some specific criterion, the main aim is usually to get back on track in case of process disturbances and to minimize the costs associated with its effects. For this purpose, an optimal combination of applicable interventions has to be selected from a set of potential ones: Typically, both *rescheduling* as well as *process variations* (i.e. dynamic switches from one process variant to another one) have to be considered in the resolution of real-world problems.

---

[1] This is a revised and extended version of the paper originally published in the Proceedings of the 3rd IEEE Conference on Intelligent Systems, London, 2006.

J. Kuster and D. Jannach: *Extending the Resource-Constrained Project Scheduling Problem for Disruption Management*, Studies in Computational Intelligence (SCI) **109**, 43–61 (2008)
www.springerlink.com   © Springer-Verlag Berlin Heidelberg 2008

However, the currently existing applications of disruption management mainly focus on the rescheduling part of the problem: They are concerned with the mere temporal shift of activities within a schedule. Even though particularly Zhu et al. [3] introduce basic aspects of structural flexibility when considering mode alternations (i.e. the change of the durations and the amount of resources required by a specific activity) as a potential form of intervention, we claim that this is still not sufficient for the effective provision of decision support in realistic problems. Apart from the temporal shift and the parametric modification of activities, the responsible decision maker might want to insert or remove process steps, change their order or parallelize what has been planned for serial execution (or vice versa).

Research on disruption management is strongly driven by operations research and thus focusing on the application of mathematical programming. Although the use of respective methods makes it possible to identify exact optimal solutions, its main drawback compared to (only suboptimal) metaheuristic approaches is the significantly higher requirement of processing time: Therefore, the respective methods can only be applied to relatively small problems if real-time results are required [4].

The work presented herein is motivated by the findings of a study conducted in collaboration with Deutsche Lufthansa AG, regarding the elementary requirements of DM-related decision support systems (DSS): We propose a novel approach to disruption management, considering both rescheduling and process variations as potential interventions relevant in real-time DSS. For this purpose, the notion of alternative activities is introduced for the formal description of process variations and metaheuristic optimization is applied to an accordingly extended version of the Resource-Constrained Project Scheduling Problem (RCPSP). The remainder of this document is structured as follows: Sect. 2 introduces a framework for the formal description of disruption management problems and potential process modifications. Section 3 shows how respective problems can be solved: Well-established methodologies are combined with specific methods based on the concepts of evolutionary algorithms. Section 4 provides an illustrative example for the application of the proposed framework from the domain of airport ground process management and gives an overview on the results of the conducted performance evaluations. Finally, Sect. 5 summarizes the contributions of this chapter.

2 Modeling Process Interventions

This section describes how potential interventions can be described formally. With a particular focus on dynamic process variations, the concept of alternative activities is introduced and used as the basis for an extension of the RCPSP. It is also illustrated how the associated constructs can be applied for the description of typical forms of intervention.

2.1 Overview

Comprehensive disruption management has to consider both rescheduling and process variations as potential forms of repair activities. Possibilities of the former type are typically implicitly given by the definition of precedence relations and associated resource constraints. Any conceptual framework for schedule optimization can therefore form the basis for the identification of optimal rescheduling interventions. However, as far as process variations are concerned, additional modeling is usually required. Two different strategies can be distinguished:

- *External Description.* In this approach potential variations are modeled separated from the corresponding process: An intervention is described through sets of activities and constraints, which have to be added to or deleted from the originally planned process. Therefore, respective *add* and *delete* lists represent the elementary constructs for this form of representation. Consider for example a simple network of two activities a and b which are linked by a precedence constraint saying that a has to be finished at or before the start of b. If we assume that the parallelization of the activities represents a potential form of process variation, this intervention corresponds to the removal of the single precedence relation. If alternatively it shall be possible to insert an activity c between a and b, the respective intervention corresponds to (1) the addition of c to the network and (2) the replacement of the existing precedence relation by two constraints defining that a needs to be executed before c which itself is executed before b.
- *Internal Description.* In this approach potential process variations are described directly within the process model and the application of an intervention is regarded as the switch from a previously chosen process execution path to a valid alternative: *Alternative activities* represent a convenient modeling construct for the formal description of different valid process variants within a comprehensive process model. Considering the example discussed above, the former option can be described by introducing a choice point into the reference (i.e. the default version of the) process, where it is possible to select a or an alternative a_1 for execution: a_1 differs from a in not being linked with b through a precedence relation. For the latter option, another alternative a_2 is inserted, which represents the origin for the sequence a_2 before c before b.

Upon changes in the process model, respective interventions have to be updated accordingly if potential modifications are regarded separated from the processes (as with an external form of description): This synchronization represents a highly sensitive task since any mistake may cause inconsistencies in the disruption management problem. If alternatively only one model is used for the description of reference process and valid modification possibilities (as with an internal form of description), a potential drawback consists in the higher level of complexity associated with the description and maintenance of the process structure. In exchange, however, the difficulty of model synchronization can be avoided as consistency is guaranteed implicitly. Since we assume that (1) this represents a major advantage for realistic applications with flexibly and dynamically changing processes, and that (2) the increased modeling complexity can be efficiently handled through the provision of appropriate (abstract) modeling constructs, we will focus on the latter form of representation in the following.

2.2 Extending the RCPSP

The Resource-Constrained Project Scheduling Problem provides a well established framework for the resolution of scheduling problems. In the regarded context, we claim that it is perfectly suited for the resolution of the *rescheduling* part of the problem, due to the following reasons:

- Defining an RCPSP is easy and intuitive. Based on abstract constructs such as activities, resources, precedence constraints, etc. it is possible to define entities and relationships on a conceptual level. Therefore, especially intelligibility and maintainability are significantly better than for comprehensive mathematical models.
- Metaheuristic approaches can be used for optimization. Local search, tabu search, genetic algorithms, ant colonies, etc. can be used to search for solutions. By the use of such incremental search procedures the provision of good results is even possible in real-time: This corresponds to the realistic requirements of disruption management, where decisions must rather be made in short time than in a (globally) optimal manner.
- The RCPSP has been and still is studied extensively. Research particularly focuses on the (further) improvement of optimization algorithms and the extension and generalization of the respective modeling concepts.

However, as far as options of switching between different process variants are concerned, the RCPSP provides only little support. The only form in which process variations are possible is the alternation of modes in the Multimode RCPSP (MRCPSP, [5]): This generalization of the classical scheduling problem makes it possible to dynamically consider changes in durations and in the amounts of required resources. Moreover, Artigues et al. [6] and Elkhyari et al. [7] recently presented their ideas of providing the RCPSP with additional flexibility: The former focus on the dynamic insertion of activities, considering each arrival of an unexpected activity a disruption. The latter use explanations to handle over-constrained networks in dynamic scheduling problems. As regards the concept of alternative activities, only Beck et al. [8] have considered options of activity replacement in scheduling problems: Their approach is based on the association of a Probability of Existence (PEX) with any activity.

We herein introduce a method for describing alternative activities (as discussed before) within the conceptual framework of the RCPSP. For this purpose, we define the Extended Resource-Constrained Project Scheduling Problem x-*RCPSP* as a generalization of the classical problem. The basic idea of the extension is a distinction between *active* and *inactive* elements, all grouped and described in a comprehensive model, where only the former set of elements is actually considered during the scheduling process. Thus, the x-*RCPSP* is based on the introduction of an additional layer on top of the original RCPSP: Depending on the current state of element *activation*, different instances of the classical problem can be generated from the respective supersets. This way, the well-established methods which have been defined for the resolution of the RCPSP [9] can be applied for the generation of valid schedules. For the x-*RCPSP*, the aim of optimization is the identification of an optimal *activation* state as well as the identification of an optimal sequence for all *active* activities. Note that each change of the *activation* state corresponds to the selection of an alternative process execution path.

The structure of the x-*RCPSP* can be described as follows. A project (or process respectively) is defined by a set of potential activities $A^+ = \{0, 1, \ldots, a, a+1\}$. The first and the last element correspond to abstract start and end activities having a duration of 0 and no resource requirements associated. All *active* activities form a subset $A \subseteq A^+$ which implies that all *inactive* activities are contained in $A^+ \setminus A$. The activities grouped in $A^0 \subseteq A^+$ form the so-called reference process: This subset defines the default *activation* state and the preferred version of the process which is considered before a disruption occurs. The execution of the respective activities is based on a set of renewable resource types $R = \{1, \ldots, r\}$. For each type k, a constant amount of u_k units is available. As regards the description of activity dependencies, the following constructs can be used:

- *Duration Value.* Each activity i has a duration d_i associated, describing how long its execution lasts.
- *Precedence Constraints.* Activities can be ordered by use of precedence constraints: The existence of $p_{i,j}$ states that activity i has to be finished at or before the start of activity j. According to the distinction between *active* and *inactive* activities, two different sets are used for grouping precedence relations: P^+ contains all potentially relevant constraints, whereas the subset $P \subseteq P^+$ groups only those $p_{i,j}$ for which both i and j are contained in the set of *active* activities A.
- *Resource Requirements.* The relationship between activities and resource types is defined through resource requirements: An activity i requires $q_{i,k}$ units of type k throughout its execution. Q^+ combines all potential dependencies of elements in A^+ on elements in R whereas Q only comprises those requirements $q_{i,k}$ for which the associated activity i is currently *active*.

The set of *active* activities represents the most important of all subsets: Whenever it changes, the sets of *active* precedence constraints and *active* resource requirements have to be updated accordingly. Li et al. [10] have argued the necessity to consider mutual dependencies when regarding alternative resources: In accordance with their approach we base the definition of alternative activities on the subsequent constructs:

- *Activity Substitutions* X^+. If $x_{i,j}$ is contained in this set, it represents a legal from of modification to *deactivate* an activity i upon the *activation* of activity j: X^+ therefore describes elementary possibilities of deliberate activity substitution. Note that the respective relationship is not necessarily commutative since the option of replacing activity i with activity j does not automatically imply the possibility to substitute j with i.
- *Activity Dependencies* M^+. Changing the state of an activity might have an impact on other activities. Therefore the set of activity dependencies describes four types of binary and non-commutative relationships between the elements of A^+:
  - The existence of an element $m^{]{[}}_{i,j}$ in the set M^+ indicates that activity $j \in A^+$ has to be activated upon the activation of $i \in A^+$.
  - The existence of an element $m^{]|}_{i,j}$ in the set M^+ indicates that activity $j \in A^+$ has to be deactivated upon the activation of $i \in A^+$.
  - The existence of an element $m^{|[}_{i,j}$ in the set M^+ indicates that activity $j \in A^+$ has to be activated upon the deactivation of $i \in A^+$.
  - The existence of an element $m^{||}_{i,j}$ in the set M^+ indicates that activity $j \in A^+$ has to be deactivated upon the deactivation of $i \in A^+$.

The *x-RCPSP* represents a generalization of the classical RCPSP: Any instance of an *x-RCPSP* with $X^+ = M^+ = \phi$ can be converted into an equivalent RCPSP. Correspondingly, the methodologies for the resolution of the classical problem can be applied to our generalization, as soon as A is stable. Furthermore, the Extended Resource-Constrained Project Scheduling Problem generalizes the Multi-Mode RCPSP: It is possible to formulate an *x-RCPSP* as an MRCPSP without losing any information if all of the following properties hold:

$$x_{i,j} \in X^+ \Rightarrow x_{j,i} \in X^+ \tag{1}$$
$$x_{i,j}, x_{j,k} \in X^+ \Rightarrow x_{i,k} \in X^+ \tag{2}$$
$$x_{i,j} \in X^+, p_{i,k} \in P+ \Rightarrow p_{j,k} \in P^+$$
$$x_{i,j} \in X^+, p_{k,i} \in P+ \Rightarrow p_{k,j} \in P^+ \tag{3}$$
$$M^+ = \phi \tag{4}$$

Statement 1 corresponds to the requirement of any potential activity substitution being commutative: The possibility of returning to an original activation state must never be restricted to cyclic paths only. Statement 2 defines the requirement that all indirectly reachable alternatives can also be reached in a direct way: Switching from one alternative to another one must never require a detour. Statement 3 states that all exchangeable activities must have the same set of predecessors and successors associated: They take exactly the same position within the process. Statement 4 defines that the set of activity dependencies has to be empty: The number of elements contained in A is constant as it is thus not possible to insert or remove activities.

2.3 Modeling Patterns

This section illustrates how the concept of alternative activities can be used for the formal description of typical forms of process modification. In the following, mode alternation, resource alternation/capacity change, activity insertion/removal, order change and serialization/parallelization are discussed. For improved readability a simplified form of notation is used in the following: $i \rightarrow j$ defines that $p_{i,j} \in P^+$, $i > n \times k$ defines that $q_{i,k} = n \in Q^+$, $i \Rightarrow j$ defines that $x_{i,j} \in X^+$, $i \Leftrightarrow j$ defines that $x_{i,j}$, $x_{j,i} \in X^+$, $i][j$ defines that $m^{][}_{i,j} \in M^+$, $i][j$ defines that $m^{][}_{i,j} \in M^+$, $i|[j$ defines that $m^{|[}_{i,j} \in M^+$ and $i||j$ defines that $m^{||}_{i,j} \in M^+$.

Mode Alternation

An execution *mode* can be defined as a fixed combination of duration and resource requirements [5]. Activities, for which different execution modes are available, are considered *multi-mode*. A mode *alternation* corresponds to the switch from a previously chosen mode to another one. In the context of the RCPSP, it is a particular characteristic of the MRCPSP to be able to optimize the current mode selection along with the activity sequence. It has already been discussed under which circumstances the *x-RCPSP* can be converted into the more specific Multi-mode RCPSP (see Sect. 2). In this section, it is now shown how potential mode alternations can be described by use of the specific *x-RCPSP* constructs.

For this purpose, we consider an original network of three activities a, b and c, forming a sequence of alphabetical order. The mode of activity b shall be variable: Mode 1 corresponds to the original version, mode 2 to a slower but less resource-intense version and mode 3, finally, to a faster version requiring more resources. To describe the possibility of changing the execution mode of an activity, we insert one alternative activity per option into A^+: Instead of having b in the network we thus distinguish b_1, b_2 and b_3, all of which are based on the original version of the activity: The associated precedence relations are identical and differences merely concern durations and resource requirements. The execution of a mode alternation corresponds to the substitution of an activity with another one. The respective possibilities (i.e. the options of exchanging any pair of alternatives) are described in X^+. For the considered example, Table 1 compares the original with the accordingly modified network.

Resource Alternation and Capacity Change

Interventions regarding the dynamic modification of resource requirements and availabilities can also be modeled within the framework of the *x-RCPSP*:

Table 1. Mode alternation in the x-$RCPSP$

	Original	Modified network
A^+	a, b, c	a, b_1, b_2, b_3, c
P^+	$a \rightarrow b,$	$a \rightarrow b_1, a \rightarrow b_2, a \rightarrow b_3,$
	$b \rightarrow c$	$b_1 \rightarrow c, b_2 \rightarrow c, b_3 \rightarrow c$
X^+	ϕ	$b_1 \Leftrightarrow b_2, b_1 \Leftrightarrow b_3, b_2 \Leftrightarrow b_3$
M^+	ϕ	ϕ

Both the option of switching between alternative resources and the option of modifying resource capacities can be defined based on the previously described pattern of activity mode alternation.

1. The definition of alternative resources is first of all based on the introduction of an additional resource type. The possibility to execute an activity on this *or* the originally intended resource is described through the introduction of an additional activity mode. This way, full flexibility is provided in the definition of alternatives: Effects on costs, durations and resource requirements can be described per activity, for example. The drawback of the additional modeling workload can be alleviated through the provision of abstract modeling constructs.

2. The possibility of capacity change is described in a similar way: Instead of changing u_k directly, an additional (alternative) resource type is introduced, representing the available standby units. For each activity which may trigger a temporary extension of resource capacities an alternative activity is introduced, defining the relationship between the process step and the reserves. We claim that this approach represents an appropriate method for the description of respective interventions: In realistic scenarios capacity changes are typically motivated by and executed for *specific* activities. Moreover, a high level of flexibility is provided through the possibility to define dependencies per activity. Again, additional modeling workload can be eliminated through the provision of appropriate modeling constructs.

Activity Insertion/Removal

The dynamic insertion or removal of an activity represents an elementary form of potential process variation. This section illustrates how this option can be described by the use of the proposed constructs.

Again, we consider a simple sequence of three activities a, b and c as the original network. The possibility of inserting an additional process step e between b and c (or removing it from there, respectively) shall be described. For this purpose we distinguish two alternative versions of the optional activity's predecessor b : b_1 is bound with the execution (i.e. *activation*) of e whereas b_2 is bound with the omission (i.e. *deactivation*) of e. As regards precedence

Table 2. Activity insertion/removal in the x-RCPSP

	Original	Modified Network
A^+	a, b, c	a, b_1, b_2, c, e
P^+	$a \rightarrow b$,	$a \rightarrow b_1$, $a \rightarrow b_2$,
	$b \rightarrow c$	$b_1 \rightarrow e$, $e \rightarrow c$, $b_2 \rightarrow c$
X^+	$\emptyset$	$b_1 \Leftrightarrow b_2$
M^+	$\emptyset$	$b_1][e$, $b_2\|\|e$

constraints, e is executed after its predecessor b_1 and before the start of all successors of the original b. The insertion or removal of the optional activity corresponds to the switch from one alternative predecessor to the other one. The respective possibility is described through the insertion of appropriate elements into X^+. As regards the associated modifications in the set of *active* activities, M^+ is used to define respective dependencies: Activity e is activated whenever b_1 is activated and deactivated whenever b_2 is deactivated. Correspondingly, Table 2 compares the original network to a modified version in which it is possible to insert and remove activity e.

Order Change

Alternative activities can also form the basis for the description of the possibility to change the execution order of two arbitrary process steps.

If we consider a sequence of four activities a, b, c and d and if the possibility to exchange the positions of b and d shall be described, two versions of process execution can be distinguished: In one version b is considered before, in the other version b is considered after d. Correspondingly, each of the movable process steps is replaced by two alternative activities: One of these alternatives is positioned according to the original process, the other one is executed at the position of the respective counterpart. In the considered example, $b_1(d_1)$ inherits all related precedence constraints from $b(d)$ whereas $b_2(d_2)$ is attached to the predecessors and successors of $d(b)$. The possibility to perform the order change is defined in X^+ and the dependencies between b and d are described in M^+: It has to be guaranteed that always the same alternatives are *active* for both activities. Table 3 summarizes the differences between the original and the modified network which provides the possibility to change the execution order of b and d.

Serialization/Parallelization

Another common form of dynamic process variation is the serialization of what has been planned for parallel or the parallelization of what has been planned for serial execution. In the following it is described how this can be formulated within the x-RCPSP.

We consider a sequence of six activities a to f. The possibility of parallelizing the subsequence b to d with the execution of e shall be described. For this

Table 3. Order change in the x-$RCPSP$

	Original	Modified network
A^+	$a,\ b,\ c,\ d$	$a,\ b_1,\ b_2,\ c,\ d_1,\ d_2$
P^+	$a \to b,$	$a \to b_1,\ a \to d_2,$
	$b \to c,$	$b_1 \to c,\ d_2 \to c,$
	$c \to d$	$c \to d_1,\ c \to b_2$
X^+	ϕ	$b_1 \Leftrightarrow b_2,\ d_1 \Leftrightarrow d_2$
M^+	ϕ	$b_1][d_1,\ b_1]|d_2,$
		$d_1][b_1,\ d_1]|b_2,$
		$b_2][d_2,\ b_2]|d_1,$
		$d_2][b_2,\ d_2]|b_1$

Table 4. Parallelization/serialization in the x-$RCPSP$

	Original	Modified Network
A^+	$a,\ b,\ c,\ d,\ e,\ f$	$a,\ b,\ c,\ d,\ e_1,\ e_2,\ f$
P^+	$a \to b,$	$a \to b,\ a \to e_2,$
	$b \to c,$	$b \to c,$
	$c \to d,$	$c \to d,$
	$d \to e,$	$d \to e_1,\ d \to f,$
	$e \to f$	$e_1 \to f,\ e_2 \to f$
X^+	ϕ	$e_1 \Leftrightarrow e_2$
M^+	ϕ	ϕ

purpose, it is sufficient to introduce an alternative for the first activity of the latter sequence: e_1 represents the option of serial execution whereas e_2 represents the option of parallel execution. As regards precedence relations, e_1 merely replaces the original activity e, whereas e_2 is a successor of all predecessors (a) of the first activity of the former sequence (b) and a predecessor of all successors (f) of the original activity (e). Moreover, the last element of the originally preceding subsequence (d) has to be linked directly with the successor/s of the originally succeeding subsequence (f). The possibility to switch between serial and parallel execution is expressed through the definition of a bidirectional substitution possibility in X^+. No mutual dependencies have to be defined. Table 4 compares the original with the accordingly modified network.

3 Solving the Extended Model

This section describes how the extended version of the RCPSP can be solved based on an evolutionary approach. The choice of a metaheuristic optimization procedure has mainly been made for reasons of performance: Respective search methods typically can provide good results for larger problems in shorter time than exact optimization approaches of mathematical programming [11]. This feature corresponds to the realistic requirements of real-time disruption management.

Even though the main focus of this section is on the provision of decision support in the area of process disruption management in appropriate time, the possibilities of using the *x-RCPSP* for classical scheduling tasks such as the generation and optimization of an initial schedule are also described. In the following, first the notion of schedule and activity list is introduced. Then, the generation of an initial solution is discussed before finally the evolutionary approach for its incremental optimization is presented.

3.1 The Object of Optimization

The aim of disruption management is the identification of a set of interventions, which can be applied to the currently existing schedule in response to a disruption. In the context of the *x-RCPSP* it is thus necessary to optimize the *activation* state along with the activity sequence.

Schedules represent the start and end point of optimization. Basically, a *schedule* corresponds to a vector $(\beta_1, \beta_2, \ldots, \beta_n)$, grouping the starting times β_i of all *active* activities. If we consider A_t the set of activities carried out at a point in time t, a schedule is considered valid if all of the following statements are true (cf. [9]):

$$\beta_i \geq 0 \quad \forall i \in A \tag{5}$$

$$\beta_i + d_i \leq \beta_j \quad \forall p_{i,j} \in P \tag{6}$$

$$\Sigma_{i \in At} q_{i,k} \leq u_k \quad \forall k \in R, \forall t \tag{7}$$

Constraint 5 defines the domain for all starting times: No negative values are allowed. Constraint 6 makes sure that all precedence constraints are respected: The difference between the starting times of two linked activities must be equal to or greater than the duration of the preceding one. Constraint 7 finally defines that resource requirements must never exceed the available capacities.

Due to the difficulty of operating directly on time values when doing optimization [4], it is a common approach to introduce an intermediary layer of solution representation [9]: Respective forms of representation are composed of easily describable and modifiable elements and can be transformed into a corresponding schedule unambiguously. From the various potential candidates (see [9] for an overview) we decided on the use of *activity lists*: λ corresponds to a precedence feasible list sorting all *active* activities in the order they shall be considered during the generation of a schedule. Respective schemes for the conversion of λ to a final set of time values have been described by Kolisch et al. [9] and Hindi et al. [4] for example: Their approaches are based on the sequential insertion of the list's activities at the earliest possible starting time. Note that these Schedule Generation Schemes (SGS) always generate the same set of starting times for an activity list whereas one and the same schedule might be associated with various different lists.

3.2 The Initial Solution

In the context of disruption management optimization aims at the identification of the activity list which provides the best combination of final schedule and associated interventions. As the proposed approach is based on incremental improvement, the starting point of optimization is an initial activity list corresponding to a disrupted and/or yet unoptimized schedule. As regards the generation of the respective λ_0, basically two scenarios can be distinguished: Either a currently existing schedule has to be considered and optimized (as in disruption management) or no schedule is given and the reference process represents the only starting point for optimization (as in classical schedule optimization).

For disruption management, the conversion of the given schedule into an initial activity list is straight-forward: Since the existing timetable is assumed to respect all precedence requirements, it is sufficient to simply sort all elements $i \in A^+$ which have actually been considered for execution and which have not been started yet according to their starting times. If, alternatively, no existing schedule is given, a valid and feasible sequence has to be generated based on the reference process A^0. Algorithm 1 summarizes the respective procedure (cf. [4]) which can be used for the transformation of any combination of activity set and associated precedence constraints into a valid activity list λ: For this purpose, all contained activities are added sequentially. In each step, first the set of currently schedulable process steps A^* is determined (line 2): It basically consists of all activities which have not been scheduled so far and which do not have any or only previously considered predecessors. Note that P_i is used to refer to the set of all preceding activities of process step i. If A^* is empty before all elements have been added to λ, the network described by the processed sets is over-constrained and no valid activity list can be identified: The method returns without any result in line 3. Otherwise, an arbitrary element of A^* is selected and added at the end of λ (line 4). After all elements of the considered set of activities have been added, the method returns the thereby created activity list.

Algorithm 1 Generate Activity List (A, P)

1:    **repeat**
2:        $A^* \leftarrow \{i \in A | i \notin \lambda \wedge (P_i = \emptyset \vee P_i \subseteq \lambda)\}$
3:        **if** $A^* = \emptyset$ **then** return false
4:        **else** add an arbitrary element of A^* at the end of λ
5:    **until** $|\lambda| = |A|$
6:    return λ

3.3 An Evolutionary Algorithm

In an evolutionary algorithm (EA), optimization is accomplished through the continuous evolution of a population: Each generation comprises the fittest individuals of the previous generation and their children, which are generated through recombination and mutation. The main idea behind this concept is that a combination of good solutions might result in or be at least close to even better ones.

This section introduces an evolutionary algorithm for the optimization of the x-$RCPSP$. First, some general remarks are made on initial population, fitness function and selection scheme before afterwards specific versions of the crossover and mutation operators are described, which take the existence of alternative activities into account.

Initial Population, Fitness and Selection

The initial population consists of the initial solution and a certain amount of fellow solutions: All of them are deduced directly from λ_0 through the application of the mutation operator (as discussed below). This first generation therefore combines the option of not intervening at all with several possibilities of applying exactly one form of intervention.

The assessment of the quality of a population and the comparison of solutions is based on a so-called fitness function. In the context of the x-$RCPSP$, this function evaluates activity lists through their conversion into one single numeric value: Both the set of applied interventions as well as the implicitly defined schedule are considered in light of the predetermined goals of schedule optimization. Whereas most scheduling approaches for the resolution of the RCPSP focus on the minimization of total process execution time (the so-called *makespan*), disruption management is rather concerned with the implications of earliness and tardiness, costs for interventions as well as the difference from the original plan, for example.

As long as the current population does not fulfill the specified optimization criteria, a new generation is deduced from the existing one. It is composed of the fittest individuals and several children, the parents of which are selected with a probability proportional to their relative fitness.

A Crossover Operator for the x-RCPSP

Handling the potentially distinct sets of activities contained within the activity lists represents the main difficulty in the combination of two parent solutions. A procedure based on the idea that one parent λ_a prescribes the interventions to consider (i.e. the *activation* state of the child) whereas the other one λ_b defines relative priorities of the contained process steps (i.e. the list order) is summarized in the following.

First it is checked whether the activity sets associated with both lists are identical: If so, an RCPSP-related crossover operator can be applied to the lists (see [4], for example). Otherwise, an x-$RCPSP$-specific procedure is executed. The problem that one activity list shall prescribe the order of a distinct set of activities is resolved by the use of a so called *transition set* $T \subseteq X+$. This set basically describes how to convert the elements of λ_b into λ_a If X_a is the set of substitutions which led from λ_0 to λ_a, T combines all elements which either exist only in X_a or only in X_b: Note that for a successful conversion it has to be possible to invert all substitutions which are exclusive to the latter set. Based on this transition set and the prescribed order of activities, a new activity list is generated in an iterative procedure: The elements of λ_b are either appended, replaced by potential substitutes or omitted (if this omission is the result of any kind of relevant activity dependency).

A Mutation Operator for the x-$RCPSP$

As regards mutation, which is potentially applied to newly generated children in order to avoid early convergence to only local optima, again a specific version of the operator has to be introduced for the x-$RCPSP$. A procedure considering also the possibility to perform process variations (i.e. to exchange alternative activities) is briefly described in the following.

First, a random value is generated which defines whether rescheduling or a process variation shall be applied: A fixed value θ defines respective probabilities. In the former case, again an RCPSP-related method can be applied for the mere rearrangement of the elements contained within the activity list. In the latter case, an activity has to be replaced by an alternative: For this purpose a potential substitution $x_{i,j} \in X^+$ is randomly selected for an arbitrary element of λ: Activity i is deactivated and thus removed from the activity list whereas activity j is activated and thus added to the activity list. Of course all kind of dependencies have also to be considered. Note also that the exchange operation has always to result in a precedence-feasible activity list: An activity can only be inserted into the list after its last predecessor and all associated successors have to be shifted to its right-hand side.

4 Exemplary Application

This section first provides an illustrative example for the application of the previously presented framework to a realistic problem of real-time disruption management: After the introduction of the turnaround process – the most typical airport ground process – and three exemplary forms of potential intervention, it is shown how the x-$RCPSP$ can be used for its description and how the evolutionary optimization approach can be applied. Finally, along with possibilities of further improvements the results of turnaround-related as well as more generic performance evaluations are discussed.

4.1 Overview

The presented approach of disruption management can be applied to various problems in various domains. Wherever it is necessary to provide comprehensive decision support in the operative management of disruptions occurring during the execution of time- and resource-dependent processes, the respective concepts can be used as a basis for the proposal of interventions concerning rescheduling and dynamic process variations. Project management (see [1,3]), production planning (see [12, 13]), supply chain management (see [1, 14]), logistics management (see [1]) or traffic flow management represent typical examples of potential fields of application.

Another field, in which disruption management plays a particularly important role, is the domain of air traffic (see [2, 15–17]). Whereas existing applications mainly focus on aircraft and crew scheduling, we will illustrate how the proposed concepts can be applied for real-time disruption management in the context of the turnaround process. This process basically combines all activities carried out at an airport while an aircraft is on ground. Instead of considering all actually relevant process steps, a simplified version will be regarded, basically corresponding to the combination of core processes as mentioned by Carr [18]: After the plane reaches its gate or stand position, first the incoming passengers leave the aircraft. It is then fueled, cleaned and catered simultaneously before the outgoing passengers enter the plane. Finally, it leaves its position heading for the runway.

In the following, we will assume an instance of this process, in which a disruption occurs during taxi-in, prior to deboarding. This way, a departure delay is caused by the delay of the first activity and the implied shift of all succeeding process steps. For the resolution of this problem we assume the existence of three basic forms of potential process variation: First, an acceleration of deboarding can be reached through the assignment of additional busses. Second, it is possible to shorten cleaning, if in exchange the cabin is additionally inspected by the cabin crew prior to boarding. Third, fueling and boarding can be parallelized if the fire brigade is present for supervision. As regards potential options of rescheduling, respective possibilities are defined in the process structure itself.

4.2 Modeling the Turnaround Process

Along with the reference process we define the possibilities of dynamic process variations based on the patterns introduced in Sect. 2. A potential acceleration of boarding through the assignment of additional resources corresponds to the simple option of mode alternation. Shortening cleaning and inserting an additional step of inspection corresponds to a mixture of mode alternation and activity insertion. Finally, the possibility to execute boarding and fueling in parallel corresponds to a specific version of activity parallelization. The model resulting from the application of the respective patterns is summarized in Table 5.

Table 5. Formal description of the exemplary turnaround process

Set	Content
R	$Bus, Firebrigade$
A^0	$Start, Deb, Fue, Cat, Cle, Boa, End$
A^+	$Start, Deb, Deb^{Bus}, Fue, Fue^{Par}, Cat, Cle, Cle^{Red}, Ins, Boa, End$
P^+	$Start \rightarrow Deb, Start \rightarrow Deb^{Bus}, Deb \rightarrow Fue, Deb \rightarrow Fue^{Par}, Deb \rightarrow Cat, Deb \rightarrow Cle, Deb \rightarrow Cle^{Red}, Deb^{Bus} \rightarrow Fue, Deb^{Bus} \rightarrow Fue^{Par}, Deb^{Bus} \rightarrow Cat, Deb^{Bus} \rightarrow Cle, Deb^{Bus} \rightarrow Cle^{Red}, Fue \rightarrow Boa, Fue^{Par} \rightarrow End, Cat \rightarrow Boa, Cle \rightarrow Boa, Cle^{Red} \rightarrow Ins, Ins \rightarrow Boa, Boa \rightarrow End$
Q^+	$Deb > 1 \times Bus, Deb^{Bus} > 2 \times Bus, Fue^{Par} > 1 \times Firebrigade$
X^+	$Deb \Leftrightarrow Deb^{Bus}, Fue \Leftrightarrow Fue^{Par}, Cle \Leftrightarrow Cle^{Red}$
M^+	$Cle^{Red}][Ins, Cle\|Ins$

Airport Ground Process Disruption Management

The aim of disruption management is the minimization of the negative impact associated with an occurring disruption: In the context of the turnaround process, an exemplary goal might thus be the elimination of all pending delays. As regards the *initial solution*, we assume that $<Deb, Fue, Cle, Cat, Boa>$ can be extracted as λ_0. By *mutating* this activity list, a full population can be generated as starting point of optimization: Examples of respective instances are $<Deb, Cle, Fue, Cat, Boa>$ and $<Deb, Fue, Cle^{Red}, Ins, Cat, Boa>$. If none of the contained solutions fulfils a predetermined stopping criterion, a new generation is deduced from the fittest individuals: Applying the *crossover* operator on two parents $\lambda_a = <Deb^{Bus}, Fue, Cle, Cat, Boa>$ and $\lambda_b = <Deb^{Bus}, Cat, Cle^{Red}, Ins, Boa, Fue^{Par}>$ generates the child activity list $<Deb^{Bus}, Cat, Cle, Fue, Boa>$ by use of the transition set $T = \{Cle^{Red} \Rightarrow Cle, Fue^{Par} \Rightarrow Fue\}$. As soon as the goal of optimization is reached or a certain amount of time has passed, the genetic algorithm stops and returns a set of the best solutions found so far. The respective schedules implicitly describe associated interventions.

4.3 Prototype Implementation

The presented approach has been implemented in a Java-based prototype. In an exemplary setting, we considered $p = 20$ instances of the discussed version of the turnaround process: We defined the durations of the elements in A^+ to be $(0, 15, 7, 20, 20, 9, 14, 8, 1, 15, 0)$ and assumed the occurring disruption to be a shift of all process deadlines from some originally planned ending time to 0. Given the availability of 10 busses and 3 fire brigades, the unmodified reference processes initially cause an overall delay of 1,150 min. Based on the $n = 3$ provided modification possibilities, this value can be reduced to 984 min at most: Regarding dynamic process variations only, in the worst

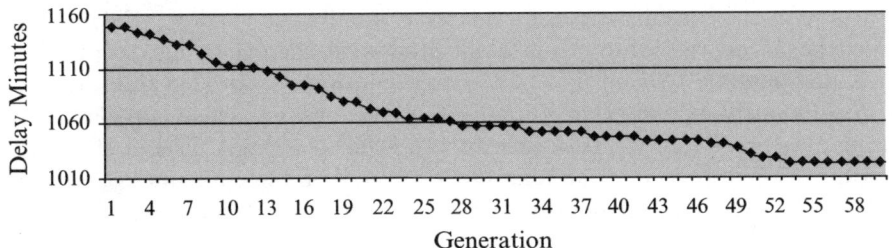

Fig. 1. Reduction of delay minutes in turnaround optimization

case $2^{n^* p} = 2^{60}$ schedules have to be evaluated for the identification of this theoretical optimum. In our heuristic approach, we considered 60 generations with 10 members each: By evaluating only 600 solutions, about 75% of the full optimization potential could be tapped within 4 s on a standard PC with 1,800 MHz and 512 MB RAM. Figure 1 illustrates an exemplary reduction of the delay minutes associated with the best known solution throughout the generations.

4.4 Further Results and Improvements

In a more generic and comprehensive approach, we have evaluated the performance of the proposed algorithms for randomly generated instances of disruption management problems. For this purpose we implemented both a program for the parameterized generation of problem instances as well as an extensive framework for the solution of DM problems in Java. Based on the former component, 320 classified test cases have been generated: Particularly the number of activities, precedence constraints and resource requirements has been varied. By use of the DM framework we then wanted to find out how much of a known optimization potential could be tapped within the first few seconds of optimization: We chose the approach of comparing the best result known after several seconds to the best result known after about half an hour of optimization, as even by use of the most powerful procedures it was not possible to identify the exact optimum solution to most of the generated instances in reasonable time. It could be observed that even for instances containing 100 activities more than 60% of the known potential could be tapped within only five seconds and that the proposed operators thus represent a good starting point for further improvements and evaluations. For further details regarding the respective evaluation please see [19].

Our most recent results indicate that particularly local forms of rescheduling can significantly improve the algorithm's performance on DM problems: Motivated by the idea of responding to a disruption right there where it takes effect, problems are resolved on a local level in the approach named Local Rescheduling (LRS): Rescheduling is regarded as an iterative process which starts with the consideration of a relatively small time window, which is identified based on the information associated with the occurring disruption. If it is not possible to identify a satisfying solution within this time frame, it is continuously extended until finally the entire search space is regarded. By using this procedure we were able to tap even more than 65% of the known optimization potential for all regarded classes of problem instances containing 1,000 activities. Further details on LRS can be found in [20].

5 Conclusions

This chapter described how both rescheduling and dynamic process variations can be considered in a comprehensive approach to disruption management. In various modeling patterns, the possibilities to describe potential forms of interventions based on the concept of alternative activities have been illustrated. An extended version of the well-known RCPSP has been introduced and a metaheuristic optimization approach based on an evolutionary algorithm has been presented. In the last chapter, the application of the introduced concepts to a real-world problem has been discussed based on the airport turnaround process before finally some remarks on the general performance of the procedures presented in this chapter have been made: The proposed algorithms are particularly powerful if they are combined with some technique of search space reduction (based on Local Rescheduling approaches, for example).

References

1. G. Yu, X. Qi, Disruption Management: Framework, Models and Applications, World Scientific Publishing, Singapore, 2004
2. J. Clausen, J. Hansen, J. Larsen, A. Larsen, Disruption Management, ORMS Today, 28: 40–43, 2001
3. G. Zhu, J.F. Bard, G. Yu, Disruption Management for Resource-Constrained Project Scheduling, Journal of the Operational Research Society, 56: 365–381, 2005
4. K.S. Hindi, H. Yang, K. Fleszar, An Evolutionary Algorithm for Resource-Constrained Project Scheduling, IEEE Transactions on Evolutionary Computation, 6: 512–518, 2002
5. S. Hartmann, Project Scheduling with Multiple Modes: A Genetic Algorithm, Annals of Operations Research, 102: 111–135, 2001
6. C. Artigues, P. Michelon, S. Reusser, Insertion Techniques for Static and Dynamic Resource Constrained Project Scheduling, European Journal of Operational Research, 149: 249–267, 2003
7. A. Elkhyari, C. Gueret, N. Jussien, Constraint Programming for Dynamic Scheduling Problems, ISS'04, Japan, 84–89, 2004
8. J.C. Beck, M.S. Fox, Constraint Directed Techniques for Scheduling with Alternative Activities, Artificial Intelligence, 121: 211–250, 2000

9. R. Kolisch, S. Hartmann, Heuristic Algorithms for Solving the Resource-Constrained Project Scheduling Problem: Classification and Computational Analysis, Project scheduling: Recent models, algorithms and applications, 147–178, Ed: J. Weglarz, Kluwer, Amsterdam, Netherlands, 1999

10. R.K.-Y. Li, R.J. Willis, Alternative Resources in Project Scheduling, Computers and Operations Research, 18: 663–669, 1991

11. J. Bautista, A. Lusa, R. Suarez, M. Mateo, R. Pastor, A. Corominas, Application of Genetic Algorithms to Assembly Sequence Planning with Limited Resources, IEEE International Symposium on Assembly and Task Planning, Porto, Portugal, 411–416, 1999

12. J. Yang, X. Qi, G. Yu, Disruption Management in Production Planning, Naval Research Logistics, 52: 420–442, 2005

13. Y. Xia, M.H. Yang, B. Golany, S. Gilbert, G. Yu, Real-Time Disruption Management in a Two-Stage Production and Inventory System, IIE Transactions, 36: 111–125, 2004

14. M. Xu, X. Qi, G. Yu, H. Zhang, C. Gao, The Demand Disruption Management Problem for a Supply Chain System with Nonlinear Demand Functions, JSSSE, 12: 82–97, 2003

15. B. Thengvall, J.F. Bard, G. Yu, Balancing User Preferences for Aircraft Schedule Recovery During Irregular Operations, IIE Transactions on Operations Engineering, 32: 181–193, 2000

16. N. Kohl, A. Larsen, J. Larsen, A. Ross, S. Tiourine, Airline Disruption Management – Perspectives, Experiences and Outlook, Technical Report, IMM, Technical University of Denmark, 2004

17. J. Clausen, A. Larsen, J. Larsen, Disruption Management in the Airline Industry – Concepts, Models and Methods, Technical Report 2005-01, IMM, Technical University of Denmark, 2005

18. F.R. Carr, Robust Decision Support Tools for Airport Surface Traffic, PhD Thesis, Massachusetts Institute of Technology, 2004

19. J. Kuster, D. Jannach, G. Friedrich, Handling Alternative Activities in Resource-Constrained Project Scheduling Problems, IJCAI 2007 – Proceedings of the Twentieth International Joint Conference on Artificial Intelligence, 1960–1965, 2007

20. J. Kuster, D. Jannach, G. Friedrich, Local Rescheduling – A Novel Approach for Efficient Response to Schedule Disruptions, IEEE Symposium on Computational Intelligence in Scheduling, 79–86, 2007

Part II

Intelligent Search and Querying

On the Evaluation of Cardinality-Based Generalized Yes/No Queries

Patrick Bosc, Nadia Ibenhssaien, and Olivier Pivert

IRISA/ENSSAT Technopole Anticipa BP 80518, 22305 Lannion Cedex France
bosc@enssat.fr, ibenhssaien@enssat.fr, pivert@enssat.fr

Summary. This chapter deals with the querying of possibilistic relational databases, by means of generalized yes/no queries whose form is: "to what extent is it possible and certain that the answer to Q satisfies property P". Here, we consider cardinality-based generalized yes/no queries (in this case, property P is about cardinality) for which a processing technique is proposed, which avoids computing all the worlds attached to the possibilistic database.

1 Introduction

In this chapter, we consider relational databases where some attribute values are imprecisely known and are represented as possibility distributions. Possibility theory [1] provides an ordinal model for uncertainty where imprecision is represented by means of a preference relation encoded by a total order over the possible situations. This approach provides a unified framework for representing precise values, as well as imprecise ones (regular sets) or vague ones (fuzzy sets), and various null value situations [2].

Let us recall that an imprecise database can be seen as a set of regular databases, called worlds, associated with a choice for each attribute value. A compact, tractable calculus valid for a subset of the relational algebra has been devised (see [3,4]). In this context, the result of a query is a possibilistic relation whose interpretations correspond to more or less possible results, equivalent to those which would have been obtained with a calculus applied to the worlds of the possibilistic database. This achievement is interesting from a methodological point of view, but the use of this type of result by a final user can be somewhat delicate. So, it becomes convenient to define queries which are more specialized to fit user needs. To meet this goal, possibilistic queries (concept initially introduced by Abiteboul [5] in the framework of null values) have been studied. Their generalized form is: "to what extent is it possible and certain that the answer to Q satisfies property P", and we have studied in [6] the case where P is: "contains a given (specified) tuple t". In this chapter, we

P. Bosc et al.: *On the Evaluation of Cardinality-Based Generalized Yes/No Queries*, Studies in Computational Intelligence (SCI) **109**, 65–79 (2008)
www.springerlink.com   © Springer-Verlag Berlin Heidelberg 2008

are interested in cardinality-based generalized yes/no queries where property P is: "contains at least (at most, exactly,...) q distinct elements" and we tackle the algorithmic aspects of their evaluation. We will see that, contrary to the previous case, we have to face complexity problems since a sequential scan of the result of Q is not sufficient. To the best of our knowledge, there does not exist any previous research work about this issue.

The structure of the chapter is the following. In Sect. 2, the notion of a possibilistic relational database is introduced. Then, the data model requested for a valid compact processing of algebraic queries is described in Sect. 3. In Sect. 4, we propose a "trial and error" algorithm to process cardinality-based generalized yes/no queries and we briefly discuss the complexity of the approach proposed. Section 5 studies what would be the impact on the algorithm/performances if the probabilistic model of uncertainty was used instead of the possibilistic one. Finally, the conclusion summarizes the contributions of the chapter and draws some lines for future works.

2 Possibilistic Databases and Worlds

In contrast to a regular database, a possibilistic relational database D may have some attributes which take imprecise values. In such a case, a possibility distribution is used to represent all the more or less acceptable candidates for the attribute. In the rest of this chapter, only finite possibility distributions are taken into account.

The first version of a possibilistic database model was introduced by Prade in the mid 80s. From a semantic point of view, a possibilistic database D can be interpreted as a set of usual databases (also called worlds), denoted by rep(D), each of which being more or less possible (one of them is supposed to correspond to the actual state of the universe modeled). This view establishes a semantic connection between possibilistic and regular databases. It is particularly interesting since it offers a canonical approach to the definition of queries addressed to possibilistic databases as will be seen later (Sect. 4). Any world Wi is obtained by choosing a candidate value in each possibility distribution appearing in D and its degree of possibility is the minimum of those of the candidates taken (according to the axioms of possibility theory).

Example 1. Let us consider the possibilistic database D involving two relations: im and pl whose respective schemas are IM(#i, ap, date, place) and PL(ap, lg, msp). Relation im describes satellite images of airplanes and each image, identified by a number (#i) (Table 1), taken on a certain location (place) a given day (date) is supposed to include a single (possibly ill-known, due to the imprecision inherent in the recognition process) airplane (ap). Relation pl gives the length (lg) and maximal speed (msp) of each airplane and is a regular (precise) relation. With the extension of im: four worlds can be drawn, since there are two candidates for date (resp. ap) in the first (resp.

Table 1. Image Relation

im	#i	ap	Date	Place
	i_1	a_1	$\{1/d_1 + 0.7/d_3\}$	p_1
	i_3	$\{1/a_3 + 0.3/a_4\}$	d_1	p_2

second) tuple of im. Each of these worlds involves relation pl which has only precise values and one of the four regular relations issued from the possibilistic relation im.

3 An Extended Possibilistic Data Model

3.1 Objective

As mentioned before, a calculus based on the processing of the query Q against worlds is intractable and a compact approach to the calculus of the answer to Q must be found out. It is then necessary to be provided with both a data model and operations which have good properties (a) the data model must be closed for the considered operations, and (b) any query (applying to the possibilistic database D) must be processed in a compact way. In addition, its result must be a compact representation of the results of this query if it were applied to all the interpretations (worlds) drawn from D, i.e., rep(Qc(D)) = Q(rep(D)), where rep(D) denotes the set of worlds associated with D and Qc stands for the query obtained by replacing the operators of Q by their compact versions. This property characterizes data models called strong representation systems.

It turns out [3] that the initial relational possibilistic model cannot comply with this property in at least two respects (notably for the selection) (a) the recovery of "missing tuples", and (b) the accounting for dependencies between candidate values. An adapted data model, which has been defined in [3], is briefly described hereafter.

3.2 Representing Possibly Missing Tuples

Because some operations (e.g. selection) filter candidate values, there is a need at the compact level for expressing that some tuples can have no representative in some worlds. A simple solution is to introduce a new attribute, denoted by N (valued in [0, 1]), which states whether or not it is legal to build worlds where no representative of the corresponding tuple is present, and, if so, the influence of this choice in terms of degree of possibility. The value of N associated with a tuple t expresses the certainty of the presence of a representative of t in any world. A tuple is denoted by a pair N/t where N equals 1 for tuples of initial possibilistic relations as well as when no candidate value has been discarded.

Table 2. Possibilistic relation 'im'

im

#i	ap	Date	Place
i_1	B-727	d_1	p_1
i_2	ATR-72	d_1	p_2
i_3	{1/B-727 + 0.7/ATR-42}	d_2	p_4
i_4	{1/B-727 + 1/B-747}	d_2	p_2

Table 3. Performing Selection on 'im'

res

#i	ap	Date	Place	N
i_1	B-727	d_1	p_1	1
i_3	B-727	d_2	p_4	0.3
i_4	B-727	d_2	p_2	0

Example 2. Let us consider the following extension of the possibilistic relation im (Table 2):

The selection based on the condition "ap = B-727"(Table 3) leads to discard the candidates which are different from this desired value. Thanks to the introduction of attribute N, the result of the selection is:

In the second tuple N is equal to 0.3, i.e, 1 minus the possibility degree attached to the most possible alternative that has been discarded. From this result, it is possible to derive the interpretation made of the single tuple $<i_1$, B-727, d_1, $p_1>$ whose degree of possibility is: $\min(1, 1 - 0.3, 1 - 0) = 0.7$.

3.3 Multiple Attribute Possibility Distributions

Another aspect of the model is related to the fact that it is sometimes necessary to express dependencies between candidate values coming from different attributes in a same tuple. This requires that the model incorporates attribute values defined as possibility distributions over several domains. This is feasible in the relational framework thanks to the concept of a nested relation. In such relations, exclusive candidates are represented as weighted tuples. Therefore, level-one relations keep their conjunctive meaning, whereas nested relations have a disjunctive interpretation.

Example 3. Let us consider the following intermediate relation int-r (Table 4) involving the nested attribute X(date, place):

This relation is associated with 12 worlds since the first tuple admits three interpretations, the second and third ones have two interpretations among which $\emptyset$ (no representative).

In order to meet the objective of a compact processing of algebraic queries, the operators must be adapted so as to accept compact relations both as inputs and outputs. It turns out that only operations such that an input tuple participates in the production of at most one element of the result, can be expected to admit a compact version. As a consequence, the intersection,

Table 4. Intermediate Relation 'int-r'

int-r	#i	ap	X date	place	N
i_1	B-727	$\{1/<d_1, p_1>+0.7/<d_1, p_2>$ $+0.4/<d_3, p_2>\}$			1
i_3	B-727	$<d_1, p_2>$			0.3
i_4	$\{0.4/B\text{-}737\}$	$\{0.3/d_3, p_2>\}$			0

the difference and the Cartesian product (then the join in the general case) are discarded and the four acceptable operators are: the selection, the projection, the fk-join (a specific join) and the union.

3.4 A brief Survey of the Four Operations

Considering the specific objective of this chapter, we will not give a detailed presentation of the operators. We limit ourselves to a brief introduction and the operations are then illustrated by an example. The interested reader may refer to [3, 4] for more details.

The three aspects of the selection are: the removal of unsatisfactory candidate values, the computation of the degree of certainty attached to each output tuple and the introduction of appropriate nested relations in the output relation if needed.

The role of the projection in the regular case is to remove undesired attributes. Here, the projection must (1) keep the duplicates in level-one relations (this is justified in [3]), (2) suppress nested relations if necessary, (3) update the possibility degrees.

Beyond selections and projections, two binary operations can be processed in a compact fashion: fk-join and union. The fk-join allows for the composition of a possibilistic relation r of schema R(W, Z), where W and Z may take imprecise values, and a regular relation s whose schema is S(W, Y) where the functional dependency $W \rightarrow Y$ holds. It consists in completing tuples of r by adding the image of the W-component. By definition, this leads to a resulting relation involving the nested relation X(W, Y), which "connects" the pairs of candidates over W and Y.

Last, the union of two independent relations whose schemas are compatible keeps all the tuples issued from the two input relations without any duplicate removal.

Example 4. Let us consider the possibilistic database composed of the relations im1(IM), im2(IM) and pl(PL) (Tables 5–8) whose respective schemas are the ones introduced in Example 1. The relations im1 and im2 are assumed to contain images of airplanes taken by two distinct satellites. Let us consider the query looking for the existence of images of airplanes whose

Table 5. Relation-Plane "Pl"

pl	ap	lg	msp
	a_1	20	1,000
	a_2	25	800
	a_4	20	1,200
	a_5	20	1,000

Table 6. Relation Image "Im1"

im1	#i	Ap	Date	Place	N
	i_1	a_3	$\{1/d_1 + 0.7/d_3\}$	p_1	1
	i_2	$\{1/a_2 + 0.7/a_1\}$	d_1	p_2	1

Table 7. Relation Image "Im2"

im2	#i	ap	Date	Place	N
	i_3	$\{1/a_4 + 1/a_5\}$	$\{0.6/d_4 + 1/d_1\}$	p_3	1

Table 8. Query-Q Results

res	#i	X			Date	Place	N
		ap	lg	msp			
	i_2	$\{0.7/<a_1,$	$20,$	$1,000>\}$	d_1	p_2	0
	i_3	$\{1/<a_4,$	$20,$	$1,200> + 1/$	$\{1/d_1\}$	p_3	0.4
		$<a_5,$	$20,$	$1,000>\}$			

maximal speed is over $900 \, \text{km h}^{-1}$ and taken by either of the two satellites at a date different from d_3 and d_4, which corresponds to the algebraic query Q: fk-join(union(select(im1, date $\notin \{d_3, d_4\}$)), select(im2, date $\notin \{d_3, d_4\}$)), select(pl, msp > 900), {ap}, {ap}). With the extensions:

We obtain the resulting relation res hereafter:

which is associated with six worlds.

4 Cardinality-Based Queries

4.1 Introducing a Post-Processing

On the basis of the definitions of the algebraic operators given above, a query can be processed in a tractable way since operations are performed in a compact fashion. However, one may wonder about the usability of the result delivered by such a query, i.e., of a compact relation as such. We think that a convenient direction is to provide users with queries which are close to their needs and whose results are easily interpretable, and then to call on (embedded) algebraic queries. Hereafter, we are interested in possibilistic queries whose most general form is: "to what extent is it possible and certain that the answer to Q satisfies property P" and which are an extension of yes/no queries. These latter are of the form: "is it true that tuple t belongs to the answer to Q?" and were studied by Abiteboul [5] in the framework of null values. In this case the answer is uncertain and it can be yes, no or maybe (instead of yes or no). Possibilistic queries have been studied in [6] in the case where P is: "contains a given (specified) tuple t". In this chapter, we are interested in cardinality-based generalized yes/no queries where property P is: "contains at least (at most, exactly, ...) q distinct elements", the answer to these queries being the possibility and the necessity (certainty) that the answer to Q contains at least (at most, exactly, ...) q distinct elements. The evaluation of such queries is based on a two-step mechanism:

1. A compact processing of the associated algebraic query, which builds a compact relation res according to the procedure depicted in the preceding section.
2. A post-processing producing the final answer (i.e., the answer to the cardinality-based query, in the form of a pair of degrees Π/N).

4.2 Problem Raised by Cardinality-Based Queries

The post-processing of the compact result of Q entails determining the possibility attached to worlds involving a certain number of elements in order to compute the degrees Π/N of the event "to what extent is it possible and certain that the answer to Q contains at least (at most, exactly, ...) q distinct elements?". In fact, if we are able to compute the possibility in each case ("at least", "at most", "exactly", ...), we can also compute the necessity. Indeed, the necessity that the answer to Q contains at least (respectively at most) q distinct elements is 1 – the possibility that the answer to Q contains less than (respectively more than) q distinct elements. Moreover, the necessity that the answer to Q contains exactly q distinct elements is 1 – the possibility that the answer to Q contains less than or more than q distinct elements.

The problem is that some tuples of the resulting relation res can produce representatives which are duplicates. According to the case considered ("at least", "at most" or "exactly"), duplicates do not have the same impact. It turns out that, in all cases, the procedure attached to the post-processing must rely on a "trial and error" technique. Let us illustrate this through two examples.

Example 5. Let us consider the following relation res = {<{1/a_1 + 0.6/a_2}, b>/0.3, <a_1, b>/1} which is assumed to be the result of the compact processing of an algebraic query Q. The degree of possibility that the answer to Q contains at least two different tuples cannot be obtained by taking the most possible representative of the two tuples of res because they are identical (<a_1, b>). Thus, the representatives which are duplicates must be identified in order to compute the exact cardinality.

Table 9. Compact Processing of Query Q

res

A	B	N
a_2	$\{1/b_3 + 0.9/b_2\}$	1
a_2	b_3	1
$\{1/a_2 + 0.3/a_1\}$	b_3	0.3
$\{0.9/a_4 + 0.8/a_5\}$	$\{0.5/b_1\}$	0

Example 6. Let us consider the following relation res (Table 9) which is assumed to be the result of the compact processing of an algebraic query Q: and the cardinality-based query: "to what extent is it possible and certain that the answer to Q has at most 1 element?".

One may think that the possibility degree delivered to the user would be 0 because each world contains at least a representative of the two tuples with $N = 1$. But in fact, there may be some duplicates among these two tuples (it is indeed the case here), that is why the procedure attached to the post-processing must rely on a "trial and error" technique. The question is whether the procedure should only concern the tuples with $N = 1$ or all the tuples somewhat certain. The world made only of the most possible representatives of the first two tuples has the possibility degree $\min(1, 1, 1-0.3, 1-0) = 0.7$ while the one containing also the best representative of the third tuple (which is (a_2, b_3)) has the possibility degree $\min(1, 1, 1, 1 - 0) = 1$. This latter world contains one tuple because the three representatives are duplicates and it is more possible than the first one. The "trial and error" procedure should therefore consider all the tuples somewhat certain.

The case "exactly" is similar to that of "at least". The same kind of reasoning can be done for the other cases ("less than", "more than").

4.3 The Algorithm

The algorithm proposed here is based on a "trial and error" technique. It aims at delivering the possibility degree π of the most satisfactory world with respect to the desired cardinality. Such an algorithm calculates a series of vectors $V = (x_1, x_2, \ldots, x_n)$ (n being the number of tuples in relation res) where each component x_i takes its values in a finite set E_i. Ultimately, it aims at finding the best solution, i.e., the best vector V.

A solution is a vector V which represents a world (a candidate regular relation for the possibilistic relation res resulting from the compact evaluation of Q). Its dimension is the number n of tuples in relation res. The components of vector V are precise tuples (the ith position of V is the representative tuple produced by the ith tuple of relation res). Some positions of V may be empty, which occurs when the value N in relation res is different from 1 (the corresponding tuple may have no representative in a given world). The algorithm and the corresponding data structures are the following:

```
Procedure OptimalSolution(i integer)
begin
  compute E_i;
  for x_j in E_i do
    if satisfactory(x_j) then
      memorize(x_j);
      if solutionFound then
        if better then keepSolution endif
      else if stillPossible then
        optimalSolution(i+1) enfif
      endif;
      undo(x_j);
    endif;
  endfor;
end;
```

E_i: list of the precise tuples corresponding to the possible representatives of the ith tuple from res (including ϕ if $N < 1$);

Π_i: list of the respective possibility degrees π_j of each x_j in E_i ($1 - N$ if x_j is ϕ);

V: represents a world of res (a regular relation that is a possible answer to Q);

Pos: vector of the same dimension as V, it contains the possibility degrees of the tuples of V (the possibility degree associated to V being the minimum over Pos);

Card: the cardinality of vector V (the number of tuples in V different from ϕ);

BestΠ: the possibility degree of the most possible world found until then;

satisfactory(x_j): $\pi_j >$ BestΠ (it is possible for the current solution to be better than the best one already found only if the possibility of the current candidate tuple is over BestΠ, since the overall possibility degree of a world is computed by means of a minimum, cf. keepSolution below);

memorize(x_j): V[i] ← x_j; Pos[i] ← π_j;

b ← ($x_j \neq \phi$ and the same tuple is not already in V (it does not exist k in [1..i-1] such that V[k] = x_j));

if b then Card ← Card + 1;

solutionFound:
($i = n$) and Card $\geq$ q for the case "at least",
($i = n$) and Card $>$ q for the case "more than",
($i = n$) and Card $\leq$ q for the case "at most",
($i = n$) and Card $<$ q for the case "less than",
($i = n$) and Card $=$ q for the case "exactly";

better: $\min_{k\ in[1,\ n]}$ Pos[k] $>$ BestΠ;

keepSolution: BestΠ ← $\min_{k\ in[1,\ n]}$ Pos[k];

stillPossible: Card $+ (n - i) \geq$ q for the case "at least" (there is no hope to construct a solution containing at least q tuples from the current vector if its cardinality is not at least equal to q minus the number of tuples that can still be chosen). Analogous conditions related to other comparators are straightforward.

undo(x_j): Pos[i] $\leftarrow$ 0; if b then Card $\leftarrow$ Card $-$ 1.

By construction, this algorithm checks all the possible worlds of res. Since the pruning conditions discard the worlds that are either not satisfactory or whose possibility degree is lower than the optimal already found, one has the guarantee to obtain the best world in the end.

The necessity degree which is the answer to the query "to what extent is it certain that the answer to Q has at least (resp. at most, exactly, ...) q distinct elements?" is 1 minus the possibility degree obtained by the algorithm above for the query "to what extent is it possible that the answer to Q has less than (resp. more than, less than or more than, ...) q distinct elements?".

Remarks. In order to reduce the number of the worlds to be computed, some improvements can be brought to the algorithm above:

1. When BestΠ is equal to 1 the processing can be stopped.
2. The sets E_i can be ranked in decreasing order on the possibility degrees. In that case, once an unsatisfactory x_j is found, the loop can be stopped (because the following x-values would be unsatisfactory too). However, this ordering does not prevent from computing the whole worlds in some cases (when the only satisfying world is the last one built).
3. In the cases "at least", "more than" and "exactly", we can take advantage of the number n of tuples in relation res. For instance, if the user is interested in 5 responses (q = 5) while the relation res contains only three tuples, the result is obviously 0.
4. For the cases "at most" and "less than", the algorithm can be evaluated only on the tuples somewhat certain. Let t be an imprecise tuple whose necessity degree is 0, if we add a representative of t to the current solution, the possibility degree associated to the solution can only decrease (or stay the same). Since the criterion is of the form at most or less than, it is thus a better idea not to take a representative of t, this choice being possible at degree 1. Furthermore, if relation res contains at most (resp. less than) q tuples with N > 0, the procedure based on a "trial and error" technique is not needed any more and the result is Π = 1 (because the possibility distributions are normalized for tuples somewhat certain).
5. In some cases, it is not necessary to compute the degrees Π and N but only one of them: if the possibility degree is less than 1, the necessity degree will be 0 (due to the property $\Pi(A) < 1 \Rightarrow N(A) = 0$), and vice versa, if the necessity degree is more than 0, Π will be equal to 1.

4.4 Example

This example is intended for illustrating the functioning of the algorithm. Let us consider the possibilistic relation r (Table 10): and the cardinality-based

Table 10. Possibilistic Relation R

R

A	B
a_2	$\{1/b_2 + 0.9/b_3\}$
a_2	$\{1/b_3 + 0.3/b_4\}$
a_3	$\{1/b_2 + 0.4/b_4\}$
$\{1/a_2 + 0.5/a_1 + 0.6/a_3\}$	b_1
$\{0.8/a_4 + 0.6/a_5 + 1/a_3\}$	$\{0.7/b_1\}$

Table 11. Selection Results on R

res

A	B	N
a_2	$\{1/b_2 + 0.9/b_3\}$	1
a_2	$\{1/b_3 + 0.3/b_4\}$	1
$\{1/a_2 + 0.5/a_1\}$	b_1	0.4
$\{0.8/a_4 + 0.6/a_5\}$	$\{0.7/b_1\}$	0

query: "to what extent is it possible and certain that the answer to the query $Q = \text{select}(r, A \neq a_3)$ (Table 11) has at most two distinct elements?"

The evaluation of query Q leads to the resulting relation res:

The corresponding sets E_i and Π_i (only of the tuples somewhat certain, cf. Remark 4), after ranking them in decreasing order on the possibility degrees, are:

$$E_1 = (<a_2, b_2>, <a_2, b_3>) \qquad \Pi_1 = (1, 0.9)$$
$$E_2 = (<a_2, b_3>, <a_2, b_4>) \qquad \Pi_2 = (1, 0.3)$$
$$E_3 = (<a_2, b_1>, \emptyset, <a_1, b_1>) \qquad \Pi_3 = (1, 0.6, 0.5)$$

To illustrate how the algorithm works on this example, we use a tree representation. The tree is split into two parts for space reasons.

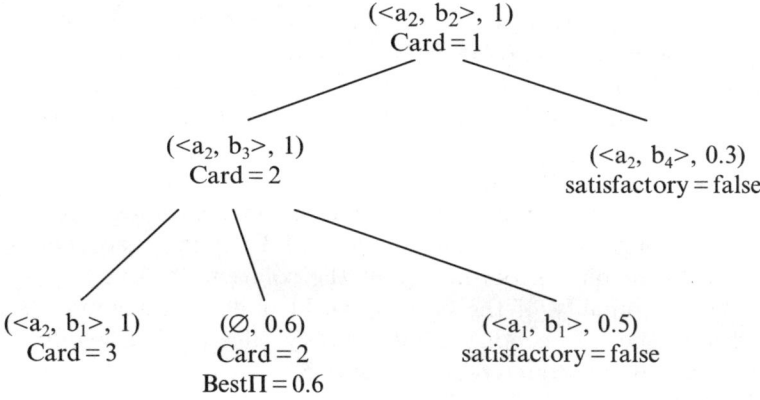

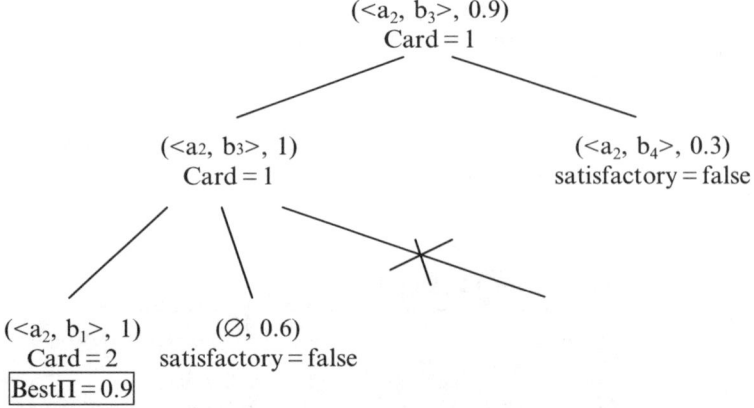

In the first part of the tree and for the node ($<a_2$, $b_4>$, 0.3), satisfactory ($<a_2$, $b_4>$) is false since at that point of the computations BestΠ = 0.6 while the solution in that branch could not be more possible than 0.3, we therefore leave the loop and come back to the calling context (the same reasoning can be done for the node ($<a_2$, $b_4>$, 0.3) of the second part of the tree.

For the node (ϕ, 0.6) of the second part of the tree, satisfactory (ϕ) is false since at that point of the computations BestΠ = 0.9. Given that the possibility degrees π_j are ranked in decreasing order, the following children of the parent of (ϕ, 0.6) are not generated.

The resulting possibility degree for the cardinality-based query above is 0.9, and the resulting necessity degree is therefore 0 (since Π < 1). In this example, we compute only five worlds instead of the 36 worlds of relation res thanks to the pruning conditions above.

4.5 About Performances

As for any of the algorithms of this family, the maximal complexity of the procedure (of Sect. 4.3) is exponential. Let us consider a relation res containing n tuples and m imprecise attributes. Let us assume that there are p candidate values per possibility distribution, and that, for each tuple, N equals 1. The time complexity in terms of recursive calls (which also corresponds to the number of computed worlds) is in $O(p^{m^*n})$, but one may expect to drastically reduce the number of computed worlds thanks to the pruning conditions.

Indeed, the algorithm described in Sect. 4.3 includes two pruning conditions: one based on the optimality of the solution (satisfactory(x_j)), the other on the cardinality of the current world under construction (stillPossible). Clearly, the first condition will be very effective if a highly possible satisfactory world is encountered early. If it is not the case (the extreme situation being that the final possibility degree equals zero), only the second pruning condition can have an effect and avoid constructing all the possible worlds of relation res. Let us notice, however, that the number of worlds attached to res is much smaller than the number of worlds attached to the initial possibilistic database, due to the reduction obtained by means of the compact evaluation of query Q. Anyway, in certain cases (when the cardinality of res is too important), due to the combinatorial nature of the algorithm, even when both pruning conditions can be used together, the complexity will still be too high to reach acceptable performances. An idea could then consist in computing only an approximate answer, i.e., an underestimation of the actual value of Π and N). For a criterion of the form "at least q", this could be done by means of a greedy algorithm scanning the resulting relation (previously ordered increasingly with respect to the possibility degree associated with each imprecise tuple) and choosing, for each imprecise tuple of res, the most possible candidate corresponding to a tuple not already chosen. Even when relation res has a "reasonable" size, such a greedy algorithm could be used as a pre-processing step in order to obtain – in a non-expensive way – a Π-value that could be used to initialize the variable BestΠ before running the trial and error procedure, so as to make the optimality-based pruning condition more effective.

5 Probabilistic Databases and Cardinality-based Generalized Yes/no Queries

5.1 Processing the Algebraic Query

A probabilistic database can also be interpreted as a weighted disjunctive set of worlds. Each world is obtained by choosing one among all the candidates pertaining to each probability distribution, and is associated with a degree corresponding to the product of the degrees tied to the candidate values appearing in it.

Concerning the database model, it has been shown in [7] that the model described in Sect. 3 constitutes also a strong representation system in a probabilistic framework, for the same set of algebraic operators (selection, projection, fk-join and union). The only difference with the possibilistic case is that we do not need an extra attribute N in order to have available the probability for an imprecise tuple to have a representative in any world. Indeed, in the probabilistic case, we know that some candidates have been discarded from a distribution when the sum of the degrees in this distribution is less than 1. In the case where a single distribution is considered, the probability attached to the situation where the tuple has no representative is equal to 1 minus the sum of the degrees attached to the remaining elements. Since the knowledge necessary to this calculus is available in the tuple, there is no need to explicitly store the degree itself in the relation. However, in order to have a generic model, attribute N can be kept even if the framework considered is the probabilistic one.

As to the processing of the algebraic query underlying a cardinality-based generalized yes/no query, it differs only slightly from the possibilistic case. Basically, what changes is that in the definitions of the operators, the minimum is replaced by the product and the maximum by the sum.

5.2 About the Post-Processing

Concerning the post-processing step, the reasoning remains in the main the same. It is practically a simple adaptation of that described in Sect. 4.3, the only change concerns the way the degrees themselves are computed.

As far as the pruning conditions are concerned, it has to be checked whether they still hold or not. The condition that concerns optimality (satisfactory(x_j)) does not hold any more in the probabilistic case: one cannot compare any more the probability of the current solution with that of the optimal solution found so far, since one has to compute all the satisfactory worlds with respect to the desired cardinality in order to sum their probability degrees (this is due to the additive nature of the probabilistic framework). The second pruning condition concerns the cardinality of the current solution. This condition still holds since it is completely independent of the degrees (and therefore, of their semantics).

The complexity of this procedure in the probabilistic case is obviously higher than in the possibilistic one, since one loses the advantage of the use of the first pruning condition. However, it remains more efficient than the one based on the computation of the worlds attached to the initial possibilistic database, due to the reduction obtained by means of the compact evaluation of query Q.

6 Conclusion

This chapter addresses the issue of querying relational databases where some attribute values are imprecise and represented by possibility distributions. An adapted model with a subset of the relational algebra has been presented (see [3, 4] for details). In this context, the result of a query is a possibilistic relation, which is not easily interpretable by a final user. This situation led us to consider a new type of queries called possibilistic queries, whose general form is: "to what extent is it possible and certain that the answer to Q satisfies property P?". In this chapter, cardinality-based generalized yes/no queries have been investigated. Their treatment is based on a two-step mechanism. The first step is the evaluation of the algebraic query involved, and the second one is a post-processing relying on a "trial and error" technique.

This work opens different lines for future research. One of them is related to the performances obtained for cardinality-based generalized yes/no queries. Clearly the approach proposed is much more efficient than a technique based on the computation of worlds but it would be interesting to assess in a more precise way the additional cost linked to the presence of imprecise data (with respect to similar queries on precise data).

References

1. Zadeh, L. A.: Fuzzy Sets as a Basis for a Theory of Possibility. Fuzzy Sets and Systems. 1 (1978) 3–28
2. Bosc, P., Prade, H.: An Introduction to the Treatment of Flexible Queries and Uncertain or Imprecise Databases. In: Uncertainty Management in Information Systems, A. Motro and P. Smets (Eds), Kluwer, Dordecht. (1997) 285–324
3. Bosc, P., Pivert, O.: Towards an Algebraic Query Language for Possibilistic Databases. 12th Conference on Fuzzy Systems (FUZZ-IEEE'03). (2003) 671–676
4. Bosc, P., Pivert, O.: About Projection-selection-join Queries Addressed to Possibilistic Relational Databases. IEEE Transactions on Fuzzy Systems. 31 (2005) 124–139
5. Abiteboul, S., Kanellakis, P., Grahne, G.: On the Representation and Querying of Sets of Possible Worlds. Theoretical Computer Science. 78 (1991) 159–187
6. Bosc, P., Duval, L., Pivert, O.: An Initial Approach to the Evaluation of Possibilistic Queries Addressed to Possibilistic Databases. Fuzzy Sets and Systems. 140 (2003) 151–166
7. Bosc, P., Pivert, O.: On a Strong Representation System for Imprecise Relational Databases. 10th International Conference on Information Processing and Management of Uncertainty in Knowledge-Based Systems (IPMU'04). Perugia, Italy. (2004) 1759–1766

Finding Preferred Query Relaxations
in Content-Based Recommenders

Dietmar Jannach

Institute for Applied Informatics, University Klagenfurt, 9020 Klagenfurt, Austria
dietmar.jannach@uni-klu.ac.at

Summary. In many content-based approaches to product recommendation, the set
of suitable items is determined by mapping the customer's needs to required product
characteristics. A 'failing query' in that context corresponds to a situation in which
none of the items in the catalog fulfills all of the customer requirements and in which
no proposal can be made. 'Query relaxation' is a common technique to recover from
such situations which aims at determining those items that fulfill as many of the
constraints as possible. This chapter proposes two new algorithms for query relaxation, which aim at resolving common shortcomings of previous approaches. The
first algorithm addresses the problem of response times for computing user-optimal
relaxations in interactive recommendation sessions. The proposed algorithm is based
on a combination of different techniques like partial evaluation of subqueries, pre-
computation of query results and compact in-memory data structures. The second
algorithm is an improvement of previous approaches to mixed-initiative failure recovery: Instead of computing all minimal 'conflicts' within the user requirements in
advance – as suggested in previous algorithms – we propose to determine *preferred*
conflicts 'on demand' and use a recent, general-purpose and fast conflict detection
algorithm for this task.[1]

1 Introduction

Recommender systems are interactive software applications that support the
online customer in his/her decision making and buying process. In *content-
based* approaches to product recommendation, the product proposals are
generated on the basis of detailed descriptions of the items in the catalog:
According to Bridge [1], case-based, utility-based, or knowledge-based recommender systems basically fall into this category. Although there may be
different techniques involved for eliciting customer requirements, ranking the
products, or finding similar items, in many implementations of such systems,
some or all of the customer's requirements are – at least initially – viewed as

---

[1] Originally published in Proceedings of the 3rd IEEE Conference on Intelligent
Systems, London, 2006.

D. Jannach: *Finding Preferred Query Relaxations in Content-Based Recommenders*, Studies
in Computational Intelligence (SCI) **109**, 81–97 (2008)
www.springerlink.com   © Springer-Verlag Berlin Heidelberg 2008

constraints that the items in the proposal have to satisfy [8]. However, when the initial selection of items is based on such a query to the catalog, situations can easily arise, in which none of the products in the catalog fulfills all of the requirements. One basic approach to recover from such situations is to search for items that fulfill as many constraints as possible, which can be achieved by incrementally eliminating one or more constraints from the query (*query relaxation*). In [8], McSherry proposes an incremental, mixed-initiative approach to query relaxation based on results that were achieved in the area of 'cooperative query answering' [3]. In his work, he maps the problem of finding items that fulfill as many constraints as possible to the problem of finding a 'maximal succeeding subquery' (XSS) of the original query. In addition, McSherry proposes the computation and utilization of 'minimal failing subqueries' (MFS) and let the user decide in an incremental process, on which part of the query he/she is willing to compromise. Nonetheless, when using the recovery algorithm described in [8], it cannot be guaranteed that the smallest possible or an optimal relaxation with respect to some function describing the 'costs' of the compromises will be found. In order to find an optimal relaxation, in general all XSSs/MFSs of the query have to be known or enumerated, a problem which was shown to be NP-hard in general [3]. In fact, even small-sized problems soon become intractable, in particular if we consider the hard real-time requirements of interactive recommender applications.

In this chapter, we propose an algorithm in which the individual subqueries of the original query are evaluated independently in advance and the set of all XSSs can be enumerated by combining these partial results without further costly query operations. The algorithm requires exactly n queries to the catalog for finding all minimal and the optimal relaxation of a query consisting of n subqueries; the additional memory requirements for storing the partial results are also limited. Our approach therefore improves existing work in the area in two directions: First, it reduces the number of database queries for computing possible relaxations to a fixed number which is required in time-bounded interactive recommender applications. In addition, in contrast to previous work in which the size of the relaxation was the main optimization criterion, our technique also supports the concept of *preferred relaxations*.

The chapter is organized as follows. After giving an introductory example in te next section, the formal foundations of the approach are summarized. We then describe our algorithm for fast enumeration of all XSSs and afterwards discuss details of the implementation and the evaluation which was done in several real-world recommender applications. The chapter ends with a discussion of previous work in the area and an outlook on future extensions.

2 Example

We will illustrate the relaxation problem and our approach with a simplified example from the domain of digital cameras. Our product database consists of the products $p1 \ldots p4$ which are characterized by different properties as shown in Fig. 1.

ID	USB	Firewire	Price	Resolution	Make
p1	true	false	400	5 MP	Canon
p2	false	true	500	5 MP	Canon
p3	true	false	200	4 MP	Fuji
p4	false	true	400	5 MP	HP

Fig. 1. Product database of digital cameras

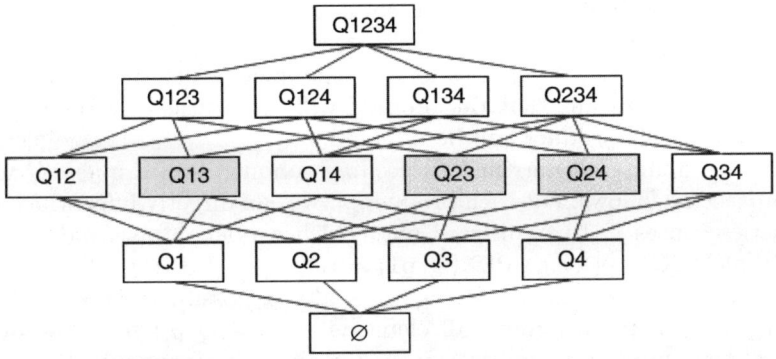

Fig. 2. Lattice of possible subqueries, minimal relaxations are printed in shaded boxes

Let us assume that the user's requirements can be expressed with the following query to the database, which unfortunately fails, given the set of available products.

$$Q \equiv \{usb = true \ (Q1), \ firewire = true \ (Q2),$$
$$price < 300 \ (Q3), \ resolution >= 5 \ MP \ (Q4)\}$$

Query relaxation will now be viewed as the problem of finding a *maximal succeeding subquery (XSS)* of Q in order to retrieve products that fulfill as many constraints as possible; the difference between the original query and an XSS is called a minimal relaxation'. In current approaches to query relaxation the original query Q is split into subqueries according to the attributes which are used in the query ($Q1$ to $Q4$ in our example). The search space in the relaxation problem is determined by the number of these subqueries, i.e., if a query can be divided into n such subqueries, there exist $(2^n - 2)$ candidates that theoretically have to be examined.[2] The lattice of the possible subqueries for our example is illustrated in Fig. 2. Testing each of the possible combinations individually is in general not possible in realistic applications because in typical implementations of such systems, each test corresponds to a query to the catalog, i.e., a database query.

---

[2] We do not have to examine the original query and the empty query.

In [9], an approach for enumerating all maximal succeeding subqueries is described in which the search space is reduced by (a) constructing the subqueries in decreasing order with respect to query length and (b) by removing subqueries of already succeeding subqueries from the remaining search space. Still, if we apply this on our example problem, only the single-element subqueries will be pruned from the search space. We therefore propose an approach, in which we can limit the number of needed catalog queries to the number of subqueries of the original query Q, which means that in our case we will only need four queries. We can achieve this by first evaluating the individual subqueries individually and then analyzing and combining the partial results in memory.

The results for the subqueries for our example are illustrated in Fig. 3: A '1' in the matrix means that the product will be returned by the subquery; '0' means that the product will be filtered out.

For determining all maximal succeeding subqueries of a query Q we can now proceed as follows. For each product p_i, we can directly infer which of the subqueries causes p_i to be filtered out, which we denote as *Product-specific Relaxation* $PSX(Q, p_i)$, e.g., $PSX(Q, p1) = \{Q2, Q3\}$, $PSX(Q, p2) = \{Q1, Q3\}$ and so forth. In fact, each $PSX(Q, p_i)$ is also a possible relaxation for the overall problem: If we remove all elements of $PSX(Q, p_i)$ from the original query, at least product p_i will satisfy the remaining requirements. Still, not all $PSX(Q, p_i)$ are minimal relaxations, but determining all minimal relaxations can be easily achieved by iterating over all $PSX(Q, p_i)$ once as follows: If the current $PSX(Q, p_i)$ is a superset of an already found relaxation, ignore it. If not, remember it and remove all those relaxations that were already found and for which the current $PSX(Q, p_i)$ is a subset.

Having determined all maximal succeeding subqueries, we can then select the one that promises to be the most useful for the customer. Without having a 'cost model' for the individual parts of the query, a suitable strategy will be to use the relaxation with the smallest cardinality. If such a cost model exists (e.g., information that users rather tend to compromise on the make than on the price), we can also determine the relaxation that minimizes this cost function. Overall, the main difference compared to previous approaches [8, 9] is that we do not search in the exponentially growing set of possible combinations of subqueries (the lattice), in which a lot of unnecessary checks are required.

ID	p1	p2	p3	p4
Q1	1	0	1	0
Q2	0	1	0	1
Q3	0	0	1	0
Q4	1	1	0	1

Product-specific relaxation for p1

Fig. 3. Evaluating the subqueries individually

3 Determining Maximal Succeeding Subqueries

We will now describe the relaxation problem and our algorithms more formally. We base our work on the formalisms introduced in [3, 8, 9].

Definition 1. *(Query): A query Q is a conjunctive query formula, i.e., $Q \equiv A_1 \wedge \ldots \wedge A_k$. Each of the A's is an atom (condition).*

In the following we denote the number of atoms of the query as $|Q|$ (query length).

Definition 2. *(Subquery): Given a query Q consisting of the conditions $A_1 \wedge \ldots \wedge A_k$, a query Q' is called asubquery of Q iff $Q' \equiv A_{s1} \wedge \ldots \wedge A_{sj}$, and $\{s_1, \ldots s_j\} \subset \{1, \ldots, k\}$*

Lemma 1. *If Q' is a subquery of Q and Q' fails, also the query Q itself must fail.*

Note that in [8] and [9], queries are split into subqueries according to the attributes that the query involves. Within our approach, however, such a specific form of partitioning is not required. Thus, in our approach it is also allowed that an individual query atom consists of a disjunction of conditions.

Valid and minimal relaxations and maximal succeeding subqueries are defined as follows.[3]

Definition 3. *(Valid relaxation): If Q is a failing query and Q' is a succeeding subquery of Q, the set of atoms of Q which are not part of Q' is called a valid relaxation of Q.*

Definition 4. *(Minimal relaxation): A valid relaxation R of a failing query Q is called minimal, if there exists no other valid relaxation R' of Q which is a subset of R.*

Maximal succeeding subqueries in the sense of [3] are directly related to minimal relaxations.

Definition 5. *(Maximal succeeding subquery – XSS): Given a failing query Q, a Maximal Succeeding Subquery XSS for Q is a non-failing subquery of Q and there exists no other query Q' which is also a non-failing subquery of Q for which holds that XSS is a subquery of Q'.*

Lemma 2. *Given a maximal succeeding subquery XSS for Q, the set of atoms of Q which are not in XSS represent a minimal relaxation R for Q.*

The approach described in this chapter aims at minimizing the number of queries to the catalog by analyzing the partial results that are obtained by evaluating the subqueries individually.

---

[3] Note that the term 'relaxation' in the context of this work means *eliminating* parts of the query rather than asking the user to revise his constraints like, e.g., in [2].

Definition 6. *(Partial query results – PQRS:) Let $P = \{p_1, \ldots, p_n\}$ be the set of products in the catalog. Given a query Q consisting of atoms $A_1, \ldots, A_k$, then $PQRS(A_i, P)$ is a function that describes the subset $P' \subseteq P$ for which condition A_i holds, $i \in \{1, \ldots, k\}$.*[4]

Lemma 3. *Given a failing query Q and a non-empty product catalog P, a relaxation R for Q always exists.*

For determining the partial query results for a query Q, exactly $|Q|$ queries to the catalog are required. Our algorithm for determining all minimal relaxations works by analyzing the possible relaxations for each product by exploiting the partial query results.

Definition 7. *(Product-specific relaxation – PSX): Let Q be a query consisting of the atoms $A_1, \ldots, A_k$, P the product catalog, and p_i an element of P. $PSX(Q, p_i)$ is defined to be a function that returns the set of atoms A_i from $A_1, \ldots, A_k$ that are not satisfied by product p_i.*

Lemma 4. *The set of atoms returned by $PSX(Q, p_i)$ is also a valid relaxation for Q.*

Determining whether an atom $a_i \in A_1, \ldots, A_k$ of a query Q is part of the product-specific relaxation for p_i can be done without further queries to the catalog by evaluating whether $p_i \in P$ is contained in the partial query result $PQRS(A_i, P)$. If p_i is not contained in $PQRS(A_i, P)$, then A_i has to be part of the product-specific relaxation $PSX(Q, p_i)$. Based on these definitions, we now describe an algorithm for determining all minimal relaxations of a query Q.

Algorithm *MinRelax* is sound and complete, i.e., it only returns minimal relaxations and it does not miss any of the minimal relaxations (Fig. 4).

Theorem 1. *Given a failing query Q and a product database P containing n products, at most n minimal relaxations can exist.*

Proof. For each product $p_i \in P$ there exists exactly one subset PSX of atoms of Q which p_i does not fulfill and which have to be definitely relaxed altogether in order to have p_i in the result set. Given n products in P, there exist exactly n such PSXs. Thus, any valid relaxation of Q has to contain all the elements of at least one of these PSXs for obtaining one of the products of P in the result set. Consequently, any relaxation which is not in the set of all PSXs of Q has to be a superset of one of the PSXs and is consequently no longer a minimal relaxation. This finally means that any minimal relaxation must be contained in the PSXs of all products and not more than $|PSX| = n$ such minimal relaxations can exist. $\square$

Proposition 1. *Algorithm MinRelax is sound and complete, i.e., it returns exactly all minimal relaxations for a failing query Q.*

---

[4] *PQRS can be represented as a matrix of zeros and ones as shown in Fig. 3*

Algorithm: MinRelax
In: A query Q, a product catalog P
Out: Set of minimal relaxations for Q

PQRS = compute the partial query results for all atoms a_i of Q
    for the product catalog Q
MinRS = $\emptyset$
forall $p_i \in$ P **do**
  PSX = compute the product-specific relaxation $PSX(Q, p_i)$ by using PQRS
  % check relaxations that were already found
  SUB = {r $\in$ MinRS|r is subquery of PSX}
  if SUB $\neq \emptyset$
       % current relaxation is superset of existing
       Continue with next p_i
  endif
  SUPER = {r $\in$ MINRS|PSX is subquery of r}
  if SUPER $\neq \emptyset$
       % remove supersets
       MinRS = MinRS\SUPER
  endif
  % store the new relaxation
  MinRS = MinRS $\cup$ PSX
endfor
return MinRS

Fig. 4. Algorithm for determining all minimal relaxations

Proof. The algorithm iteratively processes the product-specific relaxations (PSXs) for all products $p_i \in$ P. From Lemma 4 we know that all these PSXs are already valid relaxations. Minimality of the relaxations returned by Min-Relax is guaranteed by the algorithm, because (a) supersets of already discovered PSXs are ignored during result construction and (b) already discovered PSXs that are supersets of the current PSX are removed from the result set. As such, there cannot exist two relaxations R1 and R2 in the result set for which R1 is a subset of R2 or vice versa. In addition, we know from Theorem 1 that all minimal relaxations are contained in the PSXs of the products of P. Since MinRelax always processes all of these elements, it is guaranteed that none of the minimal relaxations is missed by the algorithm.

3.1 Complexity Issues

In our algorithm, the number of required executions of the typically most costly operation – querying the catalog – is equals to the number of atoms of the original query. The other operation that potentially induces relevant computation times is the determination of the subset and superset property when iterating over the products. In the theoretically worst case, at each iteration i this check has to be done for all of the previously found $i - 1$ relaxations.

This means that if we have n products, $(n * (n + 1))/2$ of such checks have to be done in the worst case. However, such checks can be efficiently done in memory and our experiments show that in realistic cases the number of actual checks that have to be made is at least an order of magnitude lower than the theoretical upper bound.

The efficiency of the algorithm with respect to the number of required queries comes at the price of a slightly increased *space complexity* for storing the partial results: For each of the individual atoms of the query, the list of matching products has to be stored. However, there exist p products and the query consists of a atoms, we need *at most $p * a$* bits for storing the raw information when using, e.g., a representation based on bit-sets.

3.2 Searching Preferred Relaxations

Up to now, we have assumed that relaxations of smaller size, i.e., those who contain fewer conditions to be removed from the query, are preferable for the user. In an interactive recommender application we therefore might pick one of the smallest relaxations and present it to the user. However, as also mentioned in [8], the users might have different preferences on which product characteristics they are more willing to compromise. Such specific preferences can be incorporated into our approach by associating *costs* (of relaxation) with each atom of the query and defining an overall cost function that for instance takes both the number of atoms in the relaxation and these individual costs into account.

Definition 8. *(Cost function) Let Q be a failing query and R a valid relaxation for Q consisting of the atoms $a_1, \ldots, a_k$. If ICOSTS is a function that associates a positive integer number with each a_i expressing the individual costs of relaxing atom a_i, the overall costs for R for Q can be described by any function COSTS(Q, R, ICOSTS) that returns a positive integer number expressing the overall costs of R. In addition it has to hold that $COSTS(Q, R', ICOSTS) < COSTS(Q, R, ICOSTS)$ if $R' \subset R$} for ensuring that adding further atoms to a relaxation does not decrease the cost value.*

Given such a cost function, we can define an ordering between the possible relaxations and describe the properties of an *optimal* relaxation.

Definition 9. *(Optimal relaxation): Given a failing query Q, a valid relaxation R for Q is said to be optimal, if there exists no other valid relaxation R' for which $COSTS(Q, R', ICOSTS) < COSTS(Q, R, ICOSTS)$.*

Determining the optimal relaxation based on our PSX-representation can be easily done by scanning the set of PSXs, evaluating the cost function individually and remembering the PSX that minimizes this cost function. Besides searching only for the optimal relaxation, it is also easily possible to determine an ordering among the relaxations for those cases, in which we want to present the user a list of possible relaxations he/she can choose from. Such an ordering can be achieved by sorting the PSX's according to their costs and by removing supersets of already found PSX's.

Note that the cost value for relaxing an individual atom of the query can come from different sources: They can be defined in advance, they could be derived from previous recommendation sessions by taking into account which compromises users typically prefer, or the user could also be directly questioned about his/her personal preferences.

4 Incremental Relaxation

Another general strategy for dealing with query failures in interactive recommender applications is to let the user incrementally decide on which attributes (i.e., constraints) he/she is willing to compromise instead of computing a complete relaxation at once. Such incremental, user-driven approaches can be implemented with the help of 'Minimal Failing Subqueries' [3,8]:

Definition 10. *(Minimal Failing Subquery – MFS): A failing subquery Q^* of a given query Q is a minimally failing subquery of Q if no proper subquery of Q^* is a failing query.*

In general, a failing query may consist of many different MFSs: In our example problem from Sect. 2, the MFSs of Q are {Q1, Q2}, {Q2, Q3}, and {Q3, Q4}. From Definition 10 we know that we have to relax at least one element from each MFS in order to get a non-empty result set.[5] In an incremental approach to relaxation, we can therefore iteratively select one of the (remaining) MFSs of the problem and let the user decide, which of the constraints should be relaxed next until a solution can be found.

It has been already shown that enumerating all MFSs of a query is in general an NP-hard problem although finding one arbitrary MFS for a query of size N can be accomplished with the help of N subsequent queries [3].

However, not all of the MFSs may be equally preferable for the user. Let us assume that the costs for the individual subqueries in our example are Q1 = 10, Q2 = 20, Q3 = 30, Q4 = 40 which means that {Q1, Q3} would be the most preferable solution. It would therefore be better to first present MFS {Q1, Q2} and then {Q2, Q3} (which includes the chance that the best relaxation {Q1, Q3} is found) rather than presenting {Q2, Q3} and {Q3, Q4}, which can only lead to non-optimal solutions in a first try.

In contrast to the approach described in [8], we propose to compute such minimal *preferred* MFSs 'on-demand' rather than computing all possible MFSs in advance, which can be a very costly operation. The computation of such preferred MFSs (conflicts) can be accomplished by adapting Junker's

---

[5] In fact, the set of all minimal relaxations corresponds to the *Hitting Set* of all MFS, compare, e.g. [10].

QuickXPlain [7] algorithm for our purposes. QuickXPlain is a general, non-intrusive technique for detecting conflicts in overconstrained problems whose main properties are that it is capable of taking preferences into account while at the other hand it minimizes the number of required *consistency checks* (i.e., queries in our case) based on a divide-and-conquer search strategy. The complexity results from [7] show for instance that in one possible configuration of the algorithm, finding a conflict of size k in n elements in the best case only needs $log_2(n/k) + 2k$ queries and $2k * log_2(n/k) + 2k$ in the worst case.[6] The adapted version of QuickXPlain is depicted in Fig. 5.

A possible algorithm for interactive relaxation that uses our adapted QuickXPlain algorithm is listed in Fig. 6. In this algorithm, the user is interactively asked to select one of the elements of the remaining conflicts and a choice point is set on every selection so that the user can revise his/her selection through backtracking and try a different path. All elements that have already been chosen for relaxation are not considered anymore when the

---

Algorithm: mfsQX
In: A failing query Q
Out: A preferred conflict of Q

A = sorted list of atoms of Q
return $mfsQI(\emptyset, A)$

function mfsQI (BG,A)
In: BG: List of atoms in background
    A: List of atoms of failing query
Out: A preferred conflict of Q

% Construct and check the current set of atoms
Query = $\wedge_{b \in BG}$ (b)
% Current branch has become inconsistent
if *query* is not successful
        return $\emptyset$
endif
if $|A| = 1$
        return 1
endif
% Split remaining atoms into two parts
C1 = $\{a_i \in A | i < (|A|/2)\}$
C2 = $A \backslash C1$
% Evaluate branches
Δ_1 = mfsQI(BG $\cup$ C1, C2)
Δ_2 = mfsQI(BG $\cup \Delta_1$, C1)
return $\Delta_1 \cup \Delta_2$

Fig. 5. Using QuickXPlain for computing preferred MFS

---

[6] This means that if there exist N atoms in the query, we will always require less than N queries for finding a conflict in realistic cases.

Algorithm: Interactive Relax
In: Sorted list of atoms A of failing query A

Query = $\bigwedge_{a \in A}$ (a)
if query is not successful
    % Compute a minimal preferred conflict
    conflict = $mfsQX(\emptyset, A)$
    remaining = conflict
    % Set up the choice points
    do |conflict| **times**
            choice = *Ask user to select an option from remaining or 'backtrack'*
            if choice = 'backtrack' **then return**
            remaining = remaining \ {choice}
            % Remove the choice and try again
            interactiveRelax(A\ {*choice*})
    end do
else
    Minimize the relaxation and compute results
    Report success and show proposal to user
    response = Ask user if result is acceptable
    if response = 'yes' **then**
        exit function
    % backtrack to last choice point
    else return
    endif
endif

Fig. 6. Basic algorithm for interactive relaxation

algorithm searches for the next conflict. The exploration of the search space is thus similar to a basic depth-first backtracking procedure (AND/OR search) for Constraint Satisfaction Problems from [11].

Note that in the algorithm in Fig. 6 we *minimize* the relaxation before the results are presented to the user, because – due to the user-driven nature of the algorithm – relaxations may not be minimal. In our example, this can happen if the user selects Q1 from the first conflict {Q1, Q2}, then Q2 from {Q2, Q3}, and finally one of the elements of the remaining conflict {Q3, Q4}. Both {Q1, Q2, Q3} and {Q1, Q2, Q4} are valid relaxations, but not minimal, since Q2 is superfluous. However, minimizing the relaxation is trivial as we only have to check for each element of the relaxation individually if we can remove it from the relaxation without making the query become a failing query again. The main purpose of the minimization procedure can be seen in the fact that more of the initial constraints of the user can be taken into account. Of course, if a recommender system also uses the elements of the relaxation for constructing user-understandable explanations, these effects of minimization have to be explained to the user adequately.

In summary, we view the algorithms described in this section as possible alternatives for computing relaxations whenever it is not possible to compute the partial query results as described in Sect. 3.

5 Evaluation

The proposed technique has been implemented and evaluated within the Advisor Suite system (see, e.g., [4, 6]), a fully knowledge-based framework for the development of interactive recommender applications. In this system, the initial set of products to be presented to the user is determined with the help of 'if-then-style' *filter rules* that relate customer requirements with product characteristics. This indirection allows us to implement a more user-oriented interaction view, such that the (non-experienced) users do not have to be questioned directly about desired product characteristics. If we consider the example from Sect. 2, we would, for instance, not ask the user about his need for 'Firewire' support, but rather try to find out what his/her mobility and connectivity requirements are.

Typical *filter rules* for our example problem could be the following:

F1: **if** high-quality-printouts are required **then**
  only recommend cameras with a resolution higher than 5MP
F2: **if** user entered price limit L **then**
  only recommend cameras that cost less than L
F3: **if** user needs high connectivity **then**
  only recommend cameras that support 'Firewire'

At run-time, when a product proposal has to be generated – which is typically done after an initial requirements elicitation phase – the filter rules are evaluated by the system: For each rule it is determined, whether the condition in the antecedent of the rule is fulfilled, i.e., whether the filter rule is *active* or not. The conclusions of those active rules are then used to construct a conjunctive query to the catalog. Note that the conclusions of the rules can contain arbitrary complex expressions on product characteristics, e.g., consisting of several conjunctions and disjunctions. This modular way of modeling recommendation rules also forms the basis for splitting up the conjunctive query into individual atoms for relaxation in a natural way.

The filter rules themselves are modeled in Advisor Suite with the help of graphical editing tools (see Fig. 7). Each rule can also be annotated with explanatory texts which are presented to the user in the explanation phase. A text can be maintained both for the case when the rule was successfully applied as well as for the case that the rule had to be relaxed. Finally, we can define an a-priori *priority value* for each rule, which corresponds to the costs of relaxing the rule in cases that no product fulfills all of the requirements. In practical applications, these priority values are defined by the domain expert

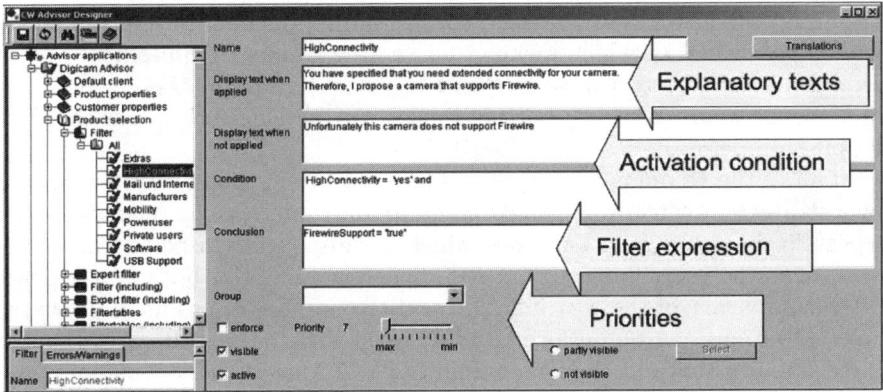

Fig. 7. Graphical editing tool for filter rules

who for instance knows that typical customers rather compromise on the manufacturer of a camera than on other characteristics.

Up to now, about twenty different recommender applications for various domains have been built with Advisor Suite and have been successfully deployed in commercial settings, which gives us in particular a good impression of the size and complexity of realistic knowledge bases:

- The number of products available in the catalog typically ranges from a few dozen to a several hundred.
- The number of filter rules remains manageable, i.e., only a few dozen rules were required in nearly all cases.
- Many of the rules are mutually exclusive with regard to their activation condition, i.e., only a smaller part of the rules is actually *active* when relaxation has to be done.

5.1 Running Times

The first aspect we have evaluated are the running times for determining (optimal) relaxations: Advisor Suite is a Java-based system that operates on top of standard relational database systems; all tests and measurements have been performed on standard desktop PCs with a 'Pentium M 2 GHz' processor and 512 MB of RAM. The most costly operation when determining relaxations in our approach are the queries that are required to compute the result sets for the individual filters, i.e., for each *active* filter, exactly one query has to be executed. The query time mostly depends on the number of products in the catalog and to a smaller extent on the complexity of the query. On average, a single query takes around 5 ms in test cases with around 500 products.

If we assume that we have a knowledge base containing 70 filter rules, and 35 of them are active – which is already a rather hard case in realistic settings – the time for computing the individual filter results is $35 \times 5 \text{ ms} = 175 \text{ ms}$.

The results of the individual filter rules are represented in memory in the form of *Java BitSets*, which means that (a) the memory requirement for the raw data is limited to *NumberOfProducts* ∗ *NumberOfActiveFilters* bits, and (b) that the analysis of the partial results can be efficiently done based on fast bit-set operations.

In all of the recommender applications that are in productive use, we follow a strategy in which we initially compute only one *optimal* relaxation with respect to size and relaxation costs, which is subsequently used to explain the proposal to the user. The time needed for determining this optimum depends on the complexity of the cost function and the number of products. However, these operations can be performed in-memory and in our test cases they required about a tenth of the query time (e.g., 17.5 ms in the above-mentioned example).

Regarding query time, we also exploited a particularity of the Advisor Suite approach of modeling *filter rules*: In many cases, the consequent of the rules contains no 'variables' (e.g., '*attribute usb must be true*'). Therefore, we can pre-compute and cache the partial results for such rules in advance, e.g., when the server is started up. In addition, these partial results can also be shared among different recommendation sessions of different users, since the partial results remain stable as long as the filter rule is not changed and the set of products is the same. Such pre-computation, however, is not possible if the consequent contains variables, like in filter rule F2 in the example above, that takes the user input with respect to allowed costs directly into account. Still, our experiences from different application domains show that the major part of the filter rules do not contain such variables, which means that the number of needed queries can be significantly reduced. Overall, even if no such pre-computation was done, in none of our test cases more than 500 ms were required for finding the optimal relaxation.

5.2 Usability Aspects

In all our fielded recommender applications, we have adopted a strategy in which we immediately compute a relaxation when we recognize that no product satisfies all constraints that were elicited in the advisory dialog. Thus, the relaxation process is not visible for the user in the first place, since in all cases one or more products will be recommended. Only when the user asks '*Why this recommendation?*', the proposal is explained with the help of the natural-language text annotations of the filter rules: 'Pro'-arguments correspond to filters that could be applied; 'con'-arguments correspond to the relaxed ones.

This initial relaxation is computed based on the a-priori priorities that were maintained by the domain expert. Since these priorities may not match the customer's actual preferences, the user is then given the possibility to interactively manipulate the priorities: In particular, the user can choose one or more of the relaxed filter rules and state that these rules should not be relaxed. After that, a new relaxation is computed and an alternative proposal is made. Of course, if the user enforces the application of too many filter rules, an empty result can be the consequence. In such situations, the user can *undo* his/her decisions and evaluate other alternatives.

For one of the applications built with Advisor Suite a detailed study was performed [12] in order to evaluate the usability of the overall system. The study of an application hosted on Austria's largest e-Commerce site (with respect to daily visits) included the analysis of more than 100,000 recommendation sessions and about 1,600 feedback forms. Although this study was not primarily concerned with the relaxation facilities, we could learn in particular from the feedback forms that the system's capability to produce explanations was the feature that was appreciated most by the users. Even more, from the interaction logs we could see that many visitors made use of the possibility to evaluate different alternatives of relaxations, in cases when their original requirements could not be fulfilled.

Note that in the fielded applications we did not allow the users to *fine tune* the priorities by themselves, e.g., by assigning a value between 1 and 10 for each rule, because an in-house assessment showed that such a task may be too complex for most users and it is also complicated to explain the effects of changing these priorities to the user. However, our future work includes the incorporation of *self-adapting* priorities, i.e., a mechanism that tracks the users' behavior over a given time frame and learn the typical user preferences.

6 Summary

In this chapter we have shown how optimal relaxations for unsuccessful queries in the context of interactive, content-based recommenders can be efficiently computed by pre-evaluating and analyzing the individual subqueries of the failing query. The proposed approach has been implemented in a domain-independent framework for building knowledge-based recommender systems and was evaluated with the help of several real-world recommender applications. Our measurements showed that even hard test cases can be successfully solved within the tight time frames that we have to deal with in interactive recommendation sessions. An empirical evaluation suggests that relaxation and explanation features are well appreciated by the online users as long as no complex interaction sequences are required.

Our work is based on the formalisms and relaxation approach also used by Godfrey [3] in the context of 'Co-operative Query Answering'. The goal in that research field is to establish a basis for building database systems that are capable of returning *more informative* answers than only 'yes' or 'no' to the users' queries. In principle, the algorithms presented in [3] could also be applied for our specific purposes in the recommendation domain. Nonetheless, the search algorithms for XSSs from [3] do not exploit partial pre-computations for adequate run-time behavior; furthermore, no ranking of relaxations based on preferences is possible, i.e., the ranking is restricted to cardinality only.

The work from [8] and [9] is similar to ours with respect to the overall goal; the approaches presented in this chapter can be seen as algorithmic improvements that take specific characteristics of the domain into account, e.g., the limited number of records in the catalog, for guaranteeing short response times. In addition, *user preferences* can be directly taken into account for optimization purposes. Compared with [8] and [9], we also claim that our approach is more flexible with regard to how the query can be split into subqueries in a natural way, which is done in [8] on the basis of query attributes only.

Our future work will include further research towards 'self-adapting' systems, where priorities and preferred relaxations can be *learned* from different sources of knowledge like past user behavior. An recent extension of our approach, which allows us to determine optimal relaxations that comprise *at-least-n* products, can be found in [5].

References

1. D. Bridge. Product recommendation systems: A new direction. In R. Weber and C. Wangenheim, editors, Workshop Programme at fourth International Conference on Case-Based Reasoning, pages 79–86, 2001
2. D. Bridge. Towards conversational recommender systems: a dialogue grammar approach. In D. W. Aha, editor, Proceedings of the EWCBR-02 Workshop on Mixed Initiative CBR, pages 9–22, 2002
3. P. Godfrey. Minimization in cooperative response to failing database queries. International Journal of Cooperative Information Systems, 6(2):95–149, 1997
4. D. Jannach. Advisor suite – a knowledge-based sales advisory system. In R. Lopez de Mantaras and L. Saitta, editors, Proceedings of the European Conference on Artificial Intelligence, pages 720–724, Valencia, Spain, 2004. IOS Press
5. D. Jannach. Techniques for fast query relaxation in content-based recommender systems. In C. Freksa, M. Kohlhase, and K. Schill, editors, KI 2006 – 29th German Conference on AI, pages 49–63, Bremen, Germany, 2006. Springer LNAI 4314
6. D. Jannach and G. Kreutler. A knowledge-based framework for the rapid development of conversational recommenders. In X. Zhou, S. Su, M. Papazoglou, M. Orlowska, and K. Jeffery, editors, Proceedings of the Fifth International Conference on Web Information Systems – WISE 2004, pages 390–402, Brisbane, 2004. Springer LNCS 3306
7. U. Junker. Quickxplain: Preferred explanations and relaxations for over-constrained problems. In Proceedings of the National Conference on Artificial Intelligence AAAI'04, pages 167–172, San Jose, 2004. AAAI Press
8. D. McSherry. Incremental relaxation of unsuccessful queries. In P. Funk and P.A. Gonzalez Calero, editors, Proceedings of the European Conference on Case-Based Reasoning, pages 331–345, 3155, 2004. Springer LNAI
9. D. McSherry. Maximally successful relaxations of unsuccessful queries. In Proceedings of the 15th Conference on Artificial Intelligence and Cognitive Science, pages 127–136, Castlebar, Ireland, 2004

10. R. Reiter. A theory of diagnosis from first principles. Artificial Intelligence, 32(1):57–95, 1987
11. E. Tsang. Foundations of Constraint Satisfaction. Academic, UK, 1993
12. M. Zanker and C. Russ. Geizhals.at: vom Preisvergleich zur e-commerce Serviceplattform. in: S. M. Salment and M. Gröschel, editors, Handbuch Electronic Customer Care, 2004

Imprecise Analogical and Similarity Reasoning about Contextual Information

Christos Anagnostopoulos and Stathes Hadjiefthymiades

Pervasive Computing Research Group, Communications Network Laboratory,
University of Athens, Greece
bleu@di.uoa.gr, shadj@di.uoa.gr

Summary. Conceptual modeling is viewed as a promising means to represent contextual knowledge, which may be enriched with semantics. Such modeling is capable of describing context, as well as, reasoning about it. Moreover, contextual reasoning is attained taking into consideration similarity-based approaches. This article proposes approximate reasoning about similarity among pieces of context using ontological modeling, description logics representation, and fuzzy logic inference rules. We report contextual similarity and fuzzy reasoning on top of logic based context semantics. Special emphasis is placed on similarity and analogical reasoning about context.

1 Introduction

Conceptual modeling is used in order to describe concepts in a syntactic and, more interestingly, in a semantic manner. Moreover, contextual information (*context*) can be represented by hierarchical structured concepts belonging to epistemic ontologies (taxonomies of concepts). One of the major aspects of this modeling is the similarity measurement between concepts thus the similarity assessment between pieces of context has to be examined. Agents try to identify whether two concepts are semantically similar based on semantic information (e.g., generalization relations and constraints). Therefore, different conceptual modeling techniques support, inevitably, different similarity metrics. The more semantic information a conceptual model supports, the more precise the similarity measurement becomes.

Semantics is the key enabler for a reasoning process to conclude how similar and/or compatible two concepts are. The uncertainty, which arises by the logical comparison of two pieces of context, can be interpreted as the similarity measure between the corresponding concepts. Context reasoning leads to context classification provided that, context can be expressed as concepts asserted in epistemic ontologies. We propose and implement an approximate reasoning process for inferring knowledge about quite *similar* and *compatible*

C. Anagnostopoulos and S. Hadjiefthymiades: *Imprecise Analogical and Similarity Reasoning about Contextual Information*, Studies in Computational Intelligence (SCI) **109**, 99–119 (2008)
www.springerlink.com   © Springer-Verlag Berlin Heidelberg 2008

pieces of context. The most similar asserted context with the current context (*actual* context) could be approximately interpreted as the most representative one. Specifically, let the formula $p \rightarrow q$ be asserted as knowledge and let q be the actual context then, we measure how much q is similar and compatible with p. This idea has been partly derived from [1] referring to reasoning about analogy, regarding logical and similarity-based approaches.

Conceptual modeling using conceptual graphs, as already supported by the Resource Description Framework (RDF) scheme [2] does not provide rich semantic conceptual expressions, contrary to the RDF(S) scheme [2]. The former conceptual representation describes concepts and relations (binary predicates) using the well-known scheme: *subject–predicate–object*. On the other hand, RDF(S) adds more semantics in terms of conceptual classification through transitive *generalization* relations (e.g., a concept p is more specific than a concept q, that is, p *is-a* q). Therefore, RDF(S) encompasses sets of conceptual properties that characterize concepts. Moreover, Description Logic (DL) [2] enriches semantic conceptual expressions with quantification and universal constraints over relations. Moreover, DL (implemented by the OWL-DL standard [2]) supports conceptual reasoning. Consequently, the reasoned information might provide a more spherical view over the similarity measurement between concepts. Hence, generalization relations could be considered as necessary information in order to infer how similar and analogous two concepts are.

One may consider that, conceptual similarity involves an assessment based on what is known about concepts (pieces of context). Such knowledge is well expressed in *ontology* that describes the conceptualisation of the world. Ontology describes taxonomies of concepts, created by generalization relations, including axioms and constraints over relations. It describes facts that are assumed to be always true and it is able to conclude facts that are previously unasserted (unclassified). In order to reason about conceptual similarity, the taxonomical structure of concepts and the enhanced semantics that supports axioms over such taxonomies have to be taken into consideration.

The article is organized as follows: a conceptual model for representing contextual information is reported in Sect. 2. In Sect. 3, we propose a method for measuring conceptual similarity while in Sect. 4 we refer to an approximate reasoning process for inferring similar and compatible pieces of context. In Sect. 5 we evaluate our method through experiments and the last two sections outline prior work, discuss conclusions, and identify future research on the area.

2 Conceptual Context Modeling

Concepts belonging to ontology can represent contextual information. Concepts are hierarchically structured forming a *conceptual taxonomy* or *taxonomy*. Specifically, taxonomy is considered as a collection of concepts organized

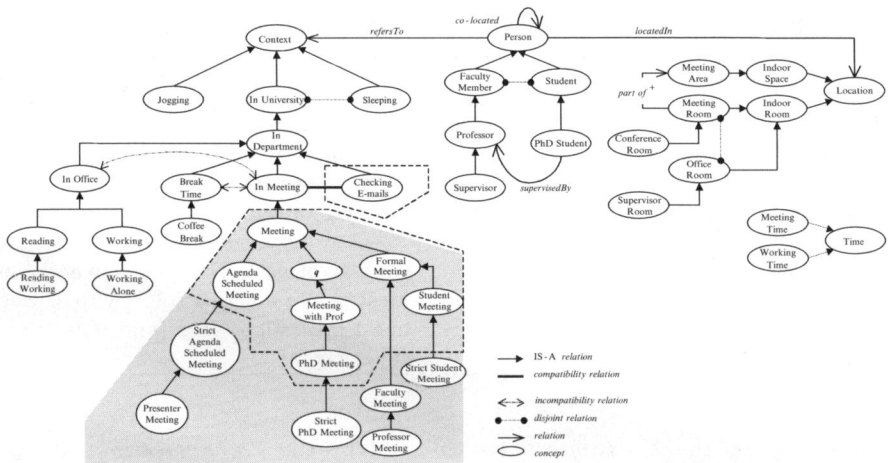

Fig. 1. Ontology: a set of conceptual taxonomies

by a partial order induced by a generalization relation, i.e., *is-a* or $\subseteq$. For instance, the concept *Meeting* is more generic than that of *Ph.D. meeting* or *Faculty meeting* (see Fig. 1), i.e., *Ph.D. meeting* $\subseteq$ *Meeting*. The hierarchy of concepts can be exploited for measuring their similarity according to their position in the taxonomy. However, in conceptual modeling, semantics is enriched with certain axioms and constraints over concepts and relations. The considered semantics is (a) the *disjoint* axiom, (b) the *closure* axiom and (c) the *compatibility* relation between concepts. The first axiom denotes that two concepts, p and q, are, by definition, disjoint, i.e., $p \subseteq \neg q$. The closure axiom defines whether quantification ($\exists$) and universal ($\forall$) constraints are applied over a relation. The compatibility relation defines whenever two concepts are characterized as compatible or not with respect to their semantic interpretation (regardless the fact that they are possibly disjoint).

In order to illustrate the meaning of such axioms, consider the concept description, in Table 1: *Ph.D. meeting* is a *Meeting* of at least ($\exists$ restriction) one Ph.D. student (restricted type of Person) with only ($\forall$ restriction) Ph.D. supervisors (restricted type of Person). Such meeting takes place only in the meeting room (restricted type of Indoor area). Consider that, Alice is a Ph.D. student and attends a Ph.D. meeting thus, the *Ph.D. meeting* concept represents Alice's actual context. The *Ph.D. meeting* concept is disjoint but compatible with the *Checking e-mails* concept; Alice might check for her e-mails during the meeting, while undertaking physical exercise, like jogging, cannot be occurred during the meeting. Hence, *Jogging* represents an incompatible concept with *Ph.D. meeting*. The compatibility relation between concepts is interpreted as the co-occurrence of their corresponding pieces of context.

In addition, consider that, Alice is co-located with other persons in a meeting room at a scheduled meeting time. Then, Alice's context is inferred to be

Table 1. Concepts description in DL syntax and natural language

DL syntax expression	Natural language
$Meeting \sqsubseteq (Context \sqcap \exists refersTo.$(Person $\sqcap \exists locatedIn.MeetingRoom \sqcap \geq 2$ $co\text{-}located \sqcap \exists co\text{-}located.Person) \sqcap$ $\exists hasTime.MeetingTime)$	*Meeting* represents the context of a person who is located in a meeting room with at least two persons at a scheduled meeting time.
$Ph.D.\ meeting \sqsubseteq (Meeting \sqcap \exists refersTo.(Ph.D.\ student \sqcap \exists locatedIn.MeetingRoom \sqcap \exists co\text{-}located.Supervisor \sqcap \forall co\text{-}located.Supervisor))$	*Ph.D. meeting* represent the *meeting* context of a Ph.D. student who is co-located at a meeting room with at least one Ph.D. supervisor (and only Ph.D. supervisors)

represented by the *Meeting* concept and not by the *Ph.D. meeting* concept. That is because, the closure axiom over the relation *co-located*, i.e., all the co-located persons must be *Ph.D. supervisors*, does not hold true in the Alice's actual context (see Table 1). Consequently, the reasoning process classifies Alice's context as a more abstract concept (*Meeting*) instead of *Ph.D. meeting*. The closure axiom results in concept classification to certain taxonomies in ontology. Concepts associated with constrained relations are more specific than those associated with unconstrained relations. According to our example, the *co-located* relation in the *Meeting* concept description is constrained by a quantification restriction, while the same relation in the *Ph.D. meeting* concept description is constrained by both restrictions (see Table 1). In this sense, specific concepts are positioned deeper in taxonomy thus, pieces of context are hierarchically structured, i.e., $context(Ph.D.\ meeting) \sqsubseteq ontext(Meeting)$.

3 Measuring Contextual Similarity

Context can be mapped into a qualitative conceptual representation. This mapping results in different levels of conceptualization of the actual world of pieces of context due to the imprecise nature of context. Levels can be represented as a hierarchy/taxonomy of concepts. Each concept belonging in taxonomy conveys specific information content [3]. Quantifying information content through conceptual modeling denotes that, the more abstract (generic) a concept the lower information content. Therefore, if there exists a unique concept in taxonomy, i.e., there are no sub-concepts its information content is null. The similarity measure between concepts quantifies the degree of equivalent information and semantics such concepts convey. Hence, such measure, called *contextual similarity*, assesses how similar two pieces of context are and it is defined in (1).

$$sim : N_+ \times \Theta \times \Theta \to [0,1] \tag{1}$$

Θ in (1) is the solution space of concepts, i.e., ontology, and N_+ is the set of positive integers. $n \in N_+$ is the level of structural complexity of concepts (i.e., composite concepts). The value of 0 denotes that the two concepts/arguments are not similar while the value of 1 indicates that the two concepts are *equivalent*. Consequently, the value of the similarity between pieces of context represented by concepts of n-level is the aggregate value of similarities of the corresponding pieces of context represented by concepts of $n - 1$ level, $n > 0$. Consider the concept $p(n)$, $q(n) \in \Theta$, where Θ is the context ontology of level n. Then, the similarity $sim(n, p(n), q(n))$ is recursively calculated in (2).

$$sim\left(n, p\left(n\right), q\left(n\right)\right) = \frac{1}{k_{n-1}} \sum_{j=1}^{k_{n-1}} sim\left(n - 1, p_j\left(n - 1\right), q_j\left(n - 1\right)\right) \qquad (2)$$

$$sim\left(n, p\left(n\right), q\left(n\right)\right) = \frac{1}{\prod\limits_{i=1}^{n} k_i} \sum_{j=1}^{k_1 \cdot \ldots \cdot k_n} sim\left(0, p\left(0\right), q\left(0\right)\right) \qquad (3)$$

The $k_{n-1} > 0$ is the number of the pieces of context $p_j(n-1)$, $q_j(n-1) \in \Theta_j$ and Θ_j is the jth taxonomy that describes context of level $n-1$. The recursive expression (2) has the solution in (3). The value of the $sim(0, p(0), q(0))$ refers to the *ground* similarity of the $p(0)$ and $q(0)$ concepts at level 0. $sim\left(\cdot, \cdot, \cdot\right)$ in (3) is proposed as a generic expression of calculating similarity based on the *ground* similarity of concepts. There are a lot of measurements that can be used to calculate the *ground* similarity [4–7]. The following sections discuss our proposed measure that takes into account enhanced context semantics. From now on, the indexing of the concept level n has been left out to keep the presentation more compact. Each concept $p(n) = p$ is assumed to be in an n-level taxonomy, $n > 0$.

3.1 Asserted Similarity Measure

Concepts are associated with transitive specialization relations ($\subseteq$) in a taxonomy Θ. Such relations denote that some concepts are more generic than other, that is, $p \subseteq q$ implies that p *is more specific than* q or p is-a q, with $p, q \in \Theta$. In this sense, the similarity measure among concepts affects their corresponding positions in taxonomy Θ. Subsequently such measure depends on the *asserted* position of a concept in Θ denoting the *explicit* knowledge of the expert. Let $U(q)$ denote the set of concepts that transitively include q, that is:

$$U\left(q\right) = \{e \in \Theta | q \subseteq e \vee e \equiv q\} \qquad (4)$$

Θ is the taxonomy that contains q starting from its most abstract concept, that is, $q \in \Theta \Leftrightarrow \exists e.\{(e \subseteq q) \vee (q \subseteq e)\}$. The most abstract a concept in Θ is the one and only concept that is *disjoint* with any other concept of different taxonomy in the ontology, that is $a \subseteq \neg b$, $a \in \Theta_i$ and $b \in \Theta_j$ with $\Theta_i \neq \Theta_j$.

We call the cardinality of $U(q)$, i.e., $|U(q)|$, as the *support* of the q concept, $sup(q)$. The *support* of q denotes the piece of information q conveys according to its position in Θ. The deeper q in Θ is the more the information q conveys.

The possible relative position of p and q belonging in taxonomy Θ is depicted in Fig. 2. It is possible that such concepts be included by common concepts. We define as *last common concept* of the p and q concepts, $lcc(p, q) = \varphi$, the last concept φ, starting from the root of Θ, that includes both p and q and holds that: $(q \subseteq \varphi) \wedge (p \subseteq \varphi) \wedge \neg(U(p) \subseteq U(q) \vee U(q) \subseteq U(p))$. Hence, φ has support $sup(\varphi) = |U(p) \cap U(q)|$ and then, $sup(p) = m + sup(\varphi)$ is the support of p, with $m = |U(p)\backslash U(\varphi)|$, and $sup(q) = n + sup(\varphi)$ is the support of q, with $n = |U(q)\backslash U(\varphi)|$. The support of φ affects significantly the asserted similarity between p and q. That is because, the higher the value of $sup(\varphi)$ the higher the common concepts that include p and q. Nonetheless, the similarity between p and q is affected by the *intermediate* concepts $I(q) \in \Theta$, with $I(q) = \{a \in \Theta | (q \subseteq a) \wedge (a \subseteq \varphi)\}$ and $I(p) \in \Theta$, with $I(p) = \{a \in \Theta | (p \subseteq a) \wedge (a \subseteq \varphi)\}$ among p and q, respectively (see Fig. 2d). The more the intermediate concepts lie among p and q the less similar q is with p. Therefore, in the Feature-Based Model defined in [5], the authors claimed that the similarity value is not only the result of *common* features

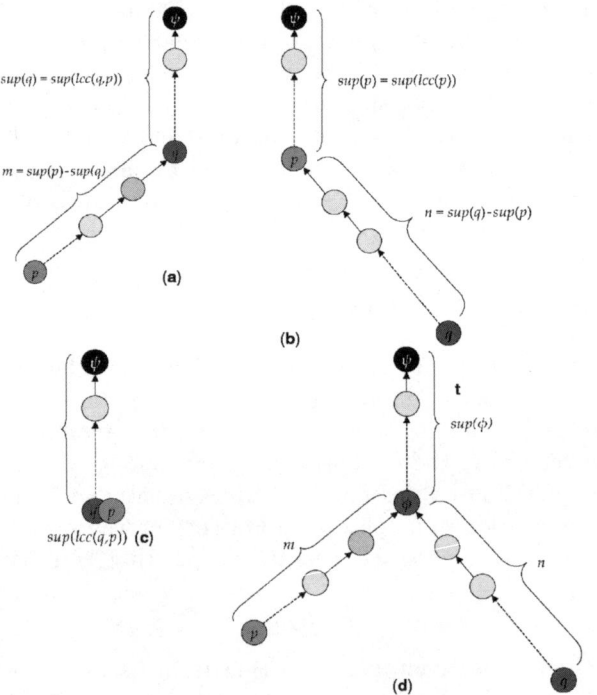

Fig. 2. Possible positions of q and p concepts in taxonomy Θ

but, also, the result of the *differences* between the two concepts. Evidently, the intermediate concepts have to be taken into account in calculating the asserted similarity.

Asserted Taxonomical Similarity

We adopt the ratio model of Tversky method [7] and define the *asserted taxonomical similarity* $sim_T(q, p)$ of two concepts q and p belonging to the Θ taxonomy in (5).

$$sim_T(q, p) = \frac{sup\left(lcc\left(q, p\right)\right)}{sup\left(lcc\left(q, p\right)\right) + a \cdot m + \beta \cdot n} \qquad (5)$$

The quantities m and n denote the support of the intermediate concepts of p and q, respectively, that is, $m = |U(p) \backslash U(lcc(q, p))|$ and $n = |U(q) \backslash U(lcc(q, p))|$. The factors α, β refer to the weights for *common* and *different* features respectively. These factors refer to the definition of an asymmetric measure but, we assume that any similarity measure is produced by symmetric measurement, hence, $\alpha + \beta = 1$. The term α, according to [5], assumes values in the range $[0, 0.5]$. The value of 0 means that the differences of q with respect to p are not sufficient in order to conclude that they are taxonomically similar, and the value of 0.5 means that the differences are necessary and sufficient to conclude such assumption. Hence, a can be determined as the fraction of the minimum length of the path from p to q in Θ as defined in (6).

$$\alpha = \frac{min\left(m, n\right)}{m + n} \qquad (6)$$

The asserted taxonomical similarity in (5) depends on the supports of p and q, thus, its values exhibit different behaviour based on the relative positions of p and q with respect to their $lcc(p, q) = \varphi$. Specifically, the $p \subseteq \varphi \subseteq \psi$ inclusion denotes that the knowledge about φ *includes* the knowledge about ψ (i.e., $\varphi \Rightarrow \psi$) without *including* any knowledge about p. Consider that the concept $p \in \Theta$ is the *asserted* representation of a user context. Let the concept $q \in \Theta$ be the *actual* representation of the current context. The *actual* representation of the unclassified context denotes the aggregation of those local pieces of context that represent the information content for q. Then, q can be *classified* according to the following cases:

1. q is a super-concept of p, i.e., $p \subseteq q$
2. q is equal to p, i.e., $p \equiv q$
3. q is a sub-concept of p, i.e., $q \subseteq p$
4. q and p are sub-concepts of their $lcc(p, q) = \varphi$

The knowledge about q indicates the degree of *belief* that the current context is an instance of the q concept. The (non-negative) degree of *belief* ($\in [0, 1]$) is monotonic over the support of a concept that is, *if* $p \subseteq \varphi \subseteq \psi$ *then*

$belief\,(q,p) \leq belief(q,\varphi) \leq belief\,(q,\psi)$ and $sup(q) \geq sup(\varphi) \geq sup(\psi)$ for the actual concept $q \in \Theta$. Such implication denotes that, the more knowledge about q, the more *belief* is accumulated to any super-concept of q. Evidently, the asserted taxonomical similarity between p and q represents the *belief* that "p is q", that is, $belief(q, p) = sim_T(q, p)$. In this sense, the relative position of p and q in Θ results in diverse attributes of such belief.

In the first case (Fig. 2a), q can be classified as a more generic concept than p, that is, p *implies* q. In this sense, $n = 0$. This means that, the support of q is $sup(q) = sup(lcc(q, p))$ and the taxonomical similarity is defined in (7) with $a = 1$ and $m = sup(p) - sup(q)$. Hence, if $sup(q)$ approaches $sup(p)$, which means that q approaches p then, the *belief* that "q is p" is a linearly monotonically increasing function over the support of q and a hyperbolically decreasing function over the support of p as illustrated in Fig. 3a.

$$sim_T\,(q,p) = \frac{sup\,(q)}{sup\,(p)} \tag{7}$$

$$sim_T\,(q,p) = \frac{sup\,(p)}{sup\,(q)} \tag{8}$$

In the second case (Fig 2c), q is classified as p, thus, the *belief* that "q is p" is 1 independently to their support and to the support of the last common concept, $sup(lcc(q, p))$. In that case, $sup(q) = sup(p)$ in (8).

In the third case (Fig 2b), q is classified as more specific concept than p, i.e., $m = 0$ (Fig. 3a.). In this sense, q implies p, to the extend that: the greater the depth of q with respect to p the lower the similarity between p and q. Hence, the belief that "q is p" is a monotonically decreasing function over the support of q. The taxonomical similarity is defined in (8), with $\beta = 1$ and $n = sup(q) - sup(p)$. The decreasing factor $sup(p)$ in (8) denotes that the more support of p the lower the decreasing rate. Moreover, $sim_T(p, q)$ decreases hyperbolically as q is a sub-concept of p. One can

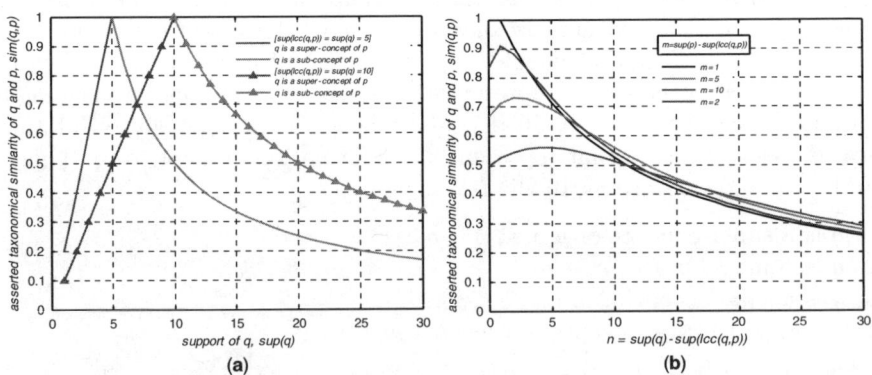

Fig. 3. The behavior of the asserted taxonomical similarity measure for the possible positions of q and p concepts is taxonomy Θ

conclude that, $\lim_{(sup(q)-sup(p))\to\infty} sim_T(p,q) = 0$. Figure 3a depicts the behaviour of $sim_T(p, q)$ for different values of $sup(p)$. A greater value of $sup(p)$ denotes that p represents more specific concept thus, $belief(q \subseteq p) > belief(q \subseteq e)$ for each $e \in \Theta$ with $q \subseteq e \subseteq p$.

An interesting analysis is presented for the fourth case in which there exists a $\varphi = lcc(p, q)$ with support $sup(\varphi)$. In that case, the $sim_T(p, q)$ is calculated in (5). The *belief* that "q is p" increases as more *intermediate* concepts $I(q)$ are accumulated between q and φ. There is a maximum value of that belief, denoting the maximum value of the similarity between p and q. This value is assumed when there is an $a \in I(q)$ concept whose $sup(a) = n/2$, and $n = sup(q) - sup(\varphi)$. As more intermediate concepts are accumulated between q and φ then, the belief that q is p monotonically decreases. That is because: for a support of $q \in [sup(\varphi), sup(a)]$, q concept is quite similar to p because it is more abstract than p with respect to φ, but on the other hand not so much dissimilar to p. For a support of $q \in [sup(a), n]$, q is more specific than p with respect to φ thus, p and q are not so similar, instead, they are *sibling concepts*. Hence, the maximum similarity between p and q is achieved when q is positioned between φ, and the sibling concept $b \in I(p)$, such that, $sup(p) < sup(b) < sup(\varphi)$.

Moreover, if $sup(\varphi) = 1$ and $m = n$, i.e., the two concepts are *siblings* with the minimum support then, their similarity is 0.5. If the support $sup(\varphi)$ increases with $m = n$, then p is also sibling to q but they seem more similar than the previous case (where $sup(\varphi) = 1$) because they share more common super-concepts, i.e., they have greater support for the last common concept. Finally, in case where $m = 1$ then, $sim(p, q)$ decreases with the support of q, as depicted in Fig. 3b. That is because, there is no intermediate concept $b \in I(p)$.

Asserted Taxonomical Similarity Based on the Disjoint Axiom

The disjoint axiom has to be taken into account in measuring asserted taxonomical similarity. If two concepts p and q are defined as *disjoint*, i.e., $p \subseteq \neg q$, then, one could claim that it is inappropriate to measure their similarity without taking into account that axiom. When the disjoint axiom is applied to p and q concepts, the subsumed concepts of p and q, respectively, are also considered as disjoint with each other. If the direct super-concepts of p and q are not declared as disjoint, but their indirect super-concepts do so, then, one has to take into account the *position* (depth) of such indirect super-concepts inside the taxonomy. Specifically, the position in taxonomy, in which the disjoint axiom is applied, generates different taxonomical measurements. Intuitively, sim_T appears to increase whenever the disjoint axiom is applied to nearer super-concepts more than distant indirect super-concepts. sim_{TD} revises sim_T by being aware of the disjoint axiom, since it subtracts the proportion of $sim_T(q, p)$ with respect to $sim_T(q_F, p_F)$, where q_F and p_F are

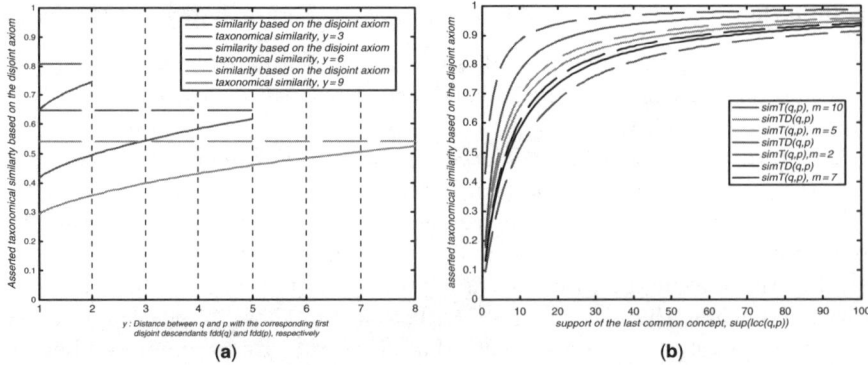

Fig. 4. The behavior of the asserted taxonomical similarity measure based on the disjoint axiom

the *first disjoint descendants* of the q and p, $q_F = fdd(q)$ and $p_F = fdd(p)$, respectively (see Fig. 4a). sim_{TD} is defined in (9).

$$sim_{TD}(q, p) = sim_T(q, p) \cdot (1 - |sim_T(q, p) - sim_T(fdd(q), fdd(p))|) \quad (9)$$

It is worth noting that, sim_{TD} cannot be measured in either RDF or RDF(S) representation schemes because they do not support the semantics of the disjoint axiom, contrary to OWL-DL.

As it is illustrated in Fig. 4a, there is a significant differentiation of sim_{TD} for diverse values of $m = sup(p) - sup(lcc(q, p))$. sim_{TD} increases as m is reduced. This means that, the longer the distance m between concepts p, q and the last common concept, $lcc(q, p)$, the smaller the overall value of sim_{TD} w.r.t disjoint axiom. Let y be the *height* between the level of p, q and the level of the first disjoint descendants, p_F, q_F, respectively (see Fig. 4a). As the value of y grows, i.e., the distance between the two levels increases, then the value of sim_{TD} decreases. That is because, the disjoint axiom is applied in higher level thus the last common concept is located higher in the taxonomy. It is worth noting that, for large values of m, sim_T and sim_{TD} assume same values.

Another dimension that has to be explored is how the support of the last common concept, $sup(lcc(p, q))$ affects the values of sim_T and sim_{TD}. In this case, the larger the $sup(lcc(p, q))$ is, the closer of sim_T and sim_{TD} become. In Fig. 4b, smaller values of $sup(lcc(p, q))$ result in broader difference between sim_T and sim_{TD}, in contrast to larger values of $sup(lcc(p, q))$. The former case denotes less similar concepts w.r.t., common super-concepts, thus, the application of the disjoint axiom plays a very significant role in taxonomical similarity. The latter case denotes very much similar concepts w.r.t., common super-concepts, thus, the impact of the disjoint axiom is negligible.

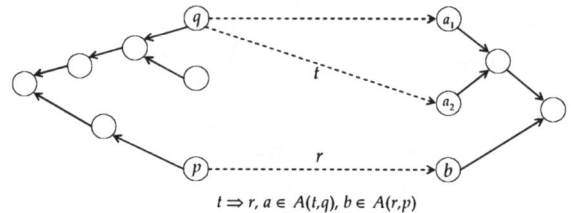

$$t \Rightarrow r, a \in A(t,q), b \in A(r,p)$$

Fig. 5. The associated concepts of the q and p concepts

Asserted Relational Similarity

The relational similarity sim_R between p and q is calculated by the asserted taxonomical similarity sim_{TD} of their associated concepts with respect to their common relations. A concept a is defined as the associated concept of the p concept through the binary relation r when $r(p, a)$. Then, the set $A(p, r)$ of the associated concepts of p through the relation r is defined as: $A(p, r) = \{a|r(p, a)\}$ (see Fig. 5). Moreover, the r and t relations can be hierarchically structured denoting that t is a specialization of r, i.e., $t \subseteq r$. Hence, two relations can be considered as similar w.r.t., a relations-taxonomy. The similarity value between r and t is denoted as $sim_T(r, t)$. In case $r = t$ then $sim_T(r, t) = 1$; otherwise $sim_T(r, t)$ derives from (5). The asserted relational similarity sim_R between p and q for a given relation r is defined in (10).

$$sim_R(q,p,r)$$
$$= \frac{\sum\limits_{a\in A(r,p)} max\left(min\left\{sim_{TD}(a,b), sim_T(r,t)\right\} | b \in A(t,q) \wedge t \subseteq r\right)}{max\left(|A(r,p)|, |A(t,q)|\right)} \qquad (10)$$

The asserted (overall) similarity, which derives from the asserted taxonomical (based on the disjoint axiom) and relational similarity, is a sum as provided in (11).

$$similarity(q,p) = \frac{1}{2}sim_{TD}(q,p) + \frac{1}{2|\Lambda|}\sum_{r\in\Lambda} sim_R(q,p,r) \qquad (11)$$

The set Λ in (11) is the set of all common most abstract relations of p and q. Such similarity is based only on semantics that asserted concepts convey within taxonomies, i.e., taxonomy of concepts, taxonomy of relations and disjoint axiom.

3.2 Analogy Measure

The asserted similarity between q and p concepts, $similarity(q, p)$ is meaningful once such concepts are in the same taxonomy, i.e., $p, q \in \Theta$ and their associated concepts belong to different taxonomies. Expressing p and q concepts in an Open World Assumption (OWA) [8] results in an adequate metric for measuring the similarity between p and q. Specifically, the knowledge that $q \to p$ can be inferred but, the knowledge that $\neg q \to \neg p$ is not implicitly inferred. On the other hand, negation is assumed as failure by expressing q and p concepts in a Closed World Assumption (CWA) [8]. DLs descriptions are expressed through an OWA thus, it is very important to take into consideration the several restrictions of the OWA over such descriptions.

The *closure* axiom can be expressed as the application of quantification ($\exists$) and universal ($\forall$) restrictions over relations. Such axiom is applied on a relation r linking a concept p with a concept a, that is, (i) for $a \in \Theta$ holds that $\exists a.r(p, a)$, i.e., p is restricted to associate with at least one a through r, and (ii) for $a \in \Theta$ holds that if $\forall a.r(p, a)$ then $a \in A(p)$, i.e., p is restricted to associate with one (not necessarily) a belonging to the set of the associated concepts of p through r. If both restrictions are applied on r, then, the closure axiom for the r relation, i.e., $\exists a.r(p, a) \wedge \forall a.r(p, a)$ with $a \in A(p)$, holds true. Concepts p and q regardless their asserted similarity, might be similar in the sense that, same restrictions are asserted over their relations. This means that, p concept is analogous to q concept due to same restrictions over common relations. Two concepts might be very similar regarding their asserted similarity but not quite analogous with respect to the closure axiom. For instance, p concept constrains its relations only with quantification restrictions, while q constrains the same relations with both types of restrictions. In that case, albeit such concepts are associated with same relations, it holds that, the q concept is more specific than the p concept. Hence, the implication $q \Rightarrow p$ holds true after the DL reasoning process.

We define as *closure concept*, $clc(p) \in \Theta$, of a concept $p \in \Theta$, the concept that constraints all relations with both types of restrictions. Let $a \in A(p, t)$ and the t relation be constrained either by quantification or universal restrictions. Then, $clc(p)$ applies both quantification and universal restrictions to the most abstract relation r, i.e., $t \subseteq r$ (see Fig. 6). Consequently, for each concept $p \in \Theta$, there is a closure concept $clc(p) \in \Theta$, that is, $clc(p) \Rightarrow p$. The *distance* d_x between p and $clc(p)$ for the restriction $x \in \{\exists, \forall\}$ is defined in (12).

$$d_x\left(p, clc\left(p\right)\right) = 1 - \min_{t|r \supseteq t}\left(max\left\{sim_T\left(t, r\right), sim_{TD}\left(a, b\right)\right\} | b \supseteq a)\right) \quad (12)$$

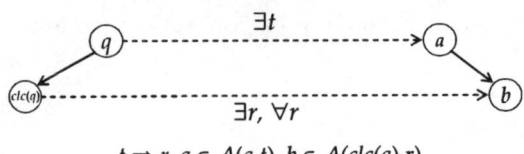

$$t \Rightarrow r,\, a \in A(q,t),\, b \in A(clc(q),r)$$

Fig. 6. The closure concept $clc(q)$ for the q concept

The concept $b \in A(pcl(c),\ r)$ is the associated concept of the closure concept of p through the most abstract relation r. Hence, the *analogy* between q and p is calculated using the distance of q and p from their corresponding closure concepts and is defined in (13). The *analogy* measure assumes value in the interval $[0,1]$. The value of 0 means analogous DL descriptions and 1 means non-analogous descriptions with respect to the closure axiom.

$$analogy\,(q,p) = \sqrt{\sum_{x \in \{\exists,\forall\}} \left(d_x\,(p, clc\,(p)) - d_x\,(q, clc\,(q))\right)^2} \qquad (13)$$

3.3 Affinity Measure

It is very important to refer to a more consolidated and holistic measure for judging both the similarity and analogy of two concepts. We define the fuzzy measure *affinity* between two concepts, which derives from the fuzzy inference of the values of the asserted similarity and the analogy for those concepts. The *affinity* measure depends to some degree on the analogy between two descriptions and to some degree on the asserted similarity of such descriptions. Let the asserted similarity and analogy between p and q concepts represent fuzzy variables. Then, the fuzzy values for the asserted similarity variable are those in Fig. 7. The corresponding fuzzy sets are *structural, semi-structural, non-structural*. Such values depend on the position of p and q concepts in the taxonomy and their relational similarity. The higher the value of similarity the more that similarity depends on the hierarchical structure of concepts in the taxonomy. Moreover, the fuzzy values for the analogy variable are those in Fig. 7. The corresponding fuzzy sets are *analogous* and *non-analogous*. An *analogous* fuzzy value for the analogy between p and q denotes that, the common relations of p and q are equivalently constrained by both types of restrictions. Consider also the fuzzy variable *affinity* between p and q concept, $affinity(p,q) \in [0,\ 1]$, with fuzzy values: *high, medium, low*. High affinity means that the asserted similarity is necessary condition that p and q concepts are similar. *Medium affinity* means that both the asserted similarity and

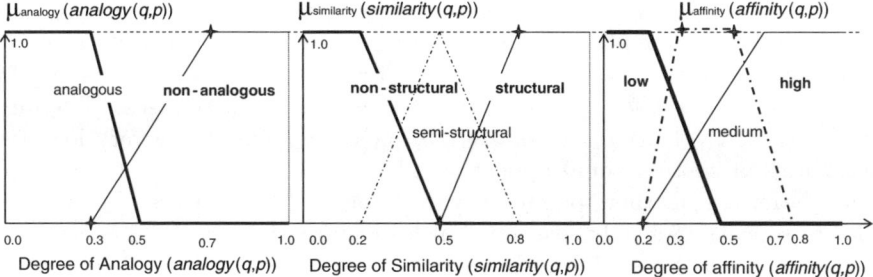

Fig. 7. Fuzzy values for the fuzzy variables: analogy, similarity and affinity measures

Table 2. Fuzzy inference rules for the affinity measure between q and p concepts

---

if **similarity** of q and p is **structural** implies **affinity** of q and p is **high**

if **similarity** of q and p is **somewhat structural** and q is **analogous** to p implies **affinity** of q and p is **high**

if **similarity** of q and p is **semi-structural** and q is **analogous** to p implies **affinity** of q and p is **medium**

if **similarity** of q and p is **non-structural** and q is **analogous** to p implies **affinity** of q and p is **low**

If **similarity** of q and p is **structural** and q is **non-analogous** to p implies **affinity** is **medium**

---

the analogy are equally considered as necessary conditions in order to outline the similarity between p and q. Finally, *low affinity* means that the analogy is regarded as being necessary condition, but not sufficient, for judging the similarity between p and q.

Therefore, the qualitative meaning of the affinity measure is the following: whenever the p and q concepts are *very* close in analogy (i.e., q is *analogous* to p, $analogy(q, p) \cong 0$), it does not strongly imply that they refer to rather similar pieces of context. On the other hand, whenever q and p are similar with respect to the similarity measure (i.e., $similarity(q, p) \cong 1$), then, there is a strong belief that they are similar, not only as analogous concepts, but also, as similar concepts referring to equivalent pieces of context. Consequently the *affinity* measure of two concepts is the fuzzy implication derived from a set of fuzzy rules associating the asserted similarity with the analogy of those concepts. A subset of the representative fuzzy rules, including concentration (e.g., *very*) and dilution (e.g., *somewhat*) operators, is depicted in Table 2.

4 Reasoning about Contextual Similarity

Reasoning about similar pieces of context means that, one has to reason about the asserted similarity and analogy measures. A concept of n-level represents a piece of context. In the following, we refer to the terms context and concept interchangeably according to the meaning. Pieces of context might relate each other through compatibility relations. Moreover, they could be declared as disjoint, but is not implied that, they are not compatible for a given world. For instance, the context $a = $ *user is attending a meeting* is defined as disjoint with the context $b = $ *user is checking her e-mails*, but these pieces of context can co-occur. Evidently, the *affinity(a,b)* measure seems to be very low, but such pieces of context could appear simultaneously. One should be aware of such relations, and, then, ought to reason about similar pieces of context in order to retrieve, not only relevant, but also compatible pieces of context.

Each concept $p_j \in \Theta$, which represents an asserted piece of context, is retrieved from the taxonomy Θ as possibly relevant and compatible with the unclassified concept q. q represents the *actual* context that has to be classified to one or more asserted concepts. However, the system could infer that, the most *relevant* concept p, i.e., that concept which assumes the highest affinity value with q; p estimates the actual context q in (14).

$$p = arg \max_{p_j \in \Theta} (affinity\,(q, p_j)) \tag{14}$$

Nonetheless, such imprecise reasoning could be revised since compatibility relations among concepts have to be taken into consideration. In that sense, we deal with *relevant* pieces of context, $\mathbf{R}$, and *compatible* pieces of context, $\mathbf{C}$, w.r.t., affinity measure and compatibility relations. Hence, the system returns pieces of context from the $\mathbf{R} \cap \mathbf{C}$ set.

Consider the position of the classified concept q, after the classification of the OWA DL reasoner RACER [9] (see Fig. 1). q is classified as a sub-concept of the $m = Meeting$ concept and, thus, disjoint with any other concept. Such classification does not allow q to be a sub-concept of the $c = Checking$ *e-mails* concept, while part of q description seems similar to the latter concept. Since c and m concepts are declared as disjoint, i.e., $c \subseteq \neg m$, then, q cannot be a sub-concept of the former one, i.e., $\neg (q \subseteq c)$. In other words, q does not extend two disjoint concepts. As a first solution, it would be better to describe the c concept with more generic concepts, hoping that $q \subseteq c$. This is not a good solution because the Θ taxonomy still contains very abstract concept descriptions loosing the specificity of the context representation. As a second solution, one could express q using even more specific concepts than those of the c description, hoping that $q \subseteq c$. Such idea sounds viable enough but two difficulties are to be addressed. Firstly, knowing exactly the actual context, i.e., q, compared with the idealistic context description c in Θ, it seems to be impossible and, secondly, even though c is considered as specific, it is also possible that Θ maintains more specific concepts than c.

Furthermore, by using default OWA DL reasoning, there is no information about the degree of inclusion. The most similar concept to q is $p = Ph.D.$ *Meeting*, which is a *Meeting*, i.e., $p \subseteq m$, something that verifies the results of our similarity measure (see Fig. 1). In addition, q is similar to c with a value of 0.5723 and q is disjoint with the latter concept, i.e., $c \subseteq \neg q$, which is undoubtedly true. On the other hand, the default DL reasoning could not take into account the compatibility relations among pieces of context. The following section copes with reasoning about compatible pieces of context.

4.1 Reasoning about Compatible Pieces of Context

The affinity value for q and m is not as high as that for any sub-concept a of m, i.e., $affinity(q, m) < affinity(q, a)$ with $a \subseteq m$. Specifically, q seems more similar to p rather than to m, i.e., $affinity(q, p) > affinity(q, m)$, because m is a more abstract description, thus, providing less information. In contrast, q is less similar to the most specific concept, $sp = Strict\ Ph.D.\ meeting$. The

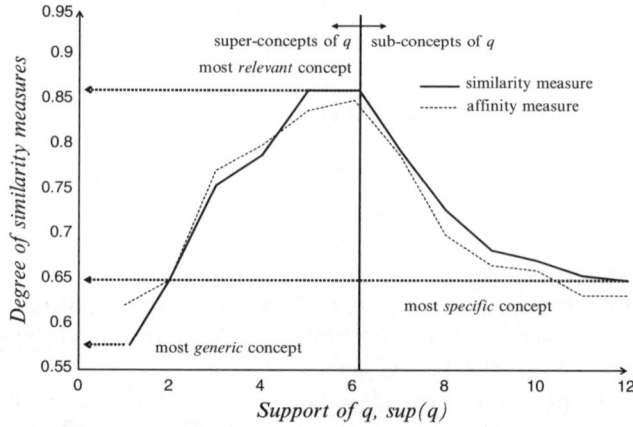

Fig. 8. The distribution of similarity measures along with the support

affinity value of any super-concept of q is lower than that of any sub-concept of q. Figure 8 depicts, the distribution of the affinity measure of the q along with the $sup(q)$, w.r.t., concepts of the taxonomy of q. The affinity value of the most specific concept, i.e., the deepest concept in the taxonomy, never drops below the minimum affinity value, which is that of the most abstract concept of the taxonomy (e.g., *Meeting* concept).

The fact that q is similar to other concepts not belonging to the *Meeting* taxonomy (e.g., *coffee break, reading,* and *checking e-mails* pieces of context), is *expected* since there is no awareness of the semantics related to compatibility relations among concepts. The similarity measurement may result in a list of relevant and compatible or incompatible pieces of context (e.g., *Meeting* and *Coffee break*).

4.2 Reasoning about Intra-Taxonomy Contexts

The affinity measure leads to those pieces of context that the actual context is classified and, quantitatively, to the degree of belief. Let p be the most relevant context to q, i.e., $p = arg \max_{p_j \in \Theta} (affinity\,(q, p_j))$, and let φ be the most generic concept that includes p, that is, $p \subseteq \varphi$. Then, we call *intra-taxonomy* concept e, the concept that belongs to the same taxonomy with p, i.e., $e \in \Theta \land p \in \Theta$. It is worth noting that, not all intra-taxonomy concepts $e \in \Theta$ are considered as relevant to q. However, one could take into consideration the behavior of the affinity measure (see Fig. 8). An intra-taxonomy concept $e \in \Theta$ is relevant to q whenever the affinity measure is greater than that of the most abstract concept φ but, lower than that of the most relevant, p. Hence, $e \in \Theta$ is relevant to q when the statement in (15) holds true.

$$affinity(q, \varphi) < affinity(q, e) < affinity(q, p) \text{ with } p \subseteq e \subseteq \varphi \quad (15)$$

In our case, $affinity(q, \varphi) < affinity(q, mp) < affinity(q, p)$ with $p \subseteq mp \subseteq q \subseteq \varphi$ and $mp = Meeting\ with\ Prof.$(intra-taxonomy concept), $p = Ph.D.$ $meeting$, and $\varphi = Meeting$. On the other hand, any intra-taxonomy concept $e \in \Theta$ which is more specific than the most relevant concept, i.e., $e \subseteq p$, assumes lower affinity value that that of p. That is because, e is more specific than p and thus, it is less similar to q since $e \subseteq p \subseteq q$. Specifically, if it holds true that $belief(p \subseteq q) > belief(e \subseteq p \subseteq q)$ then, it also holds true that, $belief(e \subseteq q) < belief(p \subseteq q)$, in other words, the concept $e \in \Theta$ is not believed to be relevant to q. According to our example $affinity(q, sp) <$ $affinity(q, p)$ thus, sp is not retrieved as a relevant context to q, with $sp = Strict\ Ph.D.\ Meeting$.

Consider those intra-taxonomy concepts $a \in \Theta$, which belong to a different sub-taxonomy of that of the most relevant concept p, i.e., for any $a \in \Theta$ there exists a concept b such that, $b = lcc(a, p)$. Hence, the only accepted relevant intra-taxonomy concept a is the one, which the statement in (16) holds true for.

$$(p \subseteq lcc\,(a, p) \subseteq q) \wedge \left(a = arg \max_{a_j \in \Theta} \left(affinity\,(q, a_j) \right) \right) \tag{16}$$

In our example, $fm = Formal\ meeting$ is retrieved as a relevant intra-taxonomy concept with $m = lcc(p, fm)$, $m = Meeting$. Hence, the accepted relevant intra-taxonomy concepts, for which the statements (15) and (16) hold true, are the dash-line encircled concepts in Fig. 1.

According to the default DL reasoning process, the only retrieved relevant concept that represents the actual context q is the $m = Meeting$ concept. Moreover, by comparing our results with those of the DL Matchmaker algorithm [10] we concluded that such algorithm counts all concepts that include m as relevant. Specifically, such algorithm considers the concept $e \in \Theta$ as $relevant$ to q such that, $e = lcc(q, m)$ with $sup(q) - sup(e) = 1$ (direct superconcept). Nonetheless, DL Matchmaker depends only on the disjoint axiom and does not take into account compatibility relations. In the worst case, since the $Checking\ E\text{-}mails$ concept would be compatible with the $Meeting$ concept, the DL Matchmaker would not consider the former as relevant to q, (the $Meeting$ concept is asserted as disjoint with every sibling concept). Compatibility among pieces of context is a special kind of relation with, inevitably, great impact on reasoning and consequently on retrieving relevant and compatible pieces of context.

5 Experimental Evaluation

We have evaluated the proposed contextual similarity and reasoning method by using abstract and specific ontologies. We evaluate the generated results using the RDF, RDF(S), and OWL-DL knowledge representations with respect to the disjoint axiom. We experiment with the behavior of our method using a relations-taxonomy and the involvement of the analogy in similarity reasoning. The proposed method is evaluated with respect to information retrieval metrics (Retrieval Status Values (RSV)). We use the standard RSVs, *precision* and *recall* [11], in different forms. Specifically, recall R is defined as the percentage of the number of retrieved and *relevant* and *compatible* concepts over the number of *relevant* and *compatible* concepts in the KB, and, precision P as the percentage of the number of retrieved and *relevant* and *compatible* concepts over the number of retrieved and *compatible* concepts. In addition the F-Measure (F) [11] is used for giving equal weight to R and P, that is $F = 2 \cdot P \cdot R \cdot (P + R)^{-1}$.

The reasoning process results in a set of *relevant* and *compatible* concepts with q context query. Figure 9a shows the P, R, and F quantities for a KB expressed in the RDF(S) scheme. The number on the horizontal axis expresses the first percentage relevance ranked concepts of the result set for that query. Such number expresses the percentage size of the result set of retrieved concepts, all of which are relevant and compatible ranked and listed in descending order according to the affinity value against q. The overall trend of precision is decreasing since the result set is constantly growing until its size reaches the total number of the concepts in the KB.

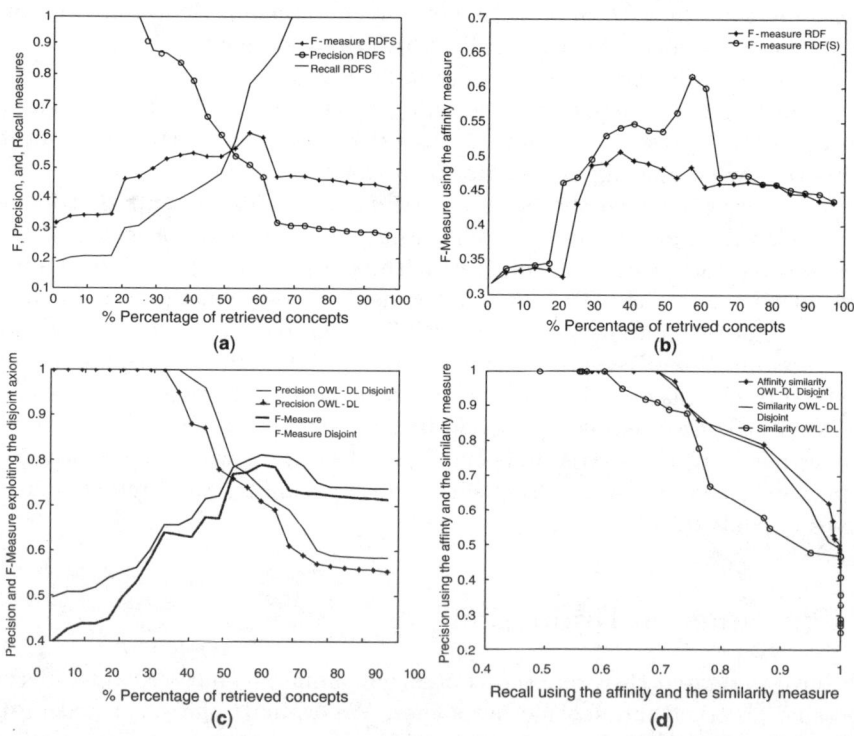

Fig. 9. Evaluation results for the proposed reasoning method

Figure 9b shows F using both RDF(S) and RDF representations. Obviously, the expressiveness in the former representation scheme assumes more precise results w.r.t., relevance ranking. Figure 9c depicts P, and F with respect to OWL-DL representation using the disjoint axiom. Evidently, the more semantic the expressions are, the more precise knowledge retrieval is achieved, since the generalization relations and the disjoint axiom are taken into account.

Figure 9d depicts P vs. R using OWL-DL representation using both the asserted similarity and affinity measures. Evidently, when analogy is taken into account in reasoning about contextual similarity, then, one obtains more precise results.

6 Related Work

The most similarity-based methods are applied on context models represented either by logic frames or objects. In [12], the author dealt with recognizing structural information from Web pages, by exploiting similarity measures over user queries. In [13], the authors discussed how to use semantic similarity algorithms in order to retrieve similar lexical pieces of context from certain lexical taxonomies. The author in [14] exploited a measuring distance over the WordNet by retrieving semantically similar meanings. In addition, similarity assessment methods and metrics used in the case-based reasoning and contextual retrieval are reported in [15].

Special interest is focused on the knowledge retrieval process dealing with highly changing contextual information. Retrieving context from ontologies appears to be a specific knowledge retrieval process. The model in [16] captures knowledge from user context (e.g., user patterns, profiles and preferences). Moreover, the authors in [17] discussed a content-based similarity-matching algorithm, which exploits the user preferences in context discovery. The authors in [18] discuss a probabilistic approach that combines logic and uncertainty theory for retrieving user context. The system, as proposed in [19], monitors user tasks and retrieves task-based user contexts. The authors in [20] proposed an algorithm that retrieves and learns user interests in the Web. Such algorithm learns the dynamics of the user interests through positive and negative relevance feedback.

7 Conclusions

In this article, we propose a method for reasoning about contextual information through similarity and analogical reasoning. The current work represents a method for knowledge retrieval based on context semantics by using crisp and approximate reasoning techniques over heterogeneous information resources. Context is represented as hierarchically structured concepts. We use the expressiveness of the DLs expressed using OWL-DL ontologies.

The proposed method retrieves the most relevant pieces of context given a context query based on similarity measures. Special focus has been given to the disjoint and closure axioms, and symmetric compatibility relations enriching context semantics. Fuzzy analogical reasoning revises the classical similarity metric denoting the significance of restrictions over contextual representations. Moreover, our method encounters not only relevant pieces of context, but also compatible ones. Evaluation experiments confirm that the semantic-based reasoning provides satisfactory results. That is because semantics is taken into consideration for retrieving relevant and compatible pieces of context. The arisen benefit of the proposed contextual similarity measure is based on exploiting semantics.

References

1. Russell, S., *Analogy by Similarity, in* David Helman (Ed.), *Analogical Reasoning*: D. Reidel, Boston, MA, 1988
2. Baader, F., Calvanese, D., McGuinness, D., Nardi, D., Patel-Schneider, P., *The Description Logic Handbook*, Cambridge University Press, Cambridge, 2003
3. Resnik, P., Using Information Content to Evaluate Semantic Similarity in a Taxonomy, *Proceedings of the International Joint Conference on Artificial Intelligence*, pp. 448–453, 1995
4. Maedche, A., Zacharias, V., Clustering Ontology-Based Metadata in the Semantic Web, *Proceedings of the Principles and Practice of Knowledge Discovery in Databases*, pp. 348–360, 2002
5. Rodriguez, M., Egenhofer, M., Determining Semantic Similarity among Entity Classes from Different Ontologies, *IEEE Transactions on Knowledge and Data Engineering*, 15(2), pp. 442–456, 2003
6. Santini, S., Jain, R., Similarity Measures, *IEEE Transactions on Pattern Analysis and Machine Intelligence*, 21(9), pp. 871–883, 1999
7. Tversky, A., Features of Similarity, *Psychological Review*, 84, pp. 327–352, 1977
8. Grimm, S., Motik, B., Closed World Reasoning in the Semantic Web Through Epistemic Operators, *Proceedings of the OWL: Experiences and Direction Workshop*, 2005
9. Haarslev, V., Moeller, R., RACER System Description, *Proceedings of the International Joint Conference on Automated Reasoning*, 2083, pp. 701–705, 2001
10. Gonzales, J., Trastour, D., Bartolini, C., Description Logics for Matchmaking of Services, *Proceedings of the Joint German/Austrian Conference on Artificial Intelligence*, pp. 139–154, 2001
11. Frakes, W., Baeza, R., *Information Retrieval: Data Structures and Algorithms*, Prentice-Hall, Englewood Cliffs, NJ, 1994
12. Cohen W., Recognizing Structure in Web Pages Using Similarity Queries, *Proceedings of the 16th National Conference of Artificial Intelligence*, pp. 59–66, 1999

13. Jiang, J., Conrath, D., Semantic Similarity Based on Corpus Statistics and Lexical Taxonomy, *Proceedings of the 10th International Conference on Research on Computational Linguistics*, pp. 19–33, 1997
14. Lewis, W., *Measuring Conceptual Distance Using WordNet: The Design of a Metric for Measuring Semantic Similarity.* The University of Arizona Working Paper in Linguistics, 12, 2002
15. Anick, P., Simoudis, E., Case-Based Reasoning and Information Retrieval: Exploring Opportunities for Technology Sharing, *American Association for Artificial Intelligence*, 1993
16. Middleton, S., DeRoure, D., Shadbolt, N., Capturing knowledge of user preferences: Ontologies in recommender systems, *Proceedings of the International Conference on Knowledge*, pp. 100–107, 2001
17. Bollacker, K., Lawrence, S., Giles, C., CireSeer: An Autonomous Web Agent for Automatic Retrieval and Identification of Interesting Publications. *Proceedings of the International Conference on Autonomous Agents*, pp. 116–123, 1998
18. Wong, S., Yao, Y., On Modeling Information Retrieval with Probabilistic Inference. *ACM Transactions on Information Systems*, 13(1) pp. 38–68, 1995
19. Leake, D., Scherle, R., Budzik, J., Hammond, K., Selecting Task Relevant Sources for Just-in-Time Retrieval, *Proceedings of the Workshop on Intelligent Information Systems*, 1999
20. Widyantoro, D., Ioerger, T., Yen, J., Learning User Interest Dynamics with a Three-Descriptor Representation, *Journal of the American Society for Information Science*, 52(3), pp. 212–225, 2000

Part III

Fuzzy Sets and Systems

A Method for Constructing V. Young's Fuzzy Subsethood Measures and Fuzzy Entropies

H. Bustince, E. Barrenechea, and M. Pagola

Departamento de Automática y Computación, Universidad Pública de Navarra, Campus de Arrosadía, s/n, 31006 Pamplona, Spain
bustince@unavarra.es, edurne.barrenechea@unavarra.es, miguel.pagola@unavarra.es

Summary. In the first part of the chapter we show the three most important axiomatizations of the concept of subsethood measure. Then we present the reasons why we focus on the definition given by V. Young. Next we study a method for constructing said measures and we analyze the conditions in which they satisfy the axioms of Sinha and Dougherty. Afterwards we study the way of obtaining fuzzy entropies that fulfill the valuation property from said subsethood measures.

1 Introduction

A fuzzy subsethood measure (also called a measure of inclusion) is a relation between fuzzy sets A and B, which indicates the degree to which A is contained in (is a subset of) B.

Traditionally, fuzzy set inclusion is defined according to Zadeh's [1] original proposal. For A and B fuzzy sets in a universe X he defined:

$$A \leq B \text{ if and only if for all } x \in X, \ \mu_A(x) \leq \mu_B(x).$$

For many researchers, this definition is too rigid and it does not do justice to the spirit of the Theory of Fuzzy Sets. Bandler and Kohout [2] call the definition of inclusion by Zadeh *an unconscious step backwards in the realm of dichotomy.* In 1980 these authors suggest the following definition: Consider two fuzzy sets A and B in a universe X, then the degree to which A is a subset of B is given by:

$$Inf_{x \in X} \jmath(\mu_A(x), \mu_B(x)),$$

where $\jmath : [0,1]^2 \rightarrow [0,1]$ is such that $\jmath(0,0) = \jmath(0,1) = \jmath(1,1) = 1$ and $\jmath(1,0) = 0$.

This fact has led many authors to study functions:

$$\sigma : \mathcal{F}(\mathcal{X}) \times \mathcal{F}(\mathcal{X}) \rightarrow [0,1],$$

H. Bustince et al.: *A Method for Constructing V. Young's Fuzzy Subsethood Measures and Fuzzy Entropies*, Studies in Computational Intelligence (SCI) **109**, 123–138 (2008)
www.springerlink.com   © Springer-Verlag Berlin Heidelberg 2008

(where $\mathcal{F}(X)$ represents the class of fuzzy sets in the universe X), such that $\sigma(A, B)$ shows how much A is contained in B.

Obviously, we must impose on σ certain conditions, (axioms), in order to achieve a good representation of the fuzzy subsethood measure.

Historically, in fuzzy literature, three axiomatizations have been given for fuzzy subsethood measures. The first one was given by Kitainik [3] in 1987. Later, in 1993 Sinha and Dougherty [4] give a collection of nine axioms, plus three additional ones. Finally, Young [5] in 1996 proposes four axioms for these measures. In 1999 Fan et al. in [6] modified one of Young's axioms.

1.1 Kitainik's Axioms

Generally speaking for this author, a fuzzy subsethood measure is considered as a fuzzy binary relation on the set of all fuzzy subsets $\mathcal{F}(X)$ of X, that satisfies the four following axioms:

(K_1) $\sigma(A, B) = \sigma(B_c, A_c)$ for all $A, B \in \mathcal{F}(X)$;

(K_2) $\sigma(A, B \wedge C) = \wedge\{\sigma(A, B), \sigma(A, C)\}$ for all $A, B, C \in \mathcal{F}(X)$;

(K_3) $\sigma(A, B) = \sigma(S(A), S(B))$ for all $A, B \in \mathcal{F}(X)$, where the fuzzy set $S(A)$ is defined as $\mu_{S(A)}(x_i) = \mu_A(s(x_i))$ with a one-to-one mapping $s : X \rightarrow X$;

(K_4) Applying σ to crisp sets, it coincides with the usual set inclusion;

The previous axiomatization was motivated by the following inter-relations between crisp sets:

$$A \subseteq B \Leftrightarrow B_c \subseteq A_c$$

$$A \subseteq (B \cap C) \Leftrightarrow (A \subseteq B) \wedge (A \subseteq C)$$

$$A \subseteq B \Leftrightarrow S(A) \subseteq S(B).$$

Dubois and Prade in [7,8] establish that the binary fuzzy relation σ can also be interpreted as an inclusion grade, (the interpretation of S for crisp sets is given in [3,8]). Due to this, we will constantly speak of fuzzy subsethood measure (or inclusion grade) instead of binary fuzzy relation.

Fodor and Yager [8,9], from the work of Kitainik, have studied the relationship existing between Kitainik's fuzzy subsethood measures and implication operators, obtaining the following theorem.

Theorem 1. *A $\mathcal{F}(\mathcal{X}) \times F(X) \rightarrow [0, 1]$ mapping σ satisfies $(K_1) - (K_4)$ if and only if there exists a contrapositive implication operator (in Fodor's sense) such that, for all A and B in $\mathcal{F}(X)$:*

$$\sigma(A, B) = Inf_{i=1}^n I(\mu_A(x_i), \mu_B(x_i)).$$

Therefore, we deduce from this theorem that the only fuzzy subsethood measure admitted in the sense of Kitainik, belongs to the group Bandler and Kohaut [2].

Kitainik also proves that there is no fuzzy subsethood measure satisfying at the same time the properties: reflexive, transitive and continuity. These properties have been much studied by different authors, as for example: Fodor and Yager [8], Kehagias and Konstantinidou [10]. In [11] Willmott studies the transitiveness of the measures of inclusion. This property has also been analyzed by Kundu in 2000 [12], etc. We will focus on the fulfillment of certain axioms and leave the study of these properties for future studies.

1.2 Sinha and Dougherty's Axioms

These authors [4] present nine axioms for the fuzzy subsethood measures. They also present three optional axioms in addition to these nine. These last three are given as optional, however, they maintain that for certain applications, they result more appropriate than some of the previous nine.

Axiom 1. $\sigma(A, B) = 1$ if and only if $A \leq B$ in Zadeh's sense.

Axiom 2. $\sigma(A, B) = 0$ if and only if $\text{Ker}(A) \cap (\text{supp}(B))_c \neq 0$, where $\text{Ker}(A) = \{x \in X | \mu_A(x) = 1\}$ and $\text{supp}(B) = \{x \in X | \mu_B(x) > 0\}$; that is, $\sigma(A, B) = 0$ if and only if exist $x \in X$ such that $\mu_A(x) = 1$ and $\mu_B(x) = 0$.

Axiom 3. If $B \leq C$, then $\sigma(A, B) \leq \sigma(A, C)$. In other words, the indicator is a non-decreasing function in the second variable.

Axiom 4. If $B \leq C$, then $\sigma(C, A) \leq \sigma(B, A)$. In other words, the indicator is a decreasing function in the first variable.

Axiom 5. $\sigma(A, B) = \sigma[S(A), S(B)]$. If we consider any one-to-one mapping $s : X \to X$. The unary shift operation is induced by $s : S : [0, 1]^X \to [0, 1]^X$ such that $\mu_{S(A)}(x) = \mu_A[s(x)]$. (That is, (K_3)).

Axiom 6. $\sigma(A, B) = \sigma(B_c, A_c)$. (That is, (K_1)).

Axiom 7. $\sigma(B \vee C, A) = \wedge \{\sigma(B, A), \sigma(C, A)\}$ for all $A, B, C \in \mathcal{F}(X)$.

Axiom 8. $\sigma(A, B \wedge C) = \wedge \{\sigma(A, B), \sigma(A, C)\}$ for all $A, B, C \in \mathcal{F}(X)$. (That is, $(K2)$).

Axiom 9. $\sigma(A, B \vee C) \geq \vee \{\sigma(A, B), \sigma(A, C)\}$ for all $A, B, C \in \mathcal{F}(X)$.

Additional axioms

Axiom 10. $\sigma(A, B) + \sigma(A, B_c) \geq 1$.

Axiom 11. If A is a refinement of B, then we require that

$$\sigma(A \vee A_c, A \wedge A_c) \leq \sigma(B \vee B_c, B \wedge B_c),$$

A_c being the complementary of the fuzzy set A with respect to the fuzzy negation c, that is, $A_c = \{(x, \mu_{A_c}(x_i) = c(\mu_A(x_i))) | x \in X\}$.

Axiom 12. If A is weakly included in B, then $\sigma(A, B) \geq \frac{1}{2}$.

Sinha and Dougherty say on their paper [4] that the reason why they impose the axioms 1 and 2 is the following: if A, B are crisp, then $\sigma(A, B) \in \{0, 1\}$. Therefore, the fuzzy subsethood measure is an extension of the concept of crisp inclusion.

Later on, Frago [13] proves that Axiom 9 is a consequence of Axiom 3. Moreover, Kitainik proves that if a fuzzy subsethood measure fulfills $(K_1) - (K_4)$, then it automatically satisfies Axioms 3 and 4 of Sinha and Dougherty. Evidently, it also satisfies Axioms 5, 6 and 8.

Due to all this, we get that Kitainik's axioms are equivalent to Sinha and Dougherty's with the only exception of Axioms 1 and 2, which Kitainik only imposes for non-fuzzy sets.

1.3 V. Young's Definition

Young [5] makes the following considerations about the previous axiomatization:

(1) From her point of view it is not fundamental to fulfill:
   if A and B are crisp, then $\sigma(A, B) \in \{0, 1\}$.
   She supports it saying that in fuzzy literature there are fuzzy subsethood measures widely used in different applications which do not fulfill that property. For example, the grade of inclusion of Goguen [14, 15]:

$$\sigma_G(A, B) = \frac{1}{n} \sum_{i=1}^{n} \wedge(1, 1 - \mu_A(x_i) + \mu_B(x_i)).$$

   In this example it is easy to see the fact that if the sets considered are crisp, then $\sigma(A, B) \in [0, 1]$. This does not mean that this measure will not give us valid information when the sets are non fuzzy. In fact, if the sets are crisp and $\sigma_G(A, B) = \frac{p}{n}$, ($n$ being the cardinal of X), then $n - p$ gives us the number of elements which belong to A and not to B.
(2) Axiom 2 is too harsh. *This axiom allows the values of A and B at one point to make $\sigma(A, B)$ equal 0. For example, if $\mu_A(x) \leq \mu_B(x)$ for all x except for one point x_0, at which $\mu_A(x_0) = 1$ and $\mu_B(x_0) = 0$, then $\sigma(A, B) = 0$, despite the fact that $\mu_A(x) \leq \mu_B(x)$ on the rest of X.*
   In addition, it shows that there exist in fuzzy literature some fuzzy subsethood measures that do not fulfill this axiom, as for example the above Goguen's inclusion grade or the fuzzy subsethood of Kosko [16, 17]:

$$\sigma_K(A, B) = \begin{cases} \dfrac{\sum_{i=1}^{n} \wedge(\mu_A(x_i), \mu_B(x_i))}{\sum_{i=1}^{n} \mu_A(x_i)}, & \text{if } A \neq 0 \\ 1, & \text{if } A = 0 \end{cases},$$

(3) Sinha and Dougherty first and later V. Young, reveal that if Axiom 2 is maintained, then $E(A) = \sigma(A \vee A_c, A \wedge A_c)$ does not fulfill the first condition of fuzzy entropy imposed by Deluca and Termini. This fact makes Sinha and Dougherty, basing themselves on the work of D. Dubois and H. Prade, propose to replace this entropy axiom with another one. However, V. Young gives an example which makes clear that the modification of the first axiom of entropy (in the sense given by Sinha and Dougherty) would lead to an important loss of information.
(4) She also shows that the relationship given by Kosko between the fuzzy subsethood measures and conditional probability disappears when considering the axiomatization of Sinha and Dougherty.

All these considerations led V. Young to give the following definition:

Definition 1. *([5]). A function*

$$\sigma_{V.Y.} : \mathcal{F}(\mathcal{X}) \times \mathcal{F}(\mathcal{X}) \to [0, 1]$$

is called a fuzzy subsethood measure, if $\sigma_{V.Y.}$ satisfies the following properties:

(a) $\sigma_{V.Y.}(A, B) = 1$ *if and only if $A \leq B$*

(b) *If $e \leq A$, then $\sigma_{V.Y.}(A, A_c) = 0$ if and only if $A = 1$, where e is the equilibrium point of the strong negation considered*

(c) *If $A \leq B \leq C$, then $\sigma_{V.Y.}(C, A) \leq \sigma_{V.Y.}(B, A)$ and if $A \leq B$, then $\sigma_{V.Y.}(C, A) \leq \sigma_{V.Y}(C, B)$*

1.4 Motivation for this Work

From our point of view, the following arguments make us focus on the study of the fuzzy subsethood measures of V. Young.

(A) The four previous considerations of V. Young.

(B) Kitainik's measures and therefore Sinha and Dougherty's have been amply studied in fuzzy literature.

(C) Due to Theorem 1 we know that in Kitainik's measures we should always take aggregation Inf, a fact that brings even Bandler and Kohout to say that the choice of this aggregation is a *harsh* criterion.

(D) We know that there exist fuzzy subsethood measures which are obtained applying the *arithmetic mean* to implication operators, such as the fuzzy subsethood measure of Goguen. Of course, this measure is not obtained by means of aggregating with the Inf implication operators, as Bandler and Kohout suggest.

All of these arguments, especially items (C) and (D) have brought us to study in the following section a mechanism for the construction of fuzzy subsethood measures of V. Young. Moreover, the arguments presented by this author have led us to analyze on the one hand the conditions in which her measures fulfill the axioms of Sinha and Dougherty and on the other, the way of obtaining fuzzy entropies that satisfy the valuation property.

2 Construction of V. Young's Subsethood Measures

The method for the construction of V. Young's subsethood measures that we present in this section is basically *aggregating (using special aggregation operators) certain functions*. For this reason we begin presenting the minimal properties that we are going to demand from the aggregation operators that we are going to use (see [18] and [19]) and then show in a proposition, the mechanism for the construction of the subsethood measures object of study. We conclude the section showing some examples.

An aggregation operator is defined by a function:

$$M : [0, 1]^n \to [0, 1]$$

for some $n \geq 2$, that satisfy at least the four following conditions:

A1. $M(x_1, \ldots, x_n) = 0$ if and only if $x_i = 0$ for all $i \in \{1, \ldots, n\}$.

A2. $M(x_1, \ldots, x_n) = 1$ if and only if $x_i = 1$ for all $i \in \{1, \ldots, n\}$.

A3. For any pair $(x_1, \ldots, x_n)$ and $(y_1, \ldots, y_n)$ of n-tuples such that $x_i, y_i \in [0, 1]$ for all $i \in \{1, \ldots, n\}$, if $x_i \leq y_i$ for all $i \in \{1, \ldots, n\}$, then $M(x_1, \ldots, x_n) \leq M(y_1, \ldots, y_n)$ i.e., M is monotonic increasing in all of its arguments.

This axiom will be denoted with $A3S$ when we demand the operator to be strictly increasing. That is: M is strictly increasing if and only if $x_i < y_i$ implies

$$M(x_1, \ldots, x_i, \ldots, x_n) < M(x_1, \ldots, y_i, \ldots, x_n), i = 1, \ldots, n.$$

A4. M is a symmetric function in all its arguments, that is,

$$M(x_1, \ldots, x_n) = M(x_{p(1)}, \ldots, x_{p(n)})$$

for any permutation p on $\{1, \ldots, n\}$.

Evidently the conditions that we impose on the aggregation operators above are more restrictive than those imposed by Dubois and Prade [20] and Klir and Folger [21] (see [18]).

In our constructions we are going to use strong negations [22] whose equilibrium point we will denote with e; that is, $e \in (0, 1)$ such that $c(e) = e$.

Proposition 1. *Let c be a strong negation such that $c(e) = e$, let $M : [0, 1]^n \to [0, 1]$ such that it satisfies A1 and A3S and let the functions*

$$g, h : [0, 1]^2 \to [0, 1], \quad such \ that$$

(1) *$g(x, y) \leq h(x, y)$ for all $x, y \in [0, 1]$*

(2) *$g(x, y) = h(x, y)$ if and only if $x \leq y$*

(3) *If $x \geq e$, then $g(x, c(x)) = 0$ if and only if $x = 1$*

(4) *If $x \leq y \leq z$, then* $\begin{cases} g(z, x) \leq g(y, x) \\ h(y, x) \leq h(z, x) \end{cases}$,

(5) *If $x \leq y$, then* $\begin{cases} g(z, x) \leq g(z, y) \\ h(z, y) \leq h(z, x) \end{cases}$,

Under these conditions $\sigma_{V.Y.} : \mathcal{F}(\mathcal{X}) \times \mathcal{F}(\mathcal{X}) \to [0, 1]$ given by

$$\sigma_{V.Y.}(A, B) = \begin{cases} 1 \ if \ h(\mu_A(x_i), \mu_B(x_i)) = 0 \ for \ all \ i \in \{1, \ldots, n\} \\ \dfrac{M_{i=1}^n(g(\mu_A(x_i), \mu_B(x_i)))}{M_{i=1}^n(h(\mu_A(x_i), \mu_B(x_i)))}, \ elsewhere \end{cases}$$

it is a fuzzy subsethood measure in the sense of V. Young.

Proof. (a) (Necessity) If $\sigma_{V.Y.} = 1$, two things can happen:
   (1) $h(\mu_A(x_i), \mu_B(x_i)) = 0$ for all x_i. By (1) we have that $g(\mu_A(x_i), \mu_B(x_i)) = 0$. By (2) it results that $\mu_A(x_i) \leq \mu_B(x_i)$ for all x_i.
   (2) If $\dfrac{M_{i=1}^n(g(\mu_A(x_i), \mu_B(x_i)))}{M_{i=1}^n(h(\mu_A(x_i), \mu_B(x_i)))} = 1$, then we have

$$M_{i=1}^n(g(\mu_A(x_i), \mu_B(x_i))) = M_{i=1}^n(h(\mu_A(x_i), \mu_B(x_i))).$$

If there exists any i for which $g(\mu_A(x_i), \mu_B(x_i)) \neq h(\mu_A(x_i), \mu_B(x_i))$, by (1) we have that $g(\mu_A(x_i), \mu_B(x_i)) < h(\mu_A(x_i), \mu_B(x_i))$. Since M satisfies $A3S$ we would have

$$M_{i=1}^n(g(\mu_A(x_i), \mu_B(x_i))) < M_{i=1}^n(h(\mu_A(x_i), \mu_B(x_i))),$$

which is a contradiction with the hypothesis. Therefore

$$g(\mu_A(x_i), \mu_B(x_i)) = h(\mu_A(x_i), \mu_B(x_i))$$

for all x_i. By (2) $\mu_A(x_i) \leq \mu_B(x_i)$ for all x_i holds.
(Sufficiency) If $\mu_A(x_i) \leq \mu_B(x_i)$ for all x_i, two things can happen:
(1) If $h(\mu_A(x_i), \mu_B(x_i)) = 0$ for all x_i, then $\sigma_{V.Y.}(A, B) = 1$.
(2) By (2) we have that

$$g(\mu_A(x_i), \mu_B(x_i)) = h(\mu_A(x_i), \mu_B(x_i)),$$

therefore

$$\sigma_{V.Y.}(A, B) = \frac{M_{i=1}^n(g(\mu_A(x_i), \mu_B(x_i)))}{M_{i=1}^n(h(\mu_A(x_i), \mu_B(x_i)))} = 1.$$

(b) If $A \geq e$, then
(Necessity) If $\sigma_{V.Y}(A, A_c) = 0$, then

$$M_{i=1}^n(g(\mu_A(x_i), c(\mu_A(x_i)))) = 0.$$

As M satisfies $A1$ we have $g(\mu_A(x_i), c(\mu_A(x_i))) = 0$ for all x_i. Because of (3) we have that $A = 1$.
(Sufficiency) If $A=1$, then by (3) we have $g(\mu_A(x_i), c(\mu_A(x_i))) = g(1,0) = 0$. As M satisfies $A1$ we have $M_{i=1}^n(g(\mu_A(x_i), c(\mu_A(x_i)))) = 0$. Obviously, $h(1,0) \neq 0$. we have to bear in mind that if it happened that $h(1,0) = 0$ then, $g(1,0) = 0 = h(1,0)$, because of (2) we would have $1 \leq 0$ which is impossible.
(c) If $A \leq B \leq C$, then
   (1) If $\sigma_{V.Y}(C, A) = 1$ by (a) we have that $C = A$, therefore $\sigma_{V.Y.}(C, A) = \sigma_{V.Y.}(B, A)$.
   (2) If $\sigma_{V.Y.}(C, A) \neq 1$, then

$$\sigma_{V.Y.}(C, A) = \frac{M_{i=1}^n(g(\mu_C(x_i), \mu_A(x_i)))}{M_{i=1}^n(h(\mu_C(x_i), \mu_A(x_i)))} \leq \frac{M_{i=1}^n(g(\mu_B(x_i), \mu_A(x_i)))}{M_{i=1}^n(h(\mu_C(x_i), \mu_A(x_i)))}$$

$$\leq \frac{M_{i=1}^n(g(\mu_B(x_i), \mu_A(x_i)))}{M_{i=1}^n(h(\mu_B(x_i), \mu_A(x_i)))} = \sigma_{V.Y.}(B, A)$$

If $A \leq B$, then

(1) If $\sigma_{V.Y.}(C, A) = 1$, then by (a) we have $C \leq A$, therefore $C \leq A \leq B$, then $\sigma_{V.Y.}(C, B) = 1$; that is $\sigma_{V.Y.}(C, A) = \sigma_{V.Y.}(C, B)$.

(2) If $\sigma_{V.Y.}(C, A) \neq 1$, then

$$\sigma_{V.Y.}(C, A) = \frac{M_{i=1}^n(g(\mu_C(x_i), \mu_A(x_i)))}{M_{i=1}^n(h(\mu_C(x_i), \mu_A(x_i)))} \leq \frac{M_{i=1}^n(g(\mu_C(x_i), \mu_B(x_i)))}{M_{i=1}^n(h(\mu_C(x_i), \mu_A(x_i)))}$$

$$\leq \frac{M_{i=1}^n(g(\mu_C(x_i), \mu_B(x_i)))}{M_{i=1}^n(h(\mu_C(x_i), \mu_B(x_i)))} = \sigma_{V.Y.}(C, B)$$

$\square$

Example 1. For a finite set X, we take the functions

$$\begin{cases} g(x, y) = \wedge(x, y) \\ h(x, y) = x \end{cases}$$

It is clear that g and h fulfill the conditions (1)–(5) of the proposition above. Therefore

$$\sigma_{V.Y.}(A, B) = \begin{cases} 1 \text{ if } \mu_A(x_i) = 0 \text{ for all } i \in \{1, \ldots, n\} \\ \dfrac{M_{i=1}^n(\wedge(\mu_A(x_i), \mu_B(x_i)))}{M_{i=1}^n(\mu_A(x_i))}, \text{ elsewhere} \end{cases}$$

it is a fuzzy subsethood measure in the sense of V. Young.

Obviously, if $M_{i=1}^n(x_1, \ldots, x_n) = \frac{1}{n}\sum_{i=1}^n x_i$ the previous expression is the fuzzy subsethood measure given by Kosko [16].

Due to [19] we know that if $\lambda > 0$, then

$$M_{i=1}^n(x_1, \ldots, x_n) = \left(\frac{1}{n}\sum_{i=1}^n x_i^\lambda\right)^{\frac{1}{\lambda}}$$

fulfills $A1$ and $A3S$. Therefore

$$\sigma_{V.Y.}(A, B) = \begin{cases} 1 \text{ if } \mu_A(x_i) = 0 \text{ for all } i \in \{1, \ldots, n\} \\ \dfrac{\left(\sum_{i=1}^n(\wedge(\mu_A(x_i), \mu_B(x_i)))^\lambda\right)^{\frac{1}{\lambda}}}{\left(\sum_{i=1}^n(\mu_A(x_i))^\lambda\right)^{\frac{1}{\lambda}}}, \text{ elsewhere} \end{cases}$$

it is a fuzzy subsethood measure in the sense of V. Young.

Example 2. For a finite set X, we take the functions

$$\begin{cases} g(x, y) = c(x) \\ h(x, y) = \vee(c(x), c(y)) \end{cases}$$

It is clear that g and h fulfill the conditions (1)–(5) of the proposition above. Therefore

$$\sigma_{V.Y.}(A,B) = \begin{cases} 1 \text{ if } A = 1 \text{ and } B = 1 \\[2mm] \dfrac{M_{i=1}^n(c(\mu_A(x_i)))}{M_{i=1}^n(\vee(c(\mu_A(x_i)), c(\mu_B(x_i))))}, & \text{elsewhere} \end{cases}$$

it is a fuzzy subsethood measure in the sense of V. Young. This example was proposed by Fan [6] as fuzzy $*$-subsethood measure.

Example 3. For a finite set X, we take the functions

$$\begin{cases} g(x,y) = \vee(c(x), \wedge(x,y)) \\[2mm] h(x,y) = \begin{cases} \vee(c(x), x) \text{ if } x \leq y \\ 1 \text{ elsewere.} \end{cases} \end{cases}$$

It is clear that g and h fulfill the conditions (1)–(5) of the proposition above. Therefore

$$\sigma_{V.Y.}(A,B) = \frac{M_{i=1}^n(\vee(c(\mu_A(x_i)), \wedge(\mu_A(x_i), \mu_B(x_i))))}{M_{i=1}^n \begin{pmatrix} \vee(c(\mu_A(x_i)), \mu_A(x_i)) \text{ if } \mu_A(x_i) \leq \mu_B(x_i) \\ 1 \text{ elsewhere} \end{pmatrix}}$$

is a fuzzy subsethood measure in the sense of V. Young. We must note that in this example the function g that we take is the implication operator of Zadeh. It is precisely the weakening of Axiom 4 by V. Young what enables the construction of this fuzzy subsethood measure from Zadeh's operator.

Example 4. For a finite set X, we take the functions

$$\begin{cases} g(x,y) = \wedge(1, 1 - x + y) \\ h(x,y) = 1 \end{cases}$$

It is clear that g and h fulfill the conditions (1)–(5) of the proposition above. Therefore

$$\sigma_{V.Y.}(A,B) = M_{i=1}^n(\wedge(1, 1 - \mu_A(x_i) + \mu_B(x_i)))$$

is a fuzzy subsethood measure in the sense of V. Young. Obviously, this example is obtained from Proposition 6 taking $h(x,y) = 1$; that is, aggregating by means of the arithmetic mean the operator of implication of Lukasiewicz.

If in the example above we take $M_{i=1}^n(x_1, \ldots, x_n) = \frac{1}{n} \sum_{i=1}^n x_i$; that is,

$$\sigma_{V.Y.}(A,B) = \frac{1}{n} \sum_{i=1}^n (\wedge(1, 1 - \mu_A(x_i) + \mu_B(x_i))$$

we get the fuzzy subsethood measure defined by Goguen in [14].

3 Fuzzy Subsethood Measure of V. Young and Axioms of D. Sinha and R. Dougherty

The result of the study of the conditions in which the constructions presented in Proposition 1 fulfill the axioms of Sinha and Dougherty is shown on Table 1.

Table 1. Proposition-1 Fulfilment

Sinha D.	V. Young
Axiom 1	It coincides with (a); that is, $\sigma_{V.Y.}$ fulfills it.
Axiom 2	It is replaced by (b); that is, If $A > e$, then $\sigma_{V.Y.}(A, A_c) = 0$ if and only if $A = 1$
Axiom 3	It coincides with the second condition of (c).
Axiom 4	It is restricted to the first condition of (c),
Axiom 5	In the conditions of Proposition 1, if M satisfies $A4$, then $\sigma_{V.Y.}$ fulfills it.
Axiom 6	In the conditions of Proposition 1, if $g(x, y) = g(c(y), c(x))$ and $h(x, y) = h(c(y), c(x)$, then, $\sigma_{V.Y.}$ fulfills it.
Axiom 7	$\sigma_{V.Y.}$ does not fulfill it. Nor does it fulfill, $\sigma_{V.Y.}(B \vee C, A) \leq \wedge(\sigma_{V.Y.}(B, A), \sigma_{V.Y.}(A, C))$
Axiom 8	$\sigma_{V.Y.}$ fulfills the following inequality: $\sigma_{V.Y.}(A, B \wedge C) \leq \wedge(\sigma_{V.Y.}(A, B), \sigma_{V.Y.}(A, C))$
Axiom 9	fulfills it.
Axiom 10	In the conditions of Proposition 1, if $g(x, y) + g(x, c(y)) \geq 1$ and $M_{i=1}^{n}(x_1, \cdots, x_n) + M_{i=1}^{n}(1 - x_1, \cdots, 1 - x_n) \geq 1$ then $\sigma_{V.Y}$ fulfills it.

Table 1. (*Continued*)

Axiom 11	$\{$ fulfills it.
Axiom 12	In the conditions of Proposition 1, if $g(x,y) \geq y$; $g(x,y) = g(c(y), c(x))$ and M is idempotent, then $\sigma_{V.Y.}$ fulfills it.

4 Fuzzy Entropy and Fuzzy Subsethood Measure in the Sense of V. Young

We know that a measure of fuzzy entropy assesses the amount of vagueness, or fuzziness, in a fuzzy set. In 1972 Deluca and Termini [23] formalize the properties of fuzzy entropy through the following axioms:

Definition 2. *A real function* $E : \mathcal{F}(\mathcal{X}) \to [0,1]$ *is called an entropy on* $\mathcal{F}(X)$, *if* E *satisfies the following properties:*

(E1) $E(A) = 0$ *if and only if* A *is nonfuzzy;*
(E2) $E(A) = 1$ *if and only if* $A = e$;
(E3) $E(A) \leq E(B)$ *if* A *refines* B; *that is,* $\mu_A(x) \leq \mu_B(x)$ *when* $\mu_B(x) \leq e$ *and* $\mu_A(x) \geq \mu_B(x)$ *when* $\mu_B(x) \geq e$;
(E4) $E(A) = E(A_c)$.

Many such entropy functions have been defined in the literature. A sample can be found in Dubois and Prade [7].

In 1983, Ebanks [24] presented the following definition of fuzzy entropy.

Definition 3. *A real function* $E : \mathcal{F}(\mathcal{X}) \to [0,1]$ *is called an entropy on* $\mathcal{F}(X)$, *if* E *has the following properties:*

(E1) $E(A) = 0$ *if and only if* A *is nonfuzzy;*
(E2) $E(A)$ *is maximum if and only if* $A = e$, *where* e *is the equilibrium point of the strong negation considered;*
(E3) $E(A) \leq E(B)$ *if* A *refines* B; *that is,* $\mu_A(x) \leq \mu_B(x)$ *when* $\mu_B(x) \leq e$ *and* $\mu_A(x) \geq \mu_B(x)$ *when* $\mu_B(x) \geq e$;
(E4) $E(A) = E(A_c)$;
(E5) *(property of valuation)* $E(A \vee B) + E(A \wedge B) = E(A) + E(B)$.

Ebanks gave the following necessary and sufficient conditions on functions so that they satisfy the requirements (E1)–(E5) for discrete fuzzy sets:

Theorem 2. *Let* $E : \mathcal{F}(X) \to R^+$ *and* $\mu_i = \mu_A(x_i)$ *for all* i. *Then* E *satisfies* (E1)–(E5) *if and only if* E *has the form* $E(A) = \sum_{i=1}^{n} g(\mu_i)$ *for some functions* $g : [0,1] \to R^+$ *that satisfy:*

(G1) $g(0) = g(1) = 0$; $g(x) > 0$ *for all* $x \in (0,1)$;
(G2) $g(x) < g(e)$ *for all* $x \in [0,1] - e$;
(G3) g *is nondecreasing on* $[0, e)$ *and nonincreasing on* $(e, 1]$;
(G4) $g(x) = g(c(x))$ *for all* $x \in [0,1]$, *where* c *is a strong negation such that* $c(e) = e$.

As V. Young recalls in [5], to relate fuzzy subsethood measure with fuzzy entropy, Kosko [16] proposes the following expression: given a fuzzy subsethood measure σ the entropy E generated by σ is defined as

$$E(A) = \sigma(A \vee A_c, A \wedge A_c), \quad \text{for all } A \in \mathcal{F}(\mathcal{X}).$$

As we have said before, this was one of the reasons which led V. Young to impose item (b) and the first condition of item (c) in Definition 1, in such a way that she obtains the following result:

Theorem 3. *([5]) If* $\sigma_{V.Y.}$ *is a fuzzy subsethood measure on* X, *then* E *defined by*

$$E_{V.Y.}(A) = \sigma_{V.Y.}(A \vee A_c, A \wedge A_c) \quad \text{for all } A \in \mathcal{F}(\mathcal{X})$$

is a fuzzy entropy measure on X.

Evidently, from functions g and h of Proposition 1 we have that

$E_{V.Y.}(A)$

$$= \begin{cases} 1 \text{ if } h(\vee(\mu_A(x_i), c(\mu_A(x_i))), \wedge(\mu_A(x_i), c(\mu_A(x_i)))) = 0 \text{ for all } i \in \{1, \ldots, n\} \\ \dfrac{M_{i=1}^n(g(\vee(\mu_A(x_i), c(\mu_A(x_i))), \wedge(\mu_A(x_i), c(\mu_A(x_i)))))}{M_{i=1}^n(h(\vee(\mu_A(x_i), c(\mu_A(x_i))), \wedge(\mu_A(x_i), c(\mu_A(x_i)))))}, \text{ elsewhere} \end{cases}$$

is a fuzzy entropy.

It is easy to see that many of the properties demanded from the functions g in Proposition 1 and the conditions on Table 1 are the same that are usually imposed on the implication operators I. The following Lemma allows us to say even more about these functions g.

Lemma 1. *Let* n *be a whole positive finite number, let* c *be any strong negation and let the function*

$$g : [0,1]^2 \to [0,1] \quad \text{such that}$$

(1') *If* $x \geq e$, *then* $g(x, c(x)) = 0$ *if and only if* $x = 1$;
(2') *If* $x \leq y \leq z$, *then* $g(z, x) \leq g(y, x)$;
(3') *If* $x \leq y$, *then* $g(z, x) \leq g(z, y)$;
(4') $g(x, y) = 1$ *if and only if* $x \leq y$,
then the function

$$G : [0,1] \to [0,1], \quad \text{given by}$$

$$G(x) = \tfrac{1}{n} g(\vee(x, c(x)), \wedge(x, c(x)))$$

satisfies the following properties:

(G1) $G(0) = G(1) = 0; G(x) > 0$ *for all* $x \in (0,1)$;
(G2) $G(x) < G(e)$ *for all* $x \in [0,1] - e$;
(G3) G *is nondecreasing on* $[0,e)$ *and nonincreasing on* $(e,1]$;
(G4) $G(x) = G(c(x))$ *for all* $x \in [0,1]$.

Proof. (G1). Bearing in mind (1') we have

$$G(0) = \tfrac{1}{n} g(1,0) = 0$$

$$G(1) = \tfrac{1}{n} g(1,0) = 0$$

Now let us see that $G(x) > 0$ for all $x \in (0,1)$. Let us suppose this is not so; that is, there exists a $x' \in (0,1)$ such that $G(x') = 0$. By definition we have that $G(x') = 0$, then

$$0 = g(\vee(x', c(x')), c(\vee(x', c(x')))).$$

Two things can happen:
(a) If $\vee(x', c(x')) \geq e$, then because of (1') we have $\vee(x', c(x')) = 1$ therefore $x' = 1$ or $x' = 0$ which is impossible, because $x' \in (0,1)$.
(b) If $\vee(x', c(x')) < e$, then $x' < e$ and $c(x') < e$, which is impossible because if $x' < e$, then $c(x') > e$ and if $c(x') < e$, then $x' > e$.
Therefore $G(x) > 0$ for all $x \in (0,1)$.
(G2). Evidently, $G(e) = \tfrac{1}{n} g(e,e)$. Let $x \in [0,1] - e$. Two things can happen:
(a) If $x < e$, then $x < e < c(x)$ by the properties (2') and (3') we have $G(x) = \tfrac{1}{n} g(c(x), x) \leq \tfrac{1}{n} g(e,x) \leq \tfrac{1}{n} g(e,e) = G(e)$.
(b) If $x > e$, then $c(x) < e < x$, therefore $G(x) = \tfrac{1}{n} g(x, c(x)) \leq \tfrac{1}{n} g(e, c(x)) \leq \tfrac{1}{n} g(e,e) = G(e)$.
Therefore for all $x \in [0,1] - e$ we have $G(x) \leq G(e)$. Let us see now that $G(x) < G(e)$. Let us suppose that it is not true; that is, there exists at least one $x' \in [0,1] - e$ such that $G(x') = G(e)$; that is, $G(x') = \tfrac{1}{n} g(\vee(x', c(x')), c(\vee(x', c(x')))) = \tfrac{1}{n} g(e,e) = \tfrac{1}{n}$, then $g(\vee(x', c(x')), c(\vee(x', c(x')))) = 1$, due to the condition (4') we have that $\vee(x', c(x')) \leq c(\vee(x', c(x'))) = \wedge(x', c(x'))$, then $\vee(x', c(x')) = \wedge(x', c(x'))$, therefore $x' = e$ which is a contradiction, therefore $G(x) < G(e)$ for all $x \in [0,1] - e$.
(G3). Let $x_1, x_2 \in [0,e)$ such that $x_1 \leq x_2 < e$, then $x_1 \leq x_2 < e < c(x_2) \leq c(x_1)$. Bearing in mind the conditions (2') and (3') we have $G(x_1) = \tfrac{1}{n} g(c(x_1), x_1) \leq \tfrac{1}{n} g(c(x_2), x_1) \leq \tfrac{1}{n} g(c(x_2), x_2) = G(x_2)$.
Let $x_1, x_2 \in (e,1]$ such that $e < x_1 \leq x_2$, then $c(x_2) \leq c(x_1) < e < x_1 \leq x_2$. In these conditions, considering the conditions (2') and (3') we have $G(x_1) = \tfrac{1}{n} g(x_1, c(x_1)) \geq \tfrac{1}{n} g(x_2, c(x_1)) \geq \tfrac{1}{n} g(x_2, c(x_2)) = G(x_2)$.
(G4). Evidently because c is strong negation.

□

In the following theorem we show the conditions in which we can construct fuzzy entropies that satisfy the valuation property from V. Young subsethood measures and therefore from the constructions developed in Proposition 1.

Theorem 4. *Let X be the referential set that is finite, non-empty such that $|X| = n$. In the same conditions as in Proposition 1, if $M_{i=1}^n (x_1, \ldots, x_n) = \frac{1}{n} \sum_{i=1}^n x_i$ and $h(x, y) = 1$ for all $x, y \in [0, 1]$. Then*

$$E_{V.Y.}(A) = \sigma_{V.Y.}(A \vee A_c, A \wedge A_c)$$

is a fuzzy entropy that satisfies the property of valuation.

Proof. We only need to bear in mind the lemma above and the theorem of Ebanks.

□

5 Conclusions and Future Line of Work

From a critical revision of the three most important axiomatizations that exist in fuzzy literature for fuzzy subsethood measures, we have concluded that the arguments shown by V. Young to replace Axiom 2 of Sinha and Dougherty make the study of her definition very attractive.

We have seen that the measures of this author can be obtained aggregating functions of $[0, 1]^2$ in $[0, 1]$. Some of these functions with properties similar to the ones usually demanded from implication operators. We have also proven the following:

(a) From her fuzzy subsethood measures we can construct fuzzy entropies. (We have seen that if we take the arithmetic mean aggregation, the entropy satisfies the property of valuation).
(b) A characteristic of her measures is that they always fulfill Axiom 9; that is,

$$\sigma_{DI}(A, B \vee C) \geq \vee\{\sigma_{DI}(A, B), \sigma_{DI}(A, C)\}$$

for all $A, B, C \in \mathcal{F}(X)$.
(c) They also fulfill the inequality: $\sigma_{DI}(A, B \wedge C) \leq \wedge\{\sigma_{DI}(A, B), \sigma_{DI}(A, C)\}$.

These facts suggest to us that in the future we must study the construction of fuzzy inclusion measures de V. Young, from M aggregations that satisfy $A1 - A4$ and I implication operators. That is, study constructions of the form:

$$\sigma(A, B) = M_{i=1}^n (I(\mu_A(x_i), \mu_B(x_i))).$$

Naturally, it will depend on the properties we demand from M and I for the functions σ to be fuzzy subsethood measures and to satisfy certain axioms.

We should focus on analyzing the influence of the conditions demanded from implication operators in order for the new measures to satisfy Sinha and Dougherty's axioms.

Another objective we have in mind is to analyze the $*-$measures of inclusion of Fan et al. We know [6] that the fuzzy inclusions that these authors define are the same as V. Young's changing the condition:

(c) If $A \leq B \leq C$, then $\sigma_{V.Y.}(C, A) \leq \sigma_{V.Y.}(B, A)$ and if $A \leq B$, then $\sigma_{V.Y.}(C, A) \leq \sigma_{V.Y}(C, B)$, by the condition:

(c) If $A \leq B \leq C$, then $\sigma_*(C, A) \leq \sigma_*(B, A)$ and $\sigma_*(C, A) \leq \sigma_*(C, B)$.

We think that we can present a method of construction similar to that of Proposition 1 for these measures and likewise analyze on the one hand the manner of constructing fuzzy entropies from them and on the other analyze the conditions that must be met by M and g in order for said measures to satisfy the axioms of Sinha and Dougherty.

References

1. L.A. Zadeh, *Fuzzy sets*, Information Control, 8: 338–353, 1965.
2. W. Bandler and L. Kohout, *Fuzzy power sets and fuzzy implication operators*, Fuzzy Sets and Systems, 4: 13–30, 1980.
3. L. Kitainik, *Fuzzy inclusions and fuzzy dichotomous decision procedures*, in J. Kacprzyk, S. Orlovski (Eds.), Optimization Models Using Fuzzy Sets and Possibility Theory, Reidel, Dordrecht pp. 154–170, 1987.
4. D. Sinha and E.R. Dougherty, *Fuzzification of set inclusion: Theory and applications*, Fuzzy Sets and Systems 55: 15–42, 1993.
5. V.R. Young, *Fuzzy subsethood*, Fuzzy Sets and Systems 77: 371–384, 1996.
6. J. Fan, X. Xie and J. Pei, *Subsethood measures: new definitions*, Fuzzy Sets and Systems, 106: 201–209, 1999.
7. D. Dubois and H. Prade, *Fuzzy Sets and Systems: Theory and Applications*, New York: Academic, 1980.
8. J. Fodor and R. Yager, *Fuzzy sets theoretic operators*, in D. Dubois, H. Prade (Eds.), Fundamentals of Fuzzy Sets, Kluwer, Boston, MA, pp. 125–193, 2000.
9. C. Cornelis, C. Van der Donk and E. Kerre, *Sinha-Dougherty approach to the fuzzification of set inclusion revisited*, Fuzzy Sets and Systems, 134: 283–295, 2003.
10. A. Kehagias and M. Konstantinidou, *L-Fuzzy Valued Inclusion Measure, L-Fuzzy Similarity and L-Fuzzy Distance*, Fuzzy Sets ans Systems, 136: 313–332, 2003.
11. R. Willmott, *On the transitivity of containment and equivalence in fuzzy power set theory*, Internatinal Journal of Mathematical Analysis and Application, 120: 384–396, 1986.
12. S. Kundu, *A representation theorem for min-transitive fuzzy relations*, Fuzzy Sets and Systems, 109: 453–457, 2000.
13. N. Frago, *Morfologia matematica borrosa basada on operadores generalizados de Lukasiwicz: procedimiento de imagenes*, Ph.D. dissertation (in Spanish), Universidad publica de Navarra, 1996.

14. B. Kosko, *L-fuzzy sets*, Journal of Mathematical Analysis and Application, 18: 145–174, 1967.
15. B. Kosko, *The logic of inexact concepts*, Synthese, 19: 325–373, 1969.
16. B. Kosko, *Fuzziness vs. probability*, Internationl Journal of General Systerm, 17: 211–240, 1990.
17. B. Kosko, *Fuzzy entropy and conditioning*, Information Science, 40: 165–174, 1986.
18. H. Bustince, J. Montero, E. Barrenechea and M. Pagola, *Semiautoduality in a restricted family of aggregation operators*, Fuzzy Sets and Systems, 158(12): 1360–1377, 2007.
19. H. Bustince, V. Mohedano, E. Barrenechea and M. Pagola, *Definition and Construction of Fuzzy DI-subsethood Measures*, Information Sciences, 176(21): 3190–3231, 2006.
20. D. Dubois and H. Prade, *Criteria aggregation and ranking of alternatives in the framework of fuzzy theory*, in: H.J. Zimmermann, L.A. Zadeh, B. Gaines (Eds.), Fuzzy Sets and Decision Analysis, TIMS Studies in Management Science, vol. 20, pp. 209–240, 1984.
21. G. J. Klir and T. A. Forlger, *Fuzzy Sets uncertainty and information*, Prentice-Hall International, 1988.
22. E. Trillas, *Sobre funciones de negación en la teoría de conjuntos difusos*, Stochastica, III-1: 47–59, (in Spanish), 1979 Reprinted (English version) (1998) in: *Advances of Fuzzy Logic*, Eds. S. Barro et altri-Universidad de Santiago de Compostela, pp. 31–43.
23. A. Deluca and S. Termini, *A definition of a nonprobabilistic entropy in the setting of fuzzy set theory*, Inform and Control, 20: 301–312, 1972.
24. B. Ebanks, *On measures of fuzziness and their representations*, Journal of Mathematical Analysis and Application, 94: 24–37, 1983.

An Incremental Learning Structure Using Granular Computing and Model Fusion with Application to Materials Processing

George Panoutsos and Mahdi Mahfouf

Institute for Microstructural and Mechanical Process Engineering, The University of Sheffield (IMMPETUS), Sheffield, S1 3JD, UK
Department of Automatic Control and Systems Engineering, University of Sheffield, Mappin Street, Sheffield, S1 3JD, UK
g.panoutsos@sheffield.ac.uk

Summary. This chapter introduces a Neural-Fuzzy (NF) modelling structure for offline incremental learning. Using a hybrid model-updating algorithm (supervised/unsupervised), this NF structure has the ability to adapt in an additive way to new input–output mappings and new classes. Data granulation is utilised along with a NF structure to create a high performance yet transparent model that entails the core of the system. A model fusion approach is then employed to provide the incremental update of the system. The proposed system is tested against a multi-dimensional modelling environment consisting of a complex, non-linear and sparse database.

1 Introduction

Data driven Computational Intelligence (CI) models are often employed to describe processes and solve engineering problems with great success. Neural-Networks (NN) [7, 11], Fuzzy, and Neural-Fuzzy systems (NF) [3, 10] as well as Evolutionary and Genetic Algorithms (GA) [9,12] have all been used in the past to solve real-world modelling and control engineering problems. The scientific maturity of such methodologies and the demand for realistic representations of high complexity real-world engineering processes allowed new and advanced features of such systems to evolve. Such features include the ability of a model/structure to incrementally learn from new data. These structures are able to learn from an initial database (with appropriate training) but at the same time incrementally adapt to new data when these are available. Additional requirements include the system's ability to interact with the environment in a continuous fashion (i.e. life-long learning mode) and having an open structure organisation (i.e. dynamically create new modules) [5]. Several

G. Panoutsos and M. Mahfouf: *An incremental Learning Structure Using Granular Computing and Model Fusion with Application to Materials Processing*, Studies in Computational Intelligence (SCI) **109**, 139–153 (2008)
www.springerlink.com

methods have been developed so far that demonstrate some of the aspects of incremental learning systems [2, 4–6].

This chapter presents a new approach that is based on Granular Computing Neural-Fuzzy networks (GrC-NF) [10]. By using such an approach, it is possible to achieve a high single network performance and at the same time maintain a high system transparency (in contrast to black-box low transparency modelling i.e. NN). An incremental learning architecture is subsequently developed using the GrC-NF models in a cascade model-fusion manner. The entropy of the cascade models is used as the main feature that assists the data fusion process.

The proposed methodology is tested against a multidimensional MISO real industrial application; the prediction of mechanical properties of heat-treated steel is investigated. Such application involves complex databases, containing data with non-linear dynamics and high interaction between the dimensions as well as sparse data of high uncertainty (measurement noise, operator errors, etc.).

2 GrC – NF Modelling

Models elicited via the Granular Computing – Neural-Fuzzy (GrC-NF) approach, as described in [10], will form the main building blocks of the Incremental Learning (IL) system. This modelling process is realised in three steps:

1. Knowledge discovery
2. Rule-base creation
3. Model optimisation

2.1 Knowledge Discovery

Granular Computing mimics the perception and the societal instinct of humans when grouping similar items together. Grouping items together is not just a matter of proximity but also of similarity measures such as similarity in class or function. Data granulation [1, 9] is achieved by a simple and transparent two-step iterative process that involves the two following steps:

- Step 1: Find the two most 'compatible' information granules and merge them together as a new information granule containing all the information included in both original granules.
- Step 2: Repeat the process of finding the two most compatible granules and merging them, until a satisfactory data abstraction level is achieved. The abstraction level can be manually or automatically set.

The most important concept in the above process is the definition of the compatibility measure, which varies between authors. This definition can be purely geometrical (distance between granules, size of granules, volume of granules), density-driven (ratio of cardinality versus granule volume) or similarity-driven (data overlap). In this chapter, the compatibility measure is defined as a function of the distance between the granules and the information density of the newly formed granule. A mathematical representation of the compatibility criterion is described by the following three equations:

$$Compatibility = f\ (w_1 \cdot \text{Distance},\ w_2 \cdot \text{Density}) \tag{1}$$

$$\text{Distance} = \sum_{i=1}^{k} (\text{position of granule}_\text{B} - \text{position of granule}_\text{A})_i \tag{2}$$

$$\text{Density} = \frac{\text{granule cardinality}}{\text{granule volume}} = \frac{\text{no. of sub-granules}}{\prod\limits_{i=1}^{k} \text{granule length}_i} \tag{3}$$

Where w_1, w_2 are weights for balancing the distance/density requirements and k is the dimensionality of the data space. This definition of compatibility uses the distance between the granules as well as the cardinality and the granule volume (to form the density criterion). Alternative forms of the compatibility criterion may also contain the granule's size (multidimensional circumference).

The growth of the granules (from data) allows for a strong linkage between the original data set (transparency) and it permits visual monitoring of the granulation process. Considering the merging of each set of granules as some information condensation or information loss, it is possible to link information loss to the merged distance and therefore plot an information loss graph. This graph can be used on-line or off-line as a criterion for terminating the granulation process. The control of the granularity level high/low allows the development of models with a variable level of abstraction i.e. low abstraction models – low number of granules/fuzzy rules.

Figure 1, shows five (5) snapshots of a 2-dimensional data granulation. Dimension A has a range between 0–2,000 units, and Dimension B between 0–800 units. The first snapshot, shown at the top of the figure, is the representation of the raw, pre-granulated data, consisting of 3,760 data points. As the iterative granulation algorithm progresses, snapshots of the granulation process are shown, consisting of 1,000 granules (second snapshot of data), 250 granules, 25 and finally 18 granules. The added information collected during the granulation process is stored, and it consists of the cardinality and the multidimensional length and density of each granule. These data are going to be used for creating a linguistic rule-base, as it is shown in the next section.

The orientation of the granules is also of high importance as it is shown in [10]. Controlling the orientation of the formed granules, and favouring one dimension over an other (or others) can have beneficial effect to the modelling process by increasing the model's sensitivity to a particular input space.

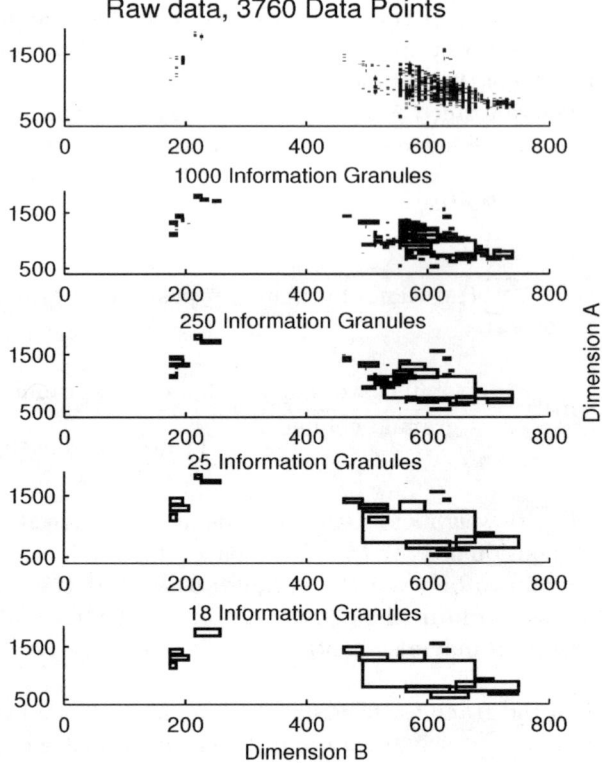

Fig. 1. Iterative information granulation, a two-dimensional example

2.2 From Information Granules to a Fuzzy Rule-Base

Consider the granulation of an input/output database provided by a multi-input single-output (MISO) system. By granulating across each input dimension individually and at the same time across the whole input space it is possible to identify relational information (rules) similar to a Mamdani Fuzzy Inference System (FIS) rule-base of the following form:

$$\text{Rule } 1: \tag{4}$$
$$\text{if } (inputA = A_1 \text{ and } inputB = B_1 \text{ and } \ldots)$$
$$\text{then } (output = O_1)$$
$$\text{Rule } 2:$$
$$\text{if } (inputA = A_2 \text{ and } inputB = B_2 \text{ and } \ldots)$$
$$\text{then } (output = O_2)$$
$$\text{M}$$

Where A_i, B_i, O_i are information granules discovered during the GrC process and i: is the number of extracted granules.

The information captured during the GrC granulation process defines the initial structure of the fuzzy rule-base, the number of rules and the initial location and width of the membership functions (MFs). For establishing the location of each MF, the centre of the each granule is considered, and the width can be determined using the multidimensional length and the density of each granule. This rule-base is then used as the initial structure of a Neural-Fuzzy system.

2.3 Model Optimisation

Using the popular 3-layer RBF Neural-Fuzzy representation [3] it is possible to optimise the model using a fast Back Error Propagation (BEP) algorithm.

Depending on the output set O_i of the consequent part of each linguistic rule (...then the output is O_i) the fuzzy system can be:

1. O: Fuzzy Set, Mamdani rule-base
2. O: Singleton, Mamdani singleton
3. O: Linear function, Takagi–Sugeno–Kang (TSK)

In this chapter, the Mamdani singleton consequent part is considered. The equations describing the Mamdani singleton Neural-Fuzzy system are given as follows:

$$y = \sum_{i=1}^{p} z_i g_i (x) \tag{5}$$

Where $g_i (x)$ is the radial basis function (RBF), defined as:

$$g_i (x) = \frac{m_i (x)}{\sum_{i=1}^{p} m_i (x)} \tag{6}$$

The NF training procedure starts by initialising the centres and widths of each fuzzy weight. This can be achieved by taking the corresponding values from the GrC rule-base structure. The NF model is then trained using an adaptive weighted back-error-propagation (BEP) algorithm [3] (Fig. 2).

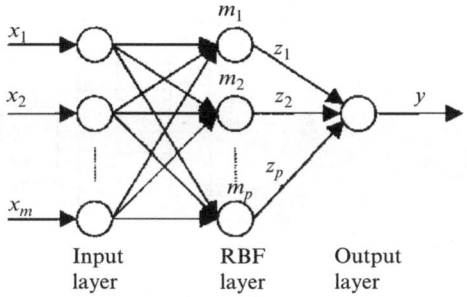

Fig. 2. General architecture of a fuzzy RBF network

3 Incremental Learning

The incremental learning/update of the system is achieved by defining an appropriate structure that is based on the NF model as described in Sect. 2. After defining the structure, two algorithms are developed, one for updating the structure (incremental learning) and the other for obtaining the model predictions.

3.1 Incremental Learning Structure

The IL model structure is based on cascaded NF models. The structure can dynamically create new sub-modules (NF networks) when new data become available. Figure 3 shows the system's architecture. The cascaded networks represent the knowledge database of the system. An information fusion algorithm is going to be used that utilises this knowledge efficiently, as will be shown in the next section.

3.2 Model Update

When new data are available the system can modify (expand) its structure to accommodate the new data, without disrupting the previously elicited model, using the algorithmic process shown in Fig. 4.

The new data are filtered by the system before they are fed to the update process. The new data vectors are compared with the existing information granules of the system and are split into *real new data* and *partially new data*. The *real new data* consist of data that belong to a totally new input space as compared to the original data, and the *partially new data* are data that belong or are close to the existing input space of the data set. Thresholds for the multidimensional granules' distance definitions have to be defined by the system designer, $Thres_{RealNew}$ and $Thres_{PartNew}$.

Each data set is handled differently by the update mechanism. The partially new data are used to perform a constrained training (fine-tuning) of the

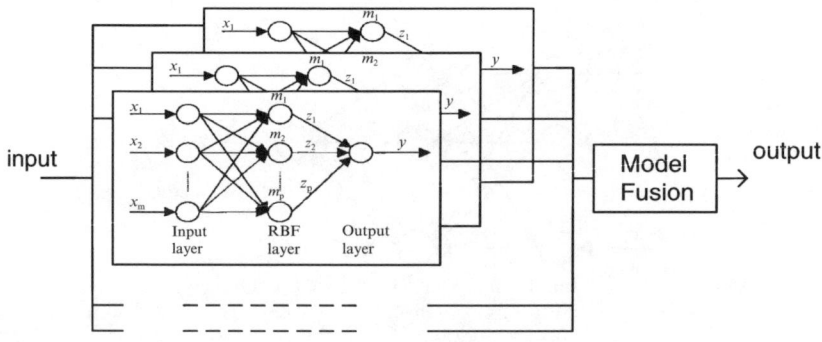

Fig. 3. IL Cascade architecture

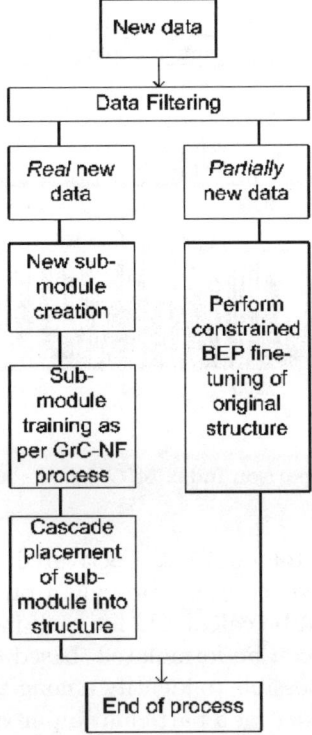

Fig. 4. The structure of the model update process

original system, so that the already existing knowledge is not disturbed. Since the input space of the partially new data is mostly covered by the system (by one or more sub-modules), there is no need to create a new module but just fine-tune the existing structure. The *"genuine"* new data are used to create a new sub-module, using the same GrC-NF modelling procedure. The new sub-module is then placed in a cascade fashion (as shown in Sect. 3.1) along with the rest of the NF sub-modules.

3.3 Model Predictions

After the *model update* process the cascade structure contains all the knowledge required by the system, the individual sub-modules cover both *old data* and *new data* input spaces. When an input vector is presented at the system an intelligent model fusion process is developed that is able to identify the appropriate sub-models for this particular input vector. The *active* sub-models provide individual predictions that are then fused together to provide the final model-prediction/answer.

Assume an input vector is presented to the system, at this stage the system does not know which sub-module (if any) is able to provide a reliable

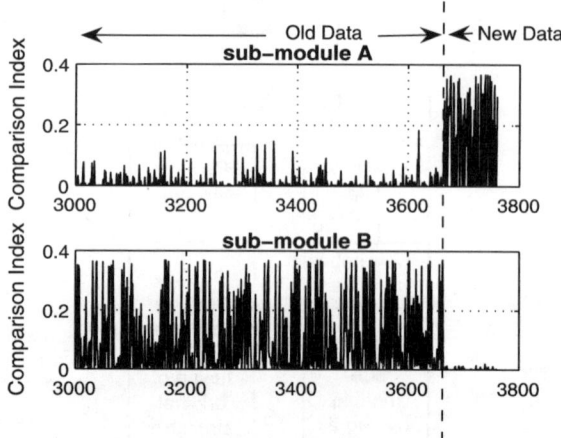

Fig. 5. Comparison index of *Old Data*, and *New Data*

answer/prediction. All sub-modules are activated and provide a single prediction each. While the NF networks provide a prediction, an entropy value
(measure of fuzziness) can be calculated for each individual network [8].

A comparative index can be formulated, based on the entropy values. As
it is shown in Fig. 5 it is possible to identify among the sub-modules which are
the ones that are more *active* for a particular input vector. Figure 5 shows how
the comparative index (entropy based) of two sub-modules varies for the *old
data* set and the *new data* set. It can be seen that the "sub-module A" produces
a higher index for the *new data* compared to the *old data*. Equivalently, "sub-
module B" produces a lower index for the *new data* compared to the *old data*.
By combining the information of both plots, it is possible to identify the sub-
module which should be credited for the correct prediction of the *new data*.

When a sub-module is very *active* to a particular input vector, it means
that the input space of the sub-module matches the input space of the given
vector and vice versa. Therefore, by calculating the entropy based comparison
index between the sub-modules it is possible to identify the ones that are more
likely to give a correct answer for a particular input vector. The selection
decision of the sub-modules is usually not very clear, i.e. when the entropy
boundaries are not very far apart, therefore it is difficult to distinguish them.
Therefore, a fuzzy decision mechanism is employed that provides a fusion
answer between the sub-modules-candidates.

If the decision variable is above or below some predefined threshold then it
is easy to assign a sub-module, if not, the answer is given by the fuzzy decision
mechanism that fuses the two answers together to provide a single value.

Following the completion of the structural design of the system, the proposed methodology is tested against a real-world industrial process, the prediction of mechanical properties of heat-treated steel.

4 Experimental Studies

Determining the optimal heat treatment regimes and weight percentage of composite materials of steel is a common but not trivial task in the steel industry. Heat treatments are frequently used in steel industry to develop the required mechanical properties of various steel grades i.e. hardening and tempering (see Fig. 6).

During the hardening stage, the material is soaked at a temperature to allow for austenitic transformation. The hardening stage is followed by quenching, which is performed using an oil or water medium. Finally, the tempering stage takes place, which aims to improve the toughness and ductility properties of the steel by heating it to specific temperatures and then air-cooling it. By predicting correctly both the optimal conditions (heat treatment) and steel composition of the steel, it is possible to obtain the required steel grade with accuracy and at a reduced cost. The modelling predictions presented in this chapter are based on both the chemical composition and the various heat treatment regimes. This is a multiple-input single-output (MISO) process that is difficult to model due to the following reasons:

- Non-linear behaviour of the process
- High interaction between the multivariable input spaces
- Measurement uncertainty of the industrial data
- High complexity of the optimisation space
- Sparse data space (see Fig. 7). The variables presented are: Ultimate Tensile Strength (UTS) – y-axis, and x-axis: Carbon (C), Nickel (Ni), Chromium (Cr) and Tempering Temperature (T.Temp). z-axis (vertical) shows the data density

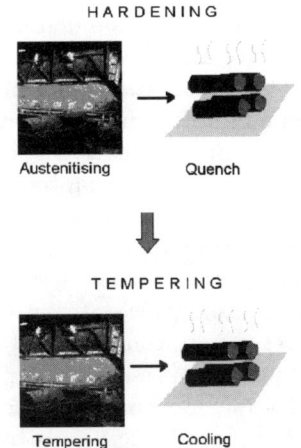

Fig. 6. Heat treatment of steel

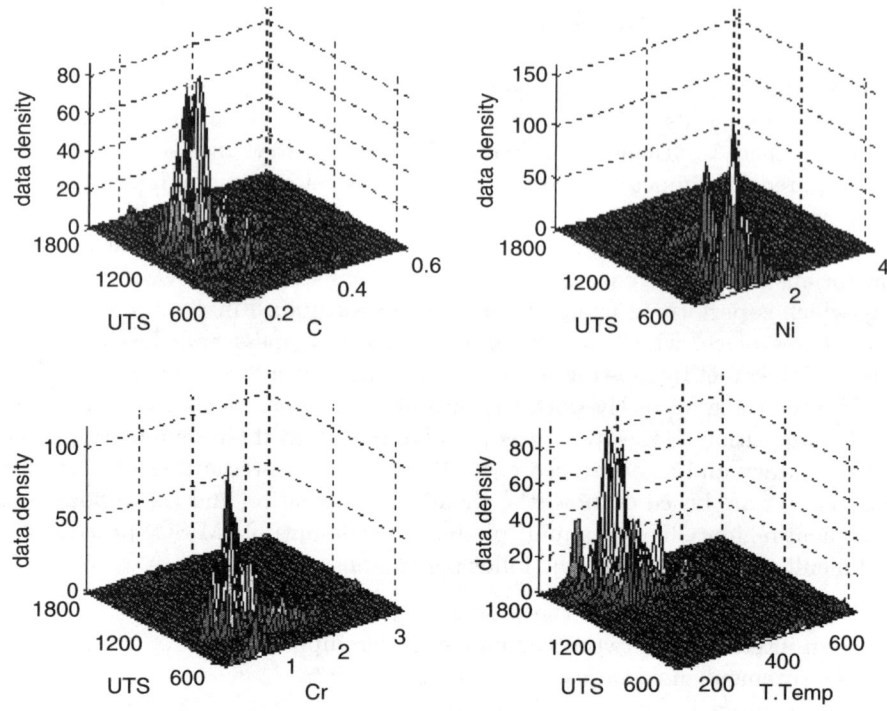

Fig. 7. Sample of the sparse data space

Black box modelling techniques [11] and grey-box modelling techniques [10] are usually employed to tackle this problem with a good level of performance. Developing such models requires a lot of effort, scientific and expert knowledge, model fine-tuning and model simulation and verification. Obtaining new data sets is an expensive and slow process in this industry, and when the new data are available, the whole modelling process has to be repeated. This can be time-consuming and it does not guaranty that the new model will retain a performance standard comparable to the old model.

Using the technique presented in this chapter, it is possible to incrementally update the initial system in order to accommodate the new data set. The new structure is then able to predict input vectors belonging to the new data set without any significant performance loss on the overall performance of the system (old data set, and old/new data set combined).

A high dimension data set, taken from the steel industry, is used for modelling purposes. Each set of points represents 15 input variables and 1 output variable. The input variables include both: (a) the chemical composition of steel (i.e. % content of C, Mn, Cr, Ni etc.) and (b) the heat treatment data (Tempering temperature, Cooling medium etc.). The output variable is the steel property that needs to be modelled/predicted, in this case the Tensile

Strength. The TS data set consists of 3,760 data vectors, which are divided as follows for the purpose of modelling:

1. Old data – training (2,747 data points)
2. Old data – validation (916 data points)
3. New data – training (72 data points)
4. New data – validation (24 data points)

Care has been taken so that the 'new data' set covers mostly an input region that is not covered by the 'old data' set (i.e. a new steel grade that is not covered by the 'old data' set). The old data set has various steel grades and the new data set contains mostly high 'Mo' data. The 'High Mo' data, are obtained from a specific steel cast that has a particular high level of Molybdenum (Mo) as compared to the rest of the data set. All data sets were cleaned for spurious or inconsistent data points and the dimensionality of the data space is 16 (15-inputs 1-output). The data space, apart from being non-linear and complex, is also very sparse. This is because these industrial data are focused towards specific grades of steel; therefore there are discontinuities in most of the input dimensions.

4.1 Initial Model Performance

The 'old' training and validation data set (sets 1 and 2) are used for training and testing the performance of the initial model. After performing data granulation to the training data set the linguistic rule-base of the system is established, as per Sect. 2.2. The model is then optimised using the BEP algorithm shown in Sect. 2.3. The model fit plots (measured vs. predicted) are shown in Fig. 8.

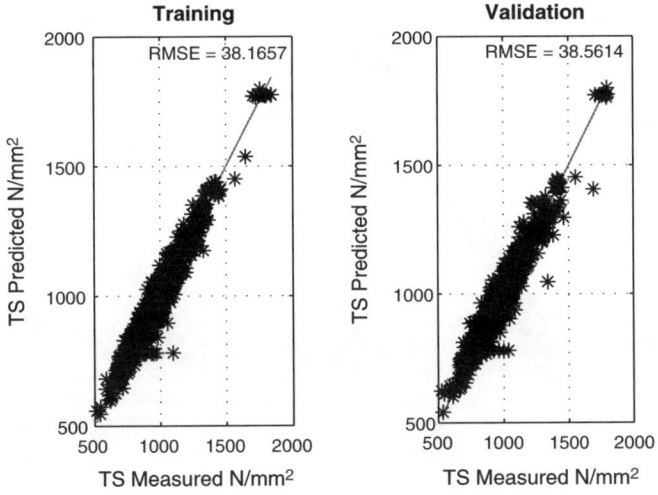

Fig. 8. Performance of the initial model

The diagonal straight line represents the '0%' error line, and ideally, if the model could predict with 100% accuracy then all data points would be on this diagonal line. The bigger the perpendicular distance to a data point the bigger the prediction error.

The performance level of the model is characterised as good by process experts and is comparable to the performance reported in [10] for a similar process.

4.2 Incremental Learning Performance

The 'new' training and validation data set is then presented to the system. The training data set is filtered by the system (as described in Sect. 3.2) and is split up into two sets named 'new data' and 'partially new data'. The partially new data are used to fine-tune the existing NF model, and the new data are used to create a new NF sub-module that is trained using the same algorithmic procedure as the initial model. The new sub-module is cascaded along with the rest of the sub-modules in the original structure.

After the incremental update procedure has finished the network is tested for its performance on the old data set as well as the new data set (training and validation). The results are shown in Fig. 9.

As seen in the model fit plot (training data sets) the structure is able to maintain a good performance, in fact similar to the one observed in the original model, but at the same time it can predict with comparable accuracy input vectors that originate from the new data set (high *Mo* data). Similar behavior is observed during the validation tests of the equivalent 'old' and 'new' data sets, as it is shown in Fig. 10. The model is able to handle correctly *unseen*

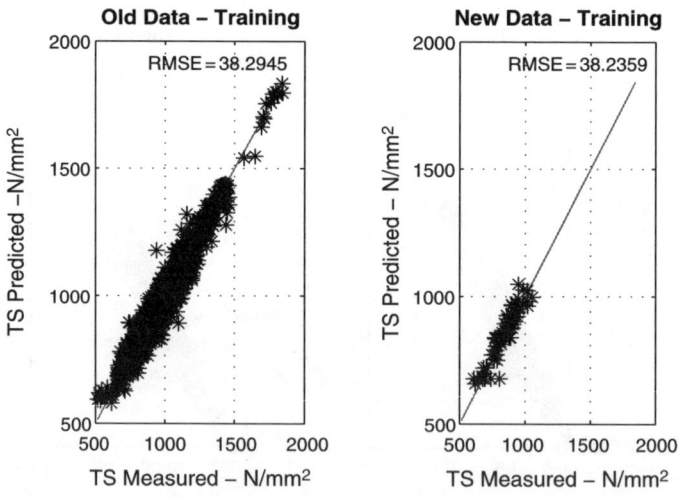

Fig. 9. IL Structure – Performance on the training data set

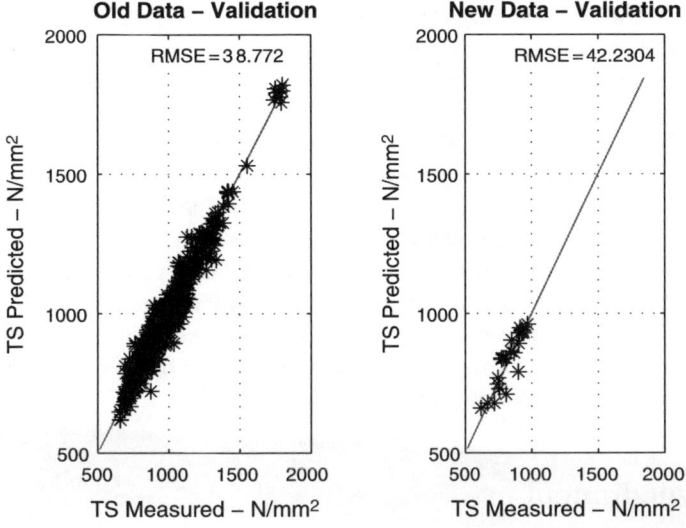

Fig. 10. IL Structure – Performance on the validation data

input data vectors; when the input MFs are adequetly excited the appropriate cascade sub-modules are activated, and via the fuzzy fusion process a single prediction is obtained with good accuracy.

5 Conclusion

In this chapter an incremental learning structure is presented, that is based on computational intelligence data-driven modeling and model fusion. The structure is based on Neural-Fuzzy models that are created using the Granular Computing – Neural-Fuzzy procedure. The cascade modeling architecture of the system and the incremental updating algorithm provide a reliable model updating routine that results in a dynamically expandable structure that does not ignore any previously gained knowledge. All knowledge is maintained (stored) in the cascade architecture and a model-fusion technique, based on the models' fuzzy entropy value, is credited for extracting the system's knowledge and obtaining individual model predictions. These predictions are subsequently intelligently fused into a single final prediction for any given input vector.

The abstraction level of the resulting models is controlled via the level of granulation of the input space. The transparency of the resulting GrC-NF models allows the visual examination of individual rules and the validation of the model knowledge/rules as compared with experts' knowledge or any relevant theory.

The incremental learning structure is tested against a real-world modeling problem obtained from a complex industrial process, the prediction of mechanical properties of heat treated steel. This process needs to be accurately modeled and the models need to be periodically updated when new data are available. It was shown that the proposed structure is able to model the given process and incrementally update the models when needed. The performance of the model on the 'old data' is maintained, and the performance on the 'new data' is comparable to the overall original performance.

Further development of this incremental learning process includes the full automation of the model updating process and the elimination of manual supervised learning (threshold definitions etc.). Additionally, the structure can be appropriately modified to meet the demands of an environment where more frequent model updates are required.

Acknowledgment

The Authors would like to thank Corus Engineering Steels (CES) Stocksbridge, Sheffield, UK, for their expert advice and for providing the heat treatment data. This work was supported by the UK Engineering and Physical Sciences Research Council (EPSRC), UK, under grand GR/R70514/01.

References

1. A Bargiela and W Pedrycz (2003) Granular Computing an Introduction. Kluwer Academic Publishers
2. G Carpenter et al. (1991) FuzzyARTMAP: A neural network architecture for incremental supervised learning of analog multi-dimensional maps. IEEE Transactions on Neural Networks, 3, 698–713
3. MY Chen and DA Linkens (2001) A systematic neuro-fuzzy modelling framework with application to material property prediction. IEEE Transactions on Systems, Man, and Cybernetics-Part B: Cybernetics, 31(5), 781–790
4. WD Cheung, J Han et al. (1996) Maintenance of Discovered Association Rules in Large Databases: An Incremental Updating Technique. ICDE, 12th Int. Conf. on Data Engineering, IEEE Computer Society, Washington DC, USA
5. N Kasabov (1998) ECOS: A framework for evolving connectionist systems and the ECO learning paradigm. Proceedings of the International Conference on Neural Information Processing, Kitakyushu, Japan: IOS, pp. 1222–1235
6. N Kasabov (2001) Evolving fuzzy neural networks for supervised/unsupervised online knowledge-based learning. IEEE Transactions on Systems Man and Cybernetics, B 31, 902–918
7. DA Linkens and L Vefghi (1997) Recognition of patient anaesthetic levels: neural network systems, principal components analysis, and canonical discriminant variates. Artificial Intelligence in Medicine 11(2), 155–173
8. NR Pal and JC Bezdek (1994) Measuring fuzzy uncertainty. IEEE Transactions on Fuzzy Systems, 2(2)

9. G Panoutsos and M Mahfouf (2005) Granular computing and evolutionary fuzzy modelling for mechanical properties of alloy steels. IFAC 2005, Proceedings of, 16th IFAC World Congress, July 4–8 2005, Prague, Czech Republic
10. G Panoutsos and M Mahfouf (2005) Discovering knowledge and modelling systems using granular computing and neurofuzzy Structures. NiSIS'05, 1st Symposium, Nature Inspired Smart Information Systems, Algarve, Portugal, 3–5 Oct. 2005
11. J Tenner (1999) Optimisation of the heat treatment of steel using neural networks Ph.D. Thesis, Department of Automatic Control and Systems Engineering, The University of Sheffield, UK
12. YY Yang, DA Linkens, and M Mahfouf (2003) Genetic algorithms and hybrid neural network modelling for aluminium stress–strain prediction. Proceedings of the Institute of Mechanical Engineers: Part I – Journal of Systems and Control Engineering, 217(I), 7–21

Switched Fuzzy Systems: Representation Modelling, Stability Analysis, and Control Design

Hong Yang[1], Georgi M. Dimirovski[2,3], and Jun Zhao[1]

[1] Northeastern University, School of Information Science & Technology, 110004 Shenyang, Liaoning, People's Republic of China
cherryyh@126.com, zhaojun@ise.neu.edu.cn
[2] SS Cyril and Methodius University, School of Electrical Engineering & Info. Technologies, 1000 Skopje, Karpos 2 BB, Republic of Macedonia
gdimirovski@dogus.edu.tr
[3] Dogus University of Istanbul, School of Engineering, Acibadem, Zeamet Sk 21, 34722 Istanbul - Kadikoy, Republic of Turkey

Summary. Stability issues for switched systems whose subsystems are all fuzzy systems, either continuous-time or discrete-time, are studied and new results derived. Innovated representation models for switched fuzzy systems are proposed. The single Lyapunov function method has been adopted to study the stability of this class of switched fuzzy systems. Sufficient conditions for quadratic asymptotic stability are presented and stabilizing switching laws of the state–dependent form are designed. The elaborated illustrative examples and the respective simulation experiments demonstrate the effectiveness of the proposed method.

1 Introduction

The large class of switched systems has attracted extensive research during the last couple of decades both as such and also in conjunction of the even larger class of hybrid systems, e.g. see [2, 8, 13, 14, 21]. For, these systems have a wide range of potential applications. For instance, such systems are widely used in the multiple operating point control systems, the systems of power transmission and distribution, constrained robotic systems, intelligent vehicle highway systems, etc. Thus switched systems represent one of the rather important types of hybrid systems. Basically all switched systems are consisted of a family of continuous-time or discrete-time subsystems and a switching rule law that orchestrates the switching among them.

Recently switched systems have been extended further to encompass switched fuzzy systems too [12, 14, 20] following the advances in fuzzy sliding mode control [4, 9, 11] although for long time it was known that ideal

H. Yang et al.: *Switched Fuzzy Systems: Representation Modelling, Stability Analysis, and Control Design*, Studies in Computational Intelligence (SCI) **109**, 155–168 (2008)
www.springerlink.com

relay switching is a time optimal control law [17]. A switched fuzzy system involves fuzzy systems among its sub-systems. The extension emanated out of the remarkable developments in theory, applications, and the industrial implementations of fuzzy control systems, e.g. see [1, 16, 18, 19, 23], which exploited Lyapunov stability theory.

It appeared, the class of switched fuzzy systems can describe more precisely both continuous and discrete dynamics as well as their interactions in complex real-world systems. In comparison to either switched or fuzzy control systems, still few stability results on switched fuzzy control systems can be found in the literature. For the continuous-time case, in [12] a combination of hybrid systems and fuzzy multiple model systems was described and an idea of the fuzzy switched hybrid control was put forward. For the discrete-time case, in [5], a fuzzy model whose subsystems are switched systems was described. In this model switching takes place simply based on state variables or time. Subsequently the same authors gave some extensions to output [6] and to guaranteed-cost [7] control designs.

In here, an innovated representation modelling of continuous-time and discrete-time switched fuzzy systems is proposed. Sufficient conditions for asymptotic stability are derived by using the method of single Lyapunov function and the parallel distributed compensation (PDC) fuzzy controller scheme as well as the stabilizing state–dependent switching laws.

Further this study is written as follows. In Sect. 2, the representation modelling problem has been explored and innovated models proposed for both continuous-time and discrete-time cases; note, also both autonomous and non-autonomous system are observed in this study. Section 3 gives a thorough presentation of the new theoretical results derived. In Sect. 4, there are presented the illustrative examples along with the respective simulation results to demonstrate the applicability and efficiency of the new theory. Thereafter, conclusion and references are given.

2 Novel Models of Switched Fuzzy Systems

In this paper, only Takagi–Sugeno $(T - S)$ fuzzy systems representing the category of data based models are considered. This representation differs from existing ones in the literature cited: each subsystem is a $T - S$ fuzzy system hence defining an entire class of switched fuzzy systems. This class inherits some essential features of hybrid systems [2, 14] and retains all the information and knowledge representation capacity of fuzzy systems [16].

2.1 The Continuous-Time Case

Consider the continuous $T - S$ fuzzy model that involves $N_{\sigma(t)}$ rules of the type as

$$R^l_{\sigma(t)} : If\ \xi_1 is\ M_{\sigma(t)1}\Lambda\ and\ \xi_p\ is\ M^l_{\sigma(t)p},$$
$$Then\ \dot{x} = A_{\sigma(t)l}x(t) + B_{\sigma(t)l}u_{\sigma(t)}(t),\quad l = 1, 2, \ldots, N_{\sigma(t)} \tag{1}$$

where $\sigma : R_+ \to M = \{1, 2, \Lambda, m\}$ is a piecewise constant function and it is representing the *switching* signal. In rule-based model (1), $R^l_{\sigma(t)}$ denotes the lth fuzzy inference rule, $N_{\sigma(t)}$ is the number of inference rules, $u(t)$ is the input variable, and the vector $x(t) = [x_1(t)\ x_2(t)\Lambda\ x_n(t)]^T \in R^n$ represents the state variables. Vector $\xi = [\xi_1\xi_2\ \Lambda\ \xi_p]$ represents the vector of rule antecedents (premises) variables. In the linear dynamic model of the rule consequent, the matrices $A_{\sigma(t)l} \in R^{n \times n}$ and $B_{\sigma(t)l} \in R^{n \times p}$ are assumed to have the appropriate dimensions.

The ith fuzzy subsystem can be represented as follows:

$$R^l_i : If\ \xi_1\ is\ M^l_{i1}\Lambda\ and\ \xi_p\ is\ M^l_{ip},$$
$$Then\quad \dot{x}(t) = A_{il}x(t) + B_{il}u_i(t),$$
$$l = 1, 2, \ldots, N_i, \quad i = 1, 2, \ldots, m. \tag{2}$$

Thus the global model of the ith fuzzy sub-system is described by means of the equation

$$\dot{x}(t) = \sum_{l=1}^{N_i} \eta_{il}(\xi(t))\,(A_{il}x(t) + B_{il}u_i(t)), \tag{3}$$

together with

$$\eta_{il}(t) = \frac{\prod\limits_{\rho=1}^{n} \mu_{M^l_\rho(t)}}{\sum\limits_{l=1}^{N_i} \prod\limits_{\rho=1}^{n} \mu_{M^l_\rho(t)}}, \tag{4a}$$

$$0 \le \eta_{il}(t) \le 1,\ \sum_{l=1}^{N_i} \eta_{il}(t) = 1, \tag{4b}$$

where $\mu_{M^l_\rho(t)}$ denotes the membership function of the fuzzy state variable x_ρ that belongs to the fuzzy set M^l_ρ.

2.2 The Discrete-Time Case

Similarly, we can define the discrete switched $T - S$ fuzzy model including $N_{\sigma(k)}$ pieces of rules

$$R^l_{\sigma(k)} : If\ \xi_1\ is\ M^l_{\sigma(k)1}\Lambda\ and\ \xi_p\ is\ M^l_{\sigma(k)p},$$
$$Then\quad x(k+1) = A_{\sigma(k)l}x(k) + B_{\sigma(k)l}u_{\sigma(k)}(k),\ l = 1, 2, \ldots, N_{\sigma(k)} \tag{5}$$

where $\sigma(k) : \{0, 1, \Lambda\} \to \{1, 2, \Lambda, m\}$ is a sequence representing switching signal $\{0, Z_+\} \to \{1, 2, \Lambda, m\}$.

In turn, the ith sub fuzzy system can be represented as follows:

$$R^l_i : If\ \xi_1\ is\ M^l_{i1}\Lambda\ and\ \xi_p\ is\ M^l_{ip},$$
$$Then\quad x(k+1) = A_{il}x(k) + B_{il}u_i(k),$$
$$l = 1, 2, \ldots, N_i,\ i = 1, 2, \Lambda\ m. \tag{6}$$

Therefore the global model of the ith sub fuzzy system is described by means of the equation

$$x(k+1) = \sum_{l=1}^{N_i} \eta_{il}(k)\, (A_{il}x(k) + B_{il}u_i(k)), \tag{7}$$

together with

$$\eta_{il}(k) = \frac{\prod\limits_{\rho=1}^{n} \mu_{M_\rho^l}(k)}{\sum\limits_{l=1}^{N_i} \prod\limits_{\rho=1}^{n} \mu_{M_\rho^l}(k)}, \qquad 0 \le \eta_{il}(k) \le 1, \; \sum_{l=1}^{N_i} \eta_{il}(k) = 1. \tag{8}$$

The representation modeling section is thus concluded.

3 New Stability Results for Switched Fuzzy Systems

First the respective definition for quadratic asymptotic stability of switched nonlinear systems, e.g. see [2, 14, 24–26], and a related lemma in conjunction with stability analysis re recalled.

Definition 3.1. *The system (1) is said to be quadratic stable if there exist a positive definite matrix P and a state-dependent switching law $\sigma = \sigma(x)$ such that the quadratic Lyapunov function $V(x(t)) = x^T(t)Px(t)$ satisfies $\frac{d}{dt}V(x(t)) < 0$ for any $x(t) \ne 0$ along the system state trajectory from arbitrary initial conditions.*

Remark 3.1. A apparently analogous definition can be stated for the discrete-time case of system (5), and it is therefore omitted.

Lemma 3.1. *Let $a_{ij_i}(1 \le i \le m, 1 \le j_i \le N_i)$ be a group of constants satisfying*

$$\sum_{i=1}^{m} a_{ij_i} < 0, \; \forall \; 1 \le j_i \le N_i.$$

Then, there exists at least one i such that

$$a_{ij_i} < 0, \; 1 \le j_i \le N_i.$$

Proof. It is trivial and therefore omitted. □

3.1 Stability of Continuous-Time Switched Fuzzy Systems

First, the novel stability result for systems (1) with $u \equiv 0$ in the fuzzy system representation is explored.

Theorem 3.1. *Suppose there exist a positive definite matrix P and constants* $\lambda_{ij_l} \geq 0$, $i = 1, 2, \ldots, m$, $j_i = 1, 2, \ldots, N_i$ *such that*

$$\sum_{i=1}^{N_i} \lambda_{ij_i}(A_{ij_i}^T P + P A_{ij_i}) < 0, \; j_i = 1, 2, \Lambda N_i, \quad (9)$$

then the system (1) is quadratic stable under the switching law:

$$\sigma(k) = \arg\min\{\bar{V}_i(x) \triangleq \max_{j_i}\{x^T(A_{ij_i}^T P + P A_{ij_i})x < 0, j_i = 1, 2, \Lambda N_i\}\} \quad (10)$$

Proof. From inequality (9) it may well be inferred that

$$\sum_{i=1}^{N_i} \lambda_{ij_i} x^T (A_{ij_i}^T P + P A_{ij_i})x < 0, \; j_i = 1, 2, \Lambda N_i \quad (11)$$

for any $x \neq 0$. Notice that (11) holds true for any $j_i \in \{1, 2, \Lambda N_i\}$ and $\lambda_{ij} \geq 0$. On the other hand, Lemma 3.1 asserts that there exists at least one i such that

$$x^T(A_{ij_i}^T P + P A_{ij_i})x < 0, \quad (12)$$

for any j_i. Thus, the switching law (10) is a well-defined one. Next, the time derivative of the respective quadratic [8, 10, 15, 19, 25] Lyapunov function $V(x(t)) = x^T(t)Px(t)$, is to be calculated:

$$\frac{d}{dt}V(x(t)) = x^T \left[\left(\sum_{l=1}^{N_i} \eta_{il} A_{il} \right)^T P + P \left(\sum_{l=1}^{N_i} \eta_{il} A_{il} \right) \right] x$$

$$= \sum_{l=1}^{N_i} \eta_{il} x^T \left[A_{il}^T P + P A_{il} \right] x$$

Notice, here $i = \sigma(x)$ is generated by means of switching law (10). By taking (4), (12) into account, one can deduce that $dV(x(t))/dt < 0, x \neq 0$. Hence system (1) is quadratic stable under switching law (10), which ends up this proof. $\square$

Now the stability result for the more important case with $u \neq 0$, is presented. It is pointed out that the parallel distributed compensation (PDC) method for fuzzy controller design [22, 23] is used for every fuzzy sub-system. It is shown in the sequel how to design controllers to achieve quadratic stability in the closed loop and under the switching law.

Namely, local fuzzy controller and system (2) have the same fuzzy inference premise variables:

$$R_{ic}^l : If \; \xi_1 \; is \; M_{i1}^l \Lambda \; and \; \xi_p \; is \; M_{ip}^l,$$
$$Then \; u_i(t) = K_{il}x(t), \; l = 1, 2, \Lambda N_i, \; i = 1, 2, \Lambda m. \quad (13)$$

Thus, the global control is

$$u_i(t) = \sum_{l=1}^{N_i} \eta_{il} K_{il} x(t). \tag{14}$$

Then the global model of the ith sub fuzzy system is described by means of the following equation:

$$\dot{x} = \sum_{l=1}^{N_i} \eta_{il}(t) \sum_{r=1}^{N_i} \eta_{ir} (A_{il} + B_{il} K_{ir}) x(t). \tag{15}$$

Theorem 3.2. *Suppose there exist a positive definite matrix P and constants $\lambda_{ij_i} \geq 0$, $i = 1, 2, \ldots, m$, $\vartheta_i = 1, 2, \ldots, N_i$ such that*

$$\sum_{i=1}^{N_i} \lambda_{ij_i} \left[(A_{ij_i} + B_{ij_i} K_{i\vartheta_i})^T P + P(A_{ij_i} + B_{ij_i} K_{i\vartheta_i}) \right] < 0, \ j_i, \vartheta_i = 1, 2, \Lambda N_i, \tag{16}$$

Then, the system (1) along with (13)–(14) is quadratic stable under the switching law:

$$\sigma(x) = \arg\min\{\bar{V}_i(x) \overset{\Delta}{=} \max_{j_i, \vartheta_i} \{x^T [(A_{ij_i} + B_{ij_i} K_{i\vartheta_i})^T P$$
$$+ P(A_{ij_i} + B_{ij_i} K_{i\vartheta_i})] x < 0, j_i, \vartheta_i = 1, 2, \Lambda N_i\}\} \tag{17}$$

Proof. It is seen from (12) that

$$\sum_{i=1}^{N_i} \lambda_{ij_i} x^T \left[(A_{ij_i} + B_{ij_i} K_{i\vartheta_i})^T P + P(A_{ij_i} + B_{ij_i} K_{i\vartheta_i}) \right] x < 0$$
$$j_l, \vartheta_l = 1, 2, \ldots, N_i. \tag{18}$$

for any $x \neq 0$. Further, it should be noted that (18) holds for any $j_i, \vartheta_i \in \{1, 2, \Lambda N_i\}$ and $\lambda_{ij_i} \geq 0$. By virtue of Lemma 3.1, there exists at least one i such that

$$x^T \left[(A_{ij_i} + B_{ij_i} K_{i\vartheta_i})^T P + P(A_{ij_i} + B_{ij_i} K_{i\vartheta_i}) \right] x < 0$$

for any j_i, ϑ_i. Thus the switching law (17) is a well-defined one. Via similar calculations as in Theorem 3.1 using Lyapunov function $V(x) = x^T(t) P x(t)$, on can find:

$$\frac{dV(x(t))}{dt} = \dot{x}^T(t) P x(t) + x^T(t) P \dot{x}(t)$$
$$= \sum_{l=1}^{N_i} \sum_{r=1}^{N_i} \eta_{il} \eta_{ir} x^T(t) \left[(A_{il} + B_{il} K_{ir})^T P + P(A_{il} + B_{il} K_{ir}) \right] x(t)$$

in which $i = \sigma(x)$ is given by law (17). Thus, system (1) along with (13)–(14) is quadratic stable under switching law (17), and this completes the proof. $\square$

3.2 Stability of Discrete-Time Switched Fuzzy Systems

Again first the case when $u = 0$ in system (5) is considered.

Theorem 3.3. *Suppose there exist a positive definite matrix P and constants $\lambda_{ij_l} \geq 0$, $i = 1, 2, \ldots, m$, $\vartheta_i = 1, 2, \ldots, N_i$ such that*

$$\sum_{i=1}^{N_i} \lambda_{ij_i}(A_{ij_i}^T P A_{i\vartheta_i} - P) < 0, j_i, \vartheta_i = 1, 2, \Lambda\, N_i, \tag{19}$$

then the system (5) is quadratic stable under the switching law:

$$\sigma(k) = \arg\min\{\bar{V}_i(k) \overset{\Delta}{=} \max_{j_i, \vartheta_i}\{x^T(A_{ij_i}^T P A_{i\vartheta_i} - P)x < 0, j_i, \vartheta_i = 1, 2, \Lambda\, N_i\}\} \tag{20}$$

Proof. Similarly as for Theorem 3.1, from (19) it is inferred that

$$\sum_{i=1}^{N_i} \lambda_{ij_i} x^T(A_{ij_i}^T P A_{i\vartheta_i} - P)x < 0, j_i, \vartheta_i = 1, 2, \Lambda\, N_i$$

for any $x \neq 0$. Then there exists at least an i such that

$$x^T(A_{ij_i}^T P A_{i\vartheta_i} - P)x < 0,$$

for any j_i, $\vartheta_i \in \{1, 2, \Lambda\, N_i\}$. Now the time derivative of Lyapunov function $V(x(t)) = x^T(t)Px(t)$ is calculated to give:

$$\Delta V(x(k)) = V(x(k+1)) - V(x(k))$$

$$= x^T(k)\left[\left(\sum_{l=1}^{N_i} \eta_{il}(k)A_{il}\right)^T P \left(\sum_{r=1}^{N_i} \eta_{i\vartheta}(k)A_{i\vartheta}\right) - P\right]x(k)$$

$$= \sum_{l=1}^{N_i} \eta_{il} \sum_{r=1}^{N_i} \eta_{i\rho} x^T(k)\left[A_{il}^T P A_{i\vartheta} - P\right]x(k)$$

Hence the system (5) is quadratic stable under switching law (20), and this completes the proof. □

The more important case with $u \neq 0$ in system (5) is considered next. And again the PDC method for fuzzy controller design is used for every fuzzy sub-system. Namely, it is observed that local fuzzy control and system (6) have the same fuzzy inference premise variables:

$$R_{ic}^l : If\ \xi_1\ is\ M_{i1}^l \Lambda\ and\ \xi_p\ is\ M_{ip}^l,$$
$$Then\ u_i(k) = K_{il}x(k),\ l = 1, 2, \Lambda\, N_i,\ i = 1, 2, \Lambda\, m. \tag{21}$$

Thus, the global control is

$$u_i(k) = \sum_{l=1}^{N_i} \eta_{il} K_{il} x(k). \tag{22}$$

Then the global model of the ith sub fuzzy system is described by:

$$x(k+1) = \sum_{l=1}^{N_i} \eta_{il}(k) \sum_{r=1}^{N_i} \eta_{i\vartheta} (A_{il} + B_{il} K_{i\vartheta}) x(k). \tag{23}$$

Theorem 3.4. *Suppose there exist a positive definite matrix P and $\lambda_{ij_l} \geq 0$, $i = 1, 2, \ldots, m$, $\vartheta_i = 1, 2, \ldots, N_i$ such that*

$$\sum_{i=1}^{N_i} \lambda_{ij_i} \left[(A_{ij_i} + B_{ij_i} K_{i\vartheta_i})^T P (A_{ip_i} + B_{ip_i} K_{iq_i}) - P \right] < 0$$

$$j_i, \vartheta_i, p_i, q_i = 1, 2, \Lambda N_i \tag{24}$$

Then, the system (5) along with (21)–(22) is quadratic stable under the switching law:

$$\sigma(k) = \arg \min \{ \bar{V}_i(k) \overset{\Delta}{=} \max_{j_i, \vartheta_i, p_i, q_i} \{ x^T [(A_{ij_i} + B_{ij_i} K_{i\vartheta_i})^T P (A_{ip_i} + B_{ip_i} K_{iq_i})$$

$$- P] x < 0, j_i, \vartheta_i, p_i, q_i = 1, 2, \Lambda N_i \} \} \tag{25}$$

Proof. It is very similar to that of Theorem 3.3 and thus omitted. □

4 Illustrative Examples and Simulation Results

Results applying the above developed theory on two examples, one for the continuous-time and one the discrete-time case, and the respective simulations (e.g. using MathWorks software [27]) are given below.

4.1 A Continuous-Time Case of Autonomous System

Consider a continuous-time switched fuzzy system described as follows:

$$R_1^1 : \; \text{If } x \text{ is } M_{11}^1, \; \text{Then } \dot{x}(t) = A_{11} x(t)$$
$$R_1^2 : \; \text{If } x \text{ is } M_{11}^2, \; \text{Then } \dot{x}(t) = A_{12} x(t)$$
$$R_2^1 : \; \text{If } y \text{ is } M_{21}^1, \; \text{Then } \dot{x}(t) = A_{21} x(t)$$
$$R_2^2 : \; \text{If } y \text{ is } M_{21}^2, \; \text{Then } \dot{x}(t) = A_{22} x(t)$$

where

$$A_{11} = \begin{bmatrix} -17 & -0.0567 \\ 4.93 & -0.983 \end{bmatrix}, \quad A_{12} = \begin{bmatrix} -10 & 0.216 \\ -0.0132 & -45.29 \end{bmatrix};$$

$$A_{21} = \begin{bmatrix} -50 & -0.042 \\ 0.008 & -10 \end{bmatrix}, \quad A_{22} = \begin{bmatrix} -40 & -0.0867 \\ 0.047 & -120 \end{bmatrix}.$$

Above, the fuzzy sets $M_{11}^1, M_{11}^2, M_{21}^1, M_{21}^2$, respectively, are represented by means of the following membership functions:

$$\mu_{11}^1(x) = 1 - \frac{1}{1 + e^{-2x}}, \quad \mu_{11}^2(x) = \frac{1}{1 + e^{-2x}};$$

$$\mu_{21}^1(y) = 1 - \frac{1}{1 + e^{(-2(y-0.3))}}, \quad \mu_{21}^2(y) = \frac{1}{1 + e^{(-2(y-0.3))}}.$$

For

$$\sum_{i=1}^{2} \lambda_{ij_i} (A_{ij_i}^T P + P A_{ij_i}) < 0, \ j_i = 1, 2$$

by choosing $\lambda_{ij_i} = 1$, one can find (e.g. ultimately by using LMI toolbox) the following P matrix

$$P = \begin{bmatrix} 0.0159 & 0.0001 \\ 0.0001 & 0.0071 \end{bmatrix}.$$

Then system is quadratic asymptotically stable under the following switching law

$$\sigma(x) = \arg\min\{\bar{V}_i(x) \triangleq \max_{j_i}\{x^T(A_{ij_i}^T P + P A_{ij_i})x < 0, j_i = 1, 2\}\}.$$

Figure 1 above depicts the obtained simulation results for the controlled evolution of system state variables when, at the initial time instant, the system is perturbed by the state vector $x(0) = [2 \ 2]^T$.

4.2 A Cases of Discrete-Time Non-Autonomous System

Now, let consider a discrete-time switched fuzzy system that is represented as follows:

$R_1^1:$ If $x(k)$ is M_{11}^1, Then $x(k+1) = A_{11}x(k) + B_{11}u(k)$,

$R_1^2:$ If $x(k)$ is M_{11}^2, Then $x(k+1) = A_{12}x(k) + B_{12}u(k)$,

$R_2^1:$ If $y(k)$ is M_{21}^1, Then $x(k+1) = A_{21}x(k) + B_{21}u(k)$,

$R_2^2:$ If $y(k)$ is M_{21}^2, Then $x(k+1) = A_{22}x(k) + B_{22}u(k)$,

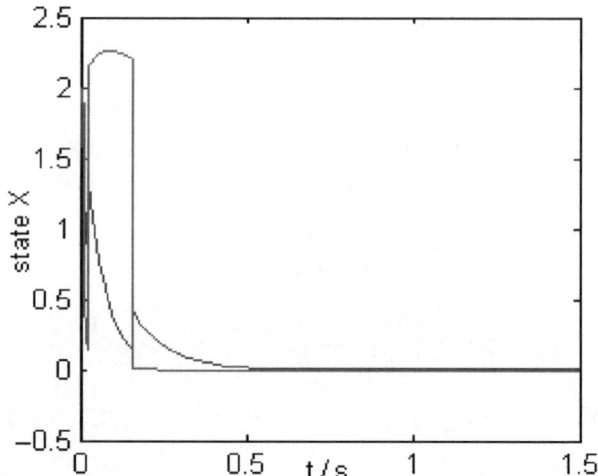

Fig. 1. The evolution time-histories for system state variables $x_1(t)$, $x_2(t)$ after the perturbation by $x(0) = [2\ 2]^T$ at the initial time instant

where

$$A_{11} = \begin{bmatrix} 0 & 1 \\ -0.0493 & -1.0493 \end{bmatrix}, \quad A_{12} = \begin{bmatrix} 0 & 1 \\ -0.0132 & -0.4529 \end{bmatrix},$$

$$A_{21} = \begin{bmatrix} 0 & 1 \\ -0.2 & -0.1 \end{bmatrix}, \quad A_{22} = \begin{bmatrix} 0.2 & 1 \\ -0.8 & -0.9 \end{bmatrix};$$

$$B_{11} = \begin{bmatrix} 0 \\ 0.4926 \end{bmatrix}, \quad B_{21} = \begin{bmatrix} 0 \\ 1 \end{bmatrix}, \quad B_{12} = \begin{bmatrix} 0 \\ 0.1316 \end{bmatrix}, \quad B_{22} = \begin{bmatrix} 0 \\ 1 \end{bmatrix}.$$

The above fuzzy sets $M_{11}^1, M_{11}^2, M_{21}^1, M_{21}^2$, respectively, are represented by means of the following membership functions:

$$\mu_{11}^1(x(k)) = 1 - \frac{1}{1 + e^{-2x(k)}}, \quad \mu_{11}^2(x(k)) = \frac{1}{1 + e^{-2x(k)}},$$

$$\mu_{21}^1(y(k)) = 1 - \frac{1}{1 + e^{(-2(y(k)-0.3))}}, \quad \mu_{21}^2(y(k)) = \frac{1}{1 + e^{(-2(y(k)-0.3))}}.$$

The state feedback gains of subsystems are obtained as

$$K_{11} = [-0.131\ -0.1148], \quad K_{12} = [-0.0623\ -2.302],$$
$$K_{21} = [1.8\ 1.9], \quad K_{22} = [-0.7\ 1.3].$$

For

$$\sum_{i=1}^{2} \lambda_{ij_i} \left[(A_{ij_i} + B_{ij_i} K_{i\vartheta_i})^T P (A_{ip_i} + B_{ij_i} K_{iq_i}) - P \right] < 0, \quad j_i, \vartheta_i, p_i, q_i = 1, 2,$$

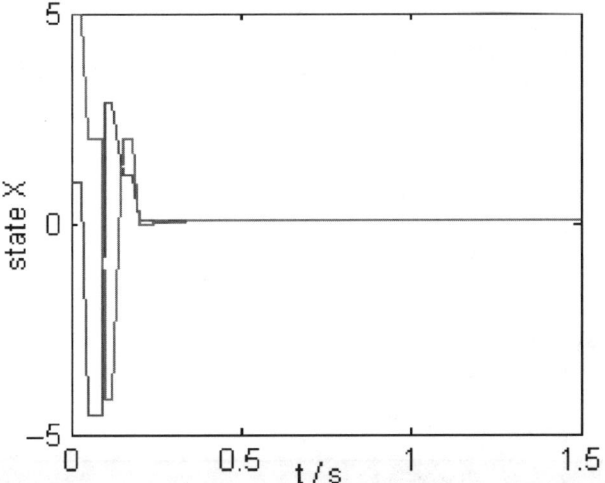

Fig. 2. The evolution time-histories for system state variables $x_1(k)$, $x_2(k)$ after the perturbation by $x(0) = [5 \ 1]^T$ at the initial time instant

by choosing $\lambda_{ij_i} = 1$, for the P matrix one can obtain the following result:

$$P = \begin{bmatrix} 0.3563 & -0.0087 \\ -0.0087 & 0.1780 \end{bmatrix}.$$

Then, the system is asymptotically stable under the following switching law

$$\sigma(k) = \arg\min\{\bar{V}_i(k) \triangleq \max_{j_i, \vartheta_i, p_i, q_i} \{x^T[(A_{ij_i} + B_{ij_i} K_{i\vartheta_i})^T P(A_{ip_i} + B_{ip_i} K_{iq_i}) - P]x < 0,$$

$$j_i, \vartheta_i, p_i, q_i = 1, 2\}\}.$$

Figure 2 below depicts the obtained simulation for the transients of controlled system state variables when, at the initial time instant, the system is perturbed by the state vector $x(0) = [5 \ 1]^T$; the sampling period was setup to $T_s = 0.05$ s.

As known from the literature, the combined switching control along with the state feedback gains does produce a reasonably varying control effort (see Fig. 3) that can be sustained by actuators.

5 Conclusion

Innovated representation models for the class of switched fuzzy systems, both continuous-time and discrete-time, based on Takagi–Sugeno fuzzy models were proposed. For both cases, new sufficient conditions for quadratic asymptotic stability of the control system with the given switching laws (Theorems 3.1

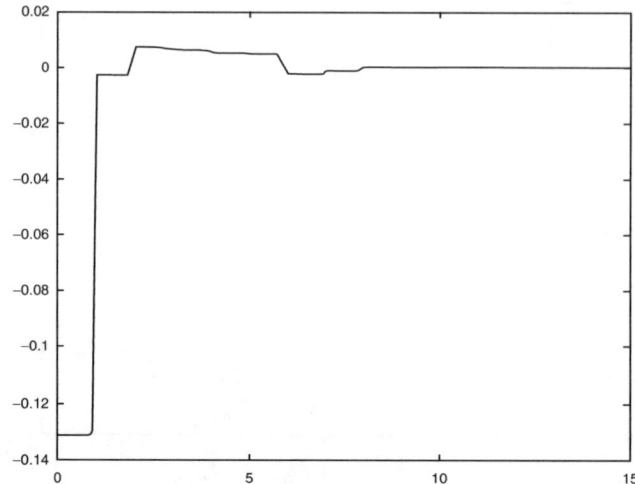

Fig. 3. The evolution time-history of the switching based state-feedback gain control of the plant system perturbed by $x(0) = [5\ 1]^T$ at the initial time instant

thru 3.4) have been derived via the common Lyapunov function approach. Following these new stability results, only the stability of a certain combination of subsystem matrices has to be checked, which is easier to carry out.

On the grounds of introducing the appropriate switching laws the stabilizing control in the state-variable dependent form has been synthesized for both these cases of fuzzy switched systems. Simulation results demonstrate that a control performance of considerable quality has been achieved in the closed loop thus promising the applicability in real-world problems.

The twofold future research is envisaged towards, firstly, reducing the conservatism of the obtained theorems and, secondly, towards deriving more sophisticated thus delicate switching based control laws.

Acknowledgments

The authors are grateful to their respective university institutions for the continuing support of their academic co-operation. In particular, Georgi M. Dimirovski would like to express his special thanks to his distinguished friends Lotfi A. Zadeh, the legend of fuzzy systems, and Janusz Kacprzyk, renown for his fuzzy based decisions support systems, for their rather useful suggestions. Also, he is thankful to the young colleague Daniel Liberson whose discussion regarding the switching control helped to clarify the presentation in this paper.

References

1. Berenji, H. R.: Fuzzy Logic Controllers. In: R. R. Yager and L. A. Zadeh (eds.): An Introduction to Fuzzy Logic Applications in Intelligent Systems. Kluwer Academic, Boston, MA (1993) 69–76.
2. Branicky, M. S.: Stability of Switched and Hybrid systems. In: Proceedings of the 34th IEEE Conference on Decision and Control, Lake Buena Vista, FL. The IEEE Press, Piscataway, NJ (1994) 3498–3503.
3. Branicky, M. S.: Multiple Lyapunov Functions and Other Analysis Tools for Switched and Hybrid Systems. IEEE Transactions on Automatic Control 43, 4 (1998) 475–482.
4. Chen, C. L., Chang, M. H.: Optimal Design of Fuzzy Sliding Mode Control: A Comparative Study. Fuzzy Sets & Systems 93 (1998) 37–48.
5. Doo, J. C., PooGyeon, P.: State Feedback Controller Design for Discrete-Time Switching Fuzzy Systems. In: Proceedings of the 41st IEEE Conference on Decision and Control, Las Vegas, NA (2002) paper A06 (1–6).
6. Doo, J. C., Seung, S. L., PooGyeon, P.: Output-Feedback H_∞ Control of Discrete-Time Switching Fuzzy Systems. In: Proceedings of the 2003 IEEE International Conference on Fuzzy Systems. The IEEE Press, Piscataway, NJ (2003) 441–446.
7. Doo, J. C., PooGyeon, P.: Guaranteed Cost Controller Design for Discrete-time Switching Fuzzy Systems. IEEE Transactions on Systems, Man & Cybernetics – Part A: Systems and Humans 34, 2 (2004) 110–119.
8. Johansson, M., Rantzer, A.: Computation of Piecewise Quadratic Lyapunov Functions for Hybrid Systems. IEEE Transactions on Automatic Control 43, 4 (1998) 555–559.
9. Li, H. X., Gatland, H. B., Green, A. W.: Fuzzy Variable Structure Control. IEEE Transactions on Systems, Man & Cybernetics – Part B: Cybernetics 27, 2 (1997) 306–312.
10. Ooba, T., Funahashi, Y.: On a common quadratic Lyapunov functions for widely distant systems. IEEE Transactions on Automatic Control 42 (1997) 1697–1699.
11. Palm, R.: Sliding Mode Fuzzy Control. In: Proceedings of the 1992 IEEE International Conference on Fuzzy Systems. The IEEE Press, Piscataway, NJ (1992) 519–526.
12. Palm, R., Driankov, D.: Fuzzy Switched Hybrid Systems – Modelling and Identification. In: Proceedings of the 1998 IEEE International Conference on Fuzzy Systems, Gaithersburg, MD. The IEEE Press, Piscataway, NJ (1998) 130–135.
13. Pettersson, S., Lennartson, B.: Stability and Robustness for Hybrid Systems. In: Proceedings of the 35th IEEE Conference on Decision and Control, Kobe, JP. The IEEE Press, Piscataway, NJ (1996) 1202–1207.
14. Savkin, A. V., Matveev, A. S.: Cyclic Linear Differential Automata: A Simple Class of Hybrid Dynamical Systems. Automatica 36, 5 (2000) 727–734.
15. Shim, H., Noh, D. J., Seo, J. H.: Common Lyapunov Function for Exponentially Stable Nonlinear Systems. Presented at the 4th SIAM Conf. on Control and Its Applications, Jacksonville, FL (1998) private communication.
16. Sugeno, M., Griffin, M. F., Bastian, A.: Fuzzy Hierarchical Control of an Unmanned Helicopter: In: Proceedings of the 5th IFSA World Congress, Seoul, KO. The International Fuzzy System Federation, Seoul, KO (1993) 1262–1265.

17. Systems and Control Encyclopaedia: Theory, Technology, Applications, M. G. Singh, Editor-in-Chief, In: Contributions on Nonlinear Systems, Vol. 5, Pergamon Press, Oxford, UK (1987) 3366–3383.
18. Takagi, T., Sugeno, M.: Fuzzy Identification of Systems and its Application to Modelling and Control. IEEE Transactions on Systems, Man & Cybernetics 15, 1 (1985) 116–132.
19. Tanaka, K., Sugeno, M.: Stability Analysis and Design of Fuzzy Control Systems. Fuzzy Set & Systems 45, 2 (1992) 135–156.
20. Tanaka, K., Masaaki, I., Hua, O. W.: Stability and Smoothness Conditions for Switching Fuzzy Systems: In: Proceedings of the 19th American Control Conference. The IEEE Press, Piscataway, NJ (2000) 2474–2478.
21. Varaiya, P. P.: Smart Cars on Smart Roads: Problems of Control. IEEE Transactions on Automatic Control 38, 2 (1993) 195–207.
22. Wang, H. O., Tanaka, K., Griffin, M.: Parallel Distributed Compensation of Nonlinear Systems by Takagi and Sugeno's Fuzzy Model. In: Proceedings of the 1995 IEEE Fuzzy System Conference FUZZ-IEEE'95, Yokahoma, JP. The IEEE Press, Piscataway, NJ (1995) 4790–4795.
23. Wang, H. O., Tanaka, K., Griffin, M.: An Approach to Fuzzy Control of Nonlinear Systems: Stability and Design Issues. IEEE Transactions on Fuzzy Systems 4, 1 (1996) 14–23.
24. Ye, H., Michael, A. N.: Stability Theory for Hybrid Dynamical Systems. IEEE Transactions on Automatic Control 43, 4 (1998) 464–474.
25. Zhao, J., Dimirovski, G. M.: Quadratic Stability for a Class of Switched Nonlinear Systems. IEEE Transactions on Automatic Control 49, 4 (2004) 574–578.
26. Dimirovski, G. M.: Lyapunov Stability in Control Synthesis Design Using Fuzzy-Logic and Neural-Networks. In: P. Borne, M. Benrejep, N. Dangoumau, and L. Lorimer (eds.): Proceedings of the 17th IMACS World Congress, Paris, July 11–15. The IMACS and Ecole Centrale de Lille, Villeneuve d'Ascq, FR (2005) Paper T5-I-01-0907, 1–8.
27. The MathWorks, Matlab Control Toolbox, Fuzzy Toolbox, LMI Tooolbox, and Simulink. The MathWorks Inc, Natick, MA (1999).

On Linguistic Summarization of Numerical Time Series Using Fuzzy Logic with Linguistic Quantifiers

Janusz Kacprzyk, Anna Wilbik, and Sławomir Zadrożny

System Research Institute, Polish Academy of Sciences, ul. Newelska 6, 01-447 Warsaw, Poland

Summary. We consider an extension to the linguistic summarization of time series data proposed in our previous papers, in particular by introducing a new protoform of the duration based summaries, that is more intuitively appealing. We summarize trends identified here with straight segments of a piecewise linear approximation of time series. Then we employ, as a set of features, the duration, dynamics of change and variability, and assume different, human consistent granulations of their values. The problem boils down to a linguistic quantifier driven aggregation of partial trends that is done via the classic Zadeh's calculus of linguistically quantified propositions. We present a modification of this calculus using the new protoform of a duration based linguistic summary. We show an application to linguistic summarization of time series data on daily quotations of an investment fund over an eight year period, accounting for the absolute performance of the fund.

1 Introduction

A linguistic data (base) summary, meant as a concise, human-consistent description of a (numerical) data set, was introduced by Yager [18] and then further developed by Kacprzyk and Yager [5], and Kacprzyk et al. [8]. The contents of a database is summarized via a natural language like expression semantics provided in the framework of Zadeh's calculus of linguistically quantified propositions [19]. Since data sets are usually large, it is very difficult for a human being to capture and understand their contents. As natural language is the only fully natural means of articulation and communication for a human being, such linguistic descriptions are the most human consistent.

In this paper we consider a specific type of data, namely time series. In this context it might be good to obtain a brief, natural language like description of trends present in the data on, e.g., stock exchange quotations, sales, etc. over a certain period of time.

Though statistical methods are widely used, we wish to derive (quasi) natural language descriptions to be considered to be an additional form of data

J. Kacprzyk et al.: *On Linguistic Summarization of Numerical Time Series Using Fuzzy Logic with Linguistic Quantifiers*, Studies in Computational Intelligence (SCI) **109**, 169–184 (2008)
www.springerlink.com   © Springer-Verlag Berlin Heidelberg 2008

description of a remarkably high human consistency. Hence, our approach is not meant to replace the classical statistical analyses but to add a new quality.

The summaries of time series we propose refer in fact to the summaries of trends identified here with straight line segments of a piece-wise linear approximation of time series. Thus, the first step is the construction of such an approximation. For this purpose we use a modified version of the simple, easy to use Sklansky and Gonzalez algorithm presented in [16].

Then we employ a set of features (attributes) to characterize the trends such as the slope of the line, the fairness of approximation of the original data points by line segments and the length of a period of time comprising the trend.

Basically the summaries proposed by Yager are interpreted in terms of the number or proportion of elements possessing a certain property. In the framework considered here a summary might look like: "Most of the trends are short" or in a more sophisticated form: "Most long trends are increasing." Such expressions are easily interpreted using Zadeh's calculus of linguistically quantified propositions. The most important element of this interpretation is a linguistic quantifier exemplified by "most." In Zadeh's [19] approach it is interpreted in terms of a proportion of elements possessing a certain property (e.g., a length of a trend) among all the elements considered (e.g., all trends).

In Kacprzyk, Wilbik and Zadrożny [9] we proposed to use Yager's linguistic summaries, interpreted in the framework of Zadeh's calculus of linguistically quantified propositions, for the summarization of time series. In our further papers (cf. Kacprzyk et al. [11–13]) we proposed, first, another type of summaries that does not use the linguistic quantifier based aggregation over the number of trends but over the time instants they take altogether. For example, such a summary can be: "Increasing trends took most of the time" or "From all increasing trends, trends of a low variability took most of the time." Such summaries do not directly fit the framework of the original Yager's approach and to overcome this difficulty we generalize our previous approach by modeling the linguistic quantifier based aggregation both over the number of trends as well over the time they take using first the Sugeno integral and then the Choquet integral [10, 14]. All these approaches have been proposed using a unified perspective given by Kacprzyk and Zadrożny [7] that is based on Zadeh's [20] protoforms.

In this paper we employ the classic Zadeh's calculus of linguistically quantified propositions. However, in comparison to our source paper (Kacprzyk et al. [11]) we propose here a new form of the protoform of duration based linguistic summaries, that is more intuitively appealing. We show results of an application to data on daily quotations of a mutual (investment) fund over an eight year period.

The paper is in line with some modern approaches to a human consistent summarization of time series – cf. Batyrshin and his collaborators [1, 2], or Chiang et al. [3] but we use a different approach.

One should mention an interesting project coordinated by the University of Aberdeen, UK, SumTime, an EPSRC Funded Project for Generating Summaries of Time Series Data[1]. Its goal is also to develop a technology for producing English summary descriptions of a time-series data set using an integration of time-series and natural language generation technology. Linguistic summaries obtained related to wind direction and speed are, cf. Sripada et al. [17]:

- WSW (West of South West) at 10–15 knots increasing to 17–22 knots early morning, then gradually easing to 9–14 knots by midnight
- During this period, spikes simultaneously occur around 00:29, 00:54, 01:08, 01:21, and 02:11 (o'clock) in these channels

They do provide a higher human consistency as natural language is used but they capture imprecision of natural language to a very limited extent. In our approach this will be overcome to a considerable extent.

2 Temporal Data and Trend Analysis

We identify trends as linearly increasing, stable or decreasing functions, and therefore represent given time series data as piecewise linear functions of some slope (intensity of an increase and decrease). These are partial trends as a global trend concerns the entire time span. There also may be trends that concern more than a window taken into account while extracting partial trends by using the Sklansky and Gonzalez [16] algorithm.

We use the concept of a uniform partially linear approximation of a time series. Function f is a uniform ε-approximation of a set of points $\{(x_i, y_i)\}$, if for a given, context dependent $\varepsilon > 0$, there holds

$$\forall i : \ |f(x_i) - y_i| \leq \varepsilon \tag{1}$$

and if f is linear, then such an approximation is a linear uniform ε-approximation.

We use a modification of the well known Sklansky and Gonzalez [16] algorithm that finds a linear uniform ε-approximation for subsets of points of a time series. The algorithm constructs the intersection of cones starting from point p_i of the time series and including a circle of radius ε around the subsequent points p_{i+j}, $j = 1, 2, \ldots$, until the intersection of all cones starting at p_i is empty. If for p_{i+k} the intersection is empty, then we construct a new cone starting at p_{i+k-1}. Figure 1a, b presents the idea of the algorithm. The family of possible solutions is indicated as a gray area. For other algorithms, see, e.g., [15].

---

[1] http://www.csd.abdn.ac.uk/research/sumtime/

First, denote:p_0 – a point starting the current cone, p_1 – the last point checked in the current cone, p_2 – the next point to be checked, Alpha_01 – a pair of angles (γ_1, β_1), meant as an interval, that defines the current cone as in Fig. 1a, Alpha_02 – a pair of angles of the cone starting at p_0 and inscribing the circle of radius ε around p_2 (cf. (γ_2, β_2) in Fig. 1a), function read_point() reads a next point of data series, function find() finds a pair

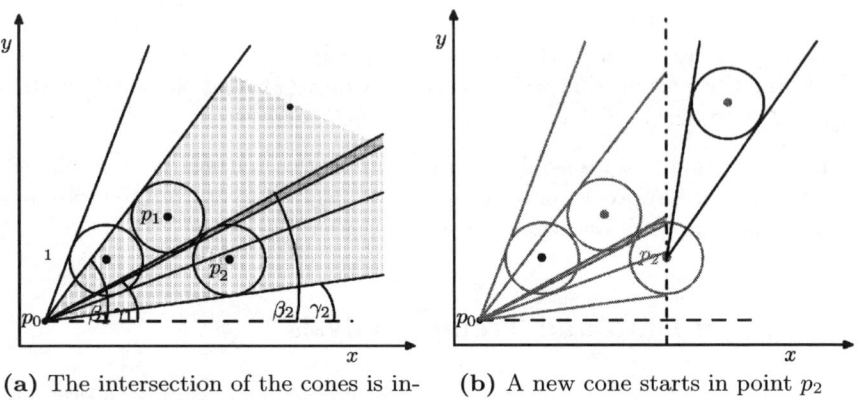

(a) The intersection of the cones is indicated by the dark grey area

(b) A new cone starts in point p_2

Fig. 1. An illustration of the algorithm for the uniform ε-approximation

```
read_point(p_0);
read_point(p_1);
while(1) {
   p_2=p_1;
   Alpha_02=find();
   Alpha_01=Alpha_02;
   do
   {
      Alpha_01 = Alpha_01 ∩ Alpha_02;

      p_1=p_2;
      read_point(p_2);
      Alpha_02=find();
} while(Alpha_01 ∩ Alpha_02 ≠ ∅);
      save_found_trend();
      p_0=p_1;
      p_1=p_2;
   }
```

Fig. 2. Pseudocode of the modified Sklansky and Gonzalez [16] algorithm for extracting trends

of angles of the cone starting at p_0 and inscribing the circle of radius ε around p_2. Then, a pseudocode of the algorithm that extracts trends is given in Fig. 2.

The bounding values of Alpha_02 (γ_2, β_2), computed by function find() correspond to the slopes of two lines tangent to the circle of radius ε around $p_2 = (x_2, y_2)$ and starting at $p_0 = (x_0, y_0)$. Thus, if $\Delta x = x_0 - x_2$ and $\Delta y = y_0 - y_2$ then:

$$\gamma_1 = arctg\left[\left(\Delta x \cdot \Delta y - \varepsilon\sqrt{(\Delta x)^2 + (\Delta y)^2 - \varepsilon^2}\right) \Big/ \left((\Delta x)^2 - \varepsilon^2\right)\right]$$

$$\gamma_2 = arctg\left[\left(\Delta x \cdot \Delta y + \varepsilon\sqrt{(\Delta x)^2 + (\Delta y)^2 - \varepsilon^2}\right) \Big/ \left((\Delta x)^2 - \varepsilon^2\right)\right]$$

The resulting linear ε-approximation of a group of points p_0, ... ,p_1 is either a single segment, chosen as, e.g., a bisector of the cone, or one that minimizes the distance (e.g., the sum of squared errors, SSE) from the approximated points, or the whole family of possible solutions, i.e., the rays of the cone.

3 Dynamic Characteristics of Trends

In our approach, while summarizing trends in time series data, we consider the following three aspects:

- Dynamics of change
- Duration
- Variability

and it should be noted that by trends we mean here global trends, concerning the entire time series (or some, probably a large, part of it), not partial trends concerning a small time span (window) taken into account in the (partial) trend extraction phase via the Sklansky and Gonzales [16] algorithm mentioned above.

In what follows we will briefly discuss these factors.

3.1 Dynamics of Change

Under the term *dynamics of change* we understand the speed of changes. It can be described by the slope of a line representing the trend, (cf. any angle η from the interval $\langle \gamma, \beta \rangle$ in Fig. 1a). Thus, to quantify dynamics of change we may use the interval of possible angles $\eta \in \langle -90; 90 \rangle$ or their trigonometrical transformation.

However it might be impractical to use such a scale directly while describing trends. Therefore we may use a fuzzy granulation in order to meet the users' needs and task specificity. The user may construct a scale of linguistic terms corresponding to various directions of a trend line as, e.g.:

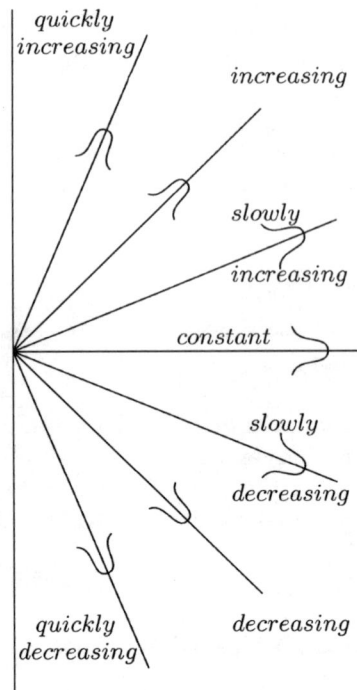

Fig. 3. A visual representation of angle granules defining the dynamics of change

– Quickly decreasing
– Decreasing
– Slowly decreasing
– Constant
– Slowly increasing
– Increasing
– Quickly increasing

Fig. 3 illustrates the lines corresponding to the particular linguistic terms.

In fact, each term represents a fuzzy granule of directions. In Batyrshin et al. [1, 2] there are presented many methods of constructing such a fuzzy granulation. The user may define a membership functions of particular linguistic terms depending on his or her needs.

We map a single value α (or the whole interval of angles corresponding to the gray area in Fig. 1b) characterizing the dynamics of change of a trend identified using the algorithm shown as a pseudocode in Fig. 2 into a fuzzy set (linguistic label) best matching a given angle. We can use, for instance, some measure of a distance or similarity, cf. the book by Cross and Sudkamp [4]. Then we say that a given trend is, e.g., "decreasing to a degree 0.8," if $\mu_{decreasing}(\alpha) = 0.8$, where $\mu_{decreasing}$ is the membership function of a fuzzy set representing "decreasing" that is a best match for angle α.

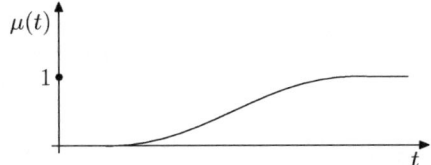

Fig. 4. An example of a membership function describing the term "long" concerning the trend duration

3.2 Duration

Duration describes the length of a single trend, meant as a linguistic variable and exemplified by a "long trend" defined as a fuzzy set whose membership function may be as in Fig. 4 where the time axis is divided into appropriate units.

The definitions of linguistic terms describing the duration depend clearly on the perspective or purpose assumed by the user.

3.3 Variability

Variability refers to how "spread out" ("vertically," in the sense of values taken on) a group of data is. The following five statistical measures of variability are widely used in traditional analyses:

- The range (maximum–minimum). Although the range is computationally the easiest measure of variability, it is not widely used, as it is based on only two data points that are extreme. This make it very vulnerable to outliers and therefore may not adequately describe the true variability.
- The interquartile range (IQR) calculated as the third quartile (the third quartile is the 75th percentile) minus the first quartile (the first quartile is the 25th percentile) that may be interpreted as representing the middle 50% of the data. It is resistant to outliers and is computationally as easy as the range.
- The variance is calculated as $\frac{\sum_i (x_i - \bar{x})^2}{n}$, where $\bar{x}$ is the mean value.
- The standard deviation – a square root of the variance. Both the variance and the standard deviation are affected by extreme values.
- The mean absolute deviation (MAD), calculated as $\frac{\sum_i |x_i - \bar{x}|}{n}$. It is not frequently encountered in mathematical statistics. This is essentially because while the mean deviation has a natural intuitive definition as the "mean deviation from the mean" but the introduction of the absolute value makes analytical calculations using this statistic much more complicated.

We propose to measure the variability of a trend as the distance of the data points covered by this trend from a linear uniform ε-approximation (cf. Sect. 2) that represents a given trend. For this purpose we propose to employ a distance between a point and a family of possible solutions, indicated as a gray cone in Fig. 1a. Equation (1) assures that the distance is definitely smaller than ε. We may use this information for the normalization. The normalized distance equals 0 if the point lays in the gray area. In the opposite case it is equal to the distance to the nearest point belonging to the cone, divided by ε. Alternatively, we may bisect the cone and then compute the distance between the point and this ray.

Similarly as in the case of dynamics of change, we find for a given value of variability obtained as above a best matching fuzzy set (linguistic label) using, e.g., some measure of a distance or similarity, cf. the book by Cross and Sudkamp [4]. Again the measure of variability is treated as a linguistic variable and expressed using linguistic terms (labels) modeled by fuzzy sets defined by the user.

4 Linguistic Data Summaries

A linguistic summary is meant as a (usually short) natural language like sentence (or some sentences) that subsumes the very essence of a set of data (cf. Kacprzyk and Zadrożny [6,7]). This data set is numeric and usually large, not comprehensible in its original form by the human being. In Yager's approach (cf. Yager [18], Kacprzyk and Yager [5], and Kacprzyk et al. [8]) the following perspective for linguistic data summaries is assumed:

- $Y = \{y_1, \ldots, y_n\}$ is a set of objects (records) in a database, e.g., the set of workers;
- $A = \{A_1, \ldots, A_m\}$ is a set of attributes characterizing objects from Y, e.g., salary, age, etc. in a database of workers, and $A_j(y_i)$ denotes a value of attribute A_j for object y_i.

A linguistic summary of a data set consists of:

- A summarizer P, i.e., an attribute together with a linguistic value (fuzzy predicate) defined on the domain of attribute A_j (e.g., "low salary" for attribute "salary").
- A quantity in agreement Q, i.e., a linguistic quantifier (e.g., most).
- Truth (validity) $\mathcal{T}$ of the summary, i.e., a number from the interval $[0,1]$ assessing the truth (validity) of the summary (e.g., 0.7); usually, only summaries with a high value of $\mathcal{T}$ are interesting.
- Optionally, a qualifier R, i.e., another attribute together with a linguistic value (fuzzy predicate) defined on the domain of attribute A_k determining a (fuzzy subset) of Y (e.g., "young" for attribute "age").

Thus, a linguistic summary may be exemplified by

$$\mathcal{T}(\text{most of employees earn low salary}) = 0.7 \qquad (2)$$

or, in a richer (extended) form, including a qualifier (e.g., young), by

$$\mathcal{T}(\text{most of young employees earn low salary}) = 0.9 \qquad (3)$$

Thus, basically, the core of a linguistic summary is a *linguistically quantified proposition* in the sense of Zadeh [19] which, for (2), may be written as

$$Qy\text{'s are } P \tag{4}$$

and for (3), may be written as

$$QRy\text{'s are } P \tag{5}$$

Then, $\mathcal{T}$, i.e., the truth (validity) of a linguistic summary, directly corresponds to the truth value of (4) or (5). This may be calculated by using either original Zadeh's calculus of linguistically quantified propositions (cf. [19]), or other interpretations of linguistic quantifiers. In the former case, the truth values (from $[0, 1]$) of (4) and (5) are calculated, respectively, as

$$\mathcal{T}(Qy\text{'s are } P) = \mu_Q \left(\frac{1}{n} \sum_{i=1}^{n} \mu_P(y_i) \right) \tag{6}$$

$$\mathcal{T}(QRy\text{'s are } P) = \mu_Q \left(\frac{\sum_{i=1}^{n} (\mu_R(y_i) \wedge \mu_P(y_i))}{\sum_{i=1}^{n} \mu_R(y_i)} \right) \tag{7}$$

where $\wedge$ is the minimum operation (more generally it can be another appropriate operation, notably a t-norm), and Q is a fuzzy set representing the linguistic quantifier in the sense of Zadeh [19], i.e., $\mu_Q : [0, 1] \longrightarrow [0, 1]$, $\mu_Q(x) \in [0, 1]$. We consider *regular non-decreasing monotone* quantifiers such that:

$$\mu(0) = 0, \quad \mu(1) = 1 \tag{8}$$
$$x_1 \leq x_2 \Rightarrow \mu_Q(x_1) \leq \mu_Q(x_2) \tag{9}$$

They can be exemplified by "most" given as in (10):

$$\mu_Q(x) = \begin{cases} 1 & \text{for } x \geq 0.8 \\ 2x - 0.6 & \text{for } 0.3 < x < 0.8 \\ 0 & \text{for } x \leq 0.3 \end{cases} \tag{10}$$

5 Protoforms of Linguistic Trend Summaries

It was shown by Kacprzyk and Zadrożny [7] that Zadeh's [20] concept of a protoform is convenient for dealing with linguistic summaries. This approach is also employed here.

Basically, a protoform is defined as a more or less abstract prototype (template) of a linguistically quantified proposition. Then, the summaries mentioned above might be represented by two types of the protoforms:

– Frequency based summaries:
  • A protoform of a short form of linguistic summaries:

$$Q \text{ trends are } P \tag{11}$$

and exemplified by:

> *Most* of trends are of a *large variability.*

- A protoform of an extended form of linguistic summaries:

$$QR \text{ trends are } P \qquad (12)$$

and exemplified by:

> *Most* of *slowly decreasing trends* are of a *large variability.*

– Duration based summaries:
  - A protoform of a short form of linguistic summaries:

$$(\text{From all trends}) \ P \text{ trends took } Q \text{ of the time} \qquad (13)$$

and exemplified by:

> (From all trends) trends *of a large variability* took *most* of the time.

- A protoform of an extended form of linguistic summaries:

$$\text{From all } R \text{ trends}, \ P \text{ trends took } Q \text{ of the time} \qquad (14)$$

and exemplified by:

> From all *slowly decreasing trends,* trends of a *large variability* took *most* of the time.

It should be noted that these summaries should be properly understood as, basically, that the (short, partial) trends, that have a large variability altogether took most of the time.

The truth values of the above types and forms of linguistic summaries will be found using the classic Zadehs calculus of linguistically quantified propositions as it is effective and efficient, and provides the best conceptual framework within which to consider a linguistic quantifier driven aggregation of partial trends that is the crucial element of our approach.

6 The Use of Zadeh's Calculus of Linguistically Quantified Propositions

Using Zadeh's [19] fuzzy logic based calculus of linguistically quantified propositions, a (proportional, nondecreasing) linguistic quantifier Q is assumed to be a fuzzy set defined in the unit interval $[0, 1]$ as, e.g. (10).

The truth values (from $[0,1]$) of (11) and (12) are calculated, respectively, as

$$T(Qy\text{'s are } P) = \mu_Q \left(\frac{1}{n} \sum_{i=1}^{n} \mu_P(y_i) \right) \tag{15}$$

$$T(QRy\text{'s are } P) = \mu_Q \left(\frac{\sum_{i=1}^{n} (\mu_R(y_i) \wedge \mu_P(y_i))}{\sum_{i=1}^{n} \mu_R(y_i)} \right) \tag{16}$$

where $\wedge$ is the minimum operation.

The computation of truth values of duration based summaries is more complicated and requires a different approach. Namely, using the Zadeh's calculus for frequency based summaries we compute a proportion of trends, that satisfy condition "trend is P," to all trends, in a simple form case. Here, for duration based summaries, while analyzing a summary "(From all trends) P trends took Q of the time" we should compute the proportion of the time during which the condition "trend is P" is satisfied to the time taken by all trends. Therefore, we should compute the time that is taken by those trends for which "trend is P" is valid. In a crisp case, when "trend is P" is either to degree 1 or 0, it is obvious, as we can use either the whole time taken by this trend or none of this time, respectively. However, what should we do if "trend is P" is to some degree? We propose to take only a part of the time defined by the degree to which "trend is P." Specifically, we compute this time as $\mu(y_i)t_{y_i}$, where t_{y_i} is the duration of trend y_i. The time taken by all trends is simply our horizon (sum of times (durations) taken by all trends). Finally, having this proportion, we may compute to which degree the proportion of the time taken by those trends which "trend is P" to all the time, is Q. A similar line of thought might be followed for the extended form of linguistic summaries.

The truth value of the short form of duration based summaries (13) is calculated as

$$T(\text{From all } y\text{'s, } Py\text{'s took } Q \text{ time}) = \mu_Q \left(\frac{1}{T} \sum_{i=1}^{n} \mu_P(y_i) t_{y_i} \right) \tag{17}$$

where T is the total time of the summarized trends and t_{y_i} is the duration of the ith trend.

The truth value of the extended form of summaries based on duration (14) is calculated as

$$T(\text{From all } Ry\text{'s, } Py\text{'s took } Q \text{ time}) = \mu_Q \left(\frac{\sum_{i=1}^{n} (\mu_R(y_i) \wedge \mu_P(y_i)) t_{y_i}}{\sum_{i=1}^{n} \mu_R(y_i) t_{y_i}} \right) \tag{18}$$

where t_{y_i} is the duration of the ith trend.

Both the fuzzy predicates P and R are assumed above to be of a rather simplified, atomic form referring to just one attribute. They can be extended to cover more sophisticated summaries involving some confluence of various, multiple attribute values as, e.g, "slowly decreasing and short."

Alternatively, we may obtain the truth values of (13) and (14) if we divide each trend which takes t_{y_i} time units into t_{y_i} trends, each lasting one time unit. For this new set of trends we use frequency based summaries with the truth values defined in (15) and (16).

7 Numerical Experiments

The method was tested on real data of daily quotations, from April 1998 to December 2006, of an investment fund that invests at most 50% of assets in shares, cf. Fig. 5, with the starting value of one share equal to PLN 10.00 and the final one equal to PLN 45.10 (PLN stands for the Polish Zloty); the minimum was PLN 6.88 while the maximum was PLN 45.15, and the biggest daily increase was PLN 0.91, while the biggest daily decrease was PLN 2.41.

It should be noted that the example shown below is meant only to illustrate the methods proposed by analyzing the absolute performance of a given investment fund. We do not deal here with a presumably more common way of analyzing an investment fund by relating its performance versus a benchmark exemplified by an average performance of a group of (similar) funds, a stock market index or a synthetic index reflecting, for instance, bond versus stock allocation.

For $\varepsilon = 0.25$ (PLN 0.25), we obtained 255 extracted trends, ranging from 2 to 71 time units (days). The histogram of duration is in Fig. 6.

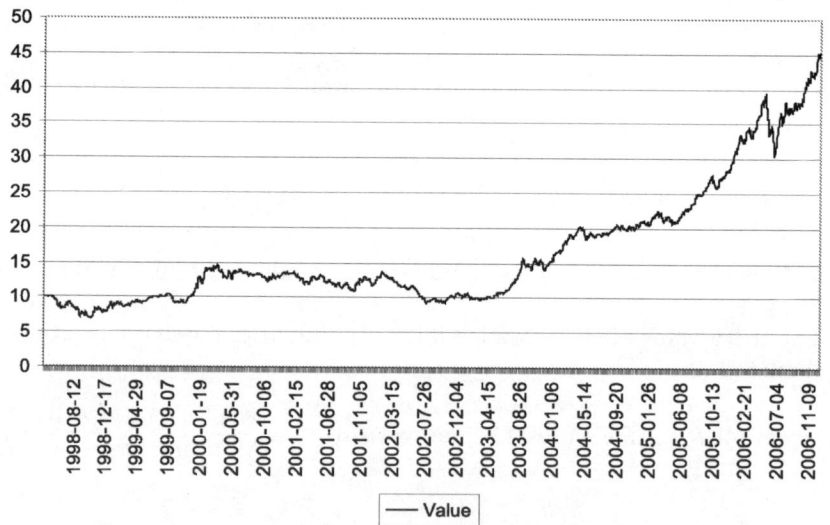

Fig. 5. A view of the original data

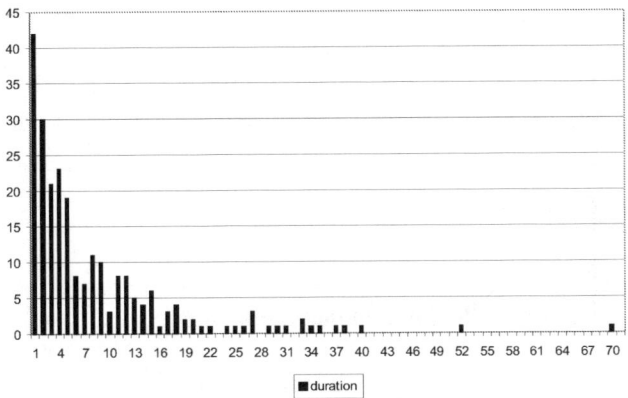

Fig. 6. Histogram of duration of trends

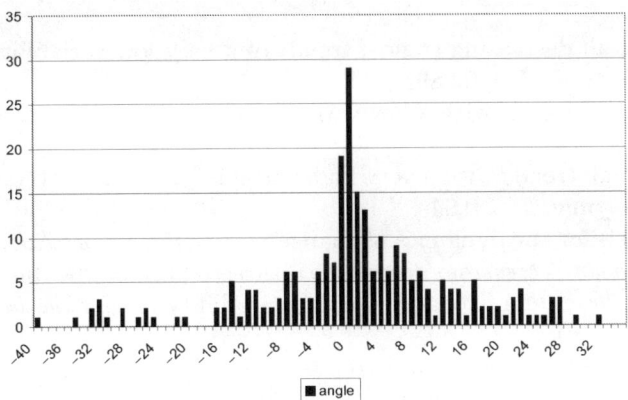

Fig. 7. Histogram of angles describing dynamic of change

Figure 7 shows the histogram of angles (dynamics of change) and the histogram of variability of trends (in %) is in Fig. 8.

Some interesting duration based summaries obtained by using the method proposed, employing the classic Zadeh's calculus of linguistically quantified propositions, and for different granulations of the dynamics of change, duration and variability, are:

– For seven labels for the dynamics of change (*quickly increasing, increasing, slowly increasing, constant, slowly decreasing, decreasing and quickly decreasing*), five labels for the duration (*very long, long, medium, short, very short*) and the variability (*very high, high, medium, low, very low*):
  • From all trends, constant trends took almost all of the time, $\mathcal{T} = 0.639$
  • From all trends, trends of a low variability took at least a half of the time are, $\mathcal{T} = 0.873$

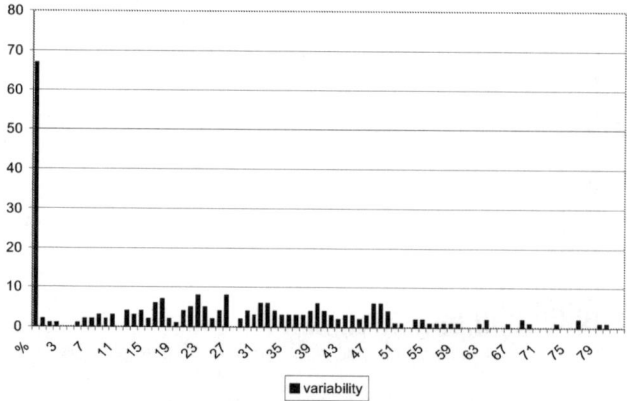

Fig. 8. Histogram of variability of trends

- From all decreasing trends, trends of a very low variability took most of the time, $\mathcal{T} = 0.989$
- From all trends with a low variability, constant trends took almost all of the time, $\mathcal{T} = 1$
- From all trends with a very high variability, constant trends took most of the time, $\mathcal{T} = 0.94$
- Five labels for the dynamics of change (*increasing, slowly increasing, constant, slowly decreasing, decreasing*), three labels for the duration (*short, medium, long*) and five labels for the variability (*very high, high, medium, low, very low*):
  - From all trends, constant trends took most of the time, $\mathcal{T} = 0.692$
  - From all trends, trends of a medium length took most of the time, $\mathcal{T} = 0.506$
  - From all decreasing trends, trends of a very low variability took most of the time, $\mathcal{T} = 0.798$
  - From all constant trends, trends of a low variability took most of the time, $\mathcal{T} = 0.5$
  - From all trends with a low variability, constant trends took most of the time, $\mathcal{T} = 0.898$

8 Concluding Remarks

First, we presented a new, more intuitively appealing, human consistent and semantically simpler protoform of the duration based summaries. Then, using this new protoform, we modified the basic Zadeh's calculus of linguistically quantified propositions that is used for a linguistic quantifier driven aggregation of partial trends that is crucial for the derivation of linguistic summaries of time series. We showed an application to linguistic summarization of time series data on daily quotations of an investment fund over an eight year period. We concentrated on the analysis of the absolute performance of the investment fund.

References

1. I. Batyrshin, On granular derivatives and the solution of a granular initial value problem, International Journal Applied Mathematics and Computer Science 12(3) (2002) 403–410.
2. I. Batyrshin, L. Sheremetov, Perception based functions in qualitative forecasting, in: I. Batyrshin, J. Kacprzyk, L. Sheremetov, L.A. Zadeh (Eds.), *Perception-Based Data Mining and Decision Making in Economics and Finance*, Springer, Berlin Heidelberg New York, 2006.
3. D.-A. Chiang, L.R. Chow, Y.-F. Wang (2000). Mining time series data by a fuzzy linguistic summary system. Fuzzy Sets and Systems, 112, 419–432.
4. V. Cross, T. Sudkamp, *Similarity and Compatibility in Fuzzy Set Theory: Assessment and Applications*, Springer, Berlin Heidelberg New York, 2002.
5. J. Kacprzyk, R.R. Yager, Linguistic summaries of data using fuzzy logic, International Journal of General Systems 30 (2001) 33–154.
6. J. Kacprzyk, S. Zadrożny, Fuzzy linguistic data summaries as a human consistent, user adaptable solution to data mining, in: B. Gabrys, K. Leiviska, J. Strackeljan (Eds.), *Do Smart Adaptive Systems Exist?*, Springer, Berlin Heidelberg New York, 2005, pp. 321–339.
7. J. Kacprzyk, S. Zadrożny, Linguistic database summaries and their protoforms: toward natural language based knowledge discovery tools, Information Sciences 173 (2005) 281–304.
8. J. Kacprzyk, R.R. Yager, S. Zadrożny, A fuzzy logic based approach to linguistic summaries of databases, International Journal of Applied Mathematics and Computer Science 10 (2000) 813–834.
9. J. Kacprzyk, A. Wilbik, S. Zadrożny, Linguistic summarization of trends: a fuzzy logic based approach, in: *Proceedings of the 11th International Conference Information Processing and Management of Uncertainty in Knowledge-Based Systems*, Paris, France, July 2–7, 2006, pp. 2166–2172.
10. J. Kacprzyk, A. Wilbik, S. Zadrożny, Linguistic summaries of time series via a quantifier based aggregation using the Sugeno integral, in: *Proceedings of 2006 IEEE World Congress on Computational Intelligence*, IEEE Press, Vancouver, BC, Canada, July 16–21, 2006, pp. 3610–3616.
11. J. Kacprzyk, A. Wilbik, S. Zadrożny, On some types of linguistic summaries of time series, in: *Proceedings of the 3rd International IEEE Conference Intelligent Systems*, IEEE Press, London, UK, 2006, pp. 373–378.
12. J. Kacprzyk, A. Wilbik, S. Zadrożny, A linguistic quantifier based aggregation for a human consistent summarization of time series, in: J. Lawry, E. Miranda, A. Bugarin, S. Li, M.A. Gil, P. Grzegorzewski, O. Hryniewicz (Eds.), *Soft Methods for Integrated Uncertainty Modelling*, Springer, Berlin Heidelberg New York, 2006, pp. 186–190.
13. J. Kacprzyk, A. Wilbik, S. Zadrożny, Capturing the essence of a dynamic behavior of sequences of numerical data using elements of a quasi-natural language,

in: *Proceedings of the 2006 IEEE International Conference on Systems, Man, and Cybernetics*, IEEE Press, Taipei, Taiwan, 2006, pp. 3365–3370.

14. J. Kacprzyk, A. Wilbik, S. Zadrożny, Linguistic summarization of time series by using the Choquet integral, in: *Proceedings of the IFSA World Congress*, 2007.

15. E. Keogh, M. Pazzani, An enhanced representation of time series which allows fast and accurate classification, clustering and relevance feedback, in: *Proc. of the 4th Int'l Conference on Knowledge Discovery and Data Mining*, New York, NY, 1998, pp. 239–241.

16. J. Sklansky, V. Gonzalez, Fast polygonal approximation of digitized curves, Pattern Recognition 12(5) (1980) 327–331.

17. S. Sripada, E. Reiter, I. Davy, SumTime-mousam: configurable marine weather forecast generator. Expert Update 6(3) (2003) 4–10.

18. R.R. Yager, A new approach to the summarization of data, Information Sciences 28 (1982) 69–86.

19. L.A. Zadeh, A computational approach to fuzzy quantifiers in natural languages, Computers and Mathematics with Applications 9 (1983) 149–184.

20. L.A. Zadeh, A prototype-centered approach to adding deduction capabilities to search engines – the concept of a protoform, in: *Proceedings of the Annual Meeting of the North American Fuzzy Information Processing Society* (NAFIPS 2002), 2002, pp. 523–525.

Part IV

Biomedical and Health Care Systems

Using Markov Models for Decision Support in Management of High Occupancy Hospital Care

Sally McClean[1], Peter Millard[2], and Lalit Garg[1]

[1] University of Ulster, Northern Ireland
   si.mcclean@ulster.ac.uk
[2] University of Westminster, London, UK
   phmillard@tiscali.co.uk

Summary. We have previously used Markov models to describe movements of patients between hospital states; these may be actual or virtual and described by a phase-type distribution. Here we extend this approach to a Markov reward model for a healthcare system with constant size. This corresponds to a situation where there is a waiting list of patients so that the total number of in-patients remains at a constant level and all admissions are from the waiting list. The distribution of costs is evaluated for any time and expressions derived for the mean cost. The approach is then illustrated by determining average cost at any time for a hospital system with two states: acute/rehabilitative and long-stay.

In addition we develop a Markov model to determine patient numbers and costs at any time where, again, there is a waiting list, so admissions are taken from this list, but we now allow a fixed growth which declines to zero as time tends to infinity. As before, the length of stay is described by a phase-type distribution, thus enabling the representation of durations and costs in each phase within a Markov framework. As an illustration, the model is used to determine costs over time for a four phase model, previously fitted to data for geriatric patients. Such an approach can be used to determine the number of patients and costs in each phase of hospital care and a decision support system and intelligent patient management tool can be developed to help hospital staff, managers and policy makers, thus facilitating an intelligent and systematic approach to the planning of healthcare and optimal use of scarce resources.

1 Introduction

Healthcare costs are increasing, as are the proportions of elderly people. This means that the care of geriatric patients is becoming an increasingly important problem, requiring careful planning and urgent attention. Old people are heavy users of hospital care largely because most people now live to die in old age. Using the Oxford Record Linkage study, [5] have shown that "generally,

S. McClean et al.: *Using Markov Models for Decision Support in Management of High Occupancy Hospital Care*, Studies in Computational Intelligence (SCI) **109**, 187–199 (2008)
www.springerlink.com   © Springer-Verlag Berlin Heidelberg 2008

hospital admissions either occurred in the years immediately before death and increased in the final year of life or were confined to that last year". With increasing longevity, old people, with their multiple medical, social and psychological problems, will inevitably place increasing demands on the health care system.

An intelligent model-based approach to the planning of healthcare, based on large-scale data routinely collected, is essential to facilitate understanding of the whole process and develop a holistic method for costing and performance measurement of hospital use. Healthcare planning should therefore include ways of predicting patient numbers and future costs of geriatric services, otherwise policies may lead to an improvement in hospital care in the short-term with a subsequent build-up of patient numbers and costs at future time points. For example, patients who are not sufficiently rehabilitated at the proper time may end up becoming long-stay and block beds that could be better utilised for acute care.

In previous work we have developed a model of patient flows within a hospital, where patients are initially admitted to an acute or rehabilitative state from which they are either discharged or die or are converted to a long-stay state [4]. Long-stay patients are discharged or die at a slower rate. Patients may be thought of as progressing through stages of acute care, rehabilitation and long-stay care where most patients are eventually rehabilitated and discharged. Thus an acute phase may be relatively quick, lasting for days or possibly weeks. A long-stay phase, on the other hand, may involve patients remaining in hospital for months, or even years. These patients may be very consuming of resources and thereby distort the performance statistics and cost implications [7].

In this chapter we discuss the use of a Markov reward model to cost the movements of patients within a hospital department. Initially we assume a constant number of beds in the department. This corresponds to a situation where there is a waiting list of patients so that the total number of in-patients remains at a constant level and all admissions are from the waiting list. Costs are assigned according to the state the patient has reached, where state here corresponds to different phases of care and recovery. Thus by assigning differential costs to the different states of the model, we may estimate the costs involved in treating people with a range of health and social problems. Using local estimates of transition rates and costings, hospital planners may thus identify cost-effective strategies. In addition we develop a Markov model to determine patient numbers and costs at any time where, again, there is a waiting list, so admissions are taken from this list, but we now allow a fixed growth in the number of beds available, which declines to zero as time tends to infinity.

The model we consider is a k-state discrete time Markov model, or phase type model [2], with costs associated with each time unit spent by each individual in each grade. Admissions to the system occur to each state to replace discharges, and we assume that the initial numbers of patients in each state of the system is known. There is also an absorbing state, usually discharge or death. Then we seek to find the distribution, in particular the means, of costs incurred in each grade of the system over time. In fact, this model is a generalization of that developed in our previous chapter [10] where we considered a 2-grade system (acute and long-stay in hospital) with no admissions. We were then only interested in the spend-down costs incurred over time if there are no further admissions; now we would like to determine the full system costs, of new admissions and current patients, for situations where there are a number of grades in hospital as well as in the community. The number of states, and associated transition parameters, are found in each case by fitting phase-type models to patient time spent in hospital and the community respectively, using maximum likelihood estimation, as previously described [3]. Here, we also extend previous work by [6] that previously fitted a model with two phases and fixed size to geriatric patient data.

This approach can be used to develop decision support system and intelligent patient management tool [1]. Since the idea of a diagnostic system was first proposed by [8] and the first experimental prototype was described by [13], numerous clinical decision support system have been developed to help clinical staff in making various clinical decisions. Using our model a decision support system is proposed which can help hospital staff, managers and policy makers in the planning and optimal use of scarce healthcare resources.

2 The Markov Model with Fixed Size

We consider patients as moving according to a discrete time Markov process where there is a cost associated with each state. The states are $S_1, \ldots, S_k$ where S_{k+1} is an absorbing state, generally discharge or death. As we have assumed previously for the two state system (k = 2), we consider a waiting list of patients so that the total number of in-patients remains at a constant level and all admissions are from the waiting list [6]. We here extend these results to any number of states; for example, [2] found four states to be appropriate for such data. Our approach here differs from previous work in that we have previously modelled the number of patients in each state whereas we here model the number of beds occupied by patients in each state, along with associated costs. We also generalise the previous results to the case where we can have any number of states. Between one time point and the next we have a probability of the bed changing from being occupied by a patient in state S_i to a patient in state S_{i+1} for $i = 1, \ldots, k - 1$. If a patient is discharged from bed S_i then the bed changes from being occupied by a patient in state S_i to being occupied by a patient in state S_1 i.e. we assume that all patients are admitted to state S_1 which typically represents an initial treatment and assessment phase.

The transition matrix of probabilities of a patient moving between one state and another between successive time points for states $S_1, \ldots, S_k$ is then:

$$\mathbf{A} = \begin{pmatrix} 1 - m_1 - d_1 & m_1 & 0 & 0 & . & . & 0 \\ 0 & 1 - m_2 - d_2 & m_2 & 0 & . & . & 0 \\ . & 0 & 0 & 1 - m_3 - d_3 & m_3 & . & . & 0 \\ & . & & . & & . & . & . \\ 0 & 0 & 0 & 0 & . & . & 1 - d_k \end{pmatrix}$$

while the transition matrix that describes changes between bed states is given by:

$$\mathbf{B} = \begin{pmatrix} 1 - m_1 & m_1 & 0 & 0 & . & . & 0 \\ d_2 & 1 - m_2 - d_2 & m_2 & 0 & . & . & 0 \\ . & d_3 & 0 & 1 - m_3 - d_3 & m_3 & . & . & 0 \\ & . & & . & & . & . & . \\ d_k & 0 & 0 & 0 & . & . & 1 - d_k \end{pmatrix}$$

Here, $m_i = \text{Prob}$ {a patient is in state S_{i+1} at time $t+1$| patient is in state S_i at time t} and $d_i = \text{Prob}$ {a patient is in state S_{i+1}, i.e. discharged, at time $t+1$| patient is in state S_i at time t} for i $= 1, \ldots, $k, i.e. m_i is the probability of moving from a grade to the next one and d_i is the probability of discharge from state S_i. We note that $\mathbf{B}$ is a stochastic matrix, representing the Markov Chain for movements between bed states, as represented in Fig. 1.

Previously we have assumed that patients are admitted according to a Poisson process [12]. This is appropriate when the hospital is not working to full capacity and there are empty beds to accommodate newly arriving patients. However, in practice, this is often not the case and newly arriving patients may have to queue for admission to hospital, typically in a waiting list. In addition, there may be some growth in bed availability over time to accommodate such queuing patients. We therefore assume in this chapter that all admissions are from the waiting list and are admitted to state S_1, typically a state where patients are assessed and may receive some initial treatment. We also assume that initially there are ν_i patients in state i for $i = 1, \ldots $k.

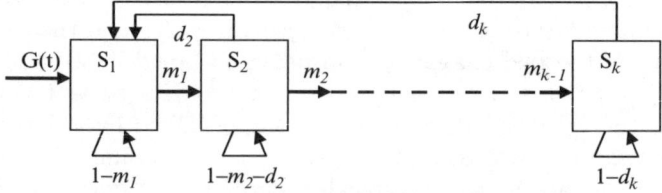

Fig. 1. Movements between bed states for the Markov model with fixed size

Then, the probability generating function (p.g.f.) is given by:

$$G(\mathbf{Z}, t) = \prod_{i=1}^{k} (1 + \mathbf{V}_i \mathbf{B}^t \mathbf{Z})^{\nu_i}. \tag{1}$$

Where $\mathbf{B}^t$ is the t-step transition matrix describing changes in bed occupancy over time t,

$\mathbf{V}_i = (0, \dots, 1, \dots, 0)$ is the row vector with 1 in the ith position, and $\mathbf{Z} = \{Z_i\}$ where $Z_i = (1 - z_i)$.

This equation represents the movement of bed occupancy between the different states $S_1, \dots, S_k$ for the initial bed occupancy distribution represented by the ν_i patients in state i for $i = 1, \dots k$ between time 0 and time t. The initial patients may therefore be regarded as being replaced by time t in one of the other states according to a multinomial distribution with transition matrix $\mathbf{V}_i \mathbf{B}^t$.

We now assume a cost c_i/time unit for a patient in state S_i. In this case, the p.g.f. is given by:

$$H(\mathbf{Z}, t) = \prod_{i=1}^{k} (1 + \mathbf{V}_i \mathbf{B}^t \mathbf{Z}_c)^{\nu_i}. \tag{2}$$

Where $\mathbf{Z}_c = \{1 - z_i^{c_i}\}$.

The cost of patients at time t is then given by:

$$\begin{aligned}
\mathbf{C}(t) &= \sum_{i=1}^{k} v_i \mathbf{V}_i \mathbf{B}^t \mathbf{c} \\
&= \sum_{i=1}^{k} \nu_i \sum_{j=1}^{k} c_j B_{ij}(t) \text{ where } \mathbf{B}^t = B_{ij}(t)
\end{aligned} \tag{3}$$

In steady state, the numbers of patients (beds) in state S_i are therefore from a multinomial distribution (N, π) where N is the total number of patients (beds) i.e.

$N = \sum_{i=1}^{k} v_i$ and $\pi = (\pi_1, \dots, \pi_k)$ is the vector of steady state probabilities of being in each (bed) state in steady state. So π is the solution of:

$$\pi \mathbf{B} = \pi.$$

Solving this equation, we obtain:

$$\pi_{i+1} = \frac{m_i}{m_{i+1} + d_{i+1}} \pi_i \text{ for } i = 1, \dots, k$$

Where we put $m_{k+1} = 0$.

Here $\pi_1 = \left\{ 1 + \frac{m_1}{m_2 + d_2} + \dots + \frac{m_{k-1}}{d_k} \right\}^{-1}$.

The corresponding costs in steady state are then given by:

$$C_\infty = \sum_{i=1}^{k} c_i \pi_i = \mathbf{c}.\boldsymbol{\pi}.$$

We can use these expressions to evaluate the number and costs of patients in each state over time and in steady state. This is illustrated in the following example.

2.1 Example 1: A Two State Model

We have previously described a two state Markov model [10] for a hospital system where the states are acute/rehabilitative and long-stay. We now extend this to a two-state system with Poisson recruitment (Fig. 2).

The transition matrix for daily transitions of patients between the states S_1 and S_2 is here:

$$\mathbf{A} = \begin{pmatrix} 1 - v - r & v \\ 0 & 1 - d \end{pmatrix}.$$

while the transition matrix for daily transitions of beds between the states S_1 and S_2 is:

$$\mathbf{B} = \begin{pmatrix} 1 - v & v \\ d & 1 - d \end{pmatrix}.$$

Also, $v = (a, l)$, where a is the initial number of acute/rehabilitative patients and l is the initial number of long-stay patients. The daily cost for the acute/rehabilitative state is c and the daily cost of the long-stay state is k.

Then the mean cost at time t is:

$$\mathbf{C}(t) = (a, l)\mathbf{B}^t \begin{pmatrix} c \\ k \end{pmatrix}$$

$$= (a, l)\left\{ \begin{pmatrix} \dfrac{d}{v+d} & \dfrac{v}{v+d} \\ \dfrac{d}{v+d} & \dfrac{v}{v+d} \end{pmatrix} + (1 - v - d)^t \begin{pmatrix} \dfrac{v}{v+d} & \dfrac{-v}{v+d} \\ \dfrac{-d}{v+d} & \dfrac{d}{v+d} \end{pmatrix} \right\} \begin{pmatrix} c \\ k \end{pmatrix}$$

$$= \frac{(a + l)(dc + vk)}{v + d} + (1 - v - d)^t \frac{(av - ld)(c - k)}{v + d}$$

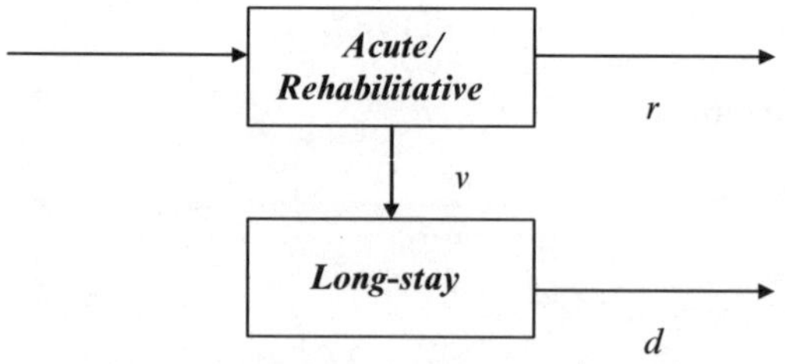

Fig. 2. A two state hospital model

The mean cost at time $t = \infty$ is thus

$$\mathbf{C}_\infty = \pi\mathbf{c}$$
$$= \frac{(a + l)(dc+vk)}{v + d}.$$

Using the data presented in Table 1, which are extracted from our previous work, we obtain, the average cost per day at a number of time points, as is illustrated in Table 2.

As we can see in Fig. 3, the steady state, which is relatively much cheaper, is attained very slowly. In steady state for these data, there are 68 patients in acute/rehabilitative care and 32 in long stay, so lower costs are achieved by

Table 1. Data for the two state model

R	V	D	A	L	C	K
0.02	0.01	0.001	100	0	£100	£50

Table 2. Costs for a number of time points for the two state model

T	1 year	2 years	5 years	10 years	∞
C(t)	£11,087	£10,204	£9,813	£9,800	£9,800

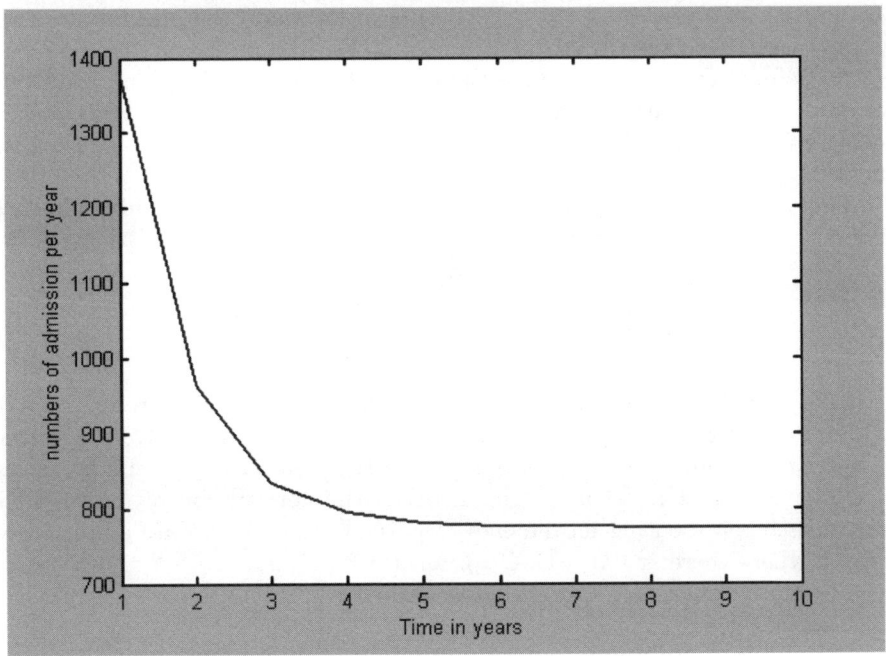

Fig. 3. Number of admissions each year

having more long-stay patients and less admissions, as is illustrated in Fig. 2. This is clearly not very desirable and illustrates the point that lowering costs may be achieved at the expense of reduced throughput and longer waiting lists. This observation at first seems rather counter-intuitive, as one might expect that high throughput would lead to lower costs. However, here the high throughput is mainly coming from short-stay patients while longer-stay patients are accumulating costs.

3 The Markov Model with Constant Growth

In this section we extend the approach of the previous section to a system, where again we let v_i be the number of patients in state S_i at time $t = 0$ for $i = 1, \ldots, k$. However, instead of a fixed number of patients (beds) in the system, new individuals enter the system, to achieve a total size of the system $T(t)$ at time t; here the total size is assumed to mean the total number of patients in states $S_1, \ldots, S_k$. The growth at time t is then given by:

$$G(t) = T(t) - T(t-1) \tag{4}$$

where $G(t)$ may be zero i.e. the size of the hospital resource available to patients is fixed, as in the previous section. This corresponds to a planned growth where management decides to expand a particular provision. Typically this growth will proceed for a fixed period of time and then decline to zero, as the target is achieved.

As before (Fig. 1), we assume that patients are admitted from a waiting list and our model essentially describes movements of beds between states, rather than patients. All new patients are admitted to state S_1, which is an initial assessment state.

The overall p.g.f. (probability generating function) of the numbers in each grade at time t is then given by:

$$G(\mathbf{Z}, t) = \left(\prod_{i=1}^{k} (1 + \mathbf{V}_i \mathbf{B}^t \mathbf{Z})^{v_i} \right) \cdot \left(\prod_{s=1}^{t} (1 + \mathbf{V}_i \mathbf{B}^{t-s} \mathbf{Z})^{G(s)} \right). \tag{5}$$

In (4), the first term corresponds to the individuals v_i who are in grade S_i at time $t = 0$, for $i = 1, \ldots, k$. By time t these patients (beds) have rearranged themselves according to a multinomial distribution with probability vector $\mathbf{V}_i \mathbf{B}^t$, as before. The second term in (4) is the convolution of multinomial terms, each representing movements of patients (beds) who are admitted at time s, where there is a growth $G(s)$, for $s = 0, \ldots, t$.

The equivalent expression for costs is then:

$$H(\mathbf{Z}, t) = \left(\prod_{i=1}^{k} (1 + \mathbf{V}_i \mathbf{B}^t \mathbf{Z}_c)^{v_i} \right) \cdot \left(\prod_{s=1}^{t} (1 + \mathbf{V}_i \mathbf{B}^{t-s} \mathbf{Z}_c)^{G(s)} \right). \tag{6}$$

The mean cost at time t is therefore now given by:

$$\mathbf{C}(t) = \sum_{i=1}^{k} v_i \mathbf{V}_i \mathbf{B}^t \mathbf{c} + \sum_{s=1}^{t} G(s) \mathbf{V}_1 \mathbf{B}^{t-s} \mathbf{c} \qquad (7)$$

As $t \to \infty$, $T(t) \to T$ and $G(t) \to 0$ we obtain:

$$\lim_{t \to \infty} \mathbf{C}(t) = C_\infty = \sum_{s=1}^{\infty} \sum_{i=1}^{k} r_{0i} \sum_{j=1}^{k} c_j \tilde{P}_{ij}(t-s) G(s)$$

$$= \left(\sum_{i=1}^{k} \nu_i + \sum_{s=1}^{\infty} G(s) \right) \left(\sum_{j=1}^{k} c_j \pi_j \right)$$

Where the first term is the total number of individuals in the system at time ∞ and the second term is the unit cost in steady state.

Here $T = \sum_{s=1}^{\infty} G(s)$ and $\lim_{t \to \infty} \sum G(t) = 0$.

3.1 Example 2: A four State Model with Exponentially Declining Growth

The four phase-type model has (Fig. 4) previously been fitted to data from a 16 year dataset by [3], who used penalised likelihood to determine the optimal number of phases in each case. The phases were identified as acute,

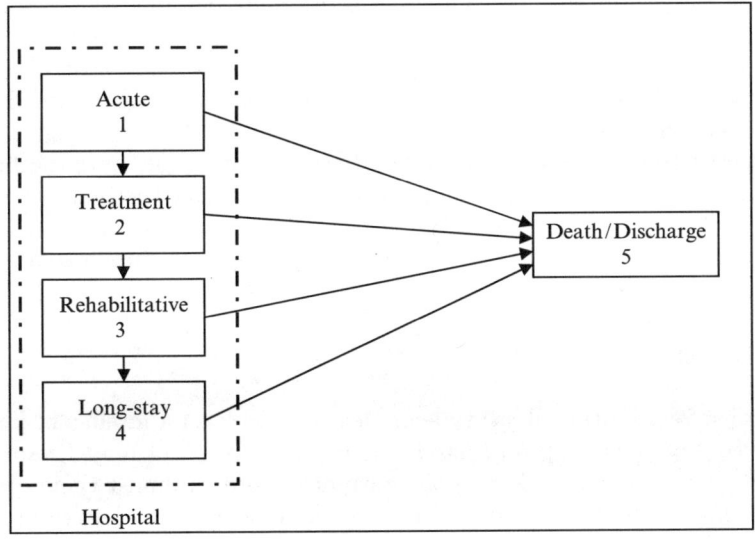

Fig. 4. The four state model

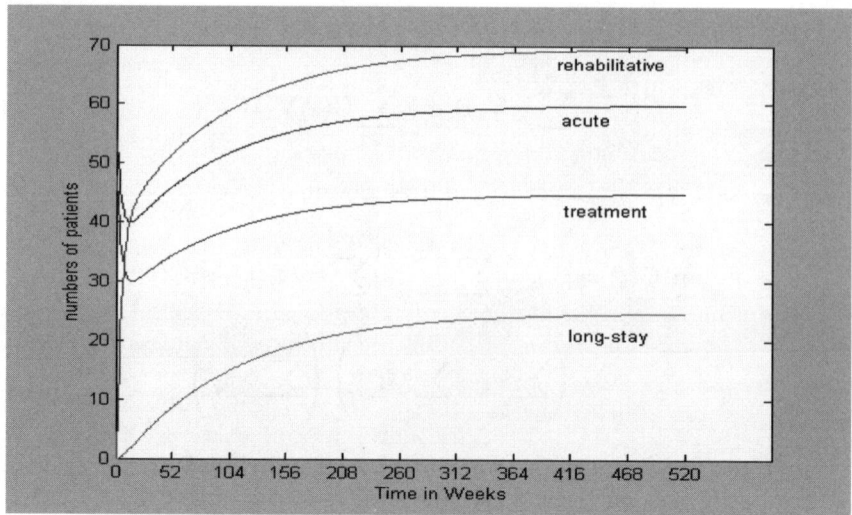

Fig. 5. The four state model

treatment, rehabilitative and long-stay (Fig. 4). From the phase-type maximum likelihood parameter estimates, we may determine transition probabilities for the Markov model. The costs have been estimated, based on current data for geriatric patients in the UK. These costs, which are indicative, are based on relative weightings of 15:10:8:6 for Acute: Treatment: Rehabilitative: Long stay.

The mean weekly numbers of patients in the hospital over time are presented in Fig. 5. Initially there are 100 acute male patients. We assume that growth occurs according to the rate $G(t) = \exp(-0.1 * t)$, leading to a total of 199 patients in steady state. Based on an average daily cost of £150 in acute care and the above ratios, the average cost per week in steady state is £143,640. All results were obtained using MATLAB. We see that, as expected, all patient numbers (and related costs) rise steeply over the first few years and then level off. The final patient numbers show the biggest group to be in rehabilitative care, with quite a few acute care patients also. There is also a significant number of patients (12.6%) in long-stay care in steady-state.

4 Data Issues

A key issue for adopting phase-type models as a vehicle for intelligent planning and costing of such hospital systems, is the estimation of model parameters from the data. In the context of the current chapter we must therefore learn parameters for the phase-type model describing a patient's length of stay in hospital (the m_i's and d_i's); in the case of the fixed growth model we also need to obtain suitable values for the growth function $G(t)$. Data on costings are also needed but may be difficult to obtain. Typically, hospital administrative data are routinely collected but can lack the detail necessary for our purposes. In particular data may be only be available as a "snapshot" (census data) of patients who have already been in hospital for a known length of time, rather than having longitudinal (cohort) data available.

Another issue is that patterns may be time heterogeneous, which means that we cannot easily combine data relating to patients from different time periods. A further aspect that we need to consider is whether data can be easily extracted from management databases. If new data collection is needed, then fitting the models can become very time-consuming and costly. In addition such complex issues as seasonality, fluctuations in demand, and weekly patterns need to be considered; typically this is achieved by analysing over an appropriate period of time to smooth out short-term effects that are not relevant to the identification of a cost-effective solution.

We have previously employed ideas from renewal theory to learn such models from census data [9] where we estimate the phase-type parameters using data on length of stay of all patients in a hospital department at a particular time-point. Such data are length-biased so we need to take account of this by modifying the likelihood terms accordingly and then employing penalised maximum likelihood to learn the parameters. However, with health services employing more powerful and complex databases, such issues of data availability are becoming increasingly less important.

5 Decision Support System

The proposed decision support system is presented in Fig. 6. It will have the following modules:

User Interface. This will work as an interface between the user and the decision support system. The user will input his/her query in the natural language and will be able to see the answer in natural language.

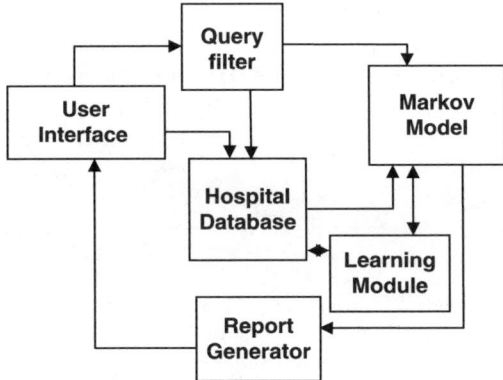

Fig. 6. Block diagram of the proposed decision support system

Query filter. This will convert the query in to the machine readable form and will extract data to run the model and will also update/append the database. according to the user input.

Hospital Database. It will contain the patient records in a database. We can use any commercial DBMS package for storing patient records. For our present system we are using the historical administrative database of the Geriatric Department of a London hospital [9].

Markov model. This is the Markov model we discussed in this chapter, which can be used to determine the number of patients and costs in each phase of hospital care.

Report generator. Report generator module will generate the report according to the answers from the Markov model module. And will convert it in to the human readable form.

We also propose a learning module to continuously evaluate and update the parameter values we used in the model to reflect the change in the database. This decision support system will help healthcare managers and policy makers in predicting the future resource requirements and therefore help in various policy decisions to ensure availability and optimum utilization of the healthcare resources at the same time minimizing the cost.

6 Conclusion

We have described discrete time Markov models that have k transient states and one absorbing state. Arrivals (admissions) to the system occur from a waiting list to replace discharges so that the total size of the system either remains constant or grows according to a pre-determined (deterministic) plan. For each of these scenarios we have attached rewards (costs) to each state at each time point and found expressions for the p.g.f. and mean costs at any time point t.

Such systems commonly occur where a number of individuals move through a graded system. We particularly focus here on a hospital system where patients move through a care system; here the states may be different stages of treatment e.g. acute, rehabilitative, long-term stay. However, the models we have been developed can also be used in other contexts, e.g. a human resource model where instead of patients we have employees and recruits replace leavers, as described in [11]. Such models can be used to develop a decision support system to assist planners who need to predict future costs under various scenarios.

References

1. A. Codrington-Virtue, T. Chaussalet, P. Millard, P. Whittlestone, and J. Kelly (2006) A system for patient management based discrete-event simulation and hierarchical clustering, *Computer-Based Medical Systems, 2006, 19th IEEE International Symposium*, Page(s):800–804

2. M.J. Faddy and S.I. McClean (1999) Analysing data on lengths of stay of hospital patients using phase-type distributions. *Applied Stochastic Models and Data Analysis*, 15, 311–317

3. M.J. Faddy and S.I. McClean (2005) Markov chain modelling for geriatric patient care. *Methods of Information in Medicine*, 44, 369–373

4. G.W. Harrison and P.H. Millard (1991) Balancing acute and long-stay care: the mathematics of throughput in departments of geriatric medicine. *Methods of Information in Medicine*, 30, 221–228

5. R.L. Himsworth and M.J. Goldacre (1999) Does time spent in hospital in the final 15 years of life increase with age at death? A population based study. *British Medical Journal*, 319, 1338–1339

6. V. Irvine, S.I. McClean, and P.H. Millard (1994) Stochastic models for geriatric inpatient behaviour. *IMA Journal of Mathematics Applied in Medicine and Biology*, 11, 207–216

7. D. Ivatts and P.H. Millard (2002) Health care modeling: why should we try? *British Journal of Healthcare Management*, 8(6), 218–222

8. R.S. Ledley and L.B. Lusted (1959) Reasoning foundations of medical diagnosis. *Science*, 130, 9–21

9. S.I. McClean and P.H. Millard (1993) Modelling in-patient bed usage in a department of geriatric medicine. *Methods of Information in Medicine*, 32, 79–81

10. S.I. McClean, B. McAlea, and P.H. Millard (1998) Using a Markov reward model to estimate spend-down costs for a geriatric department. *Journal of Operational Research Society* 10, 1021–1025

11. S.I. McClean, A.A. Papadopolou, and G. Tsaklides (2004) Discrete time reward models for homogeneous semi-Markov systems, *Communications in Statistics: Theory and Methods*, 33(3), 623–638

12. G.J. Taylor, S.I. McClean, and P.H. Millard (1999). Stochastic models of geriatric patient bed occupancy behaviour. *JRSS, Series A*, 163(1), 39–48

13. H.R. Warner, A.F. Toronto, and L. Veasy (1964) Experience with Baye's theorem for computer diagnosis of congenital heart disease. *Annals of the New York Academy of Sciences*, 115, 2–16

A Decision Support System for Measuring and Modelling the Multi-Phase Nature of Patient Flow in Hospitals

Christos Vasilakis, Elia El-Darzi, and Panagiotis Chountas

Harrow School of Computer Science, University of Westminster, HA1 3TP, UK
C.M.Vasilakis@westminster.ac.uk, eldarze, chountp@westminster.ac.uk

Summary. Multi-phase models of patient flow offer a practical but scientifically robust approach to the studying and understanding of the different streams of patients cared for by health care units. In this chapter, we put forward a decision support system that is specifically designed to identify the different streams of patient flow and to investigate the effects of the interaction between them by using readily available administrative data. The richness of the data dictate the use of data warehousing and On-Line Analytical Processing (OLAP) for data analysis and pre-processing; the complex and stochastic nature of health care systems suggested the use of discrete event simulation as the decision model. We demonstrate the application of the decision support system by reporting on a case study based on data of patients over 65 with a stroke related illness discharged by English hospitals over a year.

1 Introduction

In this chapter we propose the incorporation of discrete event simulation modelling and data warehousing techniques into a decision support system for modelling the flow of patients through hospitals and health care systems. The system can be easily applied to model different levels of health care operations. The scalability of the individual modelling components make this decision support system unique in its kind.

We focus on the analysis of patient length of stay (LoS) in hospital and bed occupancy in the care for older patients. A recent beds enquiry in the UK showed that two thirds of hospital beds are occupied by patients aged 65 and over [1]. This phenomenon is not only attributed to the higher admission rate (289 per 1,000 population for the 65+ age group as opposed to 94 per 1,000 population for the 15–64 age group) but also to the almost twice as long average LoS of this group of patients. Hence, we believe the provision of tools to aid in the analysis of hospital LoS and bed occupancy is critical to the management of these patients and to the allocation of health resources.

C. Vasilakis et al.: *A Decision Support System for Measuring and Modelling the Multi-Phase Nature of Patient Flow in Hospitals*, Studies in Computational Intelligence (SCI) **109**, 201–217 (2008)
www.springerlink.com

We also demonstrate in this chapter the viability of the proposed OLAP-enabled decision support system by applying it to real-life data. One case study is reported based on data from the English Hospital Episode Statistics database. The system is employed to support the identification of different streams of patients flowing through NHS hospitals. With the aid of the tools comprising the decision support system, the user not only can conduct extensive explorative analysis in a very efficient and timely manner but can also estimate the relevant parameters of the different streams of flow. Thereafter, it is possible to estimate the effects that different capacity constraints will have on the system and pre-test different scenarios that affect the flow of patients through it. At this level, the decision support system is capable of supporting the strategic decision making.

The rest of the chapter is organized as follows. In Sect. 2 we briefly introduce a multi-phase approach for modelling patient flow. Section 3 discusses the need for integrating data and decision modelling techniques, and illustrates the information flow in the proposed decision support system. In Sect. 4 we describe, discuss and evaluate the application of the proposed decision support system to a nationwide stoke dataset. Section 5 summarises the contribution of this chapter.

2 Measuring and Modelling the Flow of Patients

The inefficiencies of traditional methods in describing patient activity with skewed LoS distributions is well documented in the literature [2]. Simple LoS averages can offer indications but cannot accurately describe the process of care in such hospital departments as geriatric or psychiatric [3]. The complicating factor is the presence of patients with considerably longer LoS than others, in many cases in the order of months or even years. Consequently decisions on resource allocation and patient management that are based on such measures are often suboptimal [4].

Alternative methodologies that take account of the multi-phase nature of patient flow have been developed to overcome this problem [5]. McClean and Millard [6] have modelled the LoS of geriatric patients by using a two-term mixed exponential distribution. More generally, the observation that the LoS can be described in terms of mixed exponential equations have lead to the development of a flow model for modelling patient activity [7]. The different streams of patient flow can be classified as short stay (usually measured in days in the hospital care for older patients), medium stay (measured in weeks), and long stay (measured in months or years). A three-term mixed exponential model is of the form

$$N(s \geq x) = Ae^{-Bx} + Ce^{-Dx} + Ee^{-Fx} \tag{1}$$

where s is the occupancy time of a patient, x is the time in days, $N(s \geq x)$ is the total number of current patients that have been in hospital for greater than x days, and A, B, C, D, E, and F are constants, the parameters of the distribution. A specially designed program, called BOMPS, has been developed to enable the estimation of the parameters in equation (1) using the non-linear least squares method [8]. The main advantage of the software is that it uses readily available discharge data for estimating the input parameters.

Discrete event simulation models that have been developed based on the multi-phase nature of patient flow, have further extended the capabilities of the mathematical models by incorporating the stochastic nature of the system under study and the cascading effect of bed blockage in measuring the performance of alternative policies [9].

Patient LoS in hospital however, is not the only parameter to consider in describing patient activity in hospital departments [10]. Patterns of admissions and discharges and of overall occupancy play a major part in understanding the system. Weekly and seasonal variations are well reported and account for major disruptions and hospital bed shortages [11]. For example, a major seasonal variation in the pattern of admissions and discharges was the main cause behind the well publicised winter bed crisis, a cyclical phenomenon that used to appear in British hospitals, two or three weeks after the Christmas period. The rise in admissions has often been proposed as a possible explanation, however, analysis of data from a teaching hospital with bed shortages suggested delays in discharging elderly patients as a possible alternative explanation [12].

3 Integrating Data and Decision Modelling Techniques

Although LoS and bed occupancy analysis give useful insights, they also highlight the problems arising from the lack of integration between the data and decision modelling components of the decision support process [13]. Data modelling typically involves either the identification of the relationships that exist between data items, which leads to a relational data model, or the categorisation of the data items in dimensions and measures, which leads to a multi-dimensional data model. Decision modelling generally involves the development of some type of a model that is capable of representing the real life system under consideration. It has been suggested that a proper integration of data analysis and model building in a health care setting is capable of reducing the overall time required to perform a project analysis by a factor of ten [14].

The importance of data modelling is often overlooked. An analytical model should only be seen as a single activity in a chain of activities that use raw data for informing decision making. The management information value chain [15] illustrates all the activities by which information is acquired, transformed, stored, disseminated and finally presented to the decision maker and how

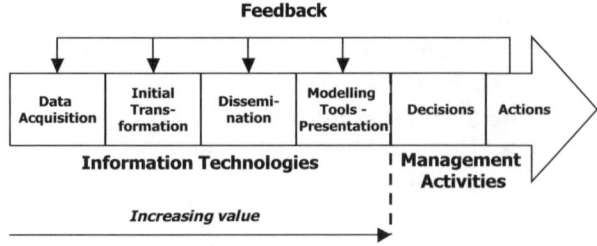

Fig. 1. The management information value chain in decision support, taken from [16]

each one of these activities increases the total value of information in an organization, Fig. 1. An analytic model must be seen as a single link in this multi-phase process that converts raw data into decisions.

The multi-dimensional data model that emerged in the past decade has been successful in analysing large volumes of data for decision making purposes [17]. Multi-dimensional databases view data as multi-dimensional data cubes that are suited for data analysis. On-Line Analytical Processing (OLAP) is a software technology that takes advantage of the data model by enabling analysts, managers and executives to gain insight into data through fast, consistent, interactive access to a wide variety of possible views of information that has been transformed from raw data to reflect the real dimensionality of the organisation as understood by the user [18]. Data are seen either as facts with associated numerical measures or as textual dimensions that characterise these facts. Dimensions are usually organised as hierarchies. Typical OLAP operations include the aggregation and de-aggregation of data along a dimension (roll-up and drill-down), the selection of specific parts of the cube (slicing), the reorientation of the dimensions in the multi-dimensional view of data on the screen (pivoting) and the displaying of values from one dimension within another one (nesting) [19]. OLAP is essentially a tool for browsing data stored in the data warehouse and it does not necessarily imply that the underlying physical model of the database is multi-dimensional.

Figure 2 illustrates the flow and transformations of information in the proposed framework of decision support. The incoming raw data may come from nationwide databases such as UK's Hospital Episode Statistics (HES), from hospital databases, and other supplementary data. Raw data are preprocessed, a process which involves loading, cleaning and adapting the data to the logical schema required by the cube building stage. After the initial transformation that can take place in an RDBMS or a spreadsheet application, raw data have acquired some clinical and business context.

The next stage involves the creation of the data model and data cubes that will facilitate the OLAP functions and it is staged on a specialised OLAP server. Data cubes are the basis for the next stage, OLAP browsing. A wide variety of client tools are available to the developers and end users, including pivot tables and graphs (incorporated in spreadsheets and internet browsers),

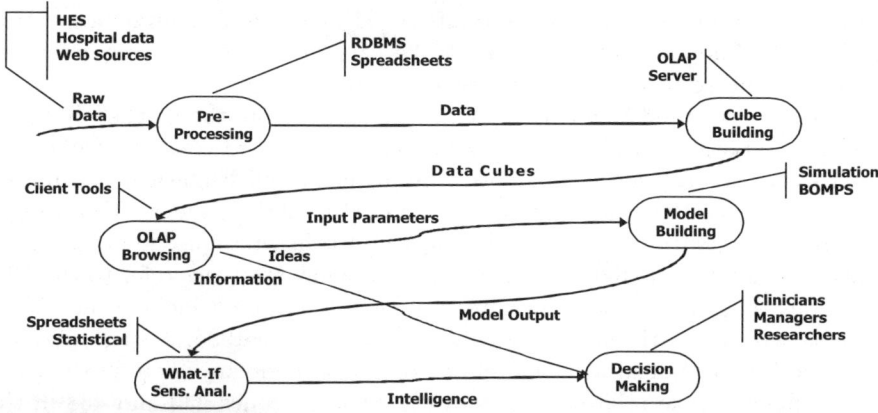

Fig. 2. Information flow in a patient activity decision support system

dedicated OLAP browsers, and query-based applications for more complex multi-dimensional data retrieval. The output of this stage is either estimations of the input parameters for the analytical models of the next stage, or information that is directed to the end users of the system.

The model building stage involves the development, instantiation and execution of some analytical or statistical models. For example, discrete event simulation or flow models that were briefly described in Sect. 2. The output of such models can then be further analysed and subjected to sensitivity analysis or provide the basis for what-if scenarios. The output of this stage provides information for intelligent decision making.

4 Application

Here we demonstrate the use of the decision support system described in the previous section on a large dataset. An additional aim is to illustrate the presence of different streams of flow in a health care system and as a result, the inadequacy of simple averages to represent LoS data. To this extent, we demonstrate how the alternative methodologies briefly described in Sect. 2 can be applied.

Stroke illness is of particular interest as its characteristics lead to a diverse hospital population in terms of clinical management, which is reflected in the LoS distribution. The neurological deficit that occurs after a stroke may be a transient phenomenon or may result in varying degrees of permanent disability. In the former cases clinical care is focused on prevention; in the latter and especially the more severe cases, the focus is on supportive after-care and multi-disciplinary rehabilitation [20]. These clinical observations pose some interesting challenges in modelling stroke related data. Moreover, the incidence/prevalence of the illness is high – stroke is the third most common cause of death in most developed countries [21]. As a result a large proportion of the NHS budget is allocated to stroke care and thus, it is important to determine and scrutinise the patterns of resource usage.

The data set used as a running example comes from the English Hospital Episode Statistics (HES) database and concerns finished consultant episodes of stroke patients, aged 65 and over, admitted by all English hospitals between April 1st, 1994 and March 31st, 1995 (148, 251 episodes). Following a HES recommendation, only records referring to the first episode of care were analysed (122, 965 spells). The LoS and occupancy statistics refer to the 105, 765 spells that had a discharge date, and therefore do not include the 17,200 patients who were still in the hospitals on 31st March 1995. A spell qualified as stroke if it contained a stroke related diagnosis code anywhere in the diagnostic chain. Stroke related diagnoses are between codes 430 and 438 in the International Classification of Diseases, Injuries and Causes of Death-Revision 9, ICD-9 [World Health Organisation 22]. No information that identified individual patients was supplied.

4.1 Stroke Data Mart Design

In this section, we outline some of the data models that underpin the OLAP environment using the formalisms defined by Thomas and Datta [23] and employed previously in modelling health data [24, 25]. Under the Thomas and Datta formalism, a cube is defined a 5-tuple $<C, A, f, d, O>$ where C is a set of characteristics, A is a set of attributes, f is a set of one-to-one mappings between a set of attributes to each characteristic, d is a boolean-valued function that partitions C into dimensions (D) and measures (M), and O is a set of partial orders [23].

The star schema for the transaction-based analysis consists of a fact table (LOS) and thirteen dimension tables, Fig. 3. The unique identifier of the spell (Spell ID) is the primary key of the fact table while the primary key of each dimension is included in the fact table as a foreign key. The dimensions are

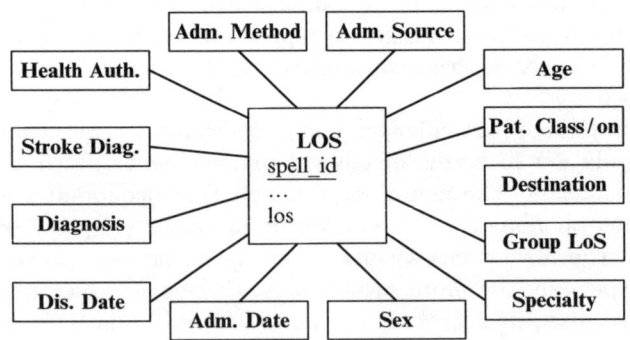

Fig. 3. LOS star schema, Stroke dataset

Admission Method, Admission Source, Age, Patient Classification, Discharge Destination, Group LoS, Finished Consultant Episode Specialty, Patient Sex, Admission Date, Discharge Date, Primary Diagnosis, Stroke Diagnosis, and Health Authority of Treatment.

A data cube, LOS, based on the above dimensional model is defined as follows:

```
C={admission_method, admission_source, age, patient_class,
destination, group_los, specialty, sex, admission_date,
discharge_date, diagnosis, stroke_diagnosis, health_authority,
spell_los};

d(admission_method)=1, d(admission_source)=1, d(age)=1,
d(patient_class)=1, d(destination)=1, d(group_los)=1,
d(specialty)=1, d(sex)=1, d(admission_date)=1,
d(discharge_date)=1, d(diagnosis)=1, d(stroke_diagnosis)=1,
d(health_authority)=1, and d(los}=0;

D={adm_method, adm_method_group, adm_source, adm_source_group,
patient_class, destination, destination_category, discharged,
group_los_week, group_los, specialty, specialty_group, sex,
adm_day, adm_week, adm_month, adm_quarter, adm_year, dis_day,
dis_week, dis_month, dis_quarter, dis_year, diagnosis,
diagnosis_group, stroke_diagnosis, stroke_diagnosis_group,
district_health_authority, regional_health_authority}
```

M={spell_id, los};
f(admission_method}={adm_method, adm_method_group},
f(admission_source}={adm_source, adm_source_group},
f(age}={age, age_group, age_lrg_group},
f(patient_class}={patient_class},
f(destination)={destination, discharged, destination_category},
f(group_los)={group_los, group_los_week, los},
f(specialty}={specialty, specialty_group},
f(sex}={sex},
f(admission_date)={adm_day, adm_week, adm_month, adm_quarter, adm_year},
f(discharge_date)={dis_day, dis_week, dis_month, dis_quarter, dis_year},
f(diagnosis)={diagnosis, diagnosis_category},
f(stroke_diagnosis)={stroke_diagnosis, stroke_diagnosis_category},
f(health_authority)={district_health_authority, regional_health_authority},
f(spell_los)={spell_id, los};
$O_{\text{admission\_method}}$={⟨adm_method, adm_category⟩},
$O_{\text{admission\_source}}$={⟨adm_source, adm_source_category⟩},
O_{age}={⟨age, age_group⟩, ⟨age_group, age_lrg_group⟩},

$O_{\text{patient\_class}} = \{\}$

$O_{\text{destination}} = \{\langle\texttt{destination, destination\_category}\rangle,$
$\langle\texttt{destination\_category, discharged}\rangle\},$

$O_{\text{group\_los}} = \{\langle\texttt{los, group\_los\_week}\rangle, \langle\texttt{group\_los\_week, group\_los}\rangle\}$ and,

$O_{\text{specialty}} = \{\langle\texttt{specialty, specialty\_group}\rangle\},$

$O_{\text{sex}} = \{\}$

$O_{\text{admission\_date}} = \{\langle\texttt{adm\_day, adm\_month}\rangle, \langle\texttt{adm\_month, adm\_quarter}\rangle,$
$\langle\texttt{adm\_quarter, adm\_year}\rangle, \langle\texttt{adm\_week, adm\_year}\rangle\},$

$O_{\text{discharge\_date}} = \{\langle\texttt{dis\_day, dis\_month}\rangle, \langle\texttt{dis\_month, dis\_quarter}\rangle,$
$\langle\texttt{dis\_quarter, dis\_year}\rangle, \langle\texttt{dis\_week, dis\_year}\rangle\},$

$O_{\text{diagnosis}} = \{\langle\texttt{diagnosis, diagnosis\_group}\rangle\},$

$O_{\text{stroke\_diagnosis}} = \{\langle\texttt{stroke\_diagnosis, stroke\_diagnosis\_group}\rangle\},$

$O_{\text{health\_authority}} = \{\langle\texttt{district\_health\_authority,}$
$\texttt{regional\_health\_authority}\rangle\}$ and,

$O_{\text{spell\_los}} = \{\}$

A very similar dimensional schema and data cube describe the periodic snapshot data model for bed occupancy analysis. Apart from the two date dimensions (admission and discharge dates) that are substituted by a single table (bed occupancy date), the remaining dimensional tables remain the same. The fact table then, OCC, becomes

```
OCC={day, adm_method, adm_source, age, patient_class, destination,
group_los, specialty, sex, diagnosis, stroke_diagnosis,
district_health_authority, admissions, discharges}.
```

The snapshot-based, data cube model designed for bed occupancy analysis is subjected to similar minor changes. Apart from merging the two date dimensional characteristics into one, the metrics of the data cube OCC are

```
M={admissions, discharges}.
```

The two star schemas and the data cube model were implemented on MS SQL Server 2000 Analysis Services. The OLAP queries were written in the MDX query language and the front-end OLAP environment was developed in MS Excel. The explorative analysis was performed using the OLAP-enabled environment including the graphs in this section, which were produced by the OLAP client.

4.2 Stroke Descriptive Statistics and Exploratory Analysis

The basic descriptive statistics of the patient LoS data are summarised in Table 1. The patient LoS distribution exhibits a standard deviation of 52.045 days over a mean of 14.29 days giving a coefficient of variation of 363%, thus indicating a very high variability in the data. The positive difference between the mean (14.29 days) and the median (7 days) LoS suggests that the data is skewed with a long tail in the distribution to the right. The large

Table 1. Descriptive statistics of LoS data, Stroke dataset

Number of records	105,765
Mean (days)	14.290
Median (days)	7
Std. deviation (days)	52.045
Skewness (days)	42.198
Kurtosis (days)	2,609.211
Min-Max (days)	0–4906
25th percentile (days)	3
75th percentile (days)	15

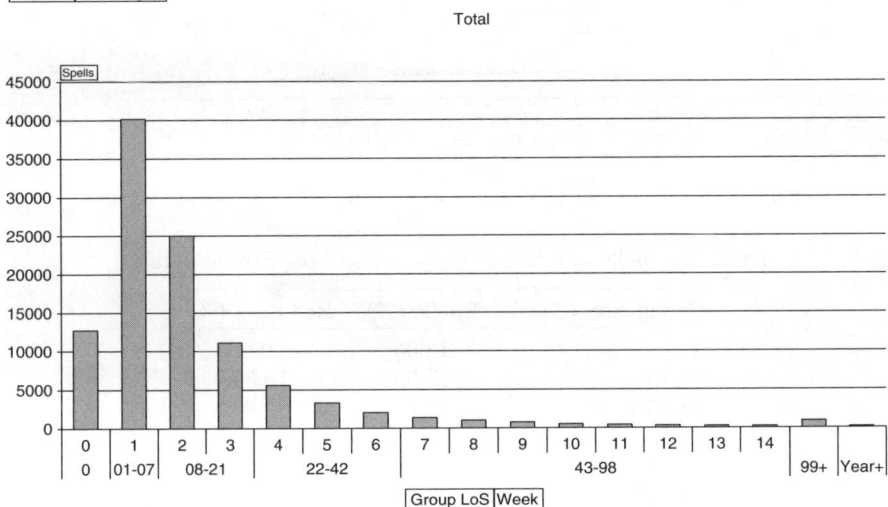

Fig. 4. Frequency distribution of LoS, Stroke dataset

positive value of skewness confirms the long right tail in the graph of the distribution. The interquartile range (IQR, the range between the 25th and 75th percentiles) is 12 days.

The frequency distribution of LoS (Fig. 4) shows a highly skewed distribution where the data initially peaks in the first week of stay in hospital indicating that the majority of patients leave hospital within the first week. There is then a very long gradual tail to the right of the distribution where there is a steady decrease in the number of patients who leave hospital with longer stays. The long tail is caused by a very small number of patients staying in hospital for a considerable amount of time – some of which occupy a bed for over a year.

The importance of taking into account the longer stay patients becomes apparent when bed occupancy is stratified, Fig. 5. This contour plot shows the daily occupancy stratified by groups of LoS. The groups refer to patients

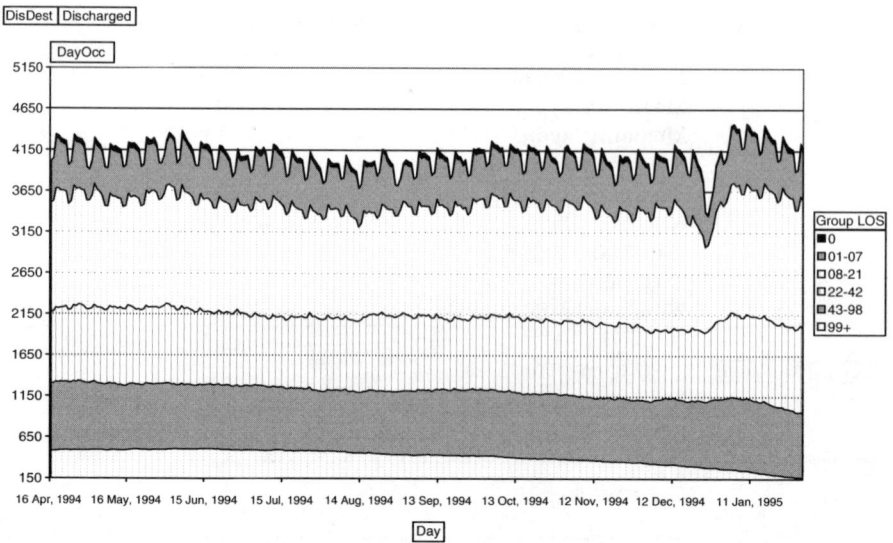

Fig. 5. Strata of daily occupancy by grouped LoS, Stroke dataset

Table 2. Spells and bed days by group LoS, Stroke dataset

Group LoS (days)	Spells (%)	Bed days (%)
0	12.08	0
01–07	37.99	10.55
08–21	34.16	30.86
22–42	10.31	21.30
43–98	4.48	18.99
99–364	0.84	8.55
365+	0.13	9.75

that stayed in hospital for 0 days, less than a week, 1–3 weeks, 4–6 weeks, 7–13 weeks, and 14+ weeks. The choice of the LoS interval for each group is based on clinical judgement. The decline towards the end of the graph in the longer stay groups is because we have excluded records of patient spells without a discharge date.

Although almost half of the patients are discharged within the first week, the majority of the beds (almost 90%) are constantly occupied by patients who stay for more than a week, Table 2. Similarly, the 1% of patients who stay for 99 days or more is responsible for almost 18% of the total bed days. Clearly, there is scope and need for a modelling technique that takes into account the above observations in the LoS distribution.

The power of the OLAP environment becomes apparent when more complex explorative analysis is required. For instance, the effect of discharge destination on LoS can be evaluated. In Fig. 6 the columns represent the average

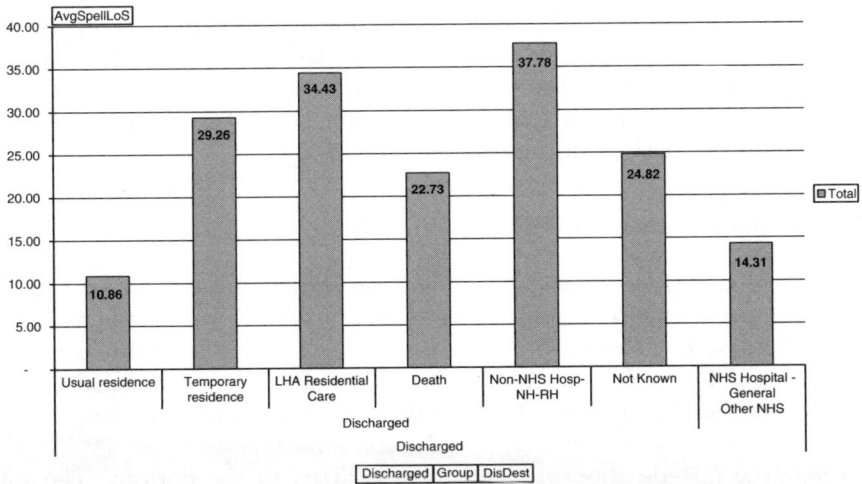

Fig. 6. Average LoS per discharge destination, Stroke dataset

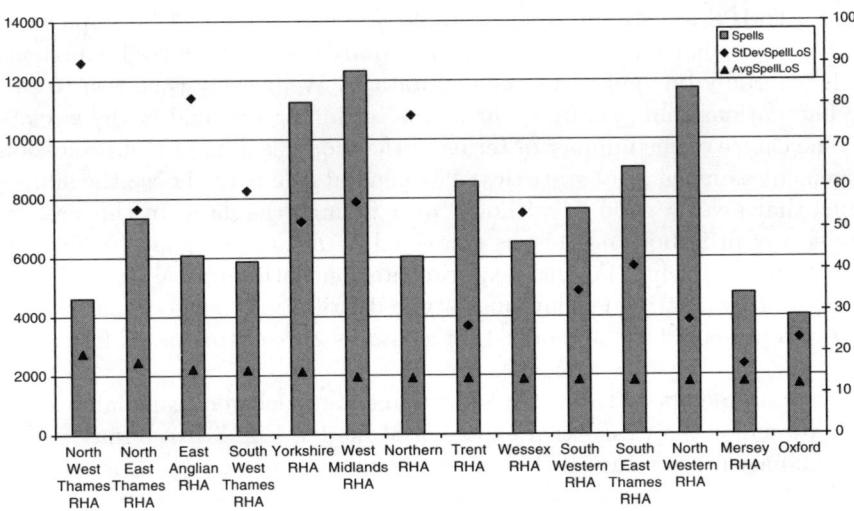

Fig. 7. Number of spells, average and standard deviation of LoS per regional health Authority, Stroke dataset

LoS per discharge destination. Clearly, there is a substantial difference in the average LoS of patients discharged to their usual residence and patients who are discharged to other destinations such as local health authority (LHA) residential care, non-NHS hospitals, and nursing and residential homes. These results are confirmed by an ANOVA test at the 95% level (p-value < 0.001).

Further analysis reveals that if outcome is simply measured by the average LoS, variations in the delivery of health care remain hidden. In Fig. 7, the number of spells is plotted on the primary y-axis (left-hand side of the plot), while the average and standard deviation of LoS is plotted on the secondary y-axis (right-hand side). Although the average LoS does not seem to vary considerably between the different regional health authorities, there is a marked difference in the variance which cannot be explained by the number of patients being treated in the regions.

4.3 Compartmental Modelling

The analysis described above is valuable in exploring and understanding various issues pertaining to the operation of the system. However, the ability to model the system not only furthers our understanding but also allows us to plan for it by testing the likely outcome of different operational scenarios. To enhance the findings of the analysis, in this section we describe the application of compartmental and simulation modelling.

First, the mixed-exponential model was fitted to the dataset. The estimated parameters after fitting the two-term and three-term exponential model to a virtual midnight census on Wednesday, 6th April 1994 using the e-fit module of BOMPS are summarised in Table 3. There were 4,157 occupied beds on that particular day. A bed census is required since the mixed-exponential model can only be applied to one-day data. A Wednesday is chosen to avoid the fluctuations that usually occur at the beginning and end of the week.

The choice of the number of terms in the model is a matter of professional judgement as much as of statistics. The general rule is to choose the simplest model that gives a good fit without "over fitting" the data. In this case, the goodness of fit in both models, as expressed by R^2, is very good (0.99642 and 0.99925 respectively). Further experimentation with survival analysis techniques and with fitting Coxian phase-type distributions suggest that the data is best represented by a model that contains three streams of flow of patients [2].

The parameters of the three-compartment model were calculated by the estimations for the three-term exponential model, Fig. 8. An admission rate of 286.2 patients per day (real admission rate per day was 289.8) is estimated on the day of the census. According to the model, these patients enter the

Table 3. Estimated parameters of 2- and 3-term mixed exponential models for census date 6th April 1994, Stroke dataset

	2-term exp. model	3-term exp. model
A	3580.728	2694.280
B	0.06902	0.09741
C	473.010	1314.213
D	0.00612	0.02766
E	0	168.415
F	0	0.00153
R^2	0.99642	0.99925

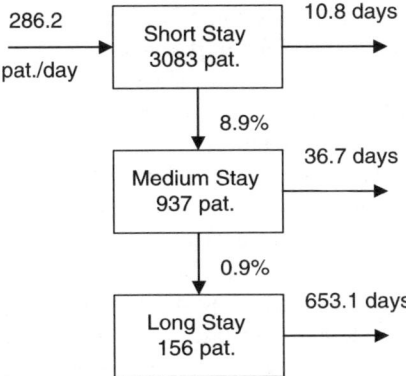

Fig. 8. Three-compartment model of patient flow, Stroke dataset

Table 4. BOMPS estimates for different discharge destinations, Stroke dataset

Discharge destination	#	First compartment Pat. (%)	First compartment LoS (days)	Second compartment Pat. (%)	Second compartment LoS (days)	Third compartment Pat. (%)	Third compartment LoS (days)
Usual residence	77,401	97.1	10.3	2.9	50.6	–	–
LHA residential care	820	93.8	27.8	6.2	97.6	–	–
Non-NHS Hosp-NH-RH	3,039	95.8	30.4	4.2	124.2	–	–
Death	14,116	39.7	3.6	57.3	17.2	3.0	180.1

LHA Local Health Authority, *NH* Nursing Home, *RH* Residential Home

short-stay compartment where they stay for 10.8 days on exponential average. Almost 9 out of 10 patients are discharged, the rest enter the medium-stay compartment where they stay for a further 36.7 days on exponential average. The majority of them are discharged, only 1 in 100 becomes a long-stay patient. The expected stay in long-stay is 653 days. The estimated number of patients in each compartment is 3,083 in short-stay, 937 in medium-stay, and 156 in long-stay. The modelled total number of patients in the system (4,176) compares favourably with the actual number of patients on the day of the census (4,157) a mere deviation of 0.46%.

Further flow rate analysis was conducted to investigate the observed differences in the average LoS for patients with different discharge destinations (recall Fig. 7). Four groups were further analysed: Usual Residence, LHA Residential Care, Non-NHS Hospitals–Nursing Homes–Residential Homes, and Death. Table 4 shows that 2-compartment models best describe all the categories of patients apart from those who died while in hospital. In this category, 4 out of 10 patients have an expected LoS of 3.6 days, while 6 out of 10 are expected to stay in the hospital for roughly 2 months (57.3 days). A small proportion (3%) stay in the hospital for half a year (180.1 days). Such a pattern of patient flow is to be expected because of the clinical progression of the disease.

In all other destinations, two streams of flow are identified. Patients who are discharged to their usual residence have the largest short-stay stream (97.1%) and a shorter LoS (10.3 days) compared to patients discharged to either LHA residential care or nursing and residential homes. These patients form a more distinct second compartment while for those patients in the short-stay stream, a longer LoS is to be expected. These delays cannot be explained just by medical reasons and it can be safely deduced that these patients occupy beds for longer periods while waiting for a suitable place to be found i.e. they may be regarded as delayed discharges.

4.4 Simulation Modelling

The compartmental model described above provides the input parameters for the discrete event simulation models [9, 12]. We first experimented with a model without capacity constraints to investigates the general behaviour of the simulations. This includes estimating the duration required for the model to reach stable state. The OLAP-enabled environment was used to facilitate the application of the Welch method to estimating the warm-up period of the simulation [26]. A warm-up period, l, of 2000 days was estimated based on 10 runs ($k = 10$), each one starting with a different random seed, Fig. 9. A 150-day moving average was required to smooth out the high frequency oscillation in the time series of the total number of patients in the system.

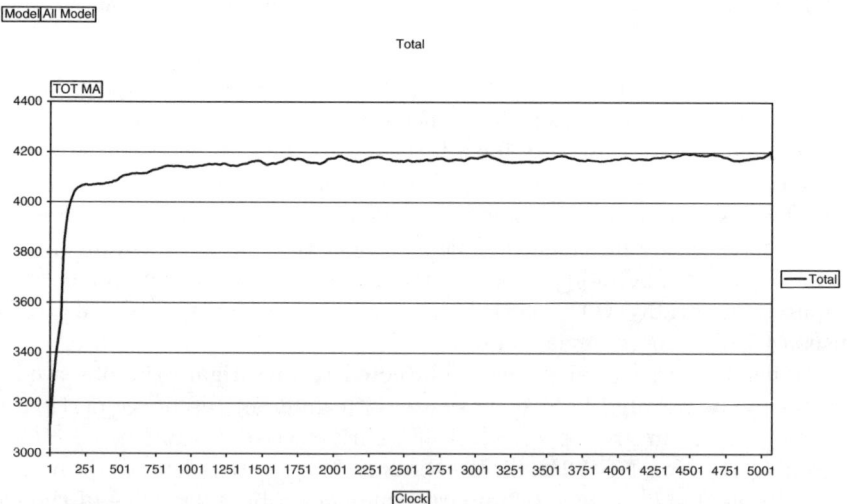

Fig. 9. Total number of patients in the system, 150-day moving average, unconstrained simulation model

The unconstrained model is also used for validation purposes. Specifically, we statistically confirmed that the point estimations for the number of patients in each compartment, after the warm-up period has been discarded, are equivalent to the estimations given by the 3-compartment model (Fig. 8). Table 5 illustrates, for each compartment, the unbiased estimators for μ and sample variance, their 95% confidence intervals (CI), and the estimates from BOMPS. One-sample t-tests at 95% CI confirm that there is no significant difference between the point estimates of simulation and those given by the three compartment model (p-values of 0.793, 0.761, and 0.119 respectively).

Having established the capacity constraints for each of the compartments, we can now introduce queues between them. The queues are used to measure blockage between the short and medium, and the medium and long-stay compartments. Different capacity constraints can be set for each compartment for conducting what-if analyses. Acute services – roughly matched with the short-stay compartment – require more spare capacity than rehabilitation and long term care services which are roughly matched with the medium and long-stay compartments. A smooth flow of patients, as expressed by low queuing and refused admission rate, is achieved only when the model operates with adequate level of emptiness in every compartment, Table 6. There is a marked increase in all the performance measures when the last two compartments operate at 95% levels, while the short-stay operates at 85% occupancy. On the contrary, a smooth level of flow is observed even when all the compartments operate at 90% levels of occupancy.

Table 5. Unconstrained model, estimates for steady-state daily occupancy ($k = 10$, $l = 2,000$, $n = 5,000$)

	Short-stay	Medium-stay	Long-stay
Mean	3,083.83	937.63	153.71
Standard Deviation	4.56	3.83	4.83
95%CI−	3,080.56	934.89	150.25
95%CI+	3,087.09	940.38	157.16
Flow model estimates	3,083.00	937.00	156.00

Table 6. Experimenting with different levels of occupancy (mean [95% CI])

Occupancy (total # beds) (%)	Refused admission (%)	Time spent in queue (days)	
		Acute to rehab	Rehab to long-stay
85 (4,913)	0 (0)	0 (0)	0.3 (0–1)
88 (4,746)	0 (0)	0 (0)	2 (0–4.5)
90 (4,640)	0.1 (0–0.2)	0 (0–0.1)	6.6 (0–17.6)
85–95–95 (4,777)	6.5 (0–17.5)	4.7 (1.7–7.6)	64.3 (29.5–99)

5 Discussion and conclusion

In this chapter we demonstrate the application of mixed-exponential, compartmental and simulation modelling techniques in modelling LoS data taken from a nationwide stroke dataset. In a separate analysis reported elsewhere [2], survival analysis was conducted and phase-type distributions were fitted to the same data. The results of these statistical techniques reached the same conclusion, that LoS data in the stroke dataset is best represented by a model that contains three streams of patient flow.

In addition, the applicability of the OLAP environment was demonstrated in terms of analysing LoS data along different dimensions. Key features of the OLAP-enabled environment that arose during this practical case are:

- The use of the first hierarchical level of the "Destination" dimension to include (or exclude) from the queries patients that are still in the system
- The use of the "Group LoS" dimension for the symmetrical treatment of LoS (as a measure and a dimension). This particular dimension allows for the generation of graphs of the distribution of LoS and the strata graphs (Figs. 4 and 5) that have proved to give particular useful insights in bed usage patterns
- The ability to generate "slices" of different daily bed censuses at the correct granularity that can then be fed to BOMPS for estimating bed usage statistics
- The implementation of Welch's method for estimating the initial transient of the steady-state simulation of patient flow

Acknowledgment

We gratefully acknowledge the guidance and support of Prof. Peter H Millard.

References

1. Shaping the Future NHS: Long Term Planning for Hospitals and Related Services. (2000) Department of Health, London
2. Vasilakis C, Marshall AH (2005) Modelling nationwide hospital length of stay: opening the black box. J Oper Res Soc 56: 862–869
3. Millard PH (1994) Current measures and their defects. In: Millard PH, McClean SI (eds.) Modelling hospital resource use: a different approach to the planning and control of health care systems. Royal Society of Medicine, London, pp. 29–37
4. Smith PC, Goddard M (2002) Performance management and operational research: a marriage made in heaven? J Oper Res Soc 53: 247–255
5. Marshall AH, Vasilakis C, El-Darzi E (2005) Length of stay-based patient flow models: recent developments and future directions. Healthc Manag Sci 8: 213–220
6. McClean SI, Millard PH (1993) Patterns of length of stay after admission in geriatric medicine: an event history approach. The Statistician 42: 263–274

7. Harrison GW, Millard PH (1991) Balancing acute and long term care: the mathematics of throughput in departments of geriatric medicine. Methods Inf Med 30: 221–228
8. Wyatt S (1995) The occupancy management and planning system (BOMPS). The Lancet 345: 243–244
9. El-Darzi E, Vasilakis C, Chaussalet T, Millard PH (1998) A simulation modelling approach to evaluating length of stay, occupancy, emptiness and bed blocking in a hospital geriatric department. Healthc Manag Sci 1: 143–149
10. Marshall AH, McClean SI (2003) Conditional phase-type distributions for modelling patient length of stay in hospital. International Transactions in Operational Research 10: 565–576
11. Winter Report 2000–2001. (2001) Department of Health, London
12. Vasilakis C, El-Darzi E (2001) A simulation study of the winter bed crisis. Healthc Manag Sci 4: 31–36
13. Koutsoukis N-S, Mitra G, Lucas C (1999) Adapting on-line analytical processing for decision modelling: the interaction of information and decision technologies. Decis Support Syst 26: 1–30
14. Isken M, Littig SJ, West M (2001) A data mart for operations analysis. J Healthc Inf Manag 15: 143–153
15. Koutsoukis NS, Dominguez-Ballesteros B, Lucas C, Mitra G (2000) A prototype decision support system for strategic planning under uncertainty. Int J Phys Distrib Logistics Manag 30: 640–660
16. Phillips RL (1994) The Management Information Value Chain. Perspectives Issue 3: Available from www.stern.nyu.edu/~abernste/teaching/Spring2001/ MIVC.htm, as of May 2007
17. Pedersen TB, Jensen CS (2001) Multidimensional Database Technology. Computer 34: 40–46
18. Codd EF, Codd SB, Salley CT (1993) Providing OLAP to User-Analysts: An IT Mandate. E.F. Codd & Associates, Sunnyvale, California
19. Vassiliadis P, Sellis T (1999) A survey of logical models for OLAP databases. ACM SIGMOD Rec 28: 64–69
20. Lee C, Vasilakis C, Kearney D, Pearse R, Millard PH (1998) The impact of the admission and discharge of stroke patients aged 65 and over on bed occupancy in English hospitals. Healthcare Manag Sci 1: 151–157
21. Gubitz G, Sandercock P (2000) Extracts from "Clinical Evidence": Acute ischaemic stroke. Br Med J 320: 692–696
22. International classification of diseases, ninth revision (ICD-9). World Health Organisation, Geneva (1977)
23. Thomas H, Datta A (2001) A conceptual model and algebra for on-line analytical processing in decision support databases. Inf Syst Res 12: 83–102
24. Vasilakis C (2003) Simulating the flow of patients: an OLAP-enabled decision support framework. Ph.D. thesis, University of Westminster
25. Vasilakis C, El-Darzi E, Chountas P (2006) An OLAP-Enabled Environment for Modelling Patient Flow. Proceedings of the 3rd IEEE Conference on Intelligent Systems (IS'06). pp. 261–266
26. Vasilakis C, El-Darzi E, Chountas P (2004) A Data Warehouse Environment for Storing and Analyzing Simulation Output Data. In: Ingalls RG, Rossetti MD, Smith JS, Peters BA (eds.) Proceedings of the 2004 Winter Simulation Conference. Institute of Electrical and Electronics Engineers, Piscataway, New Jersey, pp. 703–710

Real-Time Individuation of Global Unsafe Anomalies and Alarm Activation

Daniele Apiletti, Elena Baralis, Giulia Bruno, and Tania Cerquitelli

Dipartimento di Automatica e Informatica, Politecnico di Torino, Corso Duca degli Abruzzi 24, 10129 Torino, Italy
daniele.apiletti@polito.it, elena.baralis@polito.it,
giulia.bruno@polito.it, tania.cerquitelli@polito.it

Summary. In this chapter we present the IGUANA (real time Individuation of Global Unsafe Anomalies and Alarm activation) framework which performs real-time analysis of clinical data to assess the instantaneous risk of a patient and identify dangerous situations. The proposed approach consists of two phases. As a first step, historical data is analyzed to build a model of both normal and unsafe situations, which can be tailored to specific behaviors of a given patient clinical situation. The model exploits a risk function to characterize the risk level of a patient by analyzing his/her vital signs. Then, an online classification phase is performed. A risk label is assigned to each measure by applying the most suitable model and an alarm is triggered for dangerous situations. To allow ubiquitous analysis, this step has been developed to run on mobile devices and its performance has been evaluated on both smart phone and personal computer. Experimental results, performed on 64 records of patients affected by different diseases, show the adaptability and the efficiency of the proposed approach.

1 Introduction

Since medical applications of intelligent systems are becoming pervasive, sensor technologies may be exploited for patient monitoring to reduce hospitalization time and domicile assistance. Companies are developing non-invasive medical sensors to collect different physiological signals and create a body sensor network to monitor several health parameters (e.g., temperature, heart rate, blood pressure, oxygen saturation, serum glucose) to provide a comprehensive view of the patient's condition. An important issue in this context is the real-time analysis of physiological signals to characterize the patient condition and immediately identify dangerous situations.

The general architecture of homecare systems (see, e.g., [1, 2]) is usually composed by three subsystems: the body sensor network, the wireless local area network and the GSM network. In this scenario each individual wears a set of sensors integrated into non-invasive objects that reveal physiological

D. Apiletti et al.: *Real-Time Individuation of Global Unsafe Anomalies and Alarm Activation*, Studies in Computational Intelligence (SCI) **109**, 219–236 (2008)
www.springerlink.com

signals. These sensors are connected to the user's mobile device (e.g., smart phone, PDA) through a short range communication link (e.g., Bluetooth), in charge of transmitting recorded signals. The set of wearable sensors and the mobile device constitute the body sensor network. Possibly, the second subsystem allows the communication between the user's mobile device and the elaboration center by means of an infrastructure node (e.g., access point). Communication with the elaboration center may occur when the patient has to transfer recorded data to the system for further off-line analysis or to backup/gather historical data. Alternatively, a data transfer may occur from the elaboration center to the mobile device to send extra knowledge to improve the analysis. Finally, by means of the GSM network an alert message is immediately sent to the closest medical centre to request prompt medical intervention when a risk situation is detected.

The core of the homecare systems is the mobile device which records physiological values from wearable sensors, transmits vital signs to the elaboration center, locally elaborates/analyzes them to detect dangerous situations, and sends an alert message to request prompt medical intervention. The mobile device is also a critical point of the architecture. Since mobile devices work with different constraints (e.g., power consumption, memory, battery), the analysis of physiological signals performed on such devices requires optimized power consumption and short processing response time which are important research topics in different computer science areas.

Many efforts have been devoted to improve the hardware and the connectivity among devices (see, e.g., [1,3]) to reduce communication cost and improve device lifetime. Less attention has been devoted to the description of analysis techniques to assess the current risk level of a patient. However, the fundamental and most difficult task is the definition of efficient algorithms that automatically detect unsafe situations in real-time.

We propose the IGUANA (real time Individuation of Global Unsafe Anomalies and Alarm activation) framework which performs the real-time analysis of physiological signals to continuously monitor the risk level of a patient and detect unsafe situations. Our approach is based on data mining techniques to characterize and immediately recognize risk situations from data streams of physiological signals gathered by means of body sensors. We based our risk evaluation on a function that combines different components. Each component models a different type of deviation from the normal behavior. The IGUANA framework performs an off-line analysis of historical clinical data to build a model of both normal and unsafe situations by means of clustering techniques. This model can be tailored to specific behaviors peculiar to the clinical situation of a given patient (or patient group). The model is, then, exploited to classify real-time physiological measurements with a risk level and when a dangerous situation is detected an immediate intervention is triggered. To allow ubiquitous analysis, the classification algorithm has been developed to run on mobile devices (e.g., smart phone) and allows optimized power consumption and short processing response time.

We validated IGUANA with 64 patient records from the MIMIC-numerics database (Multi-parameter Intelligent Monitoring for Intensive Care) [4]. We analyzed four physiological signals (i.e., heart rate, systolic arterial blood pressure, diastolic arterial blood pressure, and peripheral blood oxygen saturation) which are representative of a patient's health conditions. MIMIC recordings last around 44–72 h. The experiments, performed both on smart phone and on personal computer, highlight the adaptability and the efficiency of the proposed approach.

2 The IGUANA Framework

The IGUANA framework performs real-time analysis for physiological risk assessment. To this aim, it evaluates personal health conditions by analyzing different clinical signals to identify anomalies (i.e., infrequent situations) and activate alarms in risk situations. Since conditions depend on the specific disease or patient profile, we first perform a *training phase* in which the framework automatically learns the common and uncommon behaviors by analyzing historical data. Then, an *on-line classification phase* evaluates the instantaneous risk level of the monitored patient in real-time. While the training phase is performed on a personal computer, the classification step is run on a mobile device (e.g., smart phone).

Figure 1 shows the building blocks of the IGUANA framework. The main blocks, model building and risk evaluation, are both preceded by a preprocessing phase. Preprocessing is necessary to perform data specific elaborations and to handle unacceptable physiological signal values, possibly due to sensor malfunctions.

During the training phase, given unlabeled historical clinical data, a model of safe and unsafe situations is created by means of data mining techniques. The model is tailored to specific diseases or patient profiles. Hence, different models of patients and diseases can be created. The most suitable model for

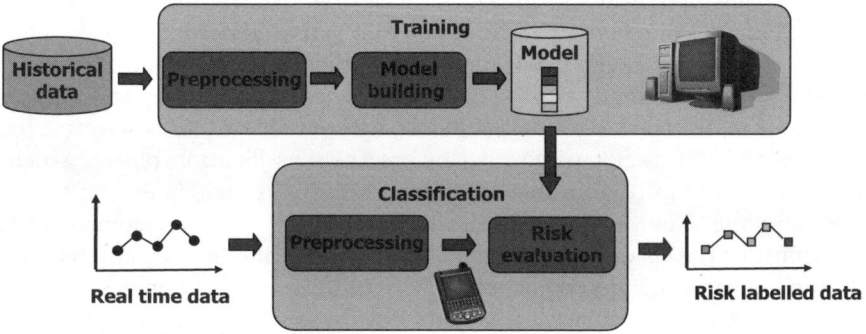

Fig. 1. Building blocks of the IGUANA framework

the current monitored patient is exploited for real-time risk evaluation. Since patient conditions depend on the contributions of several physiological signals (i.e., heart rate, blood pressure, oxygen saturation), we devise a global risk function to evaluate each time the risk indicator for the patient. The proposed risk function combines different components. Each component models a different type of deviation from the standard behavior. For example, the difference from the moving average value highlights a long term trend and the difference from the previous measure detects quick changes. Moreover, standard condition for patients may be represented by a normality band, whose upper and lower bounds are denoted as normality thresholds. Outside these thresholds, the risk of a patient additionally increases. Higher danger levels are denoted by a higher risk value. During the on-line classification phase, IGUANA processes real time streams of measurements colleted by sensors. For each measurement, the risk value is computed by means of the proposed risk function to evaluate the current patient condition. If a dangerous situation is detected an alarm (e.g., phone call, SMS) can be sent to the closest medical center to request prompt medical intervention.

2.1 Data Preprocessing

Both the model creation and the classification phases are preceded by a preprocessing phase, which aims at smoothing the effect of possibly unreliable measures performed by sensors. Preprocessing entails the following steps (1) outlier detection and removal, (2) null values (NA) handling, and (3) resampling.

Outlier detection and removal. Faulty sensors may provide unacceptable measures for the considered physiological signals. These include negative and extremely high values that cannot be reached in a living human body. The medical literature provides a validity range for each signal, whose extremes are defined NA-thresholds. Values outside such thresholds are considered outliers and substituted by null values (NA).

Null values handling. The previous step, in particular cases, may yield frequently NA-interrupted sequences of measures. During the model creation phase, a long continuous sequence of not null values is needed. Hence, before creating the model, isolated outliers (i.e., outliers preceded and followed by acceptable values) are substituted by the average of the previous and following measures. Finally, the longest subsequence without NA values is selected. This preprocessing step is not performed before the classification phase, which is applied on a stream of data incoming from sensors in real-time.

Resampling. The purpose of this step is to allow the management of the heterogeneous data sampling frequencies at which different sensors may provide their measurements. By means of a sliding window (whose size is user defined), each measurement is substituted by its average in the window. Next, to keep physiological measures synchronized, an appropriate sampling of each signal is performed.

2.2 Risk Function

The idea of quantifying the health status of a patient by means of his/her physiological signals is based on the comparison between current conditions and the common behavior derived by the analysis of previously collected data. We define as dangerous a situation in which the patient exhibits a deviation from a standard behavior described by the model. The model can be tailored to the different clinical conditions of a patient. We base our risk evaluation on a function that combines different components. Each component assesses a kind of deviation from the normal clinical behavior. The same risk function is applied to all physiological signals. However, different weights may be assigned for each signal, according to its importance in the global clinical condition evaluation and its specific physiological characteristics. The risk of each measure depends on the following components, which are depicted in Fig. 2.

- *Offset.* It is the difference between the current measurement and the moving average value. Since the moving average is the mean value of the measure in a given time window, the offset shows long term trends related to the current situation of a given patient.
- *Slope.* It is the difference between the current and previous measures. Its purpose is to detect quick changes. Hence, it shows short term trends.
- *Dist.* It is the difference between the current value and the closest of two thresholds, named "normality thresholds". Normality thresholds define a range outside which the patient risk increases because of excessively high (or low) values. Their default values are estimated as the maximum and minimum values of the moving average computed on the data analyzed during the model building phase. However, some patients may have slightly higher (or lower) values, due to particular diseases or special treatments. Hence, such thresholds should be adapted by the doctor to the specific needs of the patient.

The above risk components are combined to compute a risk value by means of the following risk function.

$$F_{risk} = |w_o \cdot Offset + w_s \cdot Slope| + w_r \cdot \sqrt{Offset^2 + Slope^2} + w_d \cdot Dist$$

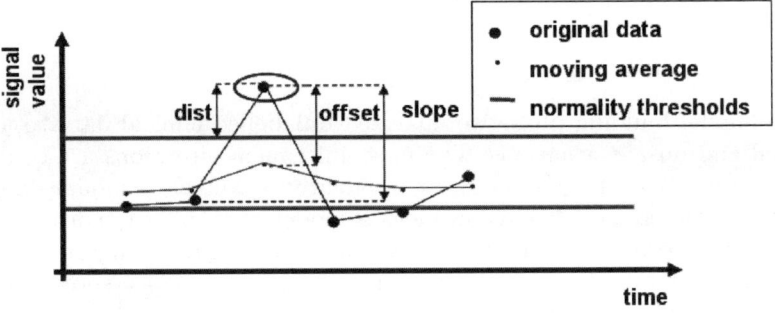

Fig. 2. Risk components

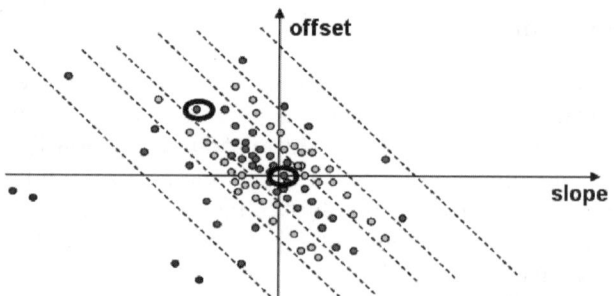

Fig. 3. Plot of the first term of the risk function

where w_o, w_s, and w_d are weight factors over the different risk components. We now analyze the contribution of each term of the risk function. The term $|w_o \cdot Offset + w_s \cdot Slope|$ is plotted in slope-offset coordinates in Fig. 3.

A higher risk is associated with the points lying in the first and third quadrant because the current value of the considered signal is far from its moving average and the current trend shows a further increase of its distance. Points in the second and fourth quadrant correspond to a lower risk value, because the signal, even if far from its moving average, shows a stabilizing trend, because its distance is decreasing. Points with the same risk value are represented on straight lines (see Fig. 3) whose slope depends on weights w_o and w_s. The first term is not sufficient to properly estimate the risk value. According to this term only, a point in the axes origin (having null slope and null offset, i.e. a measure equal to the previous one) has the same risk as a point with high opposite slope and offset values, which describes a situation where the measure is really far from the average but quickly returning to normality. The second term in (1) (i.e., $\sqrt{Offset^2 + Slope^2}$) takes into account this effect by considering the distance of a point in slope-offset coordinates from the origin. Finally, the dist contribution is added as third term. It increases the risk associated with measures outside the normality thresholds range. The weights w_s, w_o, and w_d are parametric functions of slope, offset, and dist respectively. The coefficient w_r is a function of w_o and w_s. This approach allows a wide degree of flexibility in the risk evaluation (e.g., positive and negative offset values can be easily associated with different weights).

2.3 Model Building

In the model building phase we analyzed unlabelled clinical data to identify normal and unsafe situations. The most infrequent situations are considered as representative of risky situations and are exploited to model dangerous states, while common behaviors yield a model of standard (normal) states. After turning every signal point into its corresponding risk components (offset, slope and dist), model building is performed in three steps, shown in Fig. 4 and described below.

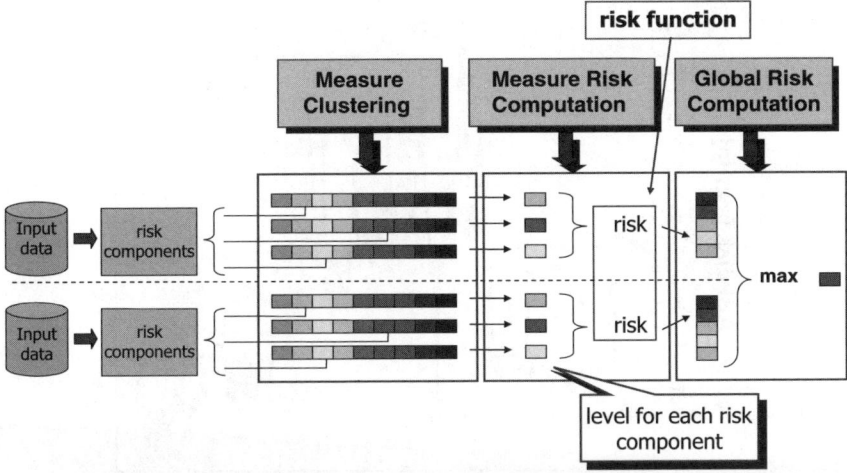

Fig. 4. Model building steps

2.4 Measure Clustering

Clustering is separately applied to the offset, slope and dist measures, to partition them in (monodimensional) clusters, characterized by homogeneous values. IGUANA is currently based on a hierarchical clustering technique, but the generality of our framework allows us to exploit any suitable clustering algorithm. Clustering algorithm selection issues are discussed in the experimental result section. At the end of this phase, for each considered sensor stream, a collection of classes for offset, slope and dist values is available, each characterized by upper and lower bounds.

2.5 Measure Risk Computation

The risk associated with each physiological signal (or measure) is computed by applying the risk function to the offset, slope and dist risk levels. Measure risk is divided into a finite number of values by means of a discretization step.

A graphical representation of the output of this step for the ABPdias (diastolic arterial blood pressure) signal is plotted in a risk diagram shown in Fig. 5. In the risk diagram points are characterized by their risk level, represented by a different shape (in ascending risk order: Crosshaped, x-shaped, and rhombus) and color (in ascending risk order: violet, red, orange). The x-axis displays the sample sequence number. Sample period is 15 s, unless differently specified. The two horizontal lines show the absolute thresholds for the signal (i.e., the maximum and minimum values allowed for any patient as a human being). Among the analyzed physiological signals, we selected ABPdias for its representative results. Hence, risk diagrams in this chapter are based on this signal and can be easily compared.

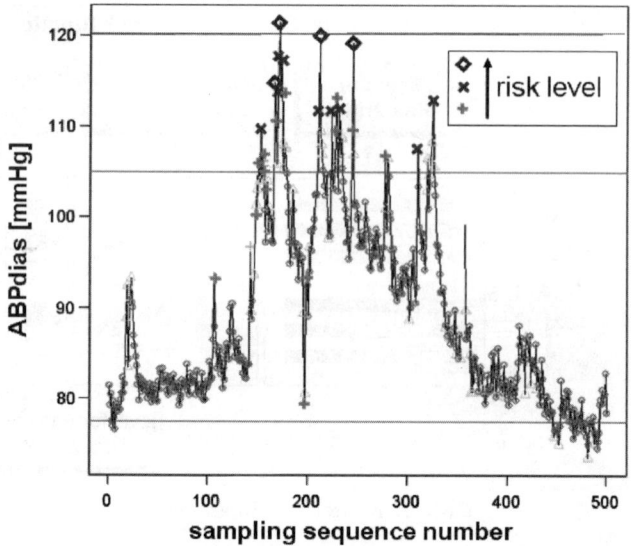

Fig. 5. ABPdias signal risk diagram

2.6 Global Risk Computation

The clinical situation of a patient depends on every physiological signal at the same instant. Since we focused our analysis on identifying unsafe anomalies which can lead to vital threat, even a single high risk measure is enough to show an unsafe trend. For this reason, the global risk associated with the clinical situation of the patient at a given instant is defined as the highest risk level among those assigned to every physiological signal at the same instant.

2.7 Classification

Real time streams of measures incoming from different sensors are initially processed separately, but the same operations are applied to all of them and at the end the results are combined into a unique value indicating the risk factor of the current clinical situation. Before starting the analysis, the value of each incoming measure is compared with user-defined absolute thresholds to determine whether it is outside a given range. If so, it is directly assigned the highest risk level. Examples of user defined thresholds for several vital signals are reported in Table 1. Next, the offset, slope and dist components are computed for the incoming measure. The needed information are (a) the previous measure, (b) the previous moving average value, and (c) the normality threshold values. A risk level is assigned to each component, by comparing the current values with the predefined classes stored in the model itself. The risk associated with each physiological signal (or measure) is computed by

Table 1. Absolute thresholds

Signal	Min	Max	Units
HR	40	150	Beats per minute
ABPsys	80	220	mmHg
ABPdias	40	120	mmHg
SpO2	90	100	%

applying the risk function to the offset, slope and dist risk levels. In this step, the user-defined weight parameters are exploited. Next, the maximum risk among all current measures is assigned as global risk level. If the obtained global risk level is above a user-defined threshold, an alarm may be triggered.[1]

3 Experimental Results

We validated our approach by means of several experiments addressing both the effect of varying different parameters of the framework (i.e., clustering algorithm selection, sliding window width, sampling frequency, risk function weights), and the performance of its current implementation.

We have identified a "standard" configuration for all parameters (i.e., clustering algorithm is hierarchical algorithm with average linkage method, sliding window size is 2 min, sampling frequency is 5 s, risk function weights are set to 1). All experiments have been performed by varying the value of a single parameter and preserving the standard configuration for all the other parameters.

3.1 Datasets

We validated the IGUANA framework with sensor measures publicly available on the Internet and collected by PhysioBank, the PhysioNet archive of physiological signals (www.physionet.org/physiobank), maintained by the Harvard-MIT Division of Health Sciences and Technology.

We analyzed 64 patient records from the MIMIC-numerics database (Multi-parameter Intelligent Monitoring for Intensive Care) [4], for whom the medical information we needed had been recorded. The patients were more than 60 years old. MIMIC-numerics is a section of the larger MIMIC database whose data is represented in the numeric format displayed in the digital instrumentation used for patient monitoring in hospitals. Since the MIMIC database collects data from bed-side ICU (Intensive Care Unit) instrumentation, the clinical situations of the patients we analyzed were extremely serious, allowing us to the test our approach in such utmost conditions. For our purpose, in the MIMIC-numerics database we chose four physiological signals considered significant of a patient's health conditions

---

[1] Actions associated with such alarms are outside the scope of this paper.

(a) heart rate (HR) [beats per minute], (b) systolic arterial blood pressure
(ABP-sys) [mmHg], (c) diastolic arterial blood pressure (ABP-dias) [mmHg],
and (d) peripheral blood oxygen saturation (SpO2) [percentage]. Original
measurements from the MIMIC-numerics database are provided every second. NA-threshold values were determined according to medical literature.
PhysioNet also provides PhysioToolkit, a software library for physiological
signal processing and analysis. We used some PhysioToolkit tools (e.g., the
rdsamp utility/command) for extracting the desired data from the databases
and during the preprocessing phase.

3.2 Clustering Algorithm Selection

We considered many clustering techniques (e.g., partitioning, hierarchical, and
density-based). We focused on partitioning and hierarchical techniques [5],
which were available in the statistical open-source environment R [6]. Partitioning algorithms, such as k-means, performed worse than hierarchical algorithms, because they clustered also normality situations in different risk levels.
Figure 6a shows this wrong behaviour. Hierarchical clustering algorithms may
use different methods to compute the inter-cluster distance. In our context,
the average linkage method yields better results than single linkage (which
forms chains of points), complete linkage, or ward. Figure 6b shows clusters
obtained by applying the average linkage method. We finally observe that the
clustering algorithm is a single, modular component of our framework. Any
suitable algorithm may be easily integrated in place of the current one.

3.3 Sliding Window Size

The size of the sliding window models the effect of the recent past on the
current situation. The longer the sliding window, the less the moving average

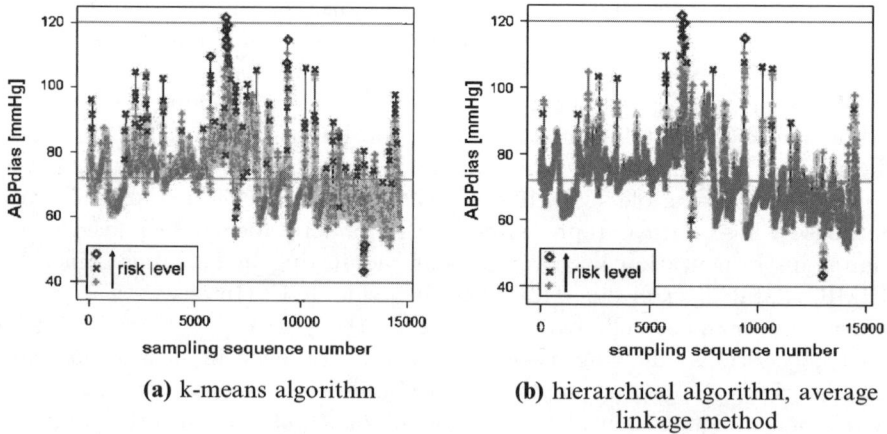

(a) k-means algorithm   (b) hierarchical algorithm, average
                                   linkage method

Fig. 6. Effect of different clustering algorithms

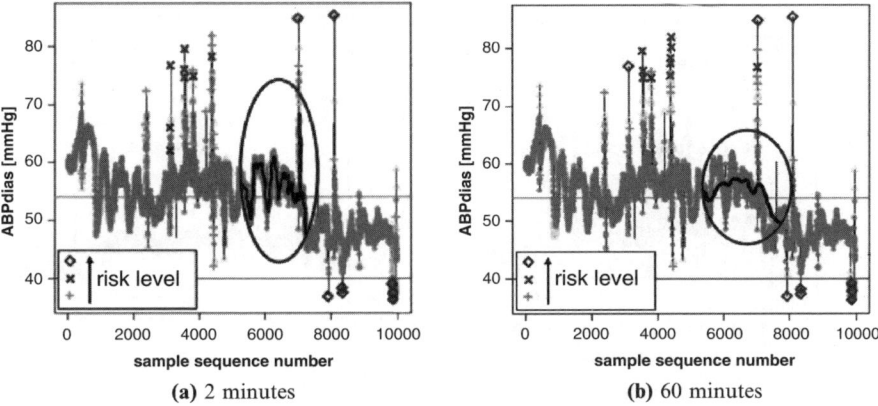

Fig. 7. Effect of the sliding window size

follows any sharp trend of a measure. Decreasing the sliding window size increases the rapid adaptation of risk evaluation to abrupt changes in a measure. The offset is the risk component affected by variations of the sliding window. It is based on the moving average value, which is strongly affected by the sliding window length. The effect of different sliding window sizes on the risk diagram is plotted in Fig. 7, with sliding window size of 2 and 60 min respectively. The variation of the sliding window size allows the IGUANA framework to adapt to different patient conditions. When a small sliding window size is considered, sudden changes of physiological values are quickly detected as potential risk conditions (see Fig. 7a). However, due to therapy side effects or a very active life style, some patients may be allowed to have quick changes in physiological values without being in danger. In this case, a longer sliding window may smooth the effect of a short, abrupt change in the context of a normal, steady situation (see Fig. 7b).

3.4 Sampling Frequency

The sampling frequency value directly affects the alarm activation delay. Every measure is assigned a risk level, which can potentially trigger an alarm. Hence, a dangerous situation can be identified within the next measure, which is in a sampling period time. For example, to identify a heart failure soon enough to have good chances of life-saving intervention by an emergency staff, the longest alarm activation delay should be 15 s. Longer delay values may be suitable for different purposes. The IGUANA framework is able to easily adapt to diverse sampling frequencies. Since sensor measures may be provided with different frequencies, in this context the adaptability of the IGUANA framework becomes essential. The effect of two different sampling frequencies on the risk diagram is highlighted in Fig. 8, where the sampling frequencies are 5 and 60 s, respectively. When the sampling frequency is high, even subtle, short

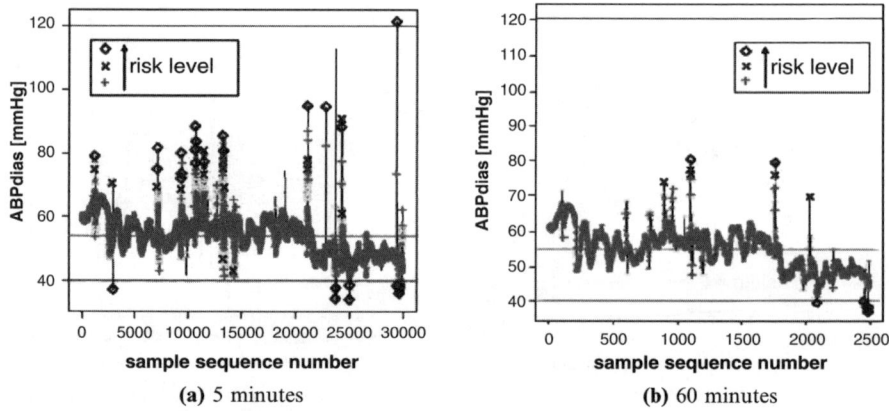

(a) 5 minutes

(b) 60 minutes

Fig. 8. Effect of the sampling frequency

anomalies are detected (see Fig. 8a), while a longer sampling period hides some of the sharpest spikes, but correctly identifies the remaining unsafe situations (see Fig. 8b).

3.5 Risk Function Weights

Risk function weights are among the most important parameters of the framework, because they directly determine the effect of each risk component (slope, offset, and dist) on the computed risk value. Hence, a physician is allowed to customize these settings according to the clinical conditions of the patient and the kind of anomalies to be detected. We report the results of some experiments performed to show the separate effect of the different risk components. All experiments are performed on the same sample dataset. They have no direct medical value, but demonstrate the adaptability of the framework to a wide range of situations. In Fig. 9 risk evaluation is only based on the offset component (slope and dist weights are set to zero). Risk rises with the distance between the measured value and the moving average. Such setting allows a physician to reveal deviations from a stationary behavior dynamically evaluated. In this case, positive or negative spikes in the signal time series are identified as dangerous situations. To separately analyze positive and negative contributions of the offset component, the offset weight w_o is set to 1 only for positive offset values in Fig. 9a, and only for negative offset values in Fig. 9b. When it is necessary to identify abrupt increases in a given measure, the slope risk component should be considered. This kind of analysis allows a physician to focus on rapid changes in the physiological behavior of the patient.

3.6 Performance

The IGUANA prototype has been developed in the R environment [6]. In our experiments, we evaluated the performance both of the off-line model

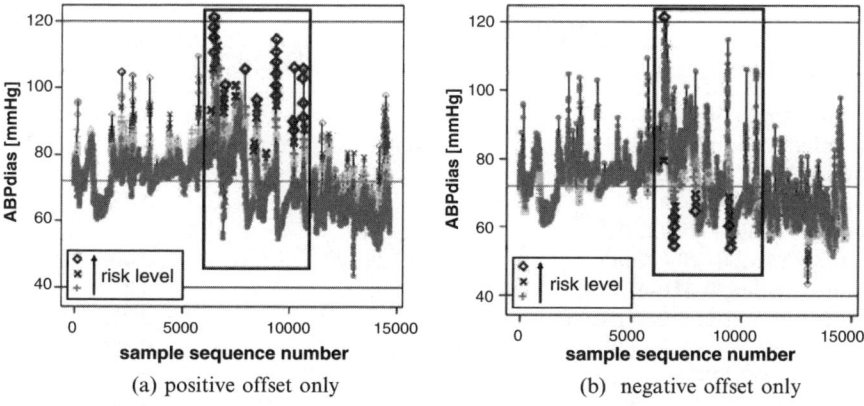

(a) positive offset only   (b) negative offset only

Fig. 9. Effect of risk function weights

building phase and of the on-line classification phase. Model building experiments have been performed on an AMD Athlon64 3200 + PC with 512 Mb main memory, Windows XP Professional operating system and R version 2.1.1. Classification experiments have been performed on an AMD AthlonXP 2000 + PC with 512 Mb main memory, Windows 2000 Professional and R version 2.1.1. For model building, we compared the performance of different clustering algorithms (i.e., partitioning and hierarchical). As expected, the k-means algorithm is about 60 times faster than the hierarchical algorithm and shows a better scalability with increasing data cardinality. However, since model creation is performed off-line, the selection of the clustering algorithm has been based on the quality of generated clusters, rather than performance. With hierarchical clustering algorithms, different methods for computing inter-cluster distance may be adopted. Different distance computation methods show a negligible effect on performance. Hence, again, the selection of the average linkage distance method was driven by cluster quality issues. Models with thousands of measures, generated by means of hierarchical clustering, are created in tens of minutes.

Performance of the classification phase is more critical, since this task is performed on-line. Furthermore, to be able to deliver real-time classification of incoming sensor data, the time requested by the classification of a single measurement set has to be less than the sampling period of the sensors.

Table 2 reports the time required for classifying a single measurement of every physiological signal. The MIMIC datasets, listed in column 1, have been identified by the name of the database (e.g., mimic), the short name of the patient's disease (e.g., angina), the patient's sex and age (e.g., m67 indicates a 67 year old man), the recording length in hours (e.g., 55 h), and the MIMIC-numerics record ID (e.g., 467n). Column 2 in Table 2 reports the total number of measurements analyzed for each patient, while column 3 reports the classification time for a single measurement. To obtain stable performance values, we separately repeated the classification step 5 times. Column 3 reports

Table 2. Classification time

Dataset	Measure samples	Classification time (ms)
mimic-angina-m67-55h-467n	51,716	0.36
mimic-bleed-m70-77h-039n	72,528	0.38
mimic-brain-m60-42h-280n	43,900	0.40
mimic-cabg-f80-57h-457n	54,060	0.38
mimic-cardio-f71-50h-293n	47,028	0.28
mimic-NA-f66-58h-276n	54,528	0.40
mimic-NA-m75-51h-474n	36,092	0.35
mimic-pulmEde-f92-71h-414n	23,488	0.36
mimic-pulmEde-m70-69h-466n	64,372	0.38
mimic-renal-f78-62h-471n	58,748	0.38
mimic-resp-f64-53h-403n	39,996	0.38
mimic-sepsis-f82-42h-269n	40,452	0.32
mimic-trauma-f92-51h-482n	40,312	0.31
mimic-valve-f77-52h-479n	49,304	0.32
mimic-resp-m90-46h-243n	43,748	0.21

classification times averaged on the 5 iterations. Classification time for a single measure is always less than 0.5 ms. The memory for the data needed by the classification process is estimated to be less than 50 bytes (supposing a 10 cluster model) for each physiological signal to be monitored.

We performed experiments with a mobile version of Iguana developed for both the Pocket PC and the Smartphone architectures.

In Fig. 10 a sample screenshot of the mobile application is presented. The instantaneous risk of each monitored vital sign is denoted as a number ranging from 1–5 while the monitored signals are denoted as ABPsys, systolic blood pressure, ABPdias, diastolic blood pressure, HR, heart beat rate, SpO2, peripheral blood oxygen saturation. To the right, the global risk is shown, together with the remaining battery power. Results are promising, since the smart phone battery proved to last many hours. Memory resources are estimated to be in the order of the hundreds of bytes for the data structures, while the complete application can be run on a 2 MB equipped Smartphone without restrictions. Since each measure requires tens of ms to be processed by the mobile application on a smart phone equipped with an OMAP850 CPU at 195 MHz, real time measure classification can be performed even at high sampling frequency. These experiments highlight both the adaptability and the efficiency of the proposed approach.

4 Related Work

Technological developments in miniaturization, diffusion of wearable medical technologies, increasing capabilities of mobile computing devices, and spreading of wireless communications improve mobile healthcare services for patients

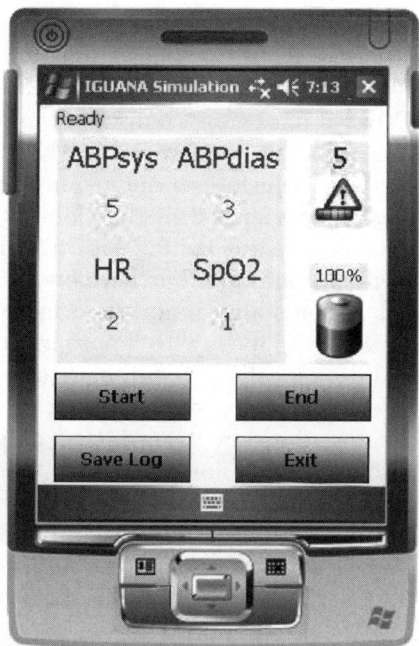

Fig. 10. Sample PocketPC screenshot

and health professionals. Furthermore, advances in health information systems and healthcare technology offer a tremendous opportunity for improving care quality while reducing cost [7]. Mobile health applications may play a key role in saving lives by allowing timely assistance, in collecting data for medical research, and in significantly cutting the cost of medical services. Non-invasive medical sensors to measure vital signs (e.g., temperature, heart rate, blood pressure, oxygen saturation, serum glucose) are currently under development [8]. These sensors may be integrated in a body sensor network to monitor various health parameters, thus providing a comprehensive view of a patient's condition. In this context dangerous situations may be timely recognized by means of real time analysis and alarms may be sent to the closest medical centre.

Several efforts have been devoted to the design of wearable medical systems [9, 10] and the reduction of power consumption of medical body sensors [11–13]. Sensor devices, integrated into intelligent wearable accessories (e.g., watches [14]), can collect physiological signals and transmit data to a mobile device.

Once the physiological data have been collected, the mobile device may send them to an elaboration centre for storage or physicians' analysis (if the elaboration centre is in the medical centre). In [15] the authors propose a monitoring system, which integrates PDA and WLAN technology. Through the WLAN, the patient's signals are transmitted in real-time to a remote central management unit, and medical staffs can access and analyse the data. Also the framework proposed in [16] is focused on a medical mobile system which performs real-time telediagnosis and teleconsultation. Patient measures are collected by a DSP-based hardware, compressed in real-time, and sent to the physicians in the hospital. The main advantage of these approaches is the simplicity of the architecture, which does not require any intelligence to the devices, since the analysis is performed in the elaboration centre. However, since this architecture introduces a delay for data transmission, it may cause a delay in detecting a critical condition. Furthermore, it requires the presence of physicians to monitor patients also in normal conditions.

If the mobile device is equipped with appropriate intelligence, it can process the physiological data locally and automatically generate alarms [17]. When the device detects a dangerous situation, it can send an alarm to the medical centre. In this way, data are transmitted only when unsafe situations occur and data compression is not strictly required. In [18] the authors concentrate on improving transmission of emergency messages, which must be reliably delivered to healthcare professionals with minimal delays and message corruption. They propose a network solution for emergency signal transmission using ad hoc wireless networks, which can be formed among patient-worn devices.

One step further towards the elaboration on mobile devices is proposed in [19] where a PDA is exploited to receive data from medical sensors and to transmit them over bandwidth-limited wireless networks. The authors specifically address the problems of managing different medical data (e.g., vital biosignals, images), developing an easy interface (for doctors) to view or acquire medical data, and supporting simultaneous data transfers over bandwidth-limited wireless links.

Many efforts have been dedicated to improving hardware and connectivity among devices [1, 3], but less attention has been devoted to investigating analysis techniques to assess the current risk level of a patient. However, the definition of efficient algorithms that automatically detect unsafe situations in real-time is a difficult task. In [10] an algorithm to discover physiological problems (e.g., cardiac arrhythmias) based on a-priori medical knowledge is proposed. Physiological time series recorded through sensors may be exploited for learning usual behavioural patterns on a long time scale. Any deviation is considered an unexpected and possibly dangerous situation. More recently, in [20] the extraction of temporal patterns from single or multiple physiological signals by means of statistical techniques (e.g., regression) is proposed. Single signal analysis provides trend descriptions such as increasing, decreasing, constant and transient. Instead, multiple signal analysis introduces a signal hierarchy and provides a global view of the clinical situation. Furthermore, a machine learning process discovers pattern templates from sequences of trends related to specific clinical events. The above mentioned solutions either are limited to specific physiological signals, or require some kind of a-priori information or fixed thresholds, or address related but different problems, such as detecting long term trends.

5 Conclusions and Future Work

IGUANA is a flexible framework that performs the real-time analysis of clinical data collected by a body sensor network. An off-line analysis is performed to build a model of both normal and unsafe situations. This model may be tailored to a specific patient disease. On-line analysis classifies each measured value assigning a risk level according to the previous model. Experimental results, performed both on personal computers and on mobile devices (e.g., smart phone) show the adaptability of the proposed approach to patients affected by different diseases and its computational efficiency. Future developments of the framework will explore different techniques to further improve the modeling phase. Two issues will be addressed (a) the analysis of the correlation among different physiological signals which contribute to a global clinical situation and (b) for specific physiological signals, the exploitation of samples in a larger time window to detect in advance dangerous situations by their early behaviors.

References

1. V. Jones, et al. Mobihealth: mobile health services based on body area networks. Technical Report TR-CTIT-06-37 Centre for Telematics and Information Technology, University of Twente, Enschede, 2006
2. D. Apiletti, E. Baralis, G. Bruno, T. Cerquitelli. IGUANA: Individuation of Global Unsafe ANomalies and Alarm activation. IEEE IS '06 – Special Session on Intelligent System for Patient Management. 267–272. September 2006, London
3. K. Lorincz, et al. Sensor networks for emergency response: Challenges and opportunities. IEEE Pervasive Computing, 3(4):16–23, 2004
4. PhysioNet – Mimic Database [Online]. Available: http://www.physionet.org/physiobank/database/mimicdb/
5. J. Han and M. Kamber. Data Mining: Concepts and Techniques. Series Editor Morgan Kaufmann Publishers. The Morgan Kaufmann Series in Data Management Systems, Jim Gray, August 2000
6. The R Project for Statistical Computing [Online]. Available: http://www.r-project.org/
7. I. Lee, et al. High-confidence medical device software and systems. Computer, 39(4):33–38, 2006
8. E. Jovanov, A. Milenkovic, C. Otto, and P. C. de Groen. A wireless body area network of intelligent motion sensors for computer assisted physical rehabilitation. Journal of NeuroEngineering and Rehabilitation, 2:6, 2005
9. F. Axisa, A. Dittimar, G. Delhomme. Smart clothes for the monitoring in real time and conditions of physiological, emotional and sensorial reaction of human. IEEE Engineering Medicine and Biology Society, 2003
10. P. Varady, Z. Benyo, and B. Benyo. An open architecture patient monitoring system using standard technologies. IEEE Transactions on Information Technology in Biomedicine, 6:95–98, 2002

11. U. Anliker, et al. AMON: a wearable multiparameter medical monitoring and alert system. IEEE Transactions on Information Technology in Biomedicine, 9:415–427, 2004
12. P.-T. Cheng, L.-M. Tsai, L.-W. Lu, and D.-L. Yang. The design of PDA-based biomedical data processing and analysis for intelligent wearable health monitoring systems. International Conference on Computer and Information Technology (CIT '04), 2004 879–884
13. P. Branche and Y. Mendelson. Signal Quality and Power Consumption of a New Prototype Reflectance Pulse Oximeter Sensor. Northeast Bioengineering Conference, Hoboken, NJ, 2005
14. http://www.skyaid.org/LifeWatch/life_watch.htm
15. Y.-H. Lin, et al. A wireless PDA-based physiological monitoring system for patient transport. IEEE Transactions on Information Technology in Biomedicine, 8(4):439–447 2004
16. H.-C. Wu, et al. A mobile system for real-time patient-monitoring with integrated physiological signal processing. IEEE BMES/EMBS Joint Conference, 2:712, 1999
17. E. Manders, B. Dawant. Data acquisition for an intelligent bedside monitoring system. IEEE Engineering in Medicine and Biology Society, 5(31):1987–1988, 1996
18. U. Varshney. Transmission of Emergency Messages in Wireless Patient Monitoring: Routing and Performance Evaluation. IEEE International Conference on System Sciences, 2006
19. S. Gupta, A. Ganz. Design considerations and implementation of a cost-effective, portable remote monitoring unit using 3G wireless data networks. IEEE EMBS, 2:3286–3289, 2004
20. S. Sharshar, L. Allart, and M. C. Chambrin. A new approach to the abstraction of monitoring data in intensive care. Lecture Notes in Computer Science, LNCS-Springer Verlag, 3581:13–22, 2005

Support Vector Machines and Neural Networks as Marker Selectors in Cancer Gene Analysis

Michalis E. Blazadonakis and Michalis Zervakis

[1]Technical University of Crete, Department of Electronic and Computer Engineering, University Campus, Chania Crete 73100, Greece
mblazad@ier.forthnet.gr

Summary. DNA micro-array analysis allows us to study the expression level of thousands of genes simultaneously on a single experiment. The problem of marker selection has been extensively studied but several aspects need yet to be addressed. In this study we add an important attribute of gene selection by considering the 'quality' of the selected markers. By the term 'quality' we refer to the property that a set of selected markers should differentiate its expression between the two possible classes of interest (positive, negative). Thus, we address the problem of selecting a small subset of representative genes that would be adequate enough to discriminate between the two classes of interest in classification, while preserving low intra-cluster and high inter-cluster distance, which eventually leads to better survival prediction.

1 Introduction

The advent of micro-array technology has given scientists a valuable tool to monitor the behavior of thousands of genes in a single experiment. The behavior of each gene is kept in a separate cell in an m by n matrix, where each row corresponds to a different gene while each column to a different experiment (sample). In a post experimental step, the expression level (behavior) of each gene is recorded in terms of a color map. Many color schemes have been used but green–black–red is the prevailing one. A green colored cell manifests that the specific gene has expressed itself *more* in the normal than in pathological situation, a red color in a cell implies exactly the opposite while a black color means that the specific gene has expressed itself in exactly the same way in both situations. Colors are translated into numbers on an open interval $[-3, +3]$ for instance, -3 means red, 0 means black and $+3$ means green.

In these kinds of experiments we encounter the problem that the number of attributes (genes or rows of the matrix) is much larger than the number of samples (columns of the matrix). Usually the number of genes is of the order of thousands while the number of columns (samples) is of the order of tenths. This problem has been extensively considered in micro-array analysis

M.E. Blazadonakis and M. Zervakis: *Support Vector Machines and Neural Networks as Marker Selectors in Cancer Gene Analysis*, Studies in Computational Intelligence (SCI) **109**, 237–258 (2008)
www.springerlink.com

through techniques that decrease the number of dimensions (genes) usually in an iterative manner. Outstanding works have been previously published addressing this problem in real domains showing encouraging results. We urge the interested reader to refer to the work of [8,18,19], while for a more technical work we site references such as [9,10,16,23]. In this study we test the ability of intelligent systems to tackle the so called "curse of dimensionality" by using intrinsic attributes of the data.

Marker (feature) selection methods can be divided into two categories namely, filter and wrapper methods. Filter methods focus on the intrinsic properties of data using various stochastic metrics such as Fisher's ratio or information gain among many others. Genes are ranked according to how they score on such a measure and the highly ranked genes, which give the highest classification accuracy, are then selected as markers. Wrapper methods on the other hand work in a recursive way, where a classifier is used to assign a relevance weight to each feature and then the feature with the lowest weight is eliminated. In the next iteration cycle weights are re-assigned and the process continues recursively, in a way that more than one features could be eliminated in each iteration. At the end of the process, the set of features achieving the highest classification accuracy is selected as the set of markers. The main difference between these two approaches is that wrapper methods work in an iterative manner where feature weights are re-evaluated and potentially changing from iteration to iteration, while in filter methods weights are evaluated once and remain stable throughout the selection process. It has been shown by various studies that the performance of wrapper methods can be superior to the performance of filter methods, we selectively refer to [11].

Concerning the desirable attributes of the selected markers, we are searching for a set of genes that would be able to predict correctly the label (class) associated with unseen samples. In addition, these genes should be able to preserve a similar expression within each class, while showing a variation on their expression levels across the two situations of interest, i.e. positive or negative state. We refer to this intrinsic property as the *quality* of the selected markers. This fact has been implicitly stated in almost every marker selection study; we selectively refer to, [1, 3, 13, 14, 18, 19]. It has also been explicitly stated by 18, 10, as well as Hastie et al. 2000. The credibility of a set of marker genes with high classification accuracy, which however is not able to show this kind of behavior, is questionable. The set of markers we are searching for should maintain a low intra-class but a high inter-class distance. Nevertheless, it is worth stressing the fact that most of the studies focus on classification accuracy ([9, 11, 23, as well as many others]) rather than on the quality of the selected markers. In this study we assess such an approach as to select marker genes that share the intrinsic characteristic of high inter-class but low intra-class distance.

Another important issue concerning marker selection has to do with the classification performance of the selected markers. It is a common practice to assess the performance of a method by its Leave One Out Cross Validation (LOOCV) error. Two types of LOOCV schemes are generally considered. The first one addresses the removal of the left-out sample before the selection of differentially expressed genes and the application of the prediction rule, while the second approach handles the removal of the left-out sample after the selection process but before the application of the prediction rule. The first is usually referred to as the External LOOCV (ELOOCV) while the second is referred to as the Internal LOOCV (ILOOCV), [2,17]. In the ILOOCV the entire training set is used in the feature selection process and the LOOCV strategy is applied after the feature selection process. On the other hand, in the ELOOCV each sample is left out before any selection process takes place and the prediction rule is then tested on that left out sample. It is obvious that ELOOCV is a more unbiased estimator of the error rate since it is totally independent of the selection process. However, ILOOCV provides a measure which can not be neglected, as it expresses the training ability of a selection rule within the training set. In other words it indicates the prediction rules that can learn or generalize better on the training set, which eventually could lead to a better generalization performance on a totally independent test set.

In this study we propose to apply a linear neuron in a wrapper manner as a marker selector trained through the Resilient Back Propagation (RPROP) method, and compare its performance with a representative wrapper method known as Recursive Feature Elimination based on Support Vector Machines (RFE-SVM) introduced by [9]. RFE-SVM has shown remarkable performance on the leukemia data set published by Golub et al. [8]. The criteria we are using to measure and compare the performance of the two approaches are ILOOCV as a measure to assess the performance of the two methods on the training set and independent test set evaluation, as a measure to assess their ability to generalize on new or unseen data. Besides accuracy metrics, we also assess their ability to select differentially expressed genes, though the Davies Bouldin (DB) index [7], as well as the survival prediction of the expression profile of the selected markers derived by the underlined methods through Kaplan–Meier curves. Our domain of application is the breast cancer data set published by [19].

2 Background Knowledge

Before we proceed it is essential to provide the basic background knowledge on support vector machines and neural networks necessary for understanding and grasping the concepts underlined on the marker selection problem.

2.1 Support Vector Machines

SVM [20] attempts to find the best separating hyperplane to distinguish between the two classes of interest, positive $(+1)$ and negative (-1). This is done by maximizing the distance $\frac{2}{\|w\|}$ between the two parallel lines $(w \cdot x) + b = 1$ and $(w \cdot x) + b = -1$, which form the margin of separation of the two classes. The final separating hyperplane passes through the middle of this margin with equation $(w \cdot x) + b = 0$. The decision function then, is a function of the form:

$$f(x) = \text{sgn}\left((w \cdot x) + b\right) \tag{1}$$

where w represents the direction vector of the hyperplane. The sign of the value returned by (1) indicates the predicted class associated with example x, while $|f(x)|$ indicates the confidence level of the resulting decision. The SVM problem can be equivalently formulated as follows:

$$minimize \ \frac{1}{2}\|w\| + C\sum_{j=1}^{n} \xi_i^2 \tag{2}$$

$$subject \ to \ y_i\left((w \cdot x_j) + b\right) \geq 1 - \xi_j, \ \xi_j \geq 0, \ j = 1, \ldots n$$

By the duality theory, a tutorial of which can be found in [6], the problem can be transformed to the following maximization problem, where λ represents the vector of Lagrange multipliers and y_i represents the label (either $+1$ or -1) of the ith

$$\underset{\lambda \in i^n}{maximize} \ \sum_{j=1}^{n} \lambda_i - \frac{1}{2}\sum_{i,j=1}^{n} \lambda_i\lambda_j y_i y_j (x_i \cdot x_j)$$

$$Subject \ to \ \begin{cases} \sum_{j=1}^{n} \lambda_j y_j = 0 \\ 0 \leq \lambda_j \leq C, j = 1, K, n \end{cases} \tag{3}$$

sample. Towards the solution of this problem, we obtain the following expression for the direction vector w:

$$w = \sum_{j=1}^{n} \lambda_j y_j \ x_j \tag{4}$$

which is actually an expansion of those training samples with non-zero λ_i, i.e. the support vectors. It can be proved that support vectors lie on the borders of the class regions (as demonstrated Fig. 1) and can be used to find b by substituting one of the support vectors to the following equation:

$$y_j\left((w \cdot x_j) + b\right) = 1. \tag{5}$$

An important issue making SVMs attractive is that they allow the use of kernels, so that the dot product in (3) can be replaced by a kernel function in the following form:

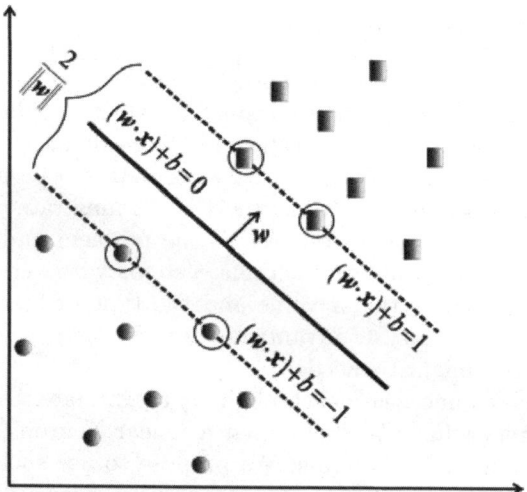

Fig. 1. Binary classification problem, showing the margin of separation between the two classes; circled points represent the support vectors

$$\underset{\lambda \in i^n}{maximize} \sum_{i=1}^{n} \lambda_i - \frac{1}{2} \sum_{i,j=1}^{n} \lambda_i \lambda_j y_i y_j k\left(\boldsymbol{x}_i, \boldsymbol{x}_j\right)$$

$$Subject\ to \begin{cases} \sum_{j=1}^{n} \lambda_j y_j = 0 \\ 0 \le \lambda_j \le C, j = 1, \dots, n \end{cases} \tag{6}$$

Besides the linear kernel in (3), other types of kernels such as polynomials of any degree, as well as Radial Basis Functions (RBF) can be used in the forms of:

$$k\left(\boldsymbol{x}, \boldsymbol{y}\right) = \left(1 + \left(\boldsymbol{x} \cdot \boldsymbol{y}\right)\right)^d$$

$$k(\boldsymbol{x}, \boldsymbol{y}) = \exp\left(-\gamma \left\|\boldsymbol{x} - \boldsymbol{y}\right\|^2\right) \tag{7}$$

2.2 The RFE-SVM Marker Selection Method

The RFE-SVM method [9] is based on SVM [20] and the idea of ranking features according to the absolute value of the components of the direction vector $\boldsymbol{w}$. As expressed in (4), each individual component of $\boldsymbol{w}$ is associated with an individual component of vector $\boldsymbol{x}$, which is the expression level of an individual feature. Thus, every feature (gene) is multiplied by a weight; the larger the absolute value of its weight, the more important that feature is according to RFE-SVM, in the sense that it contributes more to the decision function of (1). As a consequence, genes can be ranked according to the absolute value of the individual components of $\boldsymbol{w}$. A general overview of the method is given in the following steps:

3 Proposed Methodology

In this study we plan to take advantage of the fact that neural networks are open systems with many free parameters which can be adapted to the needs and peculiarities of the problem under consideration. Most frequently addressed parameters are the type of network that will be used, the number of neurons (Fig. 2), the number of layers, the learning rate and of course the training procedure that could best feet to the problem under consideration. SVMs on the other hand are more stable systems with very few parameters to be fine tuned; such as the C value and the type of kernel. Even though this generally seems to be an advantage there are case where more flexible adaptation abilities might be needed.

RFE-SVM uses a linear kernel through (1) to estimate the weight vector of the separating hyperplane. We know that a Linear Neuron (LN) can be used to approximate any linear function. We propose to use such an approach to approximate the separating hyperplane between positive and negative classes in place of the linear SVM used in the RFE-SVM method. This is applied as a linear neuron of m inputs and one output, where m corresponds to the number of genes. Considering two possible outcomes at the output layer, namely *output 0 for the negative class and 1 for the positive class*, we can use such a neuron to estimate the weight vector w of the separating hyperplane, as for step 3 in Table 1.

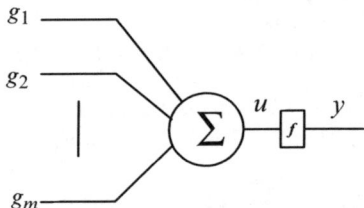

Fig. 2. A single neuron used as a marker selector

Table 1. The recursive feature elimination based on SVM (RFE-SVM) algorithm

(1)	Let m be the initial number of features
(2)	While $(m \geq 0)$
(3)	Estimate the direction vector w of the separating hyperplane using linear SVM.
(4)	Rank features according to the components of $|w|$.
(5)	Remove the feature with the smallest weight in absolute value $(m \leftarrow m - 1)$; more than one feature can be removed per iteration.
(6)	Estimate LOO accuracy of the m surviving features using a linear SVM classifier.
(7)	End While
(8)	Output as marker genes the set of surviving features which achieves maximum LOO accuracy.

3.1 Back Propagation Training

According to neural network theory the error function of a single neuron that needs to be minimized is given by:

$$E = \frac{1}{2} \sum_{j=1}^{n} (d_j - y_j)^2 \tag{8}$$

where

$$y_j = \frac{1}{1 + e^{u_j}} = f(u_j) \tag{9}$$

$$u_j = \sum_{i=1}^{m} w_i g_{ij} \tag{10}$$

and

$$f'(u_j) = y_j (1 - y_j) \tag{11}$$

where n corresponds to the number of samples, d_j represents the desirable neuron output associated with sample j and y_j is the actual output produced by the neuron for the given sample. Through a gradient descent method for the minimization of (8), w_i associated to gene g_i is updated as follows:

$$w_i(t+1) = w_i(t) + \left(-\mu \frac{\partial E}{\partial w_i}\right) = w_i(t) - \left(\mu \frac{\partial E}{\partial w_i}\right)$$

$$= w_i(t) - \sum_{j=1}^{n} \left(\frac{\partial E}{\partial y_j} \frac{\partial y_j}{\partial u} \frac{\partial u}{\partial w_i}\right) \tag{12}$$

where

$$w_i(t+1) = w_i(t) - \frac{\mu}{2} \sum_{j=1}^{n} \left(-2(d_j - y_j) y_j (1 - y_j) g_{ij}\right)$$

$$= w_i(t) + \mu \sum_{j=1}^{n} (d_j - y_j) y_j (1 - y_j) g_{ij} \tag{13}$$

$$= w_i(t) + \mu \sum_{j=1}^{n} (d_j - y_j) f'(u_j) g_{ij}$$

finally

$$w_i(t+1) = w_i(t) + \mu \sum_{j=1}^{n} e_j f'(u_j) g_{ij} \tag{14}$$

where t represents current iteration and μ is the learning rate and

$$e_j = (d_j - y_j). \tag{15}$$

3.2 Resilient Back Propagation (RPROP) Training

Neural networks often use sigmoid functions, (9), which are often called "squashing" functions since they compress the input range into an output range within [0L 1]. Sigmoid functions are characterized by the fact that their slope approaches zero as the input gets larger. This leads to the side effect of a gradient magnitude close to zero, which in turn leads to very small changes in the weights of the network, even though they might be far from their optimal values. This can be verified by (14), where when the second term of the right hand side is close to zero, the weights are updated by a very small increase or decrease.

Resilient back propagation [15] alleviates this effect by taking into consideration only the sign of the derivative and then increasing or decreasing the weight by a small factor of n^+ or n^-. Whenever the derivative of the error function E has the same sign for two consecutive iterations, the update value is increased by a factor n^+. On the other hand, whenever the derivative changes sign in two consecutive iterations, indicating that the last update was too big and the algorithm has jumped over a local minimum, the update value is decreased by a factor η^-. An overview of the algorithm is presented in Table 2. The choice of $n^+ = 1.2$ and $n^- = 0.5$ in several domains have shown very good results and these are the values used for the experiments conducted in this study. The remaining parameters Δ_0 and Δ_{max} were set to their default values 0.7 and 50 respectively [15].

Following our previous discussion, we propose to use a linear neuron trained through RPROP algorithm as the weight vector estimator in place of step 3 in Table 1. Thus, our methodology proposes to use a linear neuron instead of a linear SVM to tackle the problem of marker selection. We refer to the proposed methodology as Recursive Feature Elimination based on Linear Neuron Weights (RFE-LNW).

4 Cluster Quality Measure

Besides the gene selection methodology, it is essential to provide a measure for assessing the 'quality' of the selected markers in terms of inter- and intra-class distance. The validity or quality of features that survive at each step of the elimination process is an issue of particular interest. Based on the desirable properties of markers, the surviving genes should form well defined clusters related to the pathology states of interest. In other words, the clusters of marker genes should express small intra-class but large inter-class distance [4]. One measure to asses cluster quality is known as Davies-Bouldin index (DB) [7] that has been extensively used to assess cluster quality in various fields besides DNA micro-array analysis [5, 21, 22]. The DB index for a partition

Table 2. The Resilient back propagation (RPROP) training algorithm

---

$\forall i : \Delta_i (t) = \Delta_0$
$\forall i : \frac{\partial E}{\partial w_i} (t-1) = 0$
Repeat
  Compute gradient $\frac{\partial E}{\partial w} (t)$

  For all weights and biases
    If $\left(\frac{\partial E}{\partial w_i} (t-1) * \frac{\partial E}{\partial w_i} (t) > 0 \right)$ then
      $$\Delta_i (t) = min \left(\Delta_i (t-1) * \eta^+, \Delta_{\max} \right)$$

      $$\Delta w_i (t) = -sign \left(\frac{\partial E}{\partial w_i} (t) \right) * \Delta_i (t) \tag{16}$$

      $$w_i (t+1) = w_i (t) + \Delta w_i (t) \tag{17}$$
      $\frac{\partial E}{\partial w_i} (t-1) = \frac{\partial E}{\partial w_i} (t)$
    Else if $\left(\frac{\partial E}{\partial w_i} (t-1) * \frac{\partial E}{\partial w_i} (t) < 0 \right)$ then
      $\Delta_i (t) = \Delta_i (t-1) * \eta^-$
      $\frac{\partial E}{\partial w_i} (t-1) = 0$
    Else if $\left(\frac{\partial E}{\partial w_i} (t-1) * \frac{\partial E}{\partial w_i} (t) = 0 \right)$ then
      $\Delta w_i (t) = -sign \left(\frac{\partial E}{\partial w_i} (t) \right) * \Delta_i (t)$

      $$w_i (t+1) = w_i (t) + \Delta w_i (t) \tag{18}$$
      $\frac{\partial E}{\partial w_i} (t-1) = \frac{\partial E}{\partial w_i} (t)$
    End
  End
Until converged

U that is composed of two clusters, namely X_P corresponding to the positive class and X_N to the negative class, is given by:

$$DB (U) = \frac{\Delta (X_P) + \Delta (X_N)}{\delta (X_P, X_N)} \tag{19}$$

where $\delta (X_P, X_N)$ corresponds to inter-class distance given by:

$$\delta (X_P, X_N) = \frac{1}{|P||N|} \sum_{\substack{x \in P \\ y \in N}} d (x, y) \tag{20}$$

$d (x, y)$ is the Euclidean distance between two samples x and y and $\Delta (X_P)$ represents the intra-class distance given by:

$$\Delta (X_P) = \frac{1}{|X_P| (|X_P| - 1)} \sum_{x, y \in S} d (x, y) \tag{21}$$

with $\Delta(X_N)$ analogously defined. Optimization of the DB index minimizes intra-class distance while it maximizes inter-class distance. Therefore, smaller DB values reflect better clusters.

5 Experimental Results

The two applied methodologies were tested on the data set published by [19] on breast cancer. The data set consists of 24,481 gene expression profiles and 78 samples, 44 of which correspond to patients that remain disease free for a period of at least five years, whereas 34 correspond to patients that relapsed within a period of five years. 293 genes express missing information for all 78 patients and were removed, while other missing values were substituted using Expectation Maximization (EM) imputation [12]. In each iteration cycle, so many genes were removed as for the remaining ones to form the closest power of 2 up to 1,024 surviving genes. Then, 124 genes were removed to end up with 900 surviving features. From this point on and up to 200 surviving genes, 100 genes were eliminated in each iteration cycle up to 200 genes. From 200 surviving genes down to 100, ten genes were eliminated (per iteration) and, finally, after 100 surviving genes one gene was eliminated (per iteration) up to the end of the process. With this scalable scheme, elimination is refined as we proceed towards smaller number of surviving genes.

For the conducted experiments 150 epochs were used to train the linear neuron with all its weights and biases initialized to zero along with the parameter values given in Sect. 0. Concerning RFE-SVM the C value, (6), was set to 1,000. We emphasize that various C values were tested such as (1, 10, 100 and 1,000) and resulted in no significant differentiation on the performance of the algorithm, values of less than 1 did not give any better results.

TIGR Mev 4.0 was used as our clustering and gene expression visualization tool, JMP6 statistical software was used to derived the produced Kaplan–Meier curves in subsequent sections, while Matlab 7.01 was our implementation platform.

5.1 Internal Leave One Out Experiment

The ILOOCV is used as a measure of estimating the learning ability of the studied methodologies rather than as a measure of independent generalization performance. Besides ILOOCV in these experiments we also measure the quality of the selected features in terms of low intra but high inter class distance. The DB index is used as a measure to asses clustering performance, where the lower value reflects the better the cluster of markers.

Accuracy and quality results of the two tested methodologies are depicted in Fig. 3a,b respectively. Concering ILOOCV (Fig. 3a) both methodologies show more or less about the same performance on the average up to the point of 20 surviving features. After that point a minor advantage comes with

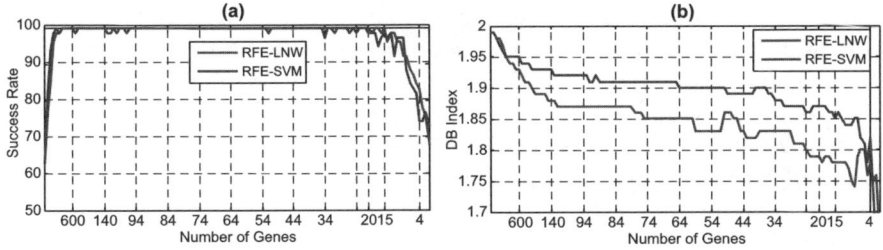

Fig. 3. ILOOCV and Quality results of the two tested methodologies

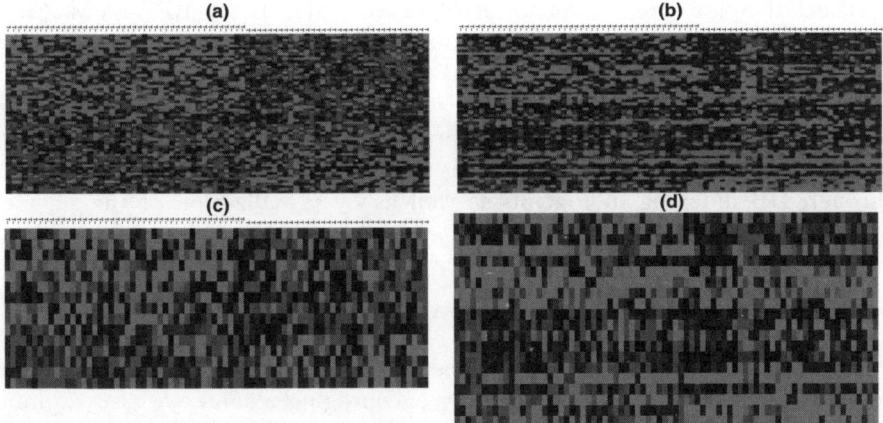

Fig. 4. Sets of markers selected by the two tested methodologies, RFE-LNW (*left*) and RFE-SVM (*right*)

RFE-LNW. Another important point is that RFE-SVM finally selects 20 genes with a 100% classification accuracy, while RFE-NNW selects a smaller number of genes (15) with a perfect classification accuracy as well. Thus, as a final conclusion on ILOOCV we state that there is minor advantage for RFE-LNW since it selects a smaller number of genes as markers.

Concerning the quality of the selected markers through out the entire feature elimination process (Fig. 3b) we notice a significant advantage of RFE-LNW over RFE-SVM. Thus in terms of cluster quality RFE-LNW produces more compact and well distinguished clusters of markers.

5.2 Expression Profile

In order to emphasize the difference of the two tested methodologies through visualization of the results we demonstrate in Fig. 4 two sets of markers selected by the two tested methodologies; genes are ranked according to Fisher's correlation. In Fig. 4a, b the expression profile of the 60 gene marker signatures selected by RFE-LNW and RFE-SVM are depicted respectively. In these two figures we notice the advantage of RFE-LNW over RFE-SVM to select more compact and distinctive clusters of marker genes. This can be verified by noticing that the 60 gene signature selected by RFE-LNW (Fig. 4a) defines more compact and well defined regions of gene expressions, where we can discriminate four different and clear areas of expression levels. These areas define the difference on the behavior of the selected markers between the two classes of interest, that is; genes that are green in negative class are red in positive and vise versa. On the other hand, the 60 gene signature selected by RFE-SVM (Fig. 4b) is showing a lower level of distinction between positive and negative class than RFE-LNW. The different areas of expression levels are not as distinct and well defined as before. This observation is also verified by the DB index, being 1.84 for RFE-LNW and 1.91 for RFE-SVM. Similar observations are also verified through (c) and (d) parts of Fig. 4 where the expression level of the final 15 and 20 markers selected through ILOOCV of RFE-LNW and RFE-SVM methods are depicted, respectively. Overall we emphasize that RFE-LNW produces more compact and distinctive clusters of gene signatures than RFE-SVM. This fact is verified by the results derived through DB indexing in Fig. 3b as well as by visualization of the selected markers, Fig. 4.

5.3 Expression Profiles as Survival Predictors

Inspired by the work of [19], as well as [18] where the expression profile was used as a survival predictor with very encouraging results, we also examine the survival prediction ability of the two expression profiles derived by the two methodologies i.e., the 15 and 20 marker signatures. For this experiment we conducted a Self Organizing Map (SOM) clustering on the derived profiles as an attempt to discover the two classes of patients that are hidden behind those two profiles. The two clusters derived by each of the two methodologies are depicted in Figs. 5 and 6 (rows correspond to genes, columns to patients).

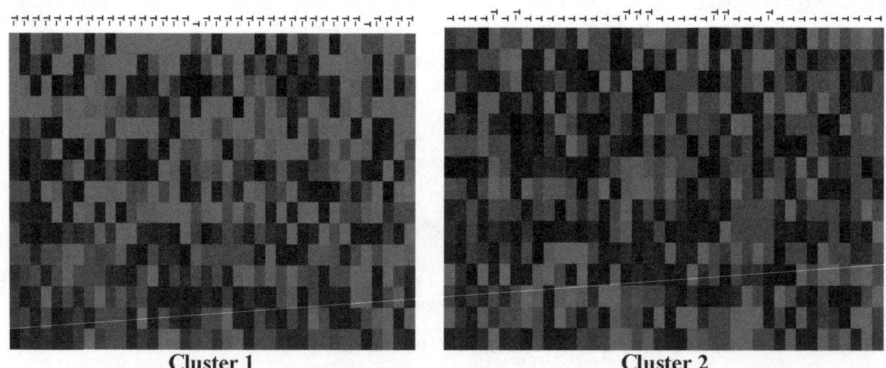

Cluster 1   Cluster 2

Fig. 5. The two clusters discovered by SOM using the 15 marker signature selected by RFE-LNW

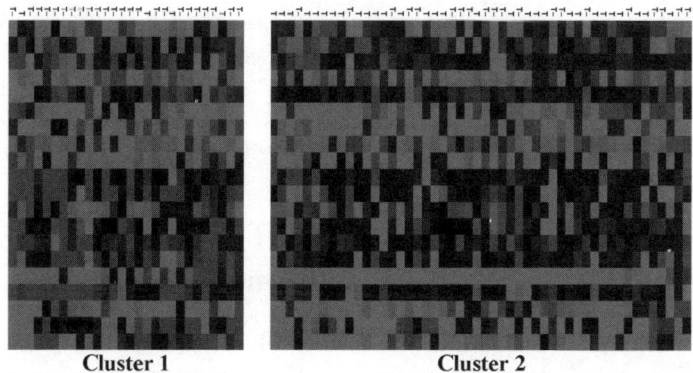

Fig. 6. The two clusters discovered by SOM using the 20 marker signature selected by RFE-SVM

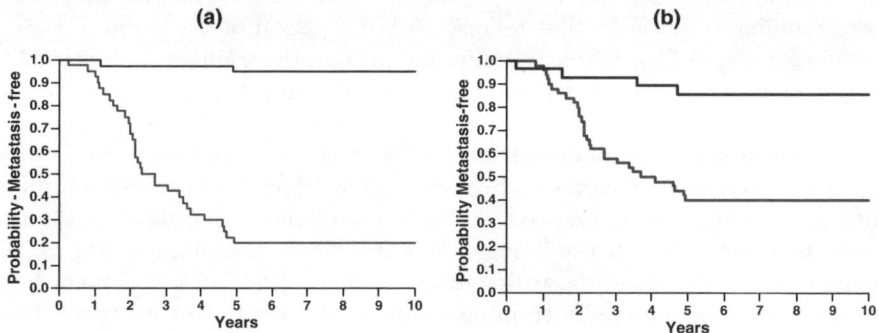

Fig. 7. Kaplan–Meier survival curves of the two tested methodologies: RFE-LNW (**a**), RFE-SVM (**b**)

The SOM was trained using 2×10^5 epochs with the Euclidean distance and average dot product similarities measures for RFE-LNW and RFE-SVM profiles, respectively. The applied similarity measures were experimentally found to produce best results on the examined expression profiles.

As a next step, we used the follow up times provided for the examples under consideration as well as the cluster label to which each example was found to belong, in order to draw the Kaplan–Meier survival curves for the corresponding two expression profiles in Fig. 7. It can be verified that the marker signature derived by RFE-LNW is a better survival predictor than the expression profile derived through RFE-SVM in Fig. 7, where blue curves correspond to good prognosis and red to poor prognosis group.

We emphasize here that the profile selected by RFE-LNW is an efficient survival predictor since the conducted SOM clustering discovered the two classes with a high accuracy of 87%, i.e. 36/44 (82%) negative samples and 32/34 (94%) positive samples were successfully clustered. On the other hand the profile derived via RFE-SVM clustered correctly the 69% of samples, i.e., 24/44 (55%) of negative samples and 30/34 (88%) of positive samples. It is worth noticing that even though the clustering of the expression profile derived by RFE-SVM is poor in accuracy compared to that derived by RFE-LNW, it is good in survival prediction since its sensitivity (percentage of true positive samples) is high.

6 Independent Test Set Evaluation

In this experiment we examine the ability of the two tested methodologies to generalize on new or unseen data. For the purpose of such an experimentation we used the independent data set published by [19] which consists of 19 samples, 7 of which are characterized negative and correspond to patients that remain metastasis free for a period of at least five years, and 12 positive corresponding to patients that relapse within a period of five years.

At each stage of the feature elimination process the training set was used to derive the weight associated with each gene and, thus, the discriminating hyperplane between the two classes. The success rate of the derived hyperplane was then tested on the independent test set. Lowest ranked genes were afterwards and as described earlier in Sect. 0 eliminated and the process continued until the gene list was exhausted. An overview of the performance of the two tested methodologies on the independent test set is visualized in Fig. 8. We demonstrate performance up to 10 surviving genes, since after this point both methods start to misclassify training samples (apparent error is greater than zero). Examining the figure above we notice that the generalization ability of RFE-LNW on unseen data is remarkable, preserving its performance within the range of 100–45 surviving features to at least at 80%. Within this interval it achieves a 94.74% success rate, which corresponds to one missed sample; the same performance is also achieved within the range of 64–62 genes (blue

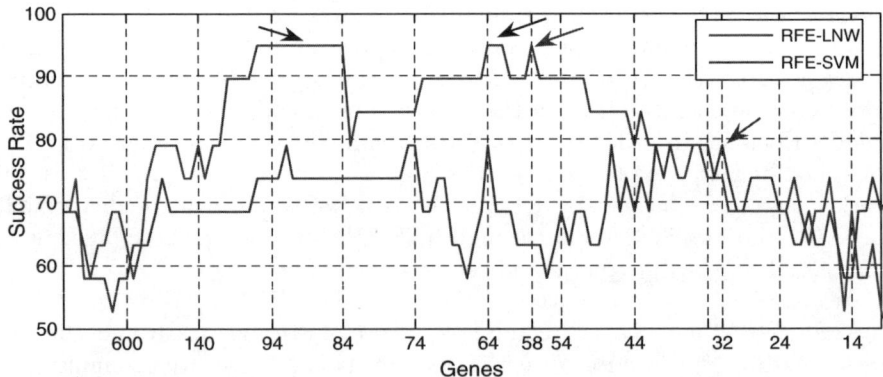

Fig. 8. Independent test set evaluation of RFE-LNW and RFE-SVM

arrows). Finally the system selects 58 genes as markers with 94.74% success rate (red arrow), since this is the smallest set of genes achieving the highest classification accuracy. On the other hand the performance of RFE-SVM is not as high, the maximum success rate achieved being 78.95% (4 missed samples). This system selects 32 genes as markers with 78.95% accuracy (green arrow). Within the range of 40–10 surviving genes the performances of the two methods are almost indistinguishable.

We emphasize that the result produced by RFE-LNW (58 markers, 94.74% success rate) in terms of accuracy is comparable to the 89.47% achieved by [19]. with 70 genes, to the 89.47% achieved by [16] with 44 genes, as well as to the 89.47% achieved by [11] with eight genes. Unfortunately in the last two studies cited above we are not given the gene name that achieved those remarkable performances. Sequence numbers and systematic gene names of the 58 marker genes selected by RFE-LNW and the 32 selected by RFE-SVM are listed on Table 3 and 4 of the appendix.

6.1 Expression Profile as a Survival Predictor

In this experiment we examine the expression profile of the marker signature selected by each of the two tested methodologies (the 58 markers selected by RFE-LNW Table 3 and the 32 selected by RFE-SVM Table 4), from the

Table 3. Sequence numbers and systematic gene names of the 58 genes selected by RFE-LNW

Sequence number	Gene name	Sequence number	Gene name
196	AB033065	10,706	NM_004953
274	X89657	10,889	AL080059
407	NM_003061	11,711	AL080109
462	NM_003079	11,993	NM_007202
1,452	NM_019886	12,259	NM_006544
1,455	Contig30646_RC	12,794	NM_016023
2,131	X05610	12,796	NM_016025
2,603	NM_003331	13,037	NM_007351
2,744	Contig44278_RC	13,270	Contig5456_RC
3,232	NM_020123	13,800	Contig47544_RC
3,670	M26880	14,968	Contig2947_RC
3,737	U72507	15,199	Contig51847_RC
3,953	Contig57447_RC	15,801	Contig39673_RC
4,966	AB018337	16,223	NM_016448
5,088	NM_012325	16,474	Contig14882_RC
5,623	NM_003674	16,669	Contig51158_RC
6,239	NM_021182	16,952	NM_015849
7,126	NM_005243	17,375	NM_018019
7,509	NM_003882	17,571	NM_018089

Table 3. (*Continued*)

Sequence number	Gene name	Sequence number	Gene name
8,024	NM_004604	17,618	NM_015910
8,071	NM_013360	17,874	AK000903
8,255	NM_005371	18,891	NM_000127
8,305	AF094508	18,898	NM_018241
8,826	Contig29820_RC	19,462	Contig58156_RC
8,982	NM_004703	20,891	BE739817_RC
9,054	NM_004721	21,350	NM_000436
9,348	Contig53480_RC	21,919	NM_001204
9,735	NM_005551	22,333	Contig54041_RC
10,477	AB002297	23,665	Contig23964_RC

Table 4. Sequence bumbers and systematic gene names of the 32 genes selected by RFE-SVM

Sequence number	Gene name	Sequence number	Gene name
196	AB033065	10,029	Contig50396_RC
447	NM_001615	10,643	NM_020974
1,148	NM_003147	10,656	NM_006398
1,348	NM_019851	10,889	AL080059
1,409	NM_001756	11,671	NM_005794
1,707	Contig14836_RC	11,780	NC_001807
1,998	NM_003283	12,297	NM_006551
2,563	Contig23399_RC	12,709	Contig11072_RC
3,828	Contig16202_RC	13,140	NM_014665
4,350	NM_002809	16,690	Contig46304_RC
4,601	Contig29617_RC	16,777	Contig10750_RC
6,747	AF131741	18,312	NM_000067
7,404	Contig31000_RC	20,265	Contig24609_RC
8,335	Contig7755_RC	20515	NM_001062
8,513	AF221520	23,239	Contig53371_RC
9,730	Contig48328_RC	23,256	Contig50950_RC

training set, on the *19 samples of the independent test set* using hierarchical clustering. We did not use SOM clustering as we did in Sect 0, since the number of samples is relatively small yielding unstable clusters, i.e. significantly different clustering results derived from run to run. In order to draw safer and more objective conclusions, hierarchical clustering is used in place of the SOM which is producing the results depicted in Fig. 9. Comparing the two methodologies through Fig. 9, we notice that in case of RFE-LNW Fig. 9a two well defined and distinctive clusters are revealed by the hierarchical clustering procedure. However, we can not state the same for the markers

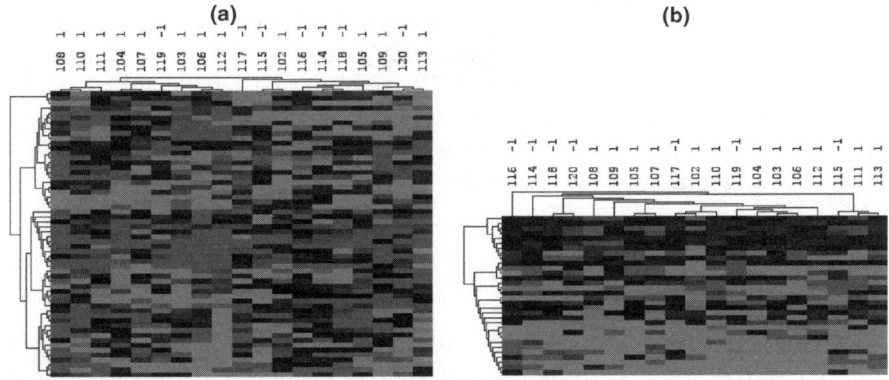

Fig. 9. Hierarchical clustering result on the independent test of the markers selected by RFE-LNW (**a**) and the corresponding Kaplan–Meier curve (**b**)

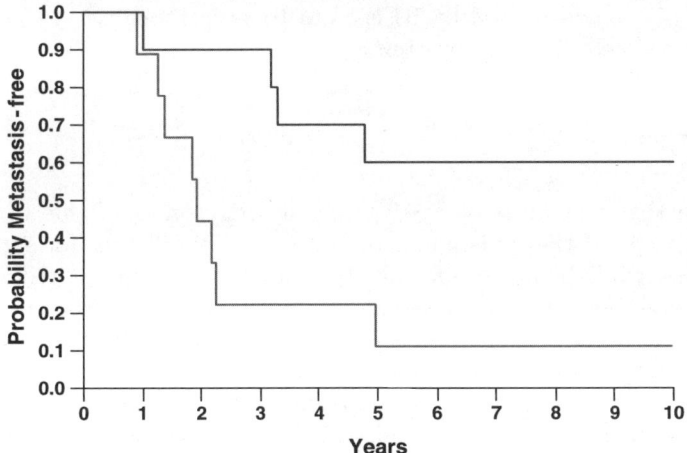

Fig. 10. Kaplan–Meier curve of the clustering result produced by RFE-LNW marker signature

derived through RFE-SVM, Fig. 9b. Clusters are not as well defined and distinctive as in the case of RFE-LNW, making our decision very subjective as to where the line that discriminates the two hidden clusters is lying. This result was derived using average dot product as a similarity measure with average linkage, we should point out that the same scenario repeated itself using any known similarity metric towards our attempt to locate two well established and distinctive clusters on the markers of RFE-SVM.

Proceeding with the clustering derived by RFE-LNW, we use the given follow up times and the cluster label each sample was assigned to, in order to derive the Kaplan–Meier survival curve depicted in Fig. 10, where we notice a significant differentiation on the survival prediction corresponding to the two clusters. One may argue that survival expectation of the good prognosis group (blue curve) is poor since it reaches only 60%, at the end. Nevertheless, the independent test set is relatively small (19 samples) to derive accurate survival estimates, and even in this case the curve still lies above the 70% up to the period of 5 years which is the most crucial one, discriminating significantly in any case the two prognostic clusters.

7 Elaboration on the Results

An open issue which has not being resolved yet is to provide a theoretical explanation of the produced results. A question of particular interest that needs to be answered is: why RFE-LNW produces more compact and distinctive clusters of markers than RFE-SVM? To answer this question we have to take into consideration the training procedures of the underlined methodologies. Equation (4), which is used by RFE-SVM to assign weights to the surviving genes can be accordingly re-written as:

$$w_i = \sum_{j=1}^{n} \lambda_j y_j g_{ij} \tag{22}$$

We notice that the value of w_i depends strongly on the value of λ_j which according to SVM theory is non-zero and probably different for support vector samples, while being zero for all remaining ones. Thus, weight w_i of gene g_i is a summation of only those samples whose λ_i is non-zero i.e. the support vectors. We stress out that this (under certain conditions) is one of the strong points of SVMs since only a small proportion of samples (the support vectors) are actually responsible for training the system. At this point we should emphasize that this fact might influence the algorithm especially in the problem of gene selection where besides accuracy we are also interested in selecting differentially expressed genes. The support vector samples lie close to the borders of the separating hyperplane (see Fig. 1) and thus there might exist cases where values of g_{ij} (expression level of gene i in sample j) are not significantly differentiated from negative to positive class, since they might lie very close to the border and reside close to each other. Taking this into account, along with the fact that support vectors should and are only a small proportion of the training samples we can address the inability of RFE-SVM to provide significantly differentiated expressed genes (on the studied domain) as an effect of the limited data samples that dominate the selection process. In essence, the requirement for differential expression is much different from that of accurate classification, which is the main objective of SVMs in the definition of the separating hyperplane. In any case, we emphasize that a more theoretical and mathematically rigorous study is still needed to address the problem which however, is beyond the scope of the present chapter.

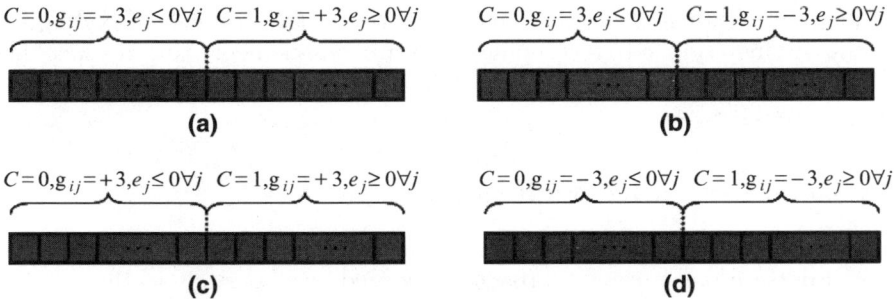

Fig. 11. Differentially versus non-differentially expressed genes

In RFE-LNW on the other hand, RPROP is used to train the linear neuron. Consider the RPROP algorithm presented in Table 2 and notice that (16) and (17) in combination with (14) can be written as:

$$w_i(t+1) = w_i(t) + \left[\mu \sum_{j=1}^{n} sign\left(e_j f'(u_j) g_{ij}\right) \right] \Delta_i(t) \qquad (23)$$

Consider also Fig. 11 which illustrates the expression level of a hypothetical gene g_i in negative ($C = 0$) and positive ($C = 1$) class respectively. In cases (a) and (b) the hypothetical gene is differentially expressed in the two classes of interest, green (negative values) in negative class and red (positive values) in positive class or vies versa. In contrast cases (c) and (d) show no differentiation on the expression level of the specific gene in the two situations of interest. Considering case (a) in combination with (23) and focusing on the negative class (green part), we notice that the term $sign\left(e_j f'(u_j) g_{ij}\right) \geq 0$ holds indeed, e_j is ≤ 0 (notice that $d = 0$ in this case and y takes values $[0 \cdots 1]$), $f'(u_j) \geq 0$ from (11) is always positive since again y ranges from $[0 \cdots 1]$ and $g_{ij} = -3$. Thus the overall weight update is positive. Now focusing on the positive class (red part) of case (a) and using the same syllogism we notice again that $sign\left(e_j f'(u_j) g_{ij}\right) \geq 0$ and the weight updata is again positive. Following about the same reasoning in case (b) of Fig. 11 one may show that (23) will produce a negative update. In contrast the summation term of the same equation in cases (c) and (d) will produce a zero update. Thus, genes that differentiate their expression between the two classes are assigned higher weights in absolute value.

The difference of the two tested methodologies as it was expected is mostly attributed to the learning procedure employed by each of the two methods. SVMs are systems which have shown remarkable performance on a variety of application domains but depending on the application and the peculiarities of the problem under consideration they have weak and strong points. The main difference of the two tested methodologies is that RFE-LNW is implicitly searching for differentially expressed genes while RFE-SVM along with the majority of wrapper methods, focuses mostly on the problem of finding the 'optimum' hyperplane to discriminate the two classes, neglecting the fact that selected genes should significantly differentiate their expression from positive to negative class.

8 Discussion and Conclusions

In this study we examine the behavior of two well known and broadly accepted pattern recognition approaches, adopted appropriately to the problem of marker selection in a micro-array experiment. Concerning Support Vector Machines, a linear kernel SVM was applied through the RFE-SVM method. In association to neural networks, an equivalent to the linear kernel, i.e. a linear neuron, was applied through RFE-LNW method. Experiments were conducted on the well known data set of breast cancer published by [19]. We emphasize that to our knowledge this is the first attempt to utilize and test the linear neuron as a gene marker selector trained using the RPROP algorithm [15], which is the key for the performance of RFE-LNW.

Experimental results demonstrate that RFE-LNW produces more compact and distinctive clusters of marker signatures, which in turn yield better survival prediction when the expression profiles of the selected markers is taken into consideration. Also RFE-LNW produced remarkable results on the accuracy performance of the independent test set, comparable to the best reported in the international bibliography. The key to the encouraging performance of RFE-LNW is that through the RPROP learning procedure it searches and finally selects markers that differentiate their expression significantly between the two classes of interest. We believe that this fact should be farther emphasized since expression profile plays a critical role to gene selection problem as it is underlined by a great number of domain expert publications, a number of which are cited in the introduction of the present study.

Based on above discussion we stress that wrapper methods will become much more valuable tools at the hands of domain experts when they explicitly search and finally succeed to select marker genes that significantly differentiate their expression across the classes of interest.

Appendix

Acknowledgements

Present work was supported by Biopattern, IST EU funded project, Proposal/ Contract no.: 508803 as well as the Hellenic Ministry of Education.

References

1. Alizadeh A, Eisen M, Davis RE, et al. (2000) Distinct substypes of diffuse large B-cell lymphoma identified by gene expression profiling. Nature, 403:503–511

2. Ambroise C, McLachlan G (2002) Selection bias in gene extraction on the basis of microarray gene-expression data. PNAS, 99:6562–6566

3. Armstrong S, Staunton J, Silverman L, et al. (2002) MLL translocations specify a distinct gene expression profile that distinguishes a unique leukemia. Nature Genetics, 30:41–47

4. Azuaje F (2002) A cluster validity frame work for genome expression data. Bionformatics, 18:319–320

5. Bandyopadhyay S, Maulik U (2001) Nonparametric genetic clustering of validity indices. IEEE Transactions on Systems, Man, and Cybernetics, 31:120–126

6. Boyd S, Vandenberghe L (2004) Convex Optimization. Oxford University Press, Oxford

7. Davie D, Bouldin, DW. (1979) A cluster separation measure. IEEE Transactions on Pattern Analysis and Machine Intelligence, PAMI, 1:224–227

8. Golub TR, Slonim DK, Tamayo P, et al. (1999) Molecular classification of cancer: class discovery and class prediction by gene expression monitoring. Science, 286:531–536

9. Guyon I, Weston J, Vapnik V (2002) Gene selection for cancer classification using support vector machines. Machine Learning, 36:389–422

10. Hestie T, Tibshirani R, Eisen MB, et al. (2000) Gene shaving as a method for identifying distinct set of genes with similar expression patterns. Journal of Genome Biology, 1(2):1–21

11. Li F, Yang Y (2005) Analysis of recursive gene selection approaches from microarray data. Bioinformatics, 21, 3741–3747

12. Little A, Rubin D (1987) Statistical Analysis with Missing Data. Wiley Series in Probability and Mathematical Statistics. Wiley, New York

13. Nutt C, Mani D, Betensky R, et al. (2003) Gene expression-based classification of malignant gliomas correlates better with survival than histological classification. Cancer Research, 63:1602–1607

14. Ramaswamy S, Ross K, Lander E, et al. (2003) A molecular signature of metastasis in primary solid tumors. Nature Genetics, 33:49–54

15. Riedmiller M, Braun H (1993) A direct adoptive method for faster backpropagation learning: The RPROP algorithm. In: Proceedings of the IEEE International Conference on Neural Networks (ICNN), 586–591

16. Shen R, Ghosh D, Chinnaiyan A, et al. (2006) Eigengene-based linear discriminant model for tumor classification using gene expression microarray data. Bioinformatics, 22:2635–2642

17. Simon R, Radmacher M, Dobbin K, et al. (2003) Pitfalls in the use of DNA microarray data for diagnostic and prognostic classification. Journal of the National Cancer Institute, 95:4–18

18. Van De Vijver MJ, He YD, Van't Veer LJ, et al. (2002) A gene expression signature as a predictor of survival in breast cancer. The New England Journal of Medicine, 347:1999–2009

19. Van't Veer LJ, Dai H, Van de Vijver, et al. (2002) Gene expression profiling predicts clinical outcome of breast cancer. Letters to Nature, 415:530–536

20. Vapnik NV (1999) The Nature of Statistical Learning Theory. Springer, Berlin Heidelberg New York

21. Vesanto J, Alhoniemi E (2000) Clustering of the self organizing map. IEEE Transactions on Neural Networks, 11:586–600

22. Wang J, Delabie J, Aashein H, Smeland E, Myklebost O (2002) Clustering of the SOM easily reveals distinct gene expression patterns: results of a reanalysis of lymphoma study. BMC Bioinformatics, 3:http://www.biomedcentral.com/1471-2105/3/36

23. Wang Y, Makedon F, Ford J, et al. (2004) HykGene: a hybrid approach for selecting marker genes for phenotype classification using microarray gene expression data. Bioinformatics, 21(8):1530–1537

An Intelligent Decision Support System in Wireless-Capsule Endoscopy

V.S. Kodogiannis[1], J.N. Lygouras[2], and Th. Pachidis[3]

[1] Centre for Systems Analysis, School of Computer Science, University
of Westminster, London, HA1 3TP, UK
kodogiv@wmin.ac.uk
[2] Laboratory of Electronics, School of Electrical and Computer Engineering,
Democritus University of Thrace, Xanthi, GR-67100, Greece
ilygour@ee.duth.gr
[3] Dept. of Petroleum and Natural Gas Technology, Kavala Institute of Technology
and Education, Kavala GR-65404, Greece
pated@otenet.gr

Summary. In this chapter, a detection system to support medical diagnosis and
detection of abnormal lesions by processing endoscopic images is presented. The
endoscopic images possess rich information expressed by texture. Schemes have been
developed to extract texture features from the texture spectra in the chromatic and
achromatic domains for a selected region of interest from each colour component
histogram of images acquired by the new M2A Swallow-able Capsule. The implementation of advanced neural learning-based schemes and the concept of fusion of
multiple classifiers dedicated to specific feature parameters have been also adopted
in this chapter. The test results support the feasibility of the proposed methodology.

1 Introduction

Medical diagnosis in clinical examinations highly relies upon physicians'
experience. For physicians to quickly and accurately diagnose a patient there
is a critical need in the area of employing computerised technologies to assist
in medical diagnosis and to access the related information. Computer-assisted
technology is certainly helpful for inexperienced physicians in making medical diagnosis as well as for experienced physicians in supporting complex
diagnosis. Such technology has become an essential tool to help physicians
in retrieving the medical information and making decisions in medical diagnosis [1]. A number of medical diagnostic decision support systems (MDSS)
based on Computational Intelligent methods have been developed to assist
physicians and medical professionals. Some medical diagnosis systems based
on computational intelligence methods use expert systems (ESs) [2] or neural
networks [3].

V.S. Kodogiannis et al.: *An Intelligent Decision Support System in Wireless-Capsule
Endoscopy*, Studies in Computational Intelligence (SCI) **109**, 259–275 (2008)
www.springerlink.com © Springer-Verlag Berlin Heidelberg 2008

Endoscopy differs from traditional medical imaging modalities in several aspects. First, Endoscopy is not based on the biophysical response of organs to X-ray or ultrasounds, but allows a direct observation of the human internal cavities via an optical device and a light source. Second, the Endoscopy investigation imposes a physical contact between the patient and the physician, and the endoscopist can assess the patient complaints before the endoscopic procedure. Finally, the patient's discomfort during the investigation prohibits repeated examinations and, with regard to the usual absence of storage system, no decision element remains available at the end of the examination; this requires that all information is gathered during a limited time period. For more than 10 years, flexible video-endoscopes have a widespread use in medicine and guide a variety of diagnostic and therapeutic procedures including colonoscopy, gastroenterology and laparoscopy [4].

Conventional diagnosis of endoscopic images employs visual interpretation of an expert physician. Since the beginning of computer technology, it becomes necessary for visual systems to "understand a scene", that is making its own properties to be outstanding, by enclosing them in a general description of an analysed environment. Computer-assisted image analysis can extract the representative features of the images together with quantitative measurements and thus can ease the task of objective interpretations by a physician expert in Endoscopy. A system capable to classify image regions to normal or abnormal will act as a second – more detailed – "eye" by processing the endoscopic video.

From the literature survey, it has been found that only a few techniques for endoscopic image analysis have been reported and they are still undergoing testing. In addition, most of the techniques were developed on the basis of features in a single domain: chromatic domain or spatial domain. Applying these techniques individually for detecting the disease patterns based on possible incomplete and partial information may lead to inaccurate diagnosis. Krishnan et al. [5] have been using endoscopic images to define features of the normal and the abnormal colon. New approaches for the characterisation of colon based on a set of quantitative parameters, extracted by the fuzzy processing of colon images, have been used for assisting the colonoscopist in the assessment of the status of patients and were used as inputs to a rule-based decision strategy to find out whether the colon's lumen belongs to either an abnormal or normal category. Endoscopic images contain rich information of texture. Therefore, the additional texture information can provide better results for the image analysis than approaches using merely intensity information. Such information has been used in CoLD (colorectal lesions detector) an innovative detection system to support colorectal cancer diagnosis and detection of precancerous polyps, by processing endoscopic images or video frame sequences acquired during colonoscopy [6]. It utilised second-order statistical features that were calculated on the wavelet transformation of each image to discriminate amongst regions of normal or abnormal tissue. A neural network based on the classic Back-propagation learning algorithm performed the classification of the features.

Intra-operative Endoscopy, although used with great success, is more invasive and associated with a higher rate of complications. Though the gastrointestinal (GI) endoscopic procedure has been widely used, doctors must be skilful and experienced to reach deep sites such as the duodenum and small intestine. The cleaning and sterilisation of these devices is still a problem leading to the desire for disposable instruments. Standard endoscopic examinations evaluate only short segments of the proximal and distal small bowel and barium follow-through has a low sensitivity and specificity of only 10% for detecting pathologies. Hence, endoscopic examination of the entire small bowel has always been a diagnostic challenge. Limitations of the diagnostic techniques in detection of the lesions located in the small bowel are mainly due to the length of the small intestine, overlying loops and intra-peritoneal location. This caused also the desire for autonomous instruments without the bundles of optical fibres and tubes, which are more than the size of the instrument itself, the reason for the objections of the patients. The use of highly integrated microcircuit in bioelectric data acquisition systems promises new insights into the origin of a large variety of health problems by providing lightweight, low-power, low-cost medical measurement devices.

At present, there is only one type of microcapsule which has been introduced recently to improve the health outcome. This first swallow-able video-capsule for the gastroenterological diagnosis has been presented by Given Imaging, a company from Israel, and its schematic diagram is illustrated in Fig. 1 [7]. The system consists of a small swallow-able capsule containing a battery, a camera on a chip, a light source, and a transmitter. The camera-capsule has a one centimetre section and a length of three centimetres so it can be swallowed with some effort. In 24 hours, the capsule is crossing the patient's alimentary canal. For the purpose of this research work, endoscopic images have been obtained using this innovative endoscopic device. They have spatial resolution of 171×151 pixels, a brightness resolution of 256 levels per colour plane (8bits), and consisted of three colour planes (red, green and blue) for a total of 24 bits per pixel.

Texture analysis is one of the most important features used in image processing and pattern recognition. It can give information about the arrangement and spatial properties of fundamental image elements. Many methods have been proposed to extract texture features, e.g. the co-occurrence matrix [8]. The definition and extraction of quantitative parameters from

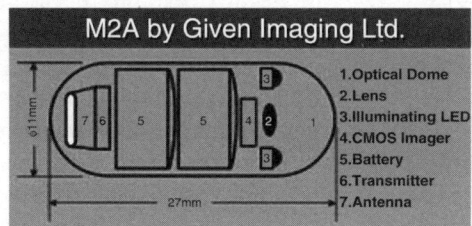

Fig. 1. Given imaging capsule

capsule endoscopic images based on texture information has been proposed. This information was initially represented by a set of descriptive statistical features calculated on the histogram of the original image. The implementation of an intelligent diagnostic system was based on various learning schemes such as Radial Basis Functions [9], Adaptive Fuzzy Logic Systems [10] and Fuzzy Inference Neural Networks [11].

The main objective of this research work which was supported by the "IVP" research project is to design a completed diagnostic software tool for the analysis and processing of wireless-capsule endoscopic images. The proposed methodology is considered in two phases. The first implements the extraction of image features while in the second phase an advanced neural network is implemented/employed to perform the diagnostic task. The extraction of quantitative parameters/features from endoscopic images based on texture information in the chromatic and achromatic domain will be considered. Emphasis has to be given to the development of a reliable but also fast pre-processing procedure. Two methodologies will be investigated. Statistical features calculated on the histogram of the original image and on its (N_{TU} – Texture spectrum) transformation. For the diagnostic part, the concept of multiple-classifier scheme is adopted, where the fusion of the individual outputs is realised using fuzzy integral. An intelligent classifier-scheme based on modified Extended Normalised Radial Basis Function (ENRBF) neural networks enhanced with split/merge issues has been also implemented while is then compared with an Radial Basis Function (RBF) network. Finally a user-interface is developed to be utilized by medical physicians, which among other functionalities integrates an intelligent diagnostic sub-system.

2 Image Features Extraction

A major component in analysing images involves data reduction which is accomplished by intelligently modifying the image from the lowest level of pixel data into higher level representations. Texture is broadly defined as the rate and direction of change of the chromatic properties of the image, and could be subjectively described as fine, coarse, smooth, random, rippled, and irregular, etc.

For this reason, we focused our attention on nine statistical measures (standard deviation, variance, skew, kurtosis, entropy, energy, inverse difference moment, contrast, and covariance) [12].

All texture descriptors are estimated for all planes in both RGB {R (Red), G (Green), B (Blue)} and HSV {H (Hue), S (Saturation), V (Value of Intensity)} spaces, creating a feature vector for each descriptor $D_i = (R_i, G_i, B_i, H_i, S_i, V_i)$. Thus, a total of 54 features (9 statistical measures x 6 image planes) are then estimated. For our experiments, we have used 70 endoscopic images related to abnormal cases and 70 images related to normal ones. Figure 2 shows samples of selected images acquired using the M2A capsule of normal and abnormal cases. Generally, the statistical measures are estimated on histograms of the original image (1st order statistics) [12].

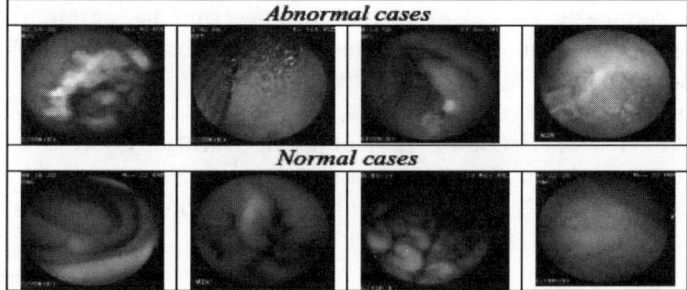

Fig. 2. Selected endoscopic images of normal and abnormal cases

However, the histogram of the original image carries no information regarding relative position of the pixels in the texture. Obviously this can fail to distinguish between textures with similar distributions of grey levels. We therefore have to implement methods which recognise characteristic relative positions of pixels of given intensity levels. An additional scheme is proposed in this study to extract new texture features from the texture spectra in the chromatic and achromatic domains, for a selected region of interest from each colour component histogram of the endoscopic images.

2.1 N_{TU} Transformation

The definition of texture spectrum employs the determination of the texture unit (TU) and texture unit number (N_{TU}) values. Texture units characterise the local texture information for a given pixel and its neighbourhood, and the statistics of all the texture units over the whole image reveal the global texture aspects. Given a neighbourhood of $\delta \times \delta$ pixels, which are denoted by a set containing $\delta \times \delta$ elements $P = \{P_0, P_1, \ldots, P_{(\delta \times \delta)-1}\}$, where P_0 represents the chromatic or achromatic (i.e. intensity) value of the central pixel and $P_i\{i = 1, 2, \ldots, (\delta \times \delta) - 1\}$ is the chromatic or achromatic value of the neighbouring pixel i, the $TU = \{E_0, E_1, \ldots, E_{(\delta \times \delta)-1}\}$, where $E_i\{i = 1, 2, \ldots, (\delta \times \delta) - 1\}$ is determined as follows:

$$E_i = \begin{cases} 0, & if \quad P_i < P_0 \\ 1, & if \quad P_i = P_0 \\ 2, & if \quad P_i > P_0 \end{cases} \tag{1}$$

The element E_i occupies the same position as the ith pixel. Each element of the TU has one of three possible values; therefore the combination of all the eight elements results in 6,561 possible TU's in total. The texture unit number (N_{TU}) is the label of the texture unit and is defined using the following equation:

$$N_{TU} = \sum_{i=1}^{(\delta \times \delta)-1} E_i \times \delta^{\iota - 1} \tag{2}$$

Where, in our case, $\delta = 3$. The texture spectrum histogram $(Hist(i))$ is obtained as the frequency distribution of all the texture units, with the abscissa showing the N_{TU} and the ordinate representing its occurrence frequency. The texture spectra of various image components {V (Value of Intensity), R (Red), G (Green), B (Blue), H (Hue), S (Saturation)} are obtained from their texture unit numbers. The statistical features are then estimated on the histograms of the N_{TU} transformations of the chromatic and achromatic planes of the image (R, G, B, H, S, V).

3 Features Evaluation

Recently, the concept of combining multiple classifiers has been actively exploited for developing highly reliable "diagnostic" systems [13]. One of the key issues of this approach is how to combine the results of the various systems to give the best estimate of the optimal result. A straightforward approach is to decompose the problem into manageable ones for several different sub-systems and combine them via a gating network. The presumption is that each classifier/sub-system is "an expert" in some local area of the feature space. The sub-systems are local in the sense that the weights in one "expert" are decoupled from the weights in other sub-networks. In this study, six sub-systems have been developed, and each of them was associated with the six planes specified in the feature extraction process (i.e. R, G, B, H, S, and V). For each sub-system, 9 statistical features have been associated with, resulting thus a total 54 features space. Each sub-system was modelled with an appropriate intelligent learning scheme. In our case, two alternative schemes have been investigated: the modified Extended Normalised Radial Basis Function (ENRBF) neural network and the Radial Basis Function Network (RBF). Such schemes provide a degree of certainty for each classification based on the statistics for each plane. The outputs of each of these networks must then be combined to produce a total output for the system as a whole as can be seen in Fig. 3.

While a usual scheme chooses one best sub-system from amongst the set of candidate sub-systems based on a winner-takes-all strategy, the current

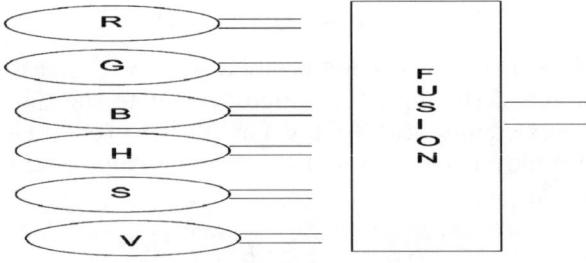

Fig. 3. Proposed fusion scheme

proposed approach runs all multiple sub-systems with an appropriate collective decision strategy. The aim in this study is to incorporate information from each feature space so that decisions are based on the whole input space. The simplest method is to take the average output from each classifier as the system output. This does not take into account the objective evidence supplied by each of the feature classifiers and confidence which we have in that classifiers results. The fuzzy integral is an alternative method that claims to resolve both of these issues by combining evidence of a classification with the systems expectation of the importance of that evidence. The fuzzy integral introduced by Sugeno and the fuzzy measures from Yager are very useful in combining information. A fuzzy measure g_λ is a set function such that:

- The fuzzy measure of an empty set is equal to zero – $g(0) = 0$
- The fuzzy measure of an entire set Q is equal to one – $g(Q) = 1$
- The fuzzy measure of set A is less than or equal to that of set B if A is a subset of $B - g(A) \le g(B)$ if $A \subset B$

This function can be interpreted as finding the maximal grade of agreement between networks' outputs and their fuzzy measures for a particular class. If the following additional property is also satisfied, the fuzzy measure is referred to as a $g_\lambda-fuzzy$ measure.

$$\forall A, B \subset Q \text{ and } A \cap B = 0, g(A \cup B) = g(A) + g(B) + \lambda g(A)g(B),$$
$$\lambda \in (-1, \infty) \tag{3}$$

where the λ measure can be given by solving the following non-linear equation

$$\lambda + 1 = \prod_{i=1}^{n} \left(1 + \lambda g^i\right) \qquad \lambda > -1 \tag{4}$$

When combining multiple NNs, let g^i denote the fuzzy measure of network i. These measures can be interpreted as quantifying how well a network properly classified the samples/patterns. They must be known and can be determined in different ways i.e., the fuzzy measure of a network could equal the ratio of correctly classified patterns during supervised training over the total number of patterns being classified. In this research, each network's fuzzy measure equalled $1 - K_i$, where K_i was network i^{th} overall testing kappa value [14].

A pattern is being classified to one of m possible output classes, c_j for $j = 1, \ldots, m$. The outputs of n different networks are being combined, where NN_i denotes the ith network. First, these networks must be renumbered/rearranged such that their a posteriori class probabilities are in descending order of magnitude for each output class j,

$$y_1(c_j) \ge y_2(c_j) \ge \ldots \ge y_n(c_j)$$

where $y_i(c_j)$ is the ith network's a posteriori class j probability. Next, each network/set of networks' $g_\lambda - fuzzy$ measure is computed for every output class j and is denoted by $g_j(A_i)$. $A_i = \{NN_1, NN_2, \ldots, NN_i\}$ is the set of the first i networks ordered correspondingly to class $j's$ associated a posteriori probabilities. These values can then be computed using the following recursive method,

- $g_j(A_1) = g_j(\{NN_1\}) = g^1$
- $g_j(A_i) = g_j(\{NN_1, \ldots, NN_i\}) = g^i + g(A_{i-1}) + \lambda g^i g(A_{i-1})$ for $1 < i < n$
- $g_j(A_n) = g_j(\{NN_1, \ldots, NN_n\}) = 1$

Finally, the fuzzy integral for each class j is defined as [15],

$$\max_{i=1}^{n} [\min[y_i(c_j), g_j(A_i)]] \tag{5}$$

The class with the largest fuzzy integral value is then chosen as the output class to which the pattern is classified. Equation 5 summarises combining multiple NNs using a fuzzy integral approach.

$$\max_{class} \left[\begin{array}{c} \max_{network} \left[\begin{array}{c} \min[y_1(c_1), g_1(A_1)] \\ \min[y_2(c_1), g_1(A_2)] \\ \ldots\ldots\ldots\ldots \\ \min[y_n(c_1), g_1(A_n)] \end{array} \right] \\ \max_{network} \left[\begin{array}{c} \min[y_1(c_2), g_2(A_1)] \\ \min[y_2(c_2), g_2(A_2)] \\ \ldots\ldots\ldots\ldots \\ \min[y_n(c_2), g_2(A_n)] \end{array} \right] \\ \ldots\ldots\ldots\ldots\ldots\ldots\ldots\ldots \\ \max_{network} \left[\begin{array}{c} \min[y_1(c_m), g_m(A_1)] \\ \min[y_2(c_m), g_m(A_2)] \\ \ldots\ldots\ldots\ldots \\ \min[y_n(c_m), g_m(A_n)] \end{array} \right] \end{array} \right] \tag{6}$$

3.1 Extended Normalised RBF Network

The first classification scheme utilised here is a modified version of the Extended Normalised Radial Basis Function Network (ENRBF) [16], which utilises a series of linear models instead of the linear combiner in an RBF network. The scheme is illustrated in Fig. 4. We propose a supervised training method for this scheme that is fully supervised and self organising in terms of structure. The method incorporates training techniques from Bayesian Ying-Yang (BYY) training which treats the problem of optimisation as one of maximising the entropy between the original non-parametric data distribution based on Kernel estimates or user specified values and the parametric distributions represented by the network. This is achieved through the derivation of a series of Expectation Maximisation (EM) update equations using a

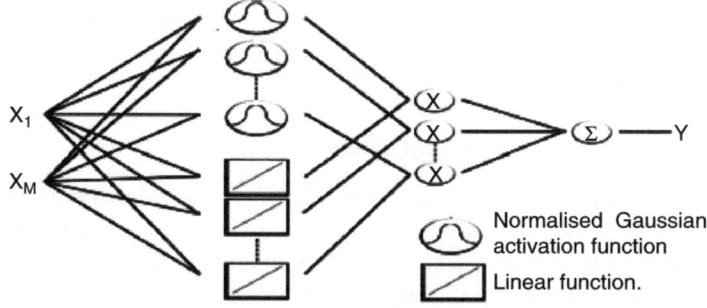

Fig. 4. ENRBF scheme

series of entropy functions as the Q function or log-likelihood function. The ENRBF network can be represented by the following equations.

$$E\left(z|x,\Theta\right) = \frac{\sum\limits_{j=1}^{K}\left(W_j^T x + c_j\right) p\left(x|j,\theta_j\right)}{\sum\limits_{j=1}^{K} p\left(x|j,\theta_j\right)} \tag{7}$$

Where z is the output of the network $z \in Z$, x is an input vector $x \in X$, $\Theta = [W, c, \theta]$ are the network parameters and $\theta = [m, \Sigma]$ are the parameters of the Gaussian activation functions given by:

$$p\left(x|j,\theta_j\right) = \exp\left\{-\tfrac{1}{2}\left(x - m_j\right)\Sigma_j^{-1}\left(x - m_j\right)\right\} \tag{8}$$

The BYY method attempts to maximise the degree of agreement between the expected value of z from the network and the true value of z from the training data. It is guaranteed to lead to a local optimum and unlike the original EM algorithm for learning the parameters of Gaussian functions this method encourages coordination between the input and output domains. Like the EM algorithm this method is also very fast in terms of the number of iterations needed for the parameters to converge.

However, as BYY is an EM based technique it is still susceptible to locally maximal values. The Split and Merge EM (SMEM) concept for Gaussian Mixture Models (GMM) proposed initially by Ueda, has been applied to the ENRBF scheme. The original SMEM algorithm is able to move neurons from over populated areas of the problem domain to underrepresented areas by merging the over populated neurons and splitting the under-populated. The use of Eigenvectors to split along the axis of maximum divergence instead of randomly as in original SMEM has been proposed recently. The SMEM algorithm suffers from the fact that before terminating all possible combinations of Split and Merge operations must be examined. Although many options can be discounted, the training time still increases exponentially with network size and again suffers from problems inherent with k-means and basic EM in that it is essentially unsupervised. In this work we incorporate the supervised nature of BYY training with improved statistical criteria for determining the neurons which poorly fit their local areas of the problem domain.

3.2 Radial Basis Function Networks

MLP network is probably the most widely used neural network paradigm. One disadvantage of this model is the difficulty in classifying a previously unknown pattern that is not classified to any of the prototypes in the training set. RBF networks train rapidly, usually orders of magnitude faster than the classic back-propagation neural networks, while exhibiting none of its training pathologies such as paralysis or local minima problems. An RBF is a function which has in-built distance criterion with respect to a centre. Such a system consists of three layers (input, hidden, output). The activation of a hidden neuron is determined in two steps: The first is computing the distance (usually by using the Euclidean norm) between the input vector and a centre c_i that represents the ith hidden neuron. Second, a function h that is usually bell-shaped is applied, using the obtained distance to get the final activation of the hidden neuron. In this case the Gaussian function G(x)

$$G(x) = \exp\left(-\frac{x^2}{\sigma^2}\right) \tag{9}$$

was used. The parameter σ is called unit width and is determined using the heuristic rule *"global first nearest-neighbour"*. It uses the uniform average width for all units using the Euclidean distance in the input space between each unit m and its nearest neighbour n. All the widths in the network are fixed to the same value σ and this results in a simpler training strategy. The activation of a neuron in the output layer is determined by a linear combination of the fixed non-linear basis functions, i.e.

$$F^*(x) = \sum_{i=1}^{M} w_i \phi_i(x) \tag{10}$$

where $\phi_i(x) = G(\|x - c_i\|)$ and w_i are the adjustable weights that link the output nodes with the appropriate hidden neurons. These weights in the output layer can then be learnt using the least-squares method. The present study adopts a systematic approach to the problem of centre selection. Because a fixed centre corresponds to a given regressor in a linear regression model, the selection of RBF centres can be regarded as a problem of subset selection. The orthogonal least squares (OLS) method can be employed as a forward selection procedure that constructs RBF networks in a rational way. The algorithm chooses appropriate RBF centres one by one from training data points until a satisfactory network is obtained. Each selected centre minimises the increment to the explained variance of the desired output, and so ill-conditioning problems occurring frequently in random selection of centres can automatically be avoided.

4 Results

The proposed approaches were evaluated using 140 clinically obtained endoscopic M2A images. For the present analysis, two decision-classes are considered: abnormal and normal. Seventy images (35 abnormal and 35 normal) were used for the training and the remaining ones (35 abnormal and 35 normal) were used for testing. The extraction of quantitative parameters from these endoscopic images is based on texture information. Initially, this information is represented by a set of descriptive statistical features calculated on the histogram of the original image. Both types of networks (i.e. ENRBF and RBF) are incorporated into a multiple classifier scheme, where the structure of each individual (for R, G, B, H, S, and V planes) classifier is composed of 9 input nodes (i.e. nine statistical features) and 2 output nodes. In a second stage, the nine statistical measures for each individual image component are then calculated though the related texture spectra after applying the (N_{TU}) transformation.

4.1 Performance of Histograms-Based Features

The multiple-classifier scheme using the ENRBF network has been trained on the six feature spaces. The network trained on the R feature space and it then achieved an accuracy of 94.28% on the testing data incorrectly classifying 2 of the normal images as abnormal and 2 abnormal as normal ones. The network trained on the G feature space misclassified 2 normal images as abnormal but not the same ones as the R space. The remaining 3 images were misclassified as normal ones. The B feature space achieved an accuracy of 94.28% on the testing data with 4 misclassifications, i.e. 3 abnormal as normal ones and the remaining one image as abnormal ones. The network trained on the H feature space achieved 91.43% accuracy on the testing data. The network trained on the S feature space achieved an accuracy of only 88.57% on the testing data. Finally, the network for the V feature space misclassified 2 normal cases as abnormal and 2 abnormal as normal ones, giving it an accuracy of 94.28% on the testing data. The fuzzy integral (FI) concept has been used here to combine the results from each sub-network and the overall system misclassified 1 normal cases as abnormal and 3 abnormal as normal ones, giving the system an overall accuracy of 94.28%. These results are illustrated in Fig. 5, while Table 1 presents the performance of individual components. It can be shown that in general the confidence levels for each correct classification is above 0.6.

In a similar way, a multi-classifier consisting of RBF networks with 9 input nodes and 2 output nodes was trained on each of the six feature spaces. The network trained on the R feature space and classified incorrectly 2 of the normal images as abnormal and 4 abnormal as normal ones. The network trained on the G feature space misclassified 3 normal images as abnormal but not the same ones as the R space. The remaining 2 images were misclassified as normal ones. The B feature space has resulted 8 misclassifications, i.e. 5 abnormal

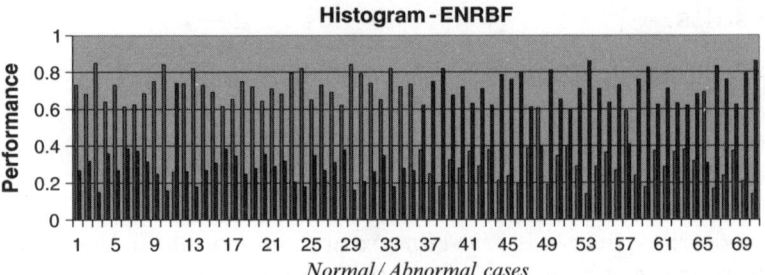

Fig. 5. Histogram-based performance

Table 1. Performances of ENRBF

ENRBF Accuracy (70 testing patterns)		
Modules	Histogram-based	N_{TU}-based
R	94.28% (4 mistakes)	92.85% (5 mistakes)
G	92.85% (5 mistakes)	97.14% (2 mistakes)
B	94.28% (4 mistakes)	95.71% (3 mistakes)
H	91.43% (6 mistakes)	94.28% (4 mistakes)
S	88.57% (8 mistakes)	91.43% (6 mistakes)
V	94.28% (4 mistakes)	97.14% (2 mistakes)
Overall	94.28% (4 mistakes)	95.71% (3 mistakes)

Table 2. Performances of RBF

RBF accuracy (70 testing patterns)		
Modules	Histogram-based	N_{TU}-based
R	91.42% (6 mistakes)	92.85% (5 mistakes)
G	92.85% (5 mistakes)	95.71% (3 mistakes)
B	88.57% (8 mistakes)	91.43% (6 mistakes)
H	90% (7 mistakes)	92.85% (5 mistakes)
S	85.71% (10 mistakes)	90.00% (7 mistakes)
V	94.28% (4 mistakes)	94.28% (4 mistakes)
Overall	88.57% (8 mistakes)	91.43% (6 mistakes)

as normal ones and the remaining 3 images as abnormal ones. The network trained on the H and S feature spaces achieved 90% and 85.71% accuracy respectively on the testing data. Finally, the network for the V feature space misclassified 3 normal cases as abnormal and one abnormal as normal one. Using the fuzzy integral (FI) concept, the overall system achieved an accuracy of 88.57%. Table 2 presents the performance of individual components. The confidence levels for each correct classification were above 0.50.

4.2 Performance of N_{TU}-Based Features

In the N_{TU}-based extraction process, the texture spectrum of the six components (R, G, B, H, S, V) have been obtained from the texture unit numbers, and the same nine statistical measures have been used in order to extract new features from each textures spectrum. A multi-classifier consisting of ENRBF networks with 9 input nodes and 2 output nodes was again trained on each of the six feature spaces. The N_{TU} transformation of the original histogram has produced a slight but unambiguous improvement in the diagnostic performance of the multi-classifier scheme. Table 1 illustrates the performances of the network in the individual components.

The ENRBF network trained on the R feature space and it then achieved an accuracy of 92.85% on the testing data incorrectly classifying 3 of the normal images as abnormal and 2 abnormal as normal ones. The network trained on the G feature space misclassified 2 normal images as abnormal but not the same ones as the R space. The B feature space achieved an accuracy of 95.71% on the testing data with 3 misclassifications, i.e. 2 abnormal as normal ones and the remaining one image as abnormal one. The network trained on the H feature space achieved 94.28% accuracy on the testing data. The network trained on the S feature space achieved an accuracy of only 91.43% on the testing data. Finally, the network for the V feature space misclassified 1 normal case as abnormal and 1 abnormal as normal one, giving it an accuracy of 97.14% on the testing data. The fuzzy integral (FI) concept has been used here to combine the results from each sub-network and the overall system provided an accuracy of 95.71%. More specifically, 1 normal case as abnormal and 2 abnormal as normal ones provide us a good indication of a "healthy" diagnostic performance. However the level of confidence in this case was slight less than the previous case (i.e. the histogram), that is 0.54 as shown in Fig. 6.

Similarly for the RBF case, the FI concept has been used to combine the results from each sub-network and the overall system misclassified 3 normal cases as abnormal and 3 abnormal as normal ones, giving the system an overall accuracy of 91.43%, despite the fact that RBF was characterised by a very fast training process. Table 2 presents the performance of individual components. The confidence levels for each correct classification were above 0.55.

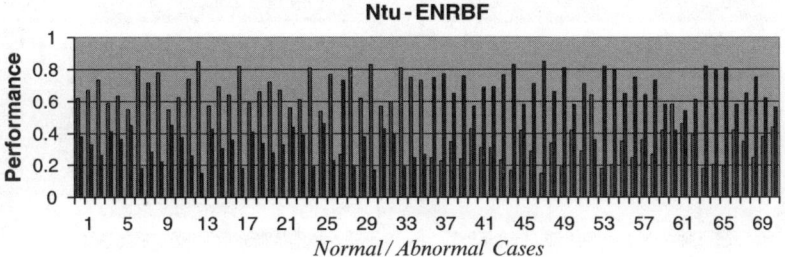

Fig. 6. N_{TU}-based performance

5 User-to-Computer Interface

There are significant differences between the requirements imposed on medical equipment, depending on the class of application. Various classes have different requirements regarding safety, reliability, cost and precision. The trend toward storing, distributing, and viewing medical images in digital form is being fuelled by two powerful forces: changes in the economic structure of the health care system and rapidly evolving technological developments.

The developed User-Interface has been designed in such a way that provides to the physician a simple and efficient tool for endoscopic imaging [17]. The physician can watch the video sequence in various speeds, select and save individual images of interest, zoom and rotate selected images as well as keeping notes on the patient history editor (bottom left side). This is illustrated in Fig. 7.

This software release includes a range of new features and enhancements for workflow efficiency and user-friendliness. It integrates Patient Management Capabilities with Diagnostic Tools. In general the benefits of the proposed Management Tool are:

Automatic diagnostic tools

• Intelligent algorithms for diagnosis of suspected cases

Productive video/individual image viewing

• Zoom, rotate
• Video playback control
• Adjustable video speed
• Video scroll bar

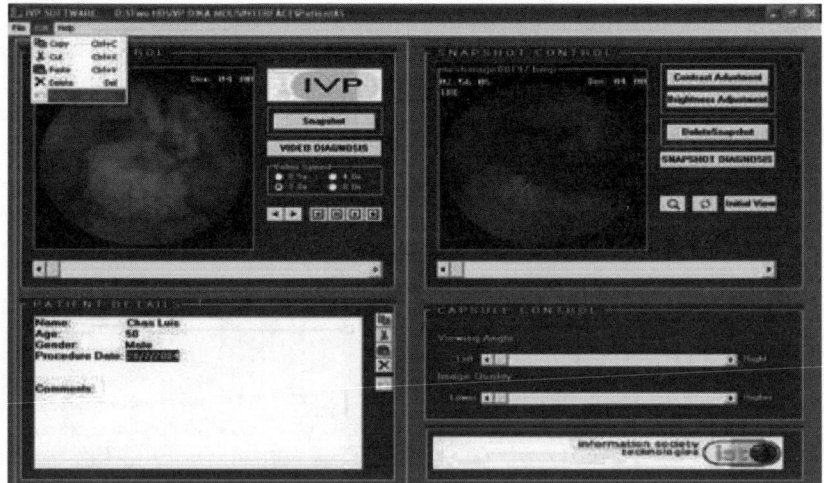

Fig. 7. Main input screen

Efficient workflow

- Patient record
- Automated video diagnosis for work load minimisation
- Storage of patient files (history record and snapshots)

Capsule control tool

- Viewing angle
- Image quality

The physician can watch the video sequence in various speeds, select and save individual images of interest, zoom and rotate selected images. In addition, the user can add/modify at any time the text on the "patient details" window using the copy, cut, paste, delete and undo buttons that are provided on the right side or alternatively the Edit Menu of the File Menu. An Intelligent algorithm already developed in MATLAB Environment has been integrated into this management scheme. The algorithm can identify possible suspicious areas. Intelligent diagnosis can be performed in two modes: video diagnosis and snapshot diagnosis:

- The video diagnosis button performs diagnosis throughout the whole video sequence and returns the results after the end on a text file on the PC.
- The snapshot diagnosis performs diagnosis only on the selected snapshot images and returns the results again on a text file on the PC.

During an endoscopic procedure the user has control over the viewing angle of the capsule as well as the image quality of the being acquired video. By changing the position of the "Viewing angle" scrollbar of the Capsule control window, he can alternate the viewing angle of the image sensor (left-right). By changing the position of the "Image quality" scrollbar of the Capsule control window, the image quality of the capsule can vary from the two extremes: lowest resolution (highest frame rate) and highest resolution (lowest frame rate).

6 Conclusions

An integrated approach on extracting texture features from wireless capsule endoscopic images has been developed. Statistical features based on texture are important features, and were able to distinguish the normal and abnormal status in the selected clinical cases. The multiple classifier approach used in this study with the inclusion of advanced neural network algorithms provided encouraging results. Two approaches on extracting statistical features from endoscopic images using the M2A Given Imaging capsule have been developed. In addition to the histogram-based texture spectrum, an alternative approach of obtaining those quantitative parameters from the texture spectra is proposed both in the chromatic and achromatic domains of the image by calculating the texture unit numbers (N_{TU}) over the histogram spectrum.

The developed intelligent diagnostic system was integrated on a user-interface environment. This user-interface was developed under the framework of the "Intracorporeal VideoProbe – IVP" European research project. This software release includes a range of new features and enhancements for workflow efficiency and user-friendliness. It also integrates Patient Management Capabilities with Powerful Diagnostic Tools.

References

1. Economou, G.P.K., Lymberopoulos, D., Karvatselou, E., Chassomeris, C.: A new concept toward computer-aided medical diagnosis – A prototype implementation addressing pulmonary diseases, IEEE Transactions on Information Technology in Biomedicine, **5** (1) 55–61 (2001)
2. Kovalerchuk, B., Vityaev, E., Ruiz, J.F.: Consistent knowledge discovery in medical; diagnosis, IEEE Engineering in Medicine and Biology Magazine, **19** (4) 26–37 (2000)
3. West, D., West, V.: Model selection for a medical diagnostic decision support system: A Breast cancer detection case, Artificial Intelligence in Medicine, **20** (3) 183–204 (2000)
4. Haga, Y., Esashi, M.: Biomedical microsystems for minimally invasive diagnosis and treatment, Proceedings of IEEE, **92** 98–114 (2004)
5. Krishnan, S., Wang, P., Kugean, C., Tjoa, M.: Classification of endoscopic images based on texture and neural network, Proceedings of 23rd Annual IEEE International Conference in Engineering in Medicine and Biology, **4** 3691–3695 (2001)
6. Maroulis, D.E., Iakovidis, D.K., Karkanis, S.A, Karras, D.A.: CoLD: A versatile detection system for colorectal lesions endoscopy video-frames, Computer Methods and Programs in Biomedicine, **70** 151–166 (2003)
7. Idden, G., Meran, G., Glukhovsky, A., Swain, P.: Wireless capsule endoscopy, Nature, 405–417 (2000)
8. Gletsos, M., Mougiakakou, S., Matsopoulos, G., Nikita, K., Nikita, A., Kelekis, D.: A computer-aided diagnostic system to characterize CT focal liver lesions: design and optimization of a neural network classifier, IEEE Transactions on Information Technology in Biomedicine, **7** (3) 153–162 (2003)
9. Wadge, E., Boulougoura, N., Kodogiannis, V.: Computer-assisted diagnosis of wireless-capsule endoscopic images using neural network based techniques, Proceedings of the 2005 IEEE International Conference on Computational Intelligence for Measurement Systems and Applications, CIMSA, 328–333 (2005)
10. Kodogiannis, V., Boulougoura, M., Wadge, E.: An intelligent system for diagnosis of capsule endoscopic images, Proceedings of the second International Conference on Computational Intelligence in Medicine and Healthcare (CIMED 2005), Portugal, 340–347 (2005)
11. Kodogiannis, V.: Computer-aided diagnosis in clinical endoscopy using neuro-Fuzzy system, IEEE FUZZ 2004, 1425–1429 (2004)
12. Boulougoura, M., Wadge, E., Kodogiannis, V., Chowdrey, H.S.: Intelligent systems for computer-assisted clinical endoscopic image analysis, Second IASTED International Conference on Biomedical Engineering, Innsbruck, Austria, 405–408 (2004)

13. Wadge, E., Kodogiannis, V.S.: Intelligent diagnosis of UTI in vivo using gas sensor arrays, International Conference on Neural Networks and Expert Systems in Medicine and HealthCare, NNESMED 2003, 93–98 (2003)
14. Kodogiannis V.S., Boulougoura M., Wadge E., Lygouras J.N.: The usage of soft-computing methodologies in interpreting capsule endoscopy, Engineering Applications in Artificial Intelligence, Elsevier, **20** 539–553 (2007)
15. Kuncheva, L.I.: Fuzzy Classifier Design, Physica, Heidelberg (2000)
16. Wadge, E.: The use of EM-Based neural network schemes for modelling and classification, PhD Thesis, Westminster University, (2005)
17. Arena, A., Boulougoura, M., Chowdrey, H.S., Dario, P., Harendt, C., Irion, K.-M., Kodogiannis, V., Lenaerts, B., Menciassi, A., Puers, R., Scherjon, C., Turgis, D.: Intracorporeal Videoprobe (IVP), in Medical and Care Computing 2, Volume 114 Studies in Health Technology and Informatics, Edited by: Bos, L. Laxminarayan, S. and Marsh, A. IOS Press, Amsterdam, Netherlands, 167–174 (2005)

Part V

Knowledge Discovery and Management

Formal Method for Aligning Goal Ontologies

Nacima Mellal, Richard Dapoigny, and Laurent Foulloy

LISTIC, Polytech'Savoie. BP 80439–74944, Annecy, France
nacima.mellal@univ-savoie.fr

Summary. Many distributed heterogeneous systems interoperate and exchange information between them. Currently, most systems are described in terms of ontologies. When ontologies are distributed, the problem of finding related concepts between them arises. This problem is undertaken by a process which defines rules to relate relevant parts of different ontologies, called "Ontology Alignment." In literature, most of the methodologies proposed to reach the ontology alignment are semi automatic or directly conducted by hand. In the present paper, we propose an automatic and dynamic technique for aligning ontologies. Our main interest is focused on ontologies describing services provided by systems. In fact, the notion of service is a key one in the description and in the functioning of distributed systems. Based on a teleological assumption, services are related to goals through the paradigm 'Service as goal achievement', through the use of ontologies of services, or precisely goals. These ontologies are called "Goal Ontologies." So, in this study we investigate an approach where the alignment of ontologies provides full semantic integration between distributed goal ontologies in the engineering domain, based on the Barwise and Seligman Information Flow (noted IF) model.

1 Introduction

In distributed environments, dynamic interaction, communication, and information exchange are highly required. Generally, systems are heterogeneous, so they need to understand what they communicate. This is known by the "Semantic nteroperability," which present a big challenge in the Artificial Intelligence area. The design of distributed systems is an important step. Recent works suggest to describe systems in terms of their goals, their functions and their physical components [3, 8]. Actually, much attention has been paid on the design of the automated control systems [2, 22], where functions and goals are fundamental to the understanding of complex systems. A functional representation of systems consists in the description of the functionality of its components (or (sub-)systems) and the relationship between them. An efficient and promising way to implement this is through the use of *Ontologies*,

N. Mellal et al.: *Formal Method for Aligning Goal Ontologies*, Studies in Computational Intelligence (SCI) **109**, 279–289 (2008)
www.springerlink.com

an explicit specification of conceptualization [9, 11]. In [20], authors have denoted a functionality of a component as a verb + noun style for representing the components activities (or actions) and its operands which needs an ontological schema for functional knowledge which specifies not only the data structure but also the conceptual viewpoint for capturing the target world. Following this approach, we associate with each goal some possible actions (at least one) in order to fulfill the intended goal [12, 21].

Usually, in distributed systems, the increasing number of goals, and thus possible combinations, requires the development of dynamic and automatic techniques for their composition and fusion. Current solutions are limited and are primarily static and manual. This needs the development of dynamic and automatic methodologies.

After a careful look at the different works and theories related to these topics, such as [16], their approach mainly builds on the IF-Map method to map ontologies in the domain of computer science departments from five UK universities. Their method is also complemented by harvesting mechanisms for acquiring ontologies, translators for processing different ontology representation formalisms and APIs for web enabled access of the generated mappings. In [14, 16, 27], first-order model theory are investigated to formalize and automatize the issues arising with semantic interoperability for which they focused on particular understandings of semantics.

To this aim, our proposed work extends the methodology proposed in [27] for communicating conceptual models (i.e., goal ontologies in the engineering domain) within distributed systems.

Multi Agent Systems (MAS) have been the subject of massive amounts of research in recent years for large scale, complex systems, distributed systems, heterogeneous systems, in open environments and adaptable systems. Software agents can be able to discover, invoke, compose, and monitor systems resources offering particular services and having particular properties, and can be able to do so with a high degree of automation if desired. It seems to be a promising candidate to support the process of goal ontologies alignment. Where the software agents represent a distributed system. For a given goal, each agent manipulates the possible goal ontologies in its system.

In the present paper, we introduce in section two the notion of goal and we define what goal ontology is in our context? In Sect. 3, we give some precise definitions of the IF model we use in our methodology. Finally, in Sect. 4, we present our methodology for aligning goal ontologies illustrating it by an algorithm.

2 Goal Ontologies

Reasoning with goal models has become an attractive and challenging topic over the past decade. The proposed model is teleological, which means that we can conceive software goals acting on data structures as determined by ends or hardware goals. Motivated by the teleological reasoning, any functional concept is described by a goal definition which is related to the intentional aspect of function [22] and some possible actions (at least one) in order to fulfill the intended goal [7, 12, 21]. The teleological basis introduced in [4–6] relates the structure and behavior of the designed system to its goals. The goal modeling requires to describe a goal representation (i.e., a conceptual structure), and to define how these concepts are related.

2.1 Goal Notion

In order to give a formal structure of a goal, we describe each service as a tuple:

$$(Actionverb, \{Property, Object\}),$$

where the action verb acts on the object's property. We distinguish two classes of objects: "Object Type" and "Object Token." Object types describe entities by their major characteristic (for example Person represents an object type), where, object tokens are specified by their identity (for example, Omar is a person). We define the couple {Property, Object}, as a "Context." When the object is an object type, the context is called "Context Type," and when it is an object token, the context is called "Context Token." We define each notion in next paragraphs.

Definition 1. *Given P, a finite set of properties, Ψ a finite set of object types. A context type is a tuple:*

$$\xi_i = (p, \psi_1, \psi_2, ..\psi_i) \tag{1}$$

where $p \in P$, denotes its property, and $\{\psi_1, \psi_2, \ldots \psi_i\} \subseteq \Psi$, a set of object type.

A similar definition holds for contexts tokens replacing object types by object tokens.

Definition 2. *A goal type is a pair (A, Ξ), where A is an action symbol and Ξ is a non-empty set of context types.*

A similar definition holds for goal tokens replacing context types by context tokens.

2.2 Goal Ontology

Links between goals are functional. These links reflect the causal dependencies between goals. In general, goals and their links are described in a hierarchy. The hierarchy illustrates the functional influence between goals, where their functional inclusion also referred to as functional part-of and denoted $\ll \subseteq \gg$.

Since we have to relate goal hierarchies through information channels, their ontological definition must include concepts together with the basic ontological axioms [28]. The ontology includes goal types as concepts and the functional part-of ordering relation as relations between goals.

Definition 3. *A goal type "γ_i" influences functionally "γ_j" iff the only way to achieve γ_j is to have already achieved γ_i, with the notation:*

$$\gamma_i \sqsubseteq \gamma_j, \tag{2}$$

Therefore, we define a goal ontology as:

Definition 4. *A goal ontology O is described by the following tuple:*

$$O = (\Gamma, \sqsubseteq_\Gamma) \tag{3}$$

where Γ is a finite set of goal types, $\sqsubseteq_\Gamma$ is a partial order on Γ.

We introduce in next section the most important definition in the IF model.

3 IF Model

3.1 IF Classification

Definition 5. *An IF Classification A is a triple $< tok(A), typ(A), | = A >$, which consists of:*

1. *a set $tok(A)$ of objects to be classified known as the instances or particulars of A that carry information,*
2. *a set $typ(A)$ of objects used to classify the instances, the types of A, and*
3. *a binary classification relation $| = A$ between $tok(A)$ and $typ(A)$ that tells one which tokens are classified as being of which types.*

The meaning of the notation a $| = A\alpha$ is "instance a is of type α in A." IF classifications are related through infomorphisms. Infomorphisms are the links between classifications.

3.2 Infomorphism

Definition 6. *Let A and B be IF classifications. An Infomorphism denoted $f =< \hat{f}, \check{f} >: A \rightleftarrows B$ is a contravariant pair of functions $\hat{f} : typ(A) \rightarrow typ(B)$ and $\check{f} : tok(B) \rightarrow tok(A)$ which satisfies the fundamental property:*

$$\check{f}(b)| = A\alpha \ iff \ b| = B \ \hat{f}(\alpha) \ for \ each \ \alpha \in \ typ(A) \ and \ b \in \ tok(b)$$

Regularities in a distributed system are expressed with IF theories and IF logics. In distributed system, IF-theories can be seen as an idealized version of the scientific laws supported by a given system.

Let A be a classification. A token $a \in tok(A)$ satisfies the constraint $\Gamma \vdash \Delta$ where (Γ, Δ) are subsets of typ(A), if a is of some types in Δ whenever a is of every type in Γ. If every token of A is constrained by (Γ, Δ), we have obviously $\Gamma \vdash_A \Delta$ and $< typ(A), \vdash_A >$ is the theory generated by A.

A theory T is said regular if for all $\alpha \in$ typ(T) and for arbitrary subsets Γ, Δ, Γ', Δ', Σ' of typ(T), the following properties hold:

- The Identity: $\alpha \vdash T\psi\alpha$
- The Weakening: if $\Gamma \vdash T\psi\Delta$ then $\Gamma, \Gamma' \vdash \Delta, \Delta'$
- The Global cut: if $\Gamma, \Gamma' \vdash \Delta, \Delta'$ for any partition[1] (Γ', Δ') of Σ', then, $\Gamma \vdash T\psi\psi\Delta$, for all if $\Gamma, \Delta \vdash T\psi typ(T)$

3.3 Local Logic

Definition 8. *A local logic* $L = <tok(L), typ(L), | = _L, \vdash _L, N_L>$ *consists of a regular*

$$IF \ theory \ th(L) = <typ(L), \vdash _L>, an \ IF \ classification \ cla(L)$$
$$= <tok(L), typ(L), | = _L>$$

and a subset $N_L \subseteq tok(L)$ *of normal tokens which satisfy all the constraints of* $th(L)$.

A token $a \in tok(L)$ is constrained by $th(L)$. Given a constraint (Γ, Δ) of $th(L)$, whenever a is of all types in Γ, then a is of some type in Δ. An IF logic L is sound if $N_L = tok(L)$. In this paper, we restrict the classification relation to normal instances, limiting ourselves to sound logics. This assumption is required to enable ontology sharing or ontology matching [15,16]. In summary, each component of a distributed system is described with a sound local logic integrating a classification and its associated theory.

$$L = <tok(L), typ(L), | = _L, \vdash _L, N_L> \tag{4}$$

Let us introduce the distributed IF structures, which base on the IF channel. IF channel models the information flow between IF classifications. The local logic is the "what" of IF, the channel is the "why"

3.4 IF Channel

Definition 9. *An IF channel consists of two classifications A1 and A2 connected through a core classification C by means of two infomorphisms f1 and f2.*

---

[1] A partition of Σ' is a pair $(\Gamma', \psi\Delta')$ of subsets Σ', such that $\Gamma' \cup \Delta' = \Sigma'$ and $\Gamma' \cap \Delta' = \phi$

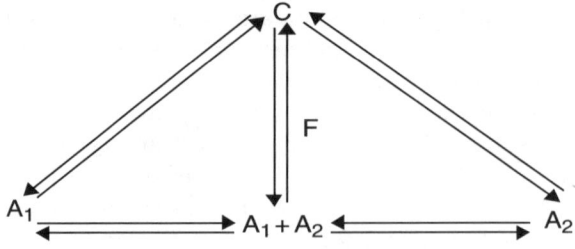

Fig. 1. The IF channel

Since local logics are inclusive concepts combining the concepts of classification and theory, they capture a more general knowledge than single classifications. Therefore there is a need to consider distributed IF logics of IF channels.

Definition 10. *Given a binary channel* $C = \{f1 : A1 \rightleftarrows C, f2 : A2 \rightleftarrows C\}$ *with a logic L on the core classification C, the distributed logic DLogC(L) of C generated by L is such as:*

$$DLogC(L) = F^{-1}[L] \tag{5}$$

The local logic on the sum $A1 + A2$ which represents the reasoning about relations among the components is also referred as the distributed logic of C generated by L while F denotes the infomorphism $F : (A1 + A2) \rightleftarrows C$. σ_1 and σ_2 are also infomorphisms, where, $\sigma_1: A_1 \rightleftarrows (A1 + A2)$ and $\sigma_2 : A_2 \rightleftarrows (A1 + A2)$, (see Fig. 1).

4 Aligning Distributed Goal Ontologies

4.1 Scenario of the Alignment

Considering a distributed system of goal ontologies, semantic interoperability turns out to find relationships (equivalence or subsumption) between concepts (e.g., goal types) which belong to different ontologies. This process is known as an ontology alignment problem. In the present work, the alignment is used to find the correspondences between goal ontologies. As a crucial topic, information exchange between functional hierarchies must occur in a semantically sound manner. Major works stem from the idea that a classification of information (types versus tokens) must exist in each of the components of a distributed system [1, 6, 18, 19]. Of particular relevance to this aim is the work in Information Flow (IF). Therefore, we follow the IF mathematical model which describes the information flow in a distributed system. A process described in [17], uses IF classifications where the classification relation is in fact a subsumption relation. We have extended this work to goal hierarchies where the classification relation expresses the functional dependency. Goal ontologies alignment is a process which describes how one or more goals of the ontology can be aligned to goal(s) of other ontologie(s) in a sound and automatic way, without altering the original ontologies. Such a process must not be considered as a simple pattern matching process, but rather an intentional alignment process since the resulting goal dependencies must respect the sum of the local logics both on the syntactic and the semantic level. Local logics express physical constraints through Gentzen sequents. Therefore there is a need to consider distributed IF logics of IF channels. The semantic integration of goal types from separate systems is achieved through a process including several steps [23]:

1. The description of each system by its possible goal ontologies.
2. The identification of the local logics of each system. To do this, IF classifications, IF theories have to be defined. Classifications of systems represent their goals, we associate goal types as *Types* of classification and context tokens as *Tokens*. The IF theories describe how different types from different systems are logically related to each other.
3. The building of the Information Channel between these systems.
4. The definition of distributed logic which permits the connection between candidate systems. Given the logic Log(C) = L on the core C, the distributed logic DLog(C) on the sum of goal hierarchies is the inverse image of Log(C) on this sum. The logic is guaranteed to be sound on those tokens of the sum that are sequences of projections of a normal token of the logic in C. We obtain sequents like relating goal(s) on remote systems to the local goal(s). The set of these couples constitute the IF theory on the core of the channel. From here it is straightforward to extend goal dependencies to dependencies between higher-level goal, and finally between distributed services.

4.2 Agent Specification

Multi-Agents Systems (MAS) models are considered as programming paradigms as well as implementation models for complex information processing systems [29]. More precisely, in distributed systems, each agent is able to locally process data and exchange only high-level information with other parts. Since an agent is anything that can be viewed as perceiving its environment through sensors and acting upon that environment through effectors [26], we exploit the high-level and dynamic nature of multi-agent interactions which is appropriate to distributed systems. The IF-based mechanism of searching for goal dependencies is typically that of distributed problem solving systems, where the component agents are explicitly designed to cooperatively achieve a given goal.

In [1], authors have described an approach to ontology negotiation between agents supporting intelligent information management. They have developed a protocol that allows agents to discover ontology conflicts and then, though incremental interpretation, clarification, and explanation, establish a common basis for communicating with each other. Our aim is not to extend this work, but to clarify the idea of using intelligent agents to facilitate the automatic alignment of goal ontologies. In this work we are mainly interested in the development of deliberative agents to implement adaptive systems in open and distributed environments. Deliberative agents are usually based on a BDI model [25], which considers agents as having certain mental attitudes, Beliefs, Desires, and Intentions (BDI). An agents beliefs correspond to information the agent has about the world (e.g., variables, facts, ...). An agents desires intuitively correspond to the goals allocated to it. An agents intentions represent desires (i.e., goals) that it has committed to achieving. A BDI architecture is intuitive and it is relatively simple to identify the process of decision-making and how to perform it. Furthermore, the notions of belief, desire and intention are easy to understand. Its main drawback lies in finding a mechanism that permits its efficient implementation. Most approaches use modal logic such as extensions to the branching time logic CTL^*, for the formalization and construction of such agents, but either they are not always completely axiomatized or they are not computationally efficient [13,24,25]. The problem lies in the great distance between the powerful logic for BDI systems and practical systems.

We have specified our proposed process using software agents. Each one is responsible of the achieving of goals in its system. The communication and the cooperation between agents may be reached by the use of IF model. For this aim, we propose the algorithm bellow.

4.3 Algorithm ≪ *Alignment of goal ontologies* ≫

```
    Input: A Goal G to achieve in System (O).
    Output: G achieved.

BEGIN
        (1): AgentO analysis the goal ontology of G; It analysis
             the ontology according to the context types of its
             goals.
          if the context types of the goals in ontology are local,
             then   G is achievable locally;
          else
                  (2): AgentO broadcasts a request to the remote
                       agents asking goals containing the context
                       types remaining in the local ontology;
                       if no agent answers; then
                        G is aborted;
                      else
                          (3): AgentO identifies the candidates
                               agents (Agents K) and analyses their
```

```
                          answers (distributed goal ontologies
                          candidates);
                  (4): Agent0 specifies goal-ontologies by
                          means of IF classifications. It
                          builds classifications of local and
                          remote goal ontologies. It generates
                          their IF theories;
                  (5): Agent0 builds the information
                          channel, by generating the core
                          classification between the requesting
                          classification and the potential
                          candidate classifications and
                          generates the infomorphisms
                          connecting them;
                  (6): Agent0 generates the IF theory on the
                          core of the channel and deduces the
                          IF logic L(Core) on the core.
                  (7): Agent0 computes the inverse image of
                          L(Core) which presents the
                          distributed logic on the sum of the
                          classifications;
                  (8): From L(Core), Agent0 selects the
                          appropriate goal according to
                          semantic constraints;
                  end-if
          end-if
END
```

In this part, we extend our approach to be applied in an multi-agents system. Each agent represents a system and communicates by exchanging information with other distant agents. For a given goal G to be achieved in system(0). Agent0 (which represents system 0) is responsible of achieving this goal. The algorithm complexity depends on the significant phases. In [23] the goal complexity is reduced to $O(n^2 * 2^p)$, n is the number of goal types, p the number of context types.

5 Conclusion

Since systems need to communicate and exchange services and information, ontologies present a good way in the description of these systems, but they are not sufficient to achieve the interaction between them in a sound manner. Formal and mathematical theories are highly required. In this paper, we have presented a formal approach for automatic and dynamic alignment of goal ontologies. Our approach is based on the mathematical model (IF model). It is well detailed in [23]. The alignment of goal ontologies expresses the functional dependencies between goals. The process of alignment proposed in this paper is specified by software agent, since it treats a distributed system.

Our future work is directed to the implementation of the proposed methodology. We plan to use the Cognitive Agent Architecture (Cougaar) with the mechanisms for building distributed agent systems [10], because it provides a rich variety of common services to simplify agent development and deployment.

References

1. Bailin S, Truszkowski W (2001) Ontology Negotiation between Agents Supporting Intelligent Information Management. In: Workshop on Ontologies in Agent Systems
2. Chittaro L, Guida G, Tasso G, Toppano E (1993) Functional and teleological knowledge in the multi modeling approach for reasoning about physical systems: A case study in diagnosis. IEEE Transactions on Systems, Man, and Cybernetics, 23(6): 1718–1751
3. Dapoigny R, Benoit E, Foulloy L (2003) Functional Ontology for Intelligent Instruments. In: Foundations of Intelligent Systems. LNAI2871, pp 88–92
4. Dapoigny R, Barlatier P, Mellal N, Benoit E, Foulloy L (2005) Inferential Knowledge Sharing with Goal Hierarchies in Distributed Engineering Systems. In: IIAI05, Pune (India), pp 590–608
5. Dapoigny R, Mellal N, Benoit E, Foulloy L (2004) Service Integration in Distributed Control Systems: an approach based on fusion of mereologies. In: IEEE Conf. on Cybernetics and Intelligent Systems (CIS'04), Singapour, December 2004, pp 1282–1287
6. Dapoigny R, Barlatier P, Benoit E, Foulloy L (2005) Formal Goal generation for Intelligent Control systems. In: 18th International Conference on Industrial & Engineering Applications of Artificial Intelligence & Expert Systems LNAI 3533, Springer, pp 712–721
7. Dardenne A, Lamsweerde A, Fickas S (1993) Goal-directed requirements acquisition. Science of Computer Programming, 20: 3–50
8. Dooley K, Skilton P, Anderson J (1998) Process knowledge bases: Facilitating reasoning through cause and effect thinking. Human Systems Management, 17(4): 281–298
9. Fikes R (1996) Ontologies: What are they, and where's the research? Principles of Knowledge Representation and Reasoning, 652–654
10. Gorton I, Haack J, McGee D, Cowell J, Kuchar O, Thomson J (2003) Evaluating Agent Architectures: Cougaar, Aglets and AAA. In: SELMAS, pp 264–278
11. Gruber T (1993) A translation approach to portable ontology specifications. Knowledge Acquisition, 5(2): 199–220
12. Hertzberg J, Thiebaux S (1994) Turning an action formalism into a alanner: A case study. Journal of Logic and Computation, 4: 617–654
13. Huber M (1999) A BDI-Theoretic Mobile Agent Architecture. In AGENTS 99 Proceedinf. of the Third Annual Conference on Autonomous Agents (1999) Seattle, WA, USA ACM, pp 236–243

14. Kalfoglou Y, Schorlemmer M (2002) Information Flow based ontology mapping. In: Proceedings 1st International Conference on Ontologies, Databases and Applications of Semantics(ODBASE'02), Irvine, CA, USA
15. Kalfoglou Y, Schorlemmer M (2003) Ontology mapping: the state of the art. The Knowledge Engineering Review, 18:1–31
16. Kalfoglou Y, Schorlemmer M (2003) IF-Map: An ontology mapping method based on information flow theory. Journal of Data Semantics, S. Spaccapietra et al. (eds.), vol. 11, Springer, Berlin Heidelberg New York
17. Kalfoglou Y, Schorlemmer M (2004) Formal Support for Representing and Automating Semantic Interoperability. In: Proceedings of the 1st European Semantic Web Symposium (ESWS'04), Heraklion, Crete
18. Kalfoglou Y, Hu B, Reynolds D (2005) On Interoperability of Ontologies for Web-based Educational Systems. In: WWW 2005 workshop on Web-based educational systems, Chiba city, Japan
19. Kent RE (2000) The information flow foundation for conceptual knowledge. In: Dynamism and Stability in Knowledge Organization. Proceedings of the Sixth International ISKO Conference. Advances in Knowledge Organization, 7: 111–117
20. Kitamura Y, Mizoguchi R (1998) Functional Ontology for Functional Understanding. In: 12th International Workshop on Qualitative Reasoning AAAI Press, pp 77–87
21. Lifschitz V (1993) A Theory of Actions. In: 9th International Joint Conference on Artificial Intelligence, M. Kaufmann (ed.), pp 432–437
22. Lind M (1994) Modeling goals and functions of complex industrial plant. Journal of Applied Artificial Intelligence, 259–283
23. Mellal N, Dapoigny R, Foulloy L (2006) The Fusion Process of Goal Ontologies using Intelligent Agents in Distributed Systems. In: International IEEE Conference on Intelligent Systems, London, UK, September, pp. 42–47
24. Myers K (1996) A Procedural Knowledge Approach to Task-Level Control. In: Proceedings of the Third International Conference on Artificial Intelligence Planning Systems, pp 158–165
25. Rao AS, Georgeff MP (1995) BDI Agents: From Theory to Practice. In: First International Conference on Multi-Agent Systems (ICMAS-95), SanFranciso, USA, pp 312–319
26. Russel S, Norvig P (1995) Artificial Intelligence, A Modern Approach, Prentice Hall, Englewood Cliffs, NJ
27. Schorlemmer M, Kalfoglou Y (2003) Using Information-Flow Theory to enable Semantic Interoperability, Technical report EDI-INF-RR-0161 University of Edinburgh
28. Varzi A (2000) Mereological commitments. Dialectica, 54: 283–305
29. Zambonelli F, Jennings NR, Wooldridge M (2003) Developing MultiAgent systems: The Gaia methodology. Journal of ACM Transactions on Software Engineering and Methodology, 12(3): 317–370

Smart Data Analysis Services

Martin Spott, Henry Abraham, and Detlef Nauck

BT Research and Venturing, Adastral Park, Ipswich IP5 3RE, UK
martin.spott@bt.com, henry.abraham@bt.com, detlef.nauck@bt.com

Summary. With corporations collecting more and more data covering virtually every aspect of their business, turning data into information and information into action is paramount for success. The current state of the business needs to be reported and based on predictions problems or opportunities can be spotted early and understood, such that actions can be pro-actively taken accordingly. We propose an architecture for a smart data analysis service that automatically builds data analysis models based on high-level descriptions of analysis goals and solution requirements, wraps the models into executable modules for a straightforward integration with operational systems, monitors the performance of the operational system and triggers the generation of a new model if the performance deteriorates or the solution requirements change.

1 Introduction

Corporations are collecting more and more data covering virtually every aspect of their business. The data concerns internal processes, suppliers, customers and probably competitors. Information derived from data can be used to improve and control internal processes, to understand, serve and target customers better. The first hurdle is quite often the large number of different systems and data sources in organisations that hinder a single view onto the entire business. Secondly, modern data analysis approaches are required to turn raw data into useful information. In the following, we will use the term *data analysis* as generic term for traditional, statistical data analysis and modern machine learning techniques likewise. The term *Intelligent Data Analysis (IDA)* can sometimes be found in literature basically meaning the same thing.

Finally, gained information will be used to make business decisions. An example is fraud detection. Based on customer data, models can be learned which are able to detect fraudulent transactions or at least mark transactions

M. Spott et al.: *Smart Data Analysis Services*, Studies in Computational Intelligence (SCI) **109**, 291–304 (2008)

with a high fraud propensity. These cases can then automatically be passed on to an advisor who checks the customer records and the latest transaction again and acts upon it.

Information gained from data will not only to be used to understand the business but also integrated with operational systems as indicated in the above example of fraud detection. Obviously, the information contained in data changes over time, which is basically a reflection of changes in the market or of internal processes. As a result, an analytics module like a fraud detector needs to be adapted on a regular basis, which is still very much a manual procedure.

The chapter describes a smart service based on our data analysis platform SPIDA (Soft Computing Platform for Intelligent Data Analysis) [1] that tackles the following problems:

1. Monitoring the performance of an operational system and trigger the generation of a new analytics module in case of performance deterioration
2. Automatic creation of a data analysis model based on user requirements
3. Wrapping the model into an executable module for integration with operational systems

The objective of our system is to keep the performance of the operational system at a high level at all times and to achieve that with no or as little human interaction as possible. Hereby, an important role is played by an intelligent wizard for the creation of data analysis models. It can create data analysis models based on given high-level user requirements. Embedded into the smart data analysis service the wizard is flexible enough to choose any analysis algorithm that meets the user requirements. Normally, such flexibility can only be achieved by human analysts. As an easier but less flexible alternative, some systems stick to the same algorithm and only change some model parameters in order to adapt to changes.

The chapter is organised as follows. We first give an overview of different application scenarios for data analysis tools or services and from those derive a generic architecture for a smart data analysis service. Afterwards, the wizard for the automatic creation of analysis processes is explained, followed by our approach to wrapping such a process into a piece of software that can be executed by an operational system.

2 Application Scenarios

In an industrial context, intelligent data analysis is normally applied in two different ways, analysts performing ad-hoc analysis on given data sets and analysis modules in operational systems executing predefined analysis tasks. Where analysts are usually looking to find an analysis solution, analysis modules are quite often the result of an analyst's work.

Analysts are either data analysis experts who act as a consultant for other parts of the business, or they are domain experts attached to the respective part of the business, know their data very well, but their data analysis know-how is limited. The first group of people is very confident in using standard data analysis or data mining tools, but the second is certainly not. Domain experts generally know what they are looking for in data, but they might not know very well, how to do that. They are usually not aware of all relevant machine learning techniques, nor which one to select for which problem and especially not, how to configure them in order to get good results.

Unfortunately, most available data analysis tools are centered around technology, which is quite independent from the actual application problem. On the other hand, business users like the domain experts mentioned above know the problem at hand, but not the technology required to solve it. Providers of data analysis solutions try to fill this gap with solutions tailored to certain industry sectors. Based on research results and the provider's knowledge gained in consulting projects, they build a set of standard analysis models as a ready-to-use approach. From the user's viewpoint, such an approach is fine as long as his problem and requirements are compatible with the provided solution. If this is not the case, the standard solution can either not be used at all, or expensive consultants are required to adapt the solution to the user's needs which might not be acceptable.

The wizard interface in SPIDA takes a different approach by allowing the user to specify the analysis problem and preferences regarding the solution at such a high level that they are easy to understand for non-experts. It then automatically generates a solution according to these specifications.

In contrast to the creative acts of human analysts, analytical modules in operational systems are traditionally static. For instance, they might be used to classify credit card transactions as fraudulent or not or to predict share prices or sales volumes on a transactional or daily basis. Usually, a human analyst would create an analysis model, which has to be integrated with the operational systems. Depending on both the analysis tool and the operational system, the integration might require software engineers to write an analysis module in the worst case, or at least a wrapper as an interface. With the introduction of data analysis functionality in databases and a standardised language for model description like Predictive Model Markup Language (PMML) [2], the integration will be quite simple in the future. Under the assumption that the analysis tool is able to create a PMML description for the model in question and the database implements the underlying analysis algorithm, the PMML description can simply be included in a PL/SQL script that will be used to analyse data in the operational system. However, it will take time before data analysis is standard in databases and a large variety of models can be transferred in that way.

In case of SPIDA, only a simple interface in the operational system is necessary that would allow the execution of any module created by the wizard.

From time to time, the underlying analysis models have to be adjusted in order to keep up with changes in the business or the market. Such adaptations are usually done manually by analysts, who produce new models based on new data. Most of the times, the underlying analysis algorithm stays the same, only parameters change. For example, the weights of a neural network or the rules in a fuzzy system might be adapted. In case of data analysis functionality in the database and a PMML model, even algorithms can be changed without compromising the integrity of operational systems. For example, it might turn out that a decision tree performs better on a new data set than a support vector machine that has been used before. By simply exchanging the PMML description of the model in the database, the new model can be integrated seamlessly. However, as mentioned earlier this is not standard, yet, and also the decision for the new algorithm and its integration is still a manual process.

Again, the wizard interface in SPIDA can automate this process by picking, configuring and implementing new algorithms without user interaction. The final decision, if a new algorithm will be uploaded into the operational system will probably still be under human supervision, but the difficult part of creating it can mostly be achieved by SPIDA.

3 Architecture for a Smart Data Analysis Service

A schema summarising all scenarios in combination with SPIDA is shown in Fig. 1. Data is fed into the operational system from a data warehouse. The data analysis module conducts data analysis for the operational system. Thereby it influences the system's performance or the performance of related business processes, because analysis results are usually used to make business decisions. For example, the data analysis module might be used to predict if a customer is going to churn and performance measures might be the relative number of churners not being flagged up by the system and the revenue lost with them. In that case, dropping system performance will be the result of an inferior analytics model. Such a drop of performance will be detected by the monitor and trigger the generation of a new analysis model. The easiest way to detect dropping performance is by comparing against predefined thresholds.

The reasons for deterioration can be manifold. In most cases, the dynamics of the market are responsible, i.e. the market has changed in a way that

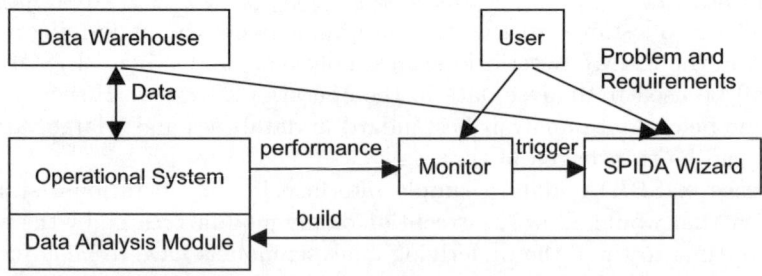

Fig. 1. Monitoring of system performance and requirements and automatic creation and integration of data analysis modules according to given requirements

cannot be detected by the analytics model, because such information is usually difficult to capture. Other reasons include changes in related business processes and data corruption.

Whenever model generation is triggered by the monitor, the wizard takes the user requirements and the latest data to build a new model. This process is being described in Sect. 4. Another reason for building a new model is a change of user requirements. For example, due to a change of company policy or for regulatory reasons, a model might be required to be comprehensible which was not the case before. If, for instance, a neural network had been used to measure the credit line of a potential borrower, we might be forced to switch to methods like decision trees or fuzzy systems which create a rule base that can be understood by managers and the regulator. We might have to show to the regulator that the model conforms to an equal opportunities act. The final step consists of turning a data analysis model or process into an executable module that can directly be called by or integrated with an operational system. We describe this step in Sect. 6.

In summary, the architecture covers the entire loop from system performance and requirements monitoring, the automatic generation of an analysis model, the build of an executable module and its integration with the operational system. All these steps are being done automatically, which, to such extent, is not possible with existing systems.

4 Automating Intelligent Data Analysis

When conducting any kind of data analysis and planning to use the results in an application we can assume to have

1. A problem definition: What is the problem we would like to solve? Do we want to predict or classify, to find groups (clustering), rules or dependencies etc?
2. Preferences regarding the solution: If the model can be adapted to new data, if it is easy to understand (rule-based, simple functions), its accuracy, execution time etc.

Experts in data analysis are well aware of these analysis problems and preferences. In the following, we assume that the data has already been prepared for analysis, i.e. it has been compiled from different data sources, if necessary, and formatted in a way that it can be used by standard data analysis tools. This process is mostly done with ETL tools (Extract, Transform, Load). An expert would then

1. Choose analysis algorithms which are suited for the problem and the given preferences
2. Pre-process data if necessary for the algorithms
3. Configure training parameters of the algorithms

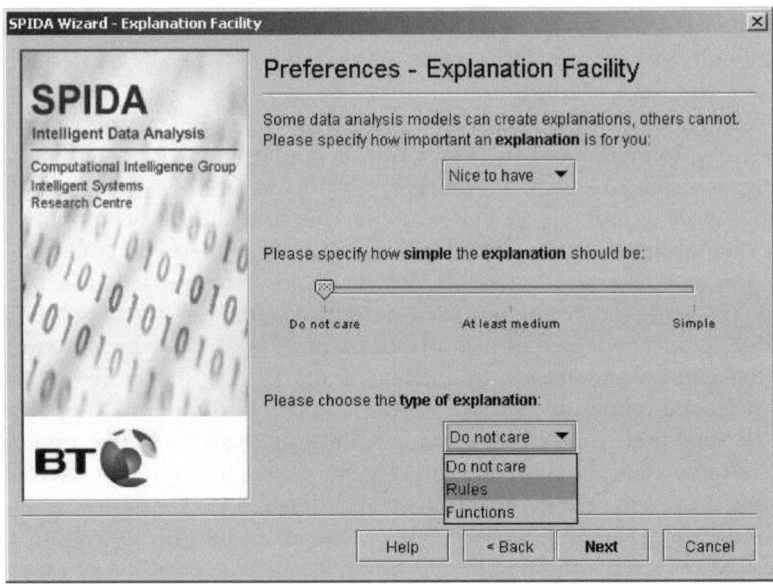

Fig. 2. Specifying preferences in the wizard, here regarding an explanation facility

4. Run the algorithms, check the performance of created models and match with preferences again
5. If improvement seems necessary and possible, go back to 3
6. Implement the model in an executable form
7. Integrate with operational system

The last two steps will be further discussed in Sect. 6.

Obviously, all of the above steps require expert knowledge. The wizard in SPIDA (Fig. 2), on the other hand, performs the steps to certain extent automatically. It engages the user in a sequence of dialogs where he can specify the analysis problem and his solution preferences as well as point to the data source. The wizard will then present all applicable analysis algorithms to the user ranked by their suitability. The user can pick the most suitable ones and trigger the creation of models. The wizard builds data analysis processes around the chosen algorithms, pre-processes the data, configures the algorithms, executes the analysis and validates the models. Information about accuracy and model complexity (e.g. the size of a generated rule base) is then available and can be incorporated in measuring the suitability of the models again. If an improvement of accuracy or model complexity according to the user requirements seems necessary and possible, the wizard will go back to Step 3, reconfigure the training algorithms and rebuild the model. The reconfigurations are a collection of expert rules. For example, if a decision tree is too complex (too many rules for a simple explanation), the maximum tree height can be reduced and a smaller tree being induced from data.

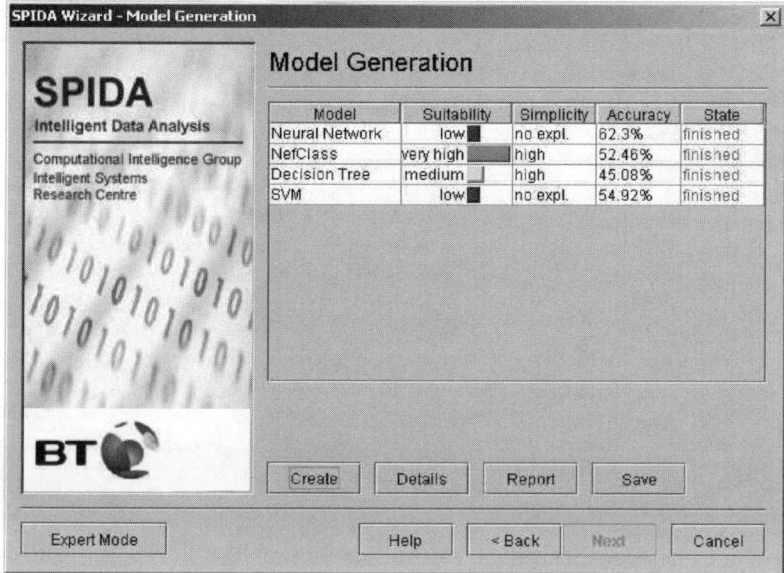

Fig. 3. Ranking of created models in the wizard

After the model building process has finished, the wizards again ranks all available models according to their suitability. Figure 3 shows four classification models with their overall suitability, simplicity of explanation (if exists) and accuracy, whereby simplicity was more important than accuracy. In this example, the user was looking for an adaptable predictive model with simple explanation. Results including a problem description, basic information about the data, model accuracy and explanations are presented in a HTML report. Figure 4 shows the part of report dealing with a fuzzy rule set as the learned model for a classifier.

Even at this stage, the user can still vary some of his requirements and finally pick the model he likes to use in his application. In the service scenario presented in this chapter, the user is taken out of the loop and the wizard automatically picks the most suitable method according to the ranking.

Figure 5 shows two typical data analysis processes automatically created by the wizard. The blocks represent data access, filters, data analysis models and visualisations. Such a process can also be manually designed, configured and executed in SPIDA's expert mode.

The wizard is based on knowledge collected from experts in data analysis. The basic idea is to match user requirements with properties of analysis methods and models. Since user requirements might look quite different from model properties, we use a a rule base to map user requirements onto desired properties, see Fig. 6. Some requirements and properties are inherently fuzzy, a user might for example ask for a model that is *easy to understand*. For this reason, the mapping rules are fuzzy as well.

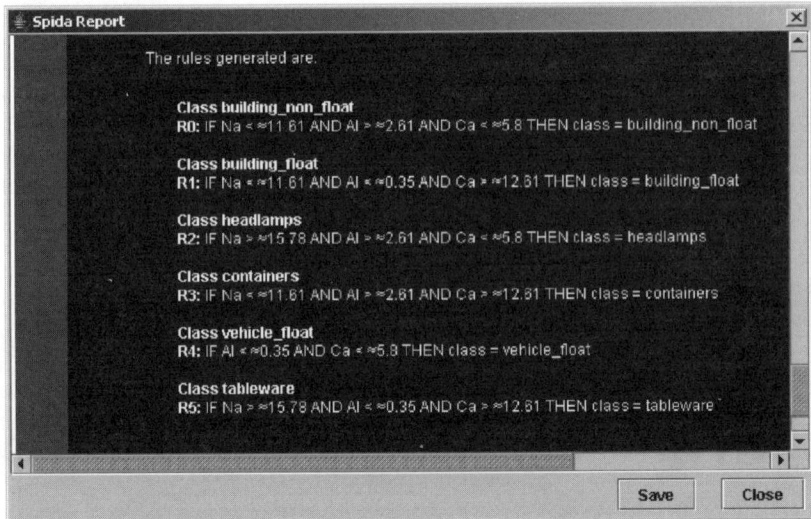

Fig. 4. An explanatory model for a classification problem in a report generated by SPIDA

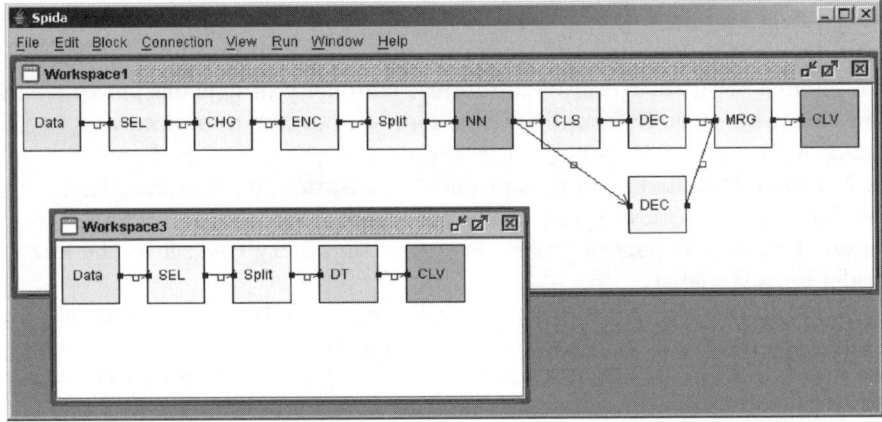

Fig. 5. Two typical data analysis processes created by the wizard with blocks for data access, filters and data analysis models

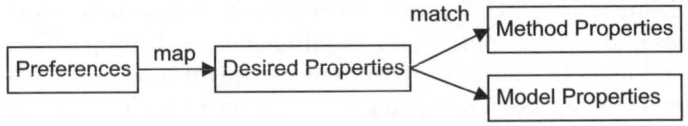

Fig. 6. Map user requirements onto desired properties and match with actual properties

In the current version of the wizard, the analysis problem and the user preferences are specified at a similar level as desired properties. They include

- Type of analysis problem (classification, function approximation, clustering, dependency analysis etc.)
- Importance of an explanation facility (do not care, nice to have, important)
- Type of explanation (do not care, rules, functions)
- Adaptability to new data (do not care, nice to have, important)
- Integration of prior knowledge (do not care, nice to have, important)
- Simplicity of an explanation
- Accuracy
- Balance importance of accuracy and simplicity

A typical mapping rule is "If simplicity preference is high and an explanation is important, the desired simplicity is medium with degree 0.6 or high with degree 1.0".

The underlying fuzzy reasoning concepts are described in detail in [3]. Using fuzzy matching techniques as briefly described in Sect. 5 and in more detail in [4], the desired properties will be matched with method and model properties of all available analysis algorithms. Thereby, method properties are the ones related to the analysis method like fuzzy methods providing a model explanation in terms of rules as opposed to a black box method like a neural network. Model properties, on the other hand, are specifically related to the model as being created by a method, e.g. the actual accuracy of a trained neural network. Naturally, model properties can only be evaluated after a model has been created (Step 4) whereas method properties can be checked at any time.

The current version of SPIDA and its wizard supports Neural Networks, Fuzzy systems, Neuro-Fuzzy systems NEFPROX and NEFCLASS, Support Vector Machines, Decision Trees, Linear Regression and Association Rules among other techniques.

Alternative approaches for the automatic creation of data analysis processes include [5–8]. The system closest to our Wizard regarding the required capabilities is the one described in [7, 8]. It uses an ontology based approach and simply describes analysis methods and pre-/post-processing methods as input/output blocks with specific interfaces. The system is built on top of the data analysis package Weka [9]. If the interfaces between two blocks match, they can be concatenated in an analysis process. If a user wants to analyse a data set, all possible analysis processes are created and executed. Once a suitable analysis process has been identified, it can be stored, re-used and shared. The authors suggest a heuristic ranking of analysis processes in order to execute only the best processes. However, they only use speed as a ranking criterion, which can be easily determined as a feature of an algorithm. More useful features about the quality of the analysis like accuracy are obviously dependent on the analysis process as well as the analysed data and are much more difficult to determine up front. Therefore, the reported results have not been very encouraging so far.

5 Measuring the Match of Requirements and Properties

We assume that the original user requirements have already been mapped onto desired properties and that information about the respective method and model properties is available. Each property is represented by a fuzzy variable that takes weighted combinations of fuzzy words as their values [3]. The desired accuracy, for example, could be 'medium (1.0) + high (1.0)' whereas a created analysis model might be accurate to the degree of 'low (0.3) + medium (0.7)'. The numbers in brackets denote the weight of the respective fuzzy word. In other words, we are looking for a model with medium or high accuracy, and the created model's accuracy is low with degree 0.3 and medium with degree 0.7. In this example, the weights for the desired accuracy sum up to 2, whereas the ones for the actual accuracy add up to 1. We interpret a combination of fuzzy words with a sum of weights greater than one as alternative fuzzy words. Rather than modeling the weights as probabilities as in [3], we assume a possibility density function [10, 11] on the fuzzy words, which allows for alternative values which, as an extreme case, could all be possible without restriction. Thereby, we define the semantics of a fuzzy word's degree of possibility as the maximum acceptable probability of a property. In the example above, we were entirely happy with an analysis model of medium or high accuracy. We therefore assigned the possibilistic weight 1 to both of them, i.e. models exhibiting the property 'low (0.0) + medium (a) + high (b)' with $a + b = 1$ are fully acceptable. In case of the requirement 'low (0.0) + medium (0.0) + high (1.0)' and the above property with $a > 0$, the weight a exceeds the possibility for 'medium' and therefore at least partially violates the requirements. Degrees of possibility can be any real number in $[0, 1]$.

In [4], we derived the following measure C for the compatibility of fuzzy requirements $\tilde{R}$ and fuzzy properties $\tilde{P}$. $\mu_{\tilde{R}}$ and $\mu_{\tilde{P}}$ denote the fuzzy membership functions of $\tilde{R}$ and $\tilde{P}$ defining the weights of fuzzy words. Using the examples above, we might have $\mu_{\tilde{P}}(\text{low}) = 0.3$, $\mu_{\tilde{P}}(\text{medium}) = 0.7$ and $\mu_{\tilde{P}}(\text{high}) = 0$.

$$C(\tilde{P}, \tilde{R}) = 1 - \sum_{\substack{x \in \mathcal{X}: \\ \mu_{\tilde{P}}(x) > \mu_{\tilde{R}}(x)}} \mu_{\tilde{P}}(x) - \mu_{\tilde{R}}(x) \tag{1}$$

Figure 7 illustrates (1) showing a fuzzy set $\tilde{R}$ for requirements on the left hand side (we used a continuous domain for better illustration), the right

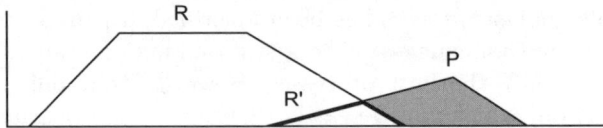

Fig. 7. Fuzzy sets of requirements $\tilde{R}$, properties $\tilde{P}$ and the intersection $\tilde{R}'$. The grey area represents incompatibility of properties with requirements

triangular function is the fuzzy set $\tilde{P}$ for properties. The right term in (1) measures the size of the grey area, which can be interpreted as a degree to which the properties violate the requirements. The size of the area is bounded by the area underneath the membership function of $\tilde{P}$ which we stipulated to be 1. That means that the right term measures the proportion of properties $\tilde{P}$ that violate the requirements and C altogether measures the area of the intersection, i.e. the compatibility of $\tilde{P}$ and $\tilde{R}$. More detailed information can be found in [4] and about matching fuzzy sets in general in [12].

6 Integration of Analytics Module with Operational System

When it comes to integrating modules with operational systems, software engineers can nowadays choose from a variety of techniques that both depend on the operational system and the module. Where some years ago, modules were quite often especially tailored for the operational system, nowadays, much more modular approaches are used that allow for using the same modules in different operational systems on the one hand, and replacing modules in operational systems on the other hand. Apart from the database approach mentioned in Sect. 2, where data analysis can directly be conducted by the database and analysis models are defined using a standard like PMML, other techniques use libraries with standard interfaces (API) and provide a respective interface in the operational system or use even less coupled techniques like web services. For our system, we decided to build executable Java libraries which can easily be used in either way.

Since we require a variety of different data analysis techniques to cover as many applications as possible, the starting point for the generation of analysis modules are data analysis platforms like SPIDA. The actual analysis model required by an operational system typically uses only a tiny fraction of such a platform's functionality. For reasons of saving space and finally costs – depending on the license model, pricing depends on the functionality being chosen – we want to restrict the module to the functions actually needed by the underlying data analysis model or process.

In case of SPIDA, we are looking to create analytics modules, which are pieces of software independent from the SPIDA platform, corresponding to a particular data analysis model. If extracted the independent piece of software can be used as a library with fixed API by the operational system. As mentioned above, changes in the analysis model are completely hidden from the operational system by the module, i.e. the module can easily be replaced if required due to changes in requirements or dropping performance.

At the heart of the integration of an analytic module in SPIDA is a general framework with which an operational system can communicate and specify its requirements and obtain an analytic module from SPIDA as a Java library.

Table 1. Compression rates of dedicated Java libraries for various mining algorithms compared to whole SPIDA package

Model	Relative size (%)
Linear Regression	33
NEFPROX	48
Decision Trees	17
Support Vector Machines	17
Neural Networks	21
NEFCLASS	47

First, the definition of the analysis problem and the user requirements are being transferred to the wizard that builds a data analysis process. This process will be transformed into an XML description. SPIDA parses the XML description and identifies the corresponding analysis components like data access, filters and the actual analysis blocks. A mapping is made between the required components and the corresponding Java classes of the platform. SPIDA's software extractor component then finds all the dependent classes for this initial set of classes. The extractor uses application knowledge and a host of other extraction techniques to find all the additional classes required to make it executable and independent from the platform. This complete set of all component classes is turned into a Jar library with suitable properties by a Jar making component of SPIDA. This Jar library is passed to the operational system, which can access the API provided in the library to control the execution of the analysis module.

As an alternative to using the wizard, XML descriptions of data analysis processes can directly be used to extract a Java library which is able to execute the given process. Furthermore, SPIDA provides a simple wrapper which can execute a given Java library. In that way we can easily create compact stand-alone packages for specific data analysis tasks.

If we think of data analysis services, we can expect many different concurrent analysis requests. Since each analysis process will create an executable object, the size of dedicated Jar libraries should be kept as small as possible in order to allow for as many users as possible. Table 1 shows the compression rates for libraries created for different data mining algorithms compared to the whole SPIDA package, which range between 47 and 17%.

7 Conclusions

As a reaction to increasing dynamics in the market place, we propose a generic architecture for a smart data analysis service for operational systems. Where current systems require manual intervention by data analysis experts if a data analysis module needs updating, our architecture automatically detects changes in performance or requirements, builds a new data analysis model and wraps the model into an executable Java library with standard API that can easily be accessed by operational systems. When building data analysis models, the wizard takes into account high-level user requirements regarding an explanation facility, simplicity of an explanation, adaptability, accuracy etc. In this way, we can vary the data analysis solution as much as possible without violating requirements. Until now, adaptive solutions only retrained models based on new data, i.e. an analysis method like a neural network was chosen once and for all and regularly adapted to new data. Switching automatically – without user interaction – to another method, e.g. to a support vector machine was not possible, let alone using user requirements for the automatic selection of the most suitable analysis method.

The data analysis platform SPIDA including the wizard has been fully implemented, also the automatic generation of executable data analysis modules. Future work includes assembling the entire architecture and trialling it in one of BT's operational Customer Relationship Management (CRM) systems.

References

1. Nauck, D., Spott, M., Azvine, B.: Spida – a novel data analysis tool. BT Technology Journal **21**(4) (2003) 104–112
2. Various: The data mining group. http://www.dmg.org
3. Spott, M.: Efficient reasoning with fuzzy words. In: Halgamuge, S.K., Wang, L., eds.: Computational Intelligence for Modelling and Predictions. Springer, Berlin Heidelberg New York, 2005
4. Spott, M., Nauck, D.: On choosing an appropriate data analysis algorithm. In: Proc. IEEE Int. Conf. on Fuzzy Systems 2005, Reno, USA, 2005 397–403
5. Wirth, R., Shearer, C., Grimmer, U., Reinartz, T.P. and Schloesser, J., Breitner, C., Engels, R., Lindner, G.: Towards process-oriented tool support for knowledge discovery in databases. In: Principles of Data Mining and Knowledge Discovery. First European Symposium, PKDD '97. Number 1263 in Lecture Notes in Computer Science, Springer, Berlin Heidelberg New York, 1997, pp. 243–253
6. Botia, J.A., Velasco, J.R., Garijo, M., Skarmeta, A.F.G.: A generic datamining system. Basic design and implementation guidelines. In: Kargupta, H., Chan, P.K., eds.: Workshop in Distributed Datamining at KDD-98, AAAI Press, New York, 1998
7. Bernstein, A., Provost, F.: An intelligent assistant for the knowledge discovery process. CeDER Working Paper IS-01-01, Center for Digital Economy Research, Leonard Stern School of Business, New York University, New York 2001
8. Bernstein, A., Hill, S., Provost, F.: Intelligent assistance for the data mining process: an ontology-based approach. CeDER Working Paper IS-02-02, Center for Digital Economy Research, Leonard Stern School of Business, New York University, New York, 2002
9. Witten, I.H., Frank, E.: Data Mining: Practical Machine Learning Tools and Techniques with JAVA Implementations. Morgan Kaufmann Publishers, San Francisco, CA, 2000

10. Gebhardt, J., Kruse, R.: The context model – an integrating view of vagueness and uncertainty. International Journal of Approximate Reasoning **9** (1993) 283–314
11. Spott, M.: Combining fuzzy words. In: Proc. IEEE Int. Conf. on Fuzzy Systems 2001, Melbourne, Australia, 2001
12. Bouchon-Meunier, B., Rifqi, M., Bothorel, S.: Towards general measures of comparison of objects. Fuzzy Sets and Systems **84**(2) (1996) 143–153

Indexing Evolving Databases for Itemset Mining

Elena Baralis, Tania Cerquitelli, and Silvia Chiusano

Dipartimento di Automatica e Informatica, Politecnico di Torino, Corso Duca degli
Abruzzi 24, 10129 Torino, Italy
elena.baralis@polito.it, tania.cerquitelli@polito.it,
silvia.chiusano@polito.it

Summary. Research activity in data mining has been initially focused on defining
efficient algorithms to perform the computationally intensive knowledge extraction
task (i.e., itemset mining). The data to be analyzed was (possibly) extracted from
the DBMS and stored into binary files. Proposed approaches for mining flat file
data require a lot of memory and do not scale efficiently on large databases. An improved memory management could be achieved through the integration of the data
mining algorithm into the kernel of the database management system. Furthermore,
most data mining algorithms deal with "static" datasets (i.e., datasets which do not
change over time). This chapter presents a novel index, called I-Forest, to support
data mining activities on evolving databases, whose content is periodically updated
through insertion (or deletion) of data blocks. I-Forest is a covering index that represents transactional blocks in a succinct form and allows different kinds of analysis.
Time and support constraints (e.g., "analyze frequent quarterly data") may be enforced during the extraction phase. The I-Forest index has been implemented into
the PostgreSQL open source DBMS and it exploits its physical level access methods.
Experiments, run for both sparse and dense data distributions, show the efficiency
of the proposed approach which is always comparable with, and for low support
threshold faster than, the Prefix-Tree algorithm accessing static data on flat file.

1 Introduction

Many real-life databases (e.g., data marts) are updated by means of blocks
of periodically incoming business information. In these databases, the content
is periodically updated through either addition of new transaction blocks or
deletion of obsolete ones. Data can be described as a sequence of incoming data
blocks, where new blocks arrive periodically or old blocks are discarded [12,13].
Examples of evolving databases are transactional data from large retail chains,
web server logs, financial stock tickers, and call detail records. Since the data
evolve overtime, algorithms have to be devised to incrementally maintain data
mining models.

E. Baralis et al.: *Indexing Evolving Databases for Itemset Mining*, Studies in Computational
Intelligence (SCI) **109**, 305–323 (2008)
www.springerlink.com

Different kinds of analysis could be performed over such data like (i) mining all available data (ii) mining only the most recent data (e.g., last month data), (iii) mining periodical data (e.g., quarterly data) and (iv) mining selected data blocks (e.g., data related to the first month of last year and the first month of this year). Consider for example transactional data from large retail chains, where every day, after shop closing, a set of transactions is added to the database [12]. In this scenario, market analysts are interested in analyzing different portions of the database to discover customer behaviors. For example, they may be interested in analyzing purchases before Christmas or during summer holidays.

Association rule mining is an expensive process that requires a significant amount of time and memory. Hence, appropriate data structures and algorithms should be studied to efficiently perform the task. Association rule mining is a two-step process: Frequent itemset extraction and association rule generation. Since the first phase is the most computationally intensive knowledge extraction task [1], research activity has been initially focused on defining efficient algorithms to perform this extraction task [1, 3, 4, 7, 10, 15, 18, 20]. The data to be analyzed is (possibly) extracted from a database and stored into binary files (i.e., flat files). Many algorithms, both memory-based [1, 3, 7, 10, 13, 16, 18] and disk-based [4, 20], are focused on specialized data structures and buffer management strategies to efficiently extract the desired type of knowledge from a flat dataset.

The wide diffusion of data warehouses caused an increased interest in coupling the extraction task with relational DBMSs. Various types of coupling (e.g., loose coupling given by SQL statements exploiting a traditional cursor interface, or tight coupling provided by optimizations in managing the interaction with the DBMS) between relational DBMSs and mining algorithms have been proposed.

A parallel effort was devoted to the definition of expressive query languages to specify mining requests. These languages often proposed extensions of SQL that allowed the specification of various types of constraints on the knowledge to be extracted (see [5] for a review of most important languages). However, the proposed architectures were at best loosely coupled with the underlying database management system [17]. The final dialogue language with the relational DBMS was always SQL.

One step further towards a tighter integration is made in [6], where techniques for optimizing queries including mining predicates have been proposed. Knowledge of the mining model is exploited to derive from mining predicates simpler predicate expressions, that can be exploited for access path selection like traditional database predicates. DBMSs exploit indices to improve the performance on complex queries. The intuition that the same approach could be "exported" to the data mining domain is the driving force of this chapter. A true integration of a novel data mining index into the PostgreSQL open source DBMS [19] is proposed and advantages and disadvantages of the proposed disk-based data structures are highlighted.

In this chapter we address itemset extraction on evolving databases. We propose an index structure, called Itemset-Forest (I-Forest), for mining data modeled as sequences of incoming data blocks. The index supports user interaction, where the user specifies different parameters for itemset extraction. It allows the extraction of the complete set of itemsets which satisfy (i) time constraints (which temporal data we are interested in) and (ii) support constraints (minimum itemset frequency).

Since the I-Forest index includes all attributes potentially needed during the extraction task, it is a covering index. Hence, the extraction can be performed by means of the index alone, without accessing the database. The data representation is complete, i.e., no support threshold is enforced during the index creation phase, to allow reusing the index for mining itemsets with any support threshold.

The I-Forest index is characterized by a set of compact structures, one for each incoming data block. Each structure provides a locally compact representation of the data block. This modular structure allows dynamically updating the I-Forest index when new data blocks arrive or old data blocks are discarded. An interesting feature of the index is its ability to represent distinct data blocks by means of different data structures. Hence, the data structure of a block can be adapted to the data distribution.

The physical organization of the I-Forest index supports efficient information retrieval during the itemset extraction task. The I-Forest index allows selective access to the needed information, thus reducing the overhead in accessing disk blocks during the extraction task. Our approach, albeit implemented into a relational DBMS, yields performance always comparable with and for low supports faster than the Prefix-tree algorithm [15] (FIMI '03 Best Implementation) on flat file.

2 Problem Statement

Let $I = \{i_1, i_2, \ldots, i_n\}$ be a set of items. A transactional data block b is a collection of transactions, where each transaction T is a set of items in I. Let b_k be the instance of b arrived at time k, where k is a time identifier. The state of a transactional database D at time t reflects all blocks received until time t. Hence, it is a finite sequence of data blocks $\{b_1, b_2, \ldots, b_n\}$, denoted as $D\,[1 \ldots t]$. $D\,[1 \ldots t]$ can be represented as a relation R, where each tuple is a triplet *(Time, TransactionId, ItemId)*. When a new block arrives, its transactions are added to R and t is updated. Analysis can be performed on sets of blocks not necessarily in sequence. We denote as Ω the set of time identifiers of the analyzed blocks (e.g., $\Omega = \{1, 2, 5\}$ means that we are interested in blocks $\{b_1, b_2, b_5\}$). $R_\Omega \subseteq R$ is the set of tuples associated to the blocks in Ω.

Given a set of constraints, the itemset extraction task is the extraction of the complete set of itemsets in R which satisfy the constraints. Constraints are among the following:

- *Time constraint*, which allows the selection of a subset of blocks in R by means of Ω.
- *Support constraint*, which defines the minimum support threshold to perform itemset extraction.

3 I-Forest Index

Itemset mining on sequences of incoming data blocks, may require a significant amount of main memory during the extraction process. By means of the I-Forest index, incoming data blocks are stored on disk in appropriate (compact) structures. During the mining phase, only the data required by the current mining process is actually loaded in main memory.

To allow different kinds of analysis and easy incremental insertion of new blocks (or deletion of obsolete ones), each data block is represented separately and independently of all others. Hence, the index is a forest of structures that represent data blocks in R. Each data block b_k in R is represented in a compact structure named *Itemset Forest block* (IF-block). A separate structure, the *Itemset Forest-Btree* (IF-Btree), links information belonging to different blocks that is accessed together during the mining process. The IF-Btree is a B-tree structure, that provides selective access to data in the IF-blocks to support efficient retrieval of information during the mining process. The complete structure of the I-Forest index is shown in Fig. 1.

Since the I-Forest index should be reusable for many extraction sessions with different constraints, its data representation is complete. No constraint (e.g., support) is enforced during the index creation phase. I-Forest is a covering index, i.e., it allows itemset extraction without accessing relation R. It includes both the time and item identifier attributes, which may be needed during the extraction task.

The I-Forest structure can be easily updated when new data blocks arrive (or old data blocks are removed). Each block b_k is stored independently

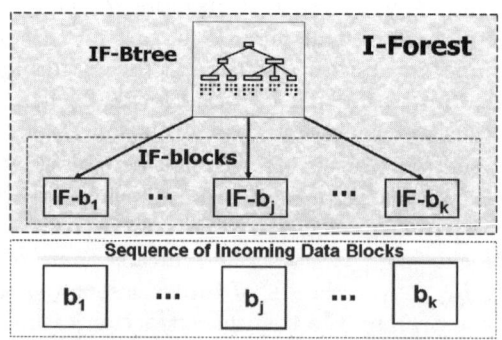

Fig. 1. I-Forest structure

by means of a IF-block, which provides a lossless and locally compact representation of the corresponding portion of R. Each IF-block may adopt a different data representation. Hence, the I-Forest index can easily adapt to skewed block distributions, where each block is characterized by a different data distribution.

3.1 I-Forest Structure

The I-Forest index includes two elements: the *Itemset Forest-blocks* (IF-blocks) and the *Itemset Forest-Btree* (IF-Btree). In the following we describe in more detail each structure.

IF-blocks. Many different compact structures could be adopted for representing IF-blocks (e.g., FP-tree [16], Inverted Matrix [10], Patricia-Tries [18], I-Tree [4]). Currently, each IF-block is represented by means of a slight variation of the I-Tree. For each (relational) data block b_k in R, an I-Tree (It_k) and an ItemList (IL_k) are built [4]. The It_k associated to b_k is a prefix tree, where each node corresponds to an item and each path encodes one or more transactions in b_k. Each node is associated with a node support value. This value is the number of transactions in b_k which contain all the items included in the subpath reaching the node. Each item is associated to one or more nodes in the same It_k. IL_k has one entry for each item in b_k, for which it reports the $block_k$-item-support value, i.e., the item frequency in b_k, and the block identifier k. The $block_k$-item-support value is obtained by adding the supports of all nodes in It_k associated to the item. The global item support value is the frequency of the item in R. This value is obtained by adding the $block_k$-item-supports of the item for each block b_k in R.

Tables 1 and 2 report (in a more succinct form than the actual relational representation) two data blocks, arrived at times t $= 1, 2$, used as a running example. Figures 2 and 3 show the structure of the corresponding It_1 and It_2

Table 1. Data block at time t $= 1$

Time	TID	Items
1	1	g,b,h,e,p,v,d
1	2	e,m,h,n,d,b
1	3	p,e,c,i,f,o,h
1	4	j,h,k,a,w,e
1	5	n,b,d,e,h
1	6	s,a,n,r,b,u,i
1	7	b,g,h,d,e,p
1	8	a,i,b
1	9	f,i,e,p,c,h
1	10	t,h,a,e,b,r
1	11	a,r,e,b,h
1	12	z,b,i,a,n,r

Table 2. Data block at time t = 2

Time	TID	Items
2	1	L
2	2	l,q
2	3	e,l,q,r
2	4	b,d,e,l,q,r
2	5	e,f,l,q,r,z
2	6	d,x,z
2	7	a,c,d,e,x,z
2	8	a,c,e,f,g,x
2	9	a,b,c,x
2	10	q,r,x
2	11	b,d,g,h,q,r,x
2	12	a,b,c,d,q,r,x,z

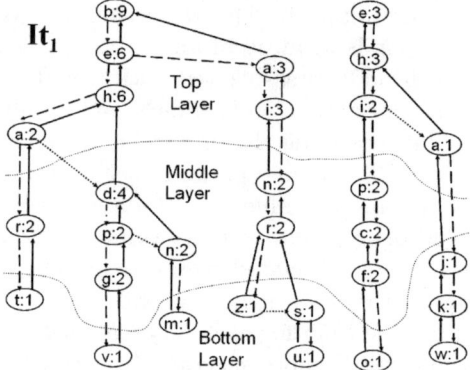

Fig. 2. IF-blocks: It_1 for block 1

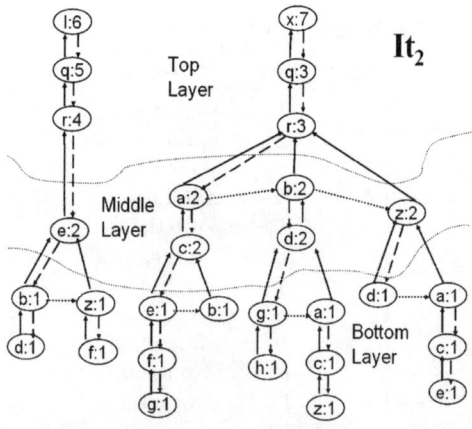

Fig. 3. IF-blocks: It_2 for block 2

associated to blocks 1 and 2 respectively. The corresponding IL_1 and IL_2 are omitted to ease readability. Consider item e in It_1. Its block$_1$-item-support is 9 (there are two nodes associated to item e, the first has node support 6 and the second 3). Furthermore, the global item support of e is 13 (its block$_1$-item-support is 9 and its block$_2$-item-support is 4).

As shown in Figs. 2 and 3, in each It_k nodes are clustered in three layers: Top, middle, and bottom [4]. Correlation analysis is exploited to store in the same disk block nodes accessed together during the mining process to reduce the number of reads. Nodes in the IF-block are linked by means of pointers which allow the retrieval from disk of the index portion required by the mining task. Each node is provided with pointers to three index blocks: (i) The disk block (and offset) including its parent (continuous edges in Figs. 2 and 3), (ii) the disk block (and offset) including its first child (dashed edges in Figs. 2 and 3), and (iii) the disk block (and offset) including its right brother (dotted edges in Figs. 2 and 3). The pointers allow both bottom-up and top-down tree traversal, thus enabling itemset extraction both with item and support constraints.

IF-Btree. It allows selective access to the IF-block disk blocks during the mining process. It has one entry for each item in relation R. The IF-Btree leaves contain pointers to all nodes in the IF-blocks. Each pointer contains the node physical address and the identifier of the data block including the node.

3.2 Updating the I-Forest Index

The proposed index supports efficient index updating when data blocks are either inserted or deleted. When a new data block b_k is available, the corresponding IF-block (It_k) is built. The IF-Btree structure is also updated by inserting pointers to the nodes in It_k. Hence, It_k data are linked to the data already in the I-Forest index. When a data block b_k is discarded, the corresponding It_k is removed and the IF-Btree is updated by removing pointers to the nodes in It_k. Functions available in the PostgreSQL library are used to update the IF-Btree.

3.3 Storage

The I-Forest index is stored in two relational tables. Table T_{Items} stores all IL_k, while table $T_{IF\text{-}blocks}$ stores every It_k. The IF-Btree is stored in a B-Tree structure. To access records in $T_{IF\text{-}blocks}$, T_{Items}, and in the IF-Btree, functions available in the access methods of PostgreSQL [19] are used. Table T_{Items} contains one record for each item that appears in each It_k. Each record contains data block identifier, item identifier, and block$_k$-item-support. Table $T_{IF\text{-}blocks}$ contains one record for each node in each It_k. Each record contains node identifier, item identifier, local node support, physical location (block number and offset within the block) of its parent, physical location (block number and offset within the block) of its first child, and physical location (block number and offset within the block) of its right brother.

4 Mining Itemsets

This section describes how itemset extraction with different constraints exploits the I-Forest index. In this chapter we address the following constraints:

- *Time constraint.* Set Ω contains the time identifiers for the selected data blocks. When the time identifier k is included in Ω, the corresponding data block b_k is analyzed. When no time constraint is enforced, all data blocks are considered in the extraction process.
- *Support constraint.* Itemsets are extracted when their frequency is higher than a minimum threshold.

Itemset extraction is performed in two steps. First, eligible items, i.e., items which satisfy all the enforced constraints, are selected. Then, the extraction process is performed. To retrieve all necessary information, both tables $T_{IF-blocks}$ and T_{Items}, and the IF-Btree are accessed by using the read functions available in the PostgreSQL access methods [19].

Enforcing constraints. The constraint enforcement step selects the items which both belong to blocks in Ω and satisfy the support constraint. These items are stored in set Λ. When no support constraint is enforced, items are inserted into Λ if they belong to at least one block in Ω. When a support constraint is enforced, we compute the joint support for each item by considering the block$_k$-item-supports for blocks in Ω. For each block b_k, the block$_k$-item-supports are stored in IL_k. Items with joint support higher than the support threshold are finally inserted into Λ.

Itemset extraction. The extraction is performed in two steps: (i) Retrieval of the necessary data from the I-Forest index and (ii) itemset extraction from the loaded data.

For each item in Λ, we read all the I-Forest paths including the item. To this aim, we retrieve the IF-Btree leaves storing the pointers to all I-Forest nodes associated to the item. Each pointer includes the node physical address and the identifier of the data block including the node. Only nodes belonging to blocks in set Ω are considered.

For each node associated to the item, all the I-Forest paths including it are retrieved from disk. Each node is read from table $T_{IF-blocks}$ by means of its physical address. The path between the node and the tree root is traversed bottom up. This path is read from table $T_{IF-blocks}$ by following the physical addresses of parent nodes. The paths in the node subtree are traversed top down. These paths are read from table $T_{IF-blocks}$ by following the physical addresses of the first child node and the right brother node.

Once read, paths are initially stored in main memory in a temporary data structure. When all paths including the current item have been retrieved, the temporary data structure is compacted in a tree structure similar to the FP-Tree [16]. Next, itemset extraction is performed by means of an adaptation of the FP-Growth algorithm [16], which is a very efficient algorithm for itemset extraction.

5 Experimental Results

We performed a variety of experiments to validate our approach, by addressing the following issues: (i) performance to generate the I-Forest index, in terms of both index creation time and index size, (ii) performance to extract frequent itemsets (by mining all available data or selected data blocks), in terms of both execution time and memory usage, (iii) effect of the DBMS buffer cache size on the hit rate, and (iv) effect of the time constraint.

We ran various experiments by considering both dense and sparse data distributions. We report here the results of the experiments on four representative datasets: Connect, Pumsb, and Kosarak datasets downloaded from UCI Machine Learning Repository [11] and the synthetic dataset T25I300D6M [2]. Connect and Pumsb [11] datasets are dense datasets, while Kosarak [11] and T25I300D6M [2] are very large and sparse. Dataset characteristics are reported in Tables 3 and 4.

To simulate block evolution, we considered three different configurations for Connect, Pumsb, and Kosarak datasets. For the first two configurations, denoted as Dataset-50 + 50 and Dataset-25 + 25 + 25 + 25, we split the original dataset respectively in two and four blocks with the same number of transactions. The last[1] configuration, denoted as Dataset-100 + 100, is composed by two identical blocks obtained by cloning the original dataset. The original dataset, denoted as Dataset-100, represents our lower bound on performance since it has no overhead due to block partitioning. To validate our approach, we compare our performance to the FIMI best implementation algorithm Prefix-tree [15] (FIMI '03 Best Implementation), a very effective state of the art algorithm for itemset extraction from flat file.

To analyze the performance of the extraction process in large evolving databases, we exploited the synthetic dataset T25I300D6M. This dataset is characterized by a sparse data distribution and a very large number of transactions (i.e., 5,999,988). We split the original dataset in 12 blocks with the same number of transactions. We analyze the performance of the extraction process when increasing the number of selected index blocks and for various support thresholds.

Both the index creation procedure and the itemset extraction algorithm are coded into the PostgreSQL open source DBMS [19]. They have been developed

---

[1] We also considered different block partitioning strategies, e.g., Dataset-75 + 25, where the first block contains 75% of the transactions and the second 25%. These experiments are not reported, since results are not significantly different.

Table 3. Characteristics of dataset configurations and corresponding index

Dataset configuration	Dataset				I-Forest index		
	Transactions	Items	AvTrSz	Size (KB)	IF-Blocks (KB)	IF-Btree (KB)	Creation time (s)
CONNECT-100	67,557	129	43	25,527	22,634	4,211	11.05
CONNECT-100 + 100	135,114	129	43	51,054	45,268	8,420	22.10
CONNECT-50	33,779	126	43	12,525	12,138	–	7.50
CONNECT-50	33,778	129	43	12,002	12,524	–	7.20
CONNECT-50 + 50	–	–	–	–	24,662	4,588	14.70
CONNECT-25	16,890	123	43	6,032	5,902	–	2.47
CONNECT-25	16,890	125	43	6,499	6,273	–	2.64
CONNECT-25	16,890	129	43	6,499	6,210	–	2.67
CONNECT-25	16,887	129	43	6,497	6,830	–	2.85
CONNECT-25 + 25 + 25 + 25	–	–	–	–	25,215	4,691	10.63
PUMSB-100	98,092	2,144	37.01	35,829	57,932	10,789	34.47
PUMSB-100 + 100	196,184	2,144	37.01	71,658	115,932	21,562	69.00
PUMSB-50	49,046	1946	37.01	17,712	30,415	–	21.50
PUMSB-50	49,046	1999	37.01	17,714	32,026	–	24.18
PUMSB-50 + 50	–	–	–	–	62,441	11,625	45.68
PUMSB-25	24,523	1787	37.01	8,756	16,592	–	8.41
PUMSB-25	24,523	1801	37.01	9,157	16,668	–	8.56
PUMSB-25	24,523	1829	37.01	8,759	17,740	–	9.04
PUMSB-25	24,523	1824	37.01	9,157	17,113	–	8.73
PUMSB-25 + 25 + 25 + 25	–	–	–	–	68,132	12,676	34.74
KOSARAK-100	1,017,029	41,244	7.9	85,435	312,647	58,401	893.81
KOSARAK-100 + 100	2,034,058	41,244	7.9	170,870	625,294	116,472	1787.50
KOSARAK-50	508,515	35,586	7.9	42,310	160,020	–	648.56
KOSARAK-50	508,515	35,139	7.9	43,125	159,458	–	651.06
KOSARAK-50 + 50	–	–	–	–	319,479	59,609	1299.62
KOSARAK-25	254,258	30,477	7.9	20,692	81,416	–	115.25
KOSARAK-25	254,258	29,666	7.9	21,167	82,199	–	109.27
KOSARAK-25	254,258	30,023	7.9	21,988	81,425	–	111.76
KOSARAK-25	254,255	29,531	7.9	21,588	81,512	–	108.80
KOSARAK-25 + 25 + 25 + 25	–	–	–	–	326,552	60,386	445.08

in ANSI C. Experiments have been performed on a 2,800 MHz Pentium IV PC with 2 GB main memory. The buffer cache of PostgreSQL DBMS has been set to the default size of 64 blocks (block size is 8 KB). All reported execution times are real times, including both system and user time, and obtained from the unix time command as in [14]. For the considered datasets, and for all configurations, the index has been generated without enforcing any support threshold.

5.1 I-Forest Index Creation and Structure

Table 3 reports the number of transactions and items, and the average transaction size (AvgTrSz) characterizing the considered dataset configurations for Connect, Pumsb, and Kosarak datasets. Table 3 also shows the size of the corresponding I-Forest index. Table 4 reports the same information for the T25I300D6M synthetic dataset. In this case, the dataset is partitioned in 12 blocks (denoted as T25I300D6M-b_i) including the same number of transactions.

Table 4. Characteristics of T25I300D6M dataset partitioning and corresponding indexes

Dataset configuration	Dataset				I-Forest index		
	Transactions	Items	AvTrSz	Size (KB)	IF-Blocks (KB)	IF-Btree (KB)	Creation time (sec)
T25I300D6M-b_1	500,000	28,381	25	163,573	711,497	–	639.90
T25I300D6M-b_2	500,000	28,402	25	166,285	711,273	–	629.18
T25I300D6M-b_3	500,000	28,401	25	178,526	712,295	–	633.63
T25I300D6M-b_4	500,000	28,409	25	178,496	712,285	–	630.62
T25I300D6M-b_5	500,000	28,408	25	178,494	711,754	–	631.87
T25I300D6M-b_6	500,000	28,398	25	178,563	711,733	–	629.49
T25I300D6M-b_7	500,000	28,392	25	178,560	711,749	–	631.27
T25I300D6M-b_8	500,000	28,399	25	178,583	712,081	–	629.09
T25I300D6M-b_9	500,000	28,386	25	178,563	711,587	–	630.95
T25I300D6M-b_{10}	500,000	28,414	25	178,541	711,689	–	626.27
T25I300D6M-b_{11}	500,000	28,399	25	178,601	712,104	–	630.64
T25I300D6M-b_{12}	499,988	28,404	25	178,533	711,664	–	631.19
T25I300D6M-b_{1-12}	5,999,988	300,000	25	2,110,318	8,541,711	397,097	7574.1

For each dataset configuration, the overall size of all IF-blocks is obtained by summing the size of the IF-blocks in the I-Forest index. The IF-Btree contains pointers to all nodes in the IF-blocks. Hence, the IF-Btree size is proportional to the total number of nodes in the IF-blocks of I-Forest.

The I-Tree internal block representation is more suitable for dense data distributions (e.g., the Connect dataset). In the I-Tree, correlated transactions are represented by a single path. Hence, in dense datasets where data are highly correlated, the I-Tree structure provides good data compression. For sparse data distributions (e.g., the Kosarak dataset), where data are weakly correlated, a lower data compression is achieved. In this case, storage of the IF-blocks requires more disk blocks. Alternative representations (e.g., Patricia-Tries [18]) may reduce the required disk space. Block partitioning increases the overall I-Forest size (including both IF-blocks and IF-Btree elements). The increase is higher for dense than for sparse data distributions, but always not significant. In dense datasets, the I-Tree compactness is partially lost due to block partitioning. Transactions correlated in the original dataset, but belonging to different data blocks, are represented as disjoint paths. In sparse datasets, most transactions are characterized by disjoint paths already in the original dataset. In particular, Connect-25 + 25 + 25 + 25 and Connect-50+50 configurations require 11.36 and 8.9% KB more than Connect-100. Pumsb-25 + 25 + 25 + 25 and Pumsb-50 + 50 configurations require 11.7 and 7.7% KB more than Pumsb-100. Instead, for the Kosarak dataset, which is sparser, Kosarak-25+25+25+25 and Kosarak-50+50 configurations require 2.16 and 4.28% KB more than Kosarak-100.

Index creation time is mainly affected by two factors: Writing the index blocks on disk and performing correlation analysis. Due to block partitioning, a larger number of nodes has to be written. However, smaller data blocks reduce the complexity of the correlation analysis task, and thus the time needed for it. Results show that the index creation time for the Dataset-25 + 25 + 25 + 25 configuration is always lower than (or at most comparable to) that for Dataset-100.

Since T25I300D6M is a very sparse dataset, block partitioning does not increase significantly the overall size of the I-Forest index. In this dataset most transactions are represented by disjoint paths already in the original dataset. The main overhead in the index size is due to the storage of the physical pointers that link nodes. However, this information is required to support the effective retrieval of correlated data during the mining process, thus reducing the number of disk reads.

5.2 Itemset Extraction Performance

To evaluate the performance of our approach, we considered the following configurations of the I-Forest index: (i) the best case, represented by Dataset-100 (no partitioning), (ii) general cases, represented by Dataset-50 + 50 and Dataset-25+25+25+25, and (iii) the worst case, represented by Dataset-100+100 (being the two data blocks identical, during the extraction phase both IF-blocks have always to be accessed).

Figures 4 and 5 report the run time for frequent itemset extraction with various support thresholds for Connect, Pumsb and Kosarak datasets, in the four configurations. Figures 4 and 5 also show the run time for the Prefix-Tree algorithm [15] on Dataset-100 (flat file). For all experiments we neglect time for writing generated itemsets. When low support thresholds are considered, our approach, albeit implemented into a relational DBMS, is always significantly faster than the Prefix-tree algorithm on flat file. When higher supports are considered, our approach yields performance comparable to the Prefix-tree

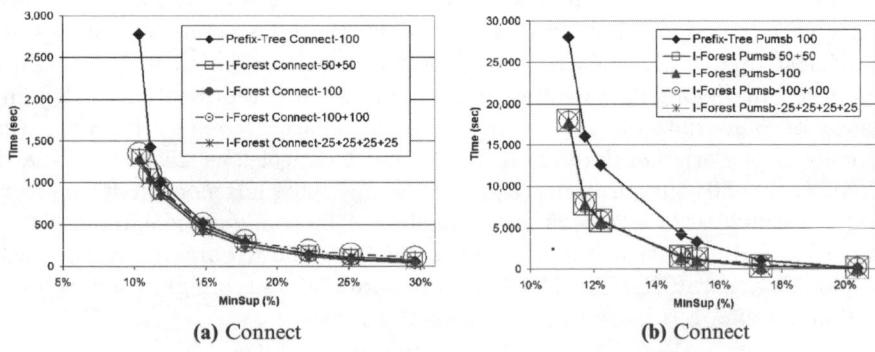

(a) Connect   (b) Connect

Fig. 4. Frequent itemset extraction time

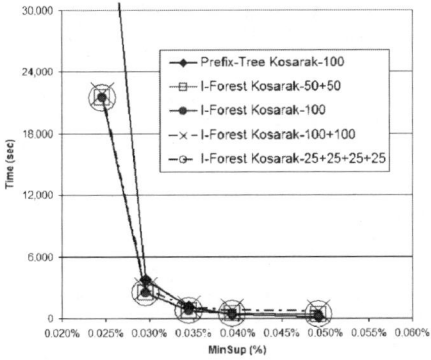

Fig. 5. Frequent itemset extraction time: Kosarak

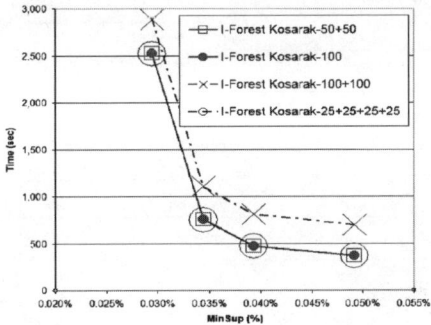

Fig. 6. Effect of block partitioning on frequent itemset extraction time: Kosarak

algorithm. When using the I-Forest index based approach, the overhead due
to data retrieval from disk is counterbalanced by an efficient memory usage
during the extraction phase. Memory usage is further discussed in the next
section.

Figures 6 and 7 analyze the overhead in frequent itemset extraction introduced by block partitioning for Connect, Pumsb and Kosarak datasets in
the four configurations. Block partitioning introduces an overhead on data retrieval since data are accessed from different IF-blocks. Experimental results
show that this overhead does not significantly affect the extraction performance for any support thresholds. In particular, the overhead due to block
partitioning is expected to (slightly) increase when mining itemsets with high
support constraint. Items occurring in many transactions are potentially correlated and they can be represented in a single index path. Block partitioning
may split this path in different blocks thus affecting data retrieval performance. On the other hand, when dealing with smaller data blocks, the index
creation procedure can provide a clever physical organization of each IF-block,
because the correlation analysis procedure may become more effective. Hence,
within each IF-block, data accessed together during the extraction phase are

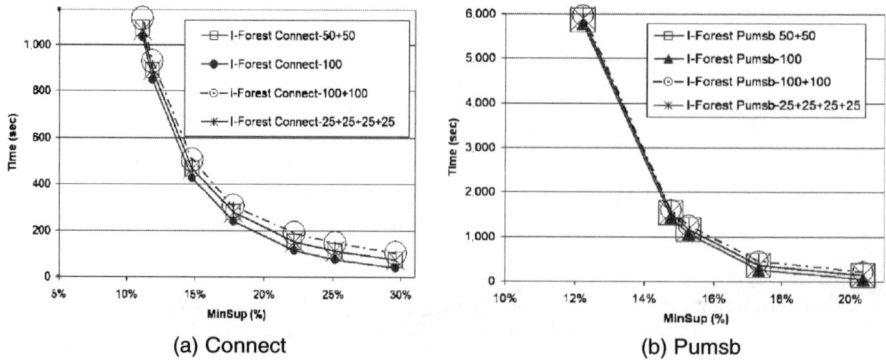

(a) Connect   (b) Pumsb

Fig. 7. Effect of block partitioning on frequent itemset extraction time

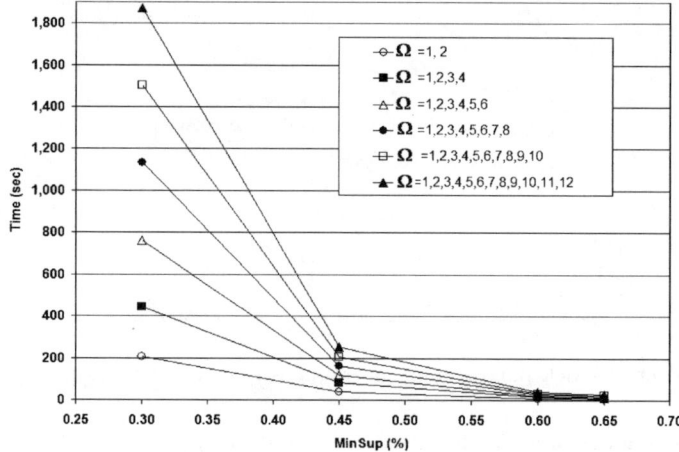

Fig. 8. Itemset extraction time with both time and support constraints: T25I300D6M

actually stored in the same (disk) block, thus reducing the I/O cost. When lower supports are considered, weakly correlated items are analyzed. These items are naturally represented in disjoint index paths.

For the configuration Dataset-100 + 100, the run time for frequent itemset extraction is higher than for all the other configurations. Recall that Dataset-100 + 100 is obtained by cloning the original dataset. Hence, to retrieve the index data, the number of disk accesses is doubled with respect to the configuration Dataset-100.

We ran experiments to evaluate the performance of itemset extraction when enforcing both time and support constraints. Experiments have been run on the (large) synthetic dataset T25I300D6M partitioned in 12 blocks. We considered six different time constraints, with increasing number of blocks to be mined (Ω lists the blocks to be mined). For each time constraint, Fig. 8 reports the run time in itemset extraction by varying the support thresholds.

The extraction time grows almost linearly with the number of analyzed blocks. The increase is mainly due to the cost for accessing data in different index blocks. The increment is more evident for lower supports. Being the original dataset sparse, most of the items have a low frequency. Hence, a large amount of data has to be retrieved when lower support thresholds are considered.

5.3 Memory Consumption

We compared the memory consumption[2] of our approach with the Prefix-Tree algorithm on flat file. We report the average memory (Fig. 9a) and the total memory (Fig. 9b) required during the extraction process. For the I-Forest index the buffer cache of the DBMS is included in the memory held by the current process. The Kosarak dataset is discussed as a representative example.

Figure 9a shows that Prefix-Tree, on average, needs a significantly larger amount of memory than our index based approach. It also requires an amount of overall memory (Fig. 9b) smaller than I-Forest only for high support values.

The I-Forest index allows the extraction algorithm to selectively load in memory only the data required for the current execution phase. Hence, both average and total required memory may become significantly smaller than that required by Prefix-Tree which stores the entire data structure in memory for the whole extraction process. For this reason, a larger memory space is available for the I-Forest extraction process, thus reducing the occurrence of memory swaps. Furthermore, since the size of PostgreSQL DBMS buffer cache is kept constant, global and average memory requirements of our approach are rather stable with decreasing support thresholds.

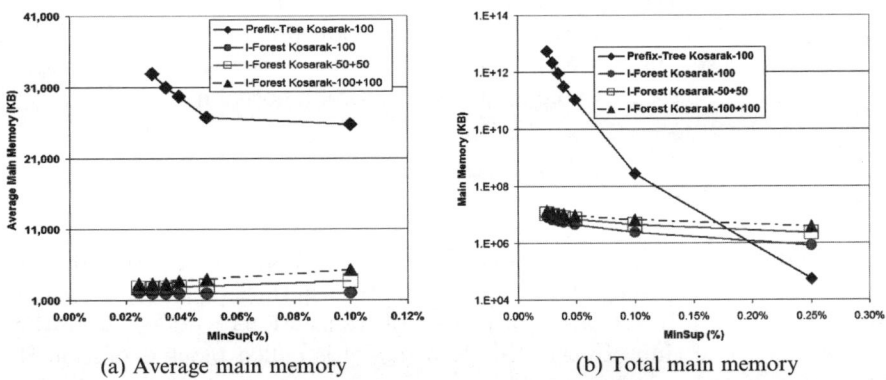

(a) Average main memory   (b) Total main memory

Fig. 9. Memory usage for Kosarak dataset

---

[2] The amount of memory required by the extraction process is read from file/proc/PID/mem (memory held by the PID process).

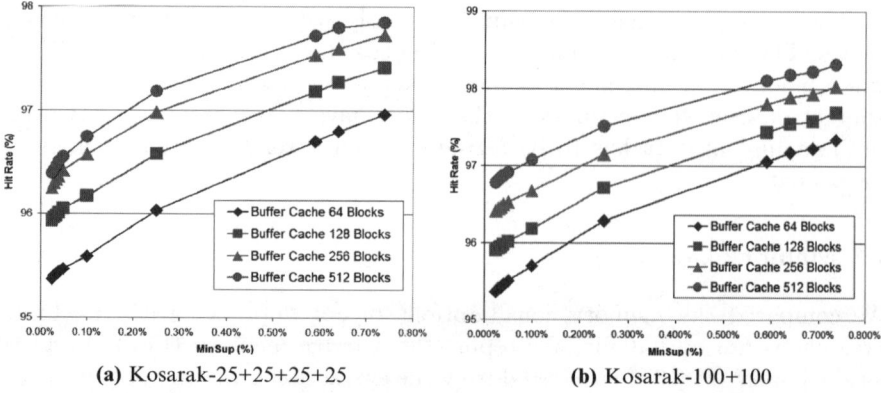

(a) Kosarak-25+25+25+25   (b) Kosarak-100+100

Fig. 10. Buffer cache and hit rate for Kosarak dataset

5.4 Buffer Cache Size

The hit rate is the ratio between the number of hits (accessed blocks already available in the buffer cache) and the total number of accesses to index blocks. In case of hit, there is no overhead caused by accessing data on disk. Hence, the I/O cost decreases when the hit rate increases.

Figure 10 shows the hit rate when accessing the PostgreSQL buffer cache during the extraction process for two different configurations of the Kosarak dataset. With the default size of the buffer cache (i.e., 64 blocks) the hit rate is quite high even when the extraction is performed with low support thresholds. When increasing the buffer cache size, the hit rate grows by about 0.5%. On the other hand, when the buffer cache size is increased, the performance of the extraction process slightly decreases (e.g., when the buffer cache is 512 blocks the CPU time required for the extraction process increases by about 4%). In this case, the decrease in I/O cost is counterbalanced by a decrease in the memory space available for the extraction algorithm.

6 Related Work

Incremental frequent pattern mining is a relevant issue in many real-life contexts. Incoming data may arrive as (i) single transactions [7] or (ii) transaction blocks [3,8,12,13]. In the latter case, a set of transactions is added to the database at the same time. Hence, the data model is called block evolution [13]. The I-Forest index structure addresses the extraction of frequent itemsets in this last context. Hence, it may be straightforwardly integrated into the framework proposed in [13].

Works in [3, 8, 12, 13, 21] address incremental frequent itemset extraction under block evolution. The FUP [8] algorithm, based on the Apriori [1] approach, requires a large number of database scans. It has been improved by

BORDERS [3], based on the notion of border sets (a set X is a border set if all its proper subsets are frequent sets but X itself is not). Border sets allow an efficient detection of new frequent sets, thus reducing the number of candidates to count and database scans. This algorithm stores the set of frequent itemsets and updates them when new data arrive. Hence, a large solution set has to be stored. Furthermore, updating the solution set may require to re-scan the whole dataset. To deal with smaller solution sets, [21] stores only the maximal itemsets, while [9] maintains the set of closed itemsets. The BORDERS approach has been improved by the ECUT (Efficient Counting Using TID-lists) algorithm [12]. To count the support of itemsets, ECUT retrieves only the relevant portion of the dataset by means of the TID-lists of 1-itemsets. The ECUT+ algorithm, also proposed in [12], materializes the TID list for itemsets with size greater than one. The number of itemsets may be very large. The heuristic solution proposed in ECUT+ is the materialization of the TID-lists of 1-itemset and 2-itemsets only. This technique improves performance with respect to ECUT, but it requires more disk space.

A different incremental technique is based on arbitrary insertions and deletions of single transactions. In this scenario, [7] proposed an extension of the FP-Tree [16], where each new transaction is inserted in the existing tree by means of a heuristic approach without exploring the whole tree. However, after several insertions the updated structure is not as compact and efficient as the corresponding FP-tree.

The I-Forest index provides a compact structure to store all information potentially required by the mining process. Furthermore, it allows selective access to this information. Hence, differently from the above approaches, it provides a flexible framework in which different types of analysis (e.g., mining only a subset of interesting blocks) can be performed. Many algorithms have been proposed to perform the computationally intensive knowledge extraction task over static databases [1, 15, 16, 18]. Since the required memory increases significantly, memory based approaches are not suitable for mining sequences of incoming data blocks. In this context, disk based approaches [4, 10, 20] are more suitable for the itemset extraction task. They exploit clever data structures to summarize the (static) database on disk. Efficient itemset extraction takes place on these ad-hoc data structures. [20] proposes B + tree based indexes to access data stored by means of either a vertical or a horizontal data representation, while [10] represents the dataset by means of an Inverted Matrix stored on disk, which is mined to extract frequent itemsets. [4] proposes an index structure, called I-Tree, to tightly integrate frequent itemset mining in PostgreSQL open source DBMS. However, the I-Tree index cannot be incrementally maintained. We exploit the I-Tree as a building block of our approach. The I-Forest index allows both incremental mining of evolving databases and smooth data evolution.

7 Conclusions and Future Work

The I-Forest is a new index structure suitable for itemset extraction from databases collections, which evolve over time through periodical insertion (or deletion) of data blocks. An algorithm to efficiently extract itemsets by exploiting the proposed index structure has been presented. Time and support constraints may be enforced to drive the extraction process. Primitives for index creation and itemset extraction have been integrated in the kernel of the PostgreSQL DBMS [19]. Experiments have been run for both sparse and dense data distributions. The experimental results show that itemset mining based on the I-Forest index is efficient both in terms of extraction time and memory usage. The performance of the proposed approach is always comparable with and for low support thresholds faster than the Prefix-Tree algorithm [11] (FIMI '03 best implementation), accessing static data on flat file.

As further extensions of this work, the following issues may be addressed: (i) *Compact structures suitable for different data distribution.* Currently, we adopted the I-Tree structure to uniformly represent all IF-blocks. Since incoming data blocks are stored on disk in independent structures, different compact structures suitable for the data distribution of each block may be exploited to optimize disk space and reduce the I/O cost. As future work we plan to include data structures appropriate for sparser data distributions (e.g., Patricia-Trie [18] and Inverted Matrix [10]). (ii) *Item constrained extraction.* Currently, the proposed algorithm based on the I-Forest index is able to extract itemsets enforcing time and support constraints. Another interesting analysis can be performed by extracting only itemsets including a selected subset of interesting items. (iii) *Integration with a mining language.* The proposed primitives, currently integrated into the kernel of the PostgreSQL DBMS, may be integrated with a query language for specifying mining requests. These low-level primitives can contribute an efficient database implementation of the basic extraction statements of the query language.

References

1. R. Agrawal and R. Srikant. Fast algorithm for mining association rules. In VLDB, 1994.
2. R. Agrawal, T. Imielinski, and A. Swami. Database mining: A performance perspective. IEEE Trans. Knowl. Data Eng., 5(6), 1993.
3. Y. Aumann, R. Feldman, and O. Lipshtat. Borders: An efficient algorithm for association generation in dynamic databases. In JIIS, vol. 12, 1999.
4. E. Baralis, T. Cerquitelli, and S. Chiusano. Index Support for Frequent Itemset Mining a Relational DBMS. In ICDE, 2005.
5. M. Botta, J.-F. Boulicaut, C. Masson, and R. Meo. A comparison between query languages for the extraction of association rules. In DaWak, 2002.
6. S. Chaudhuri, V. Narasayya, and S. Sarawagi. Efficient evaluation of queries with mining predicates. In IEEE ICDE, 2002.

7. W. Cheung and O. R. Zaiane. Incremental mining of frequent patterns without candidate generation or support constraint. In IDEAS, pp. 111–116, July 2003.
8. D. W.-L. Cheung, J. Han, V. Ng, and C. Y. Wong. Maintenance of discovered association rules in large databases: An incremental updating technique. In ICDE, pp. 106–114. IEEE Computer Society, 1996.
9. L. Dumitriu. Interactive mining and knowledge reuse for the closed-itemset incremental-mining problem. SIGKDD Explorations, 3(2):28–36, 2002.
10. M. El-Hajj and O. R. Zaiane. Inverted matrix: Efficient discovery of frequent items in large datasets in the context of interactive mining. In ACM SIGKDD, 2003.
11. FIMI. http://fimi.cs.helsinki.fi/.
12. V. Ganti, J. Gehrke, and R. Ramakrishnan. DEMON: Mining and monitoring evolving data. IEEE Trans. Knowl. Data Eng., 13(1):50–63, 2001.
13. V. Ganti, J. E. Gehrke, and R. Ramakrishnan. Mining data streams under block evolution. SIGKDD Explorations, 3(2), 2002.
14. B. Goethals and M. J. Zaki. Fimi'03: Workshop on frequent itemset mining implementations, November 2003.
15. G. Grahne and J. Zhu. Efficiently using prefix-trees in mining frequent itemsets. In FIMI, November 2003.
16. J. Han, J. Pei, and Y. Yin. Mining frequent patterns without candidate generation. In ACM SIGMOD, 2000.
17. R. Meo, G. Psaila, and S. Ceri. A tightly-coupled architecture for data mining. In IEEE ICDE, 1998.
18. A. Pietracaprina and D. Zandolin. Mining frequent itemsets using patricia tries. In FIMI, 2003.
19. Postgres. http://www.postgresql.org.
20. G. Ramesh, W. Maniatty, and M. Zaki. Indexing and data access methods for database mining. In DMKD, 2002.
21. A. Veloso, W. M. Jr., M. De Carvalho, B. Possas, S. Parthasarathy, and M. J. Zaki. Mining frequent itemsets in evolving databases. In SDM, 2002.

Likelihoods and Explanations in Bayesian Networks

David H. Glass

School of Computing and Mathematics, University of Ulster, Newtownabbey,
Co Antrim, BT37 0QB, UK
dh.glass@ulster.ac.uk

Summary. We investigate the problem of calculating likelihoods in Bayesian networks. This is highly relevant to the issue of explanation in such networks and is in many ways complementary to the MAP approach which searches for the explanation that is most probable given evidence. Likelihoods are also of general statistical interest and can be useful if the value of a particular variable is to be maximized. After looking at the simple case where only parents of nodes are considered in the explanation set, we go on to look at tree-structured networks and then at a general approach for obtaining likelihoods.

1 Introduction

In many scenarios human reasoning seems to involve producing adequate explanations of the phenomena under consideration. In many artificial intelligence applications, however, reasoning and inference can be carried out without any explicit account of explanation. This naturally raises the question as to how explanations can be extracted from such applications. This is crucial if users are to trust the reliability of the inferences made. In probabilistic systems, for example, users often find it difficult to make sense of the reasoning process unless suitable explanations are available. Unfortunately, providing an adequate account of explanation, which would be required for the automatic generation of explanations, is a notoriously difficult problem.

There has been considerable interest in how explanations can be obtained from Bayesian networks [1–8]. There are a number of reasons why Bayesian networks provide a suitable environment for finding explanations. First of all, if the structure of a network is interpreted causally then it immediately gives a framework for locating the causes (or potential causes) of particular events (or facts) which is particularly important given that causality is often taken to be a key ingredient in explanation [6, 9]. Bayesian networks also provide an obvious framework for studying statistical explanations and for ordering explanations in terms of probability.

D.H. Glass: *Likelihoods and Explanations in Bayesian Networks*, Studies in Computational Intelligence (SCI) **109**, 325–341 (2008)
www.springerlink.com © Springer-Verlag Berlin Heidelberg 2008

The foregoing discussion raises two important questions. First of all, which nodes in a Bayesian network should be included in the explanation? Should it be all the nodes in the network, or the ancestors of the evidence nodes, or a subset of the ancestors? While a full discussion of explanation in Bayesian networks is beyond the scope of this paper, it is worth noting that other researchers have addressed many of these issues or closely related topics (see for example [3–6]). Second, how should probability be used to order explanations in Bayesian networks? The main answer to this question is the Maximum A Posteriori (MAP) approach, which finds the explanation that is most probable given the evidence [1, 3, 5]. Nevertheless, this approach has been criticized [6, 10]. One difficulty is that the MAP approach can yield as the best explanation one which is negatively related to the evidence and so lowers the probability of the evidence – it is most probable in virtue of its high prior probability.

This paper considers an alternative approach. Rather than calculating the probability of an explanation given the evidence, we investigate ways to calculate the likelihood of possible explanations, i.e. the probability of the evidence given the explanation. As a result we also wish to find the explanation of maximum likelihood. The purpose of the paper is not to argue that this is a better approach than the MAP approach since this approach also has problems. Nevertheless, it is of interest to calculate likelihoods since they are certainly relevant to the concept of explanation and complementary to MAP. Likelihoods also provide statistical information that is generally of interest in investigation of data and can give important information about how a certain state of affairs can be brought about with high probability (provided the network can be interpreted causally).

The structure of the rest of the paper is as follows. In Sect. 2 we present some background information and discussion about the problem. In Sect. 3 we consider the simplest case where only the parents of the evidence nodes are considered in the explanation. Section 4 looks at calculating likelihoods in trees and Sect. 5 investigates a general approach. Preliminary results are presented in Sect. 6 and conclusions in Sect. 7.

2 Preliminaries

2.1 Bayesian Networks

Suppose V is set of nodes with each node $v \in V$ representing a random variable X_v. A Bayesian network [1, 11–13] consists of a directed acyclic graph (DAG) together with a probability distribution which satisfies the Markov condition with respect to the DAG so that each variable X_v is conditionally independent of its non-descendents given its parents $X_{pa(v)}$ in the DAG. If the conditional probability distribution for each node is given by $P(x_v|x_{pa(v)})$, the joint probability distribution can be written as,

$$P(x) = \prod_{v \in V} P(x_v|x_{pa(v)}).\tag{1}$$

In this paper only discrete variables are considered.

We now define some of the terminology that will be used in this paper (see [12,13] for further details). A sequence of nodes $[v_0, v_1, \ldots, v_m]$ is a *chain* between v_0 and v_m if there is a directed edge from v_i to v_{i+1} or from v_{i+1} to v_i for $i = 0, \ldots, m-1$. A chain between v_0 and v_m is a *directed path* from v_0 to v_m if there is a directed edge from v_i to v_{i+1} for $i = 0, \ldots, m-1$. v_0 is an *ancestor* of v_m if there is a directed path from v_0 to v_m. A *tree* is a network in which each node has at most one parent and there is only one node with no parents (the *root* node). A *singly connected* network has at most one chain between each pair of nodes. If there is more than one chain between any pair of nodes the network is *multiply connected*. We will also refer to the notion of *d-separation* which enables us to identify all the conditional independencies entailed by the Markov condition. The reader is referred to [1] for a definition and to [13] for an alternative method to identify the conditional independencies.

2.2 Definition of Problem

The basic problem to be addressed in this paper is to determine the probability of a specified configuration of a set of variables, X_T, which we shall refer to as the target set, given the possible configurations of a set of variables, X_E, which we shall refer to as the explanation set. We will denote the specified configuration of the target set as x_T and the configurations of the explanation set as x_E. Thus, we wish to calculate $P(x_T|x_E)$. We also wish to find the configuration of maximum likelihood, i.e. the configuration of the target set that maximizes the probability of the target configuration. Formally, we wish to find the configuration x_E^* of X_E such that,

$$x_E^* = \arg \max_{x_E} P(x_T|x_E).\tag{2}$$

2.3 Discussion of Problem

At first glance the problem appears to be similar to standard inferences carried out in Bayesian networks. For example, suppose evidence is entered for nodes in the network. An inference can then be performed to obtain the marginal probability of a variable given the evidence. One problem is that we wish to find the likelihood for *all* configurations of the explanation set. A simple approach would be to adapt standard inference algorithms so that they perform an inference for each configuration, but it should be possible to develop more efficient approaches. Furthermore, in the problem being considered, conditioning on the explanation set results in a loss of the factorizability of the function and so makes inference more demanding. Nevertheless, there are strong similarities with work on obtaining MAP explanations as will become clear in Sect. 5.

3 Likelihood of Parents

3.1 Explanations

3.1.1 All parents in the Explanation Eet

The simplest case occurs when the explanatory variables X_E are the parents of the target variables X_T. If X_T consists of just a single node X_t the desired configuration x_E^* can be obtained by simple inspection of the conditional probability distribution $P(x_t|x_{pa(t)})$. Furthermore, if X_E contains $X_{pa(t)}$ as a proper subset and is confined to ancestors of X_t, then variables in $X_E \backslash X_{pa(t)}$ have no effect on X_t due to the Markov condition. In this case it only makes sense to consider the projection of x_E on $x_{pa(t)}$, in which case the maximum configuration is again found by inspection of the conditional distribution.

We now consider a more general case where X_T contains multiple nodes and is given by $X_T = \{X_{t1}, \ldots X_{tm}\}$ with the explanation set consisting of the parents of these nodes, i.e. $X_E = X_{pa(T)} = \{X_{pa(t1)}, \ldots X_{pa(tm)}\}$. At this stage we consider only cases where no node in X_T has a parent which is a descendent of another node in X_T since we are considering cases where the explanation set only contains ancestors of variables in the target set. We also restrict the explanation set so that it only contains nodes not found in the target set, i.e. X_E is replaced by $X_{ER} = X_E \backslash X_T$. The likelihood of a configuration x_{ER}^* is given by,

$$P(x_T|x_{ER}) = \prod_{i \in T} P(x_i|x_{pa(i)}^T), \tag{3}$$

where $x_{pa(i)}^T$ is the configuration of the variables $X_{pa(i)}$ with any variable found in X_T constrained to take on the value specified in x_T. One way to find the configuration of maximum likelihood x_{ER}^* would be to calculate the product in (3) for all possible configurations of x_{ER}, but fortunately a more efficient approach can be found. Suppose the set of nodes T is partitioned into k groups of nodes $\{T_1, \ldots, T_k\}$ such that a node found in one group does not share any parents with any nodes except those in its own group. In order to obtain x_{ER}^* we can now consider the components corresponding to the parents of each group separately. If we denote the projection of x_{ER}^* onto the parents of group T_j as $x_{ER,j}^*$, we obtain for each j,

$$x_{ER,j}^* = \arg \max_{x_{pa(j)}^T} \prod_{i \in T_j} P(x_i|x_{pa(i)}^T). \tag{4}$$

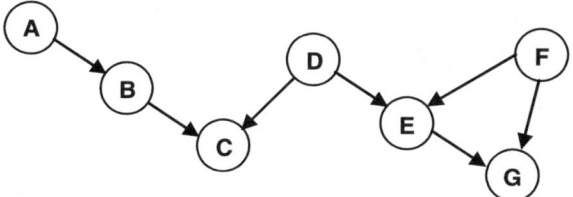

Fig. 1. The DAG for the Bayesian network considered in example 1

Example 1. Consider the Bayesian network whose DAG is shown in Fig. 1 and consider the target set $T = \{B, E, G\}$ with the configuration (b_1, e_1, g_1). We note that the target set can be partitioned as $T = \{B\} \cup \{E, G\}$ so that the parents of the nodes in each partition are distinct. Since the explanation set will then be $\{A, D, F\}$ we can obtain the configuration of maximum likelihood by selecting the value a_i of A which maximizes $P(b_1|a_i)$ and the values d_j and f_k of D and F which maximize $P(e_1|d_j, f_k) \times P(g_1|e_1, f_k)$.

3.1.2 Subset of Parents in Explanation Set

When the explanation set consists of a proper subset of the set of parents of the target set, $X_E \subset X_{pa(T)}$, things are less straightforward since the likelihood cannot be obtained by simply taking the product obtained from the relevant conditional probability tables. Part of the motivation for looking at this problem is that it raises some of the issues that will be important later in the paper. In fact, the most general solution to this problem will require the methods proposed in Sect. 5.

To illustrate the general idea we consider three scenarios, shown in Fig. 2, wherethere is only one node in the target set, node C, and two parent nodes A and B, only one of which is in the explanation set, node A. Thus, supposing C to have value c_1, we wish to calculate $P(c_1|a_i)$. In the first scenario we consider the simple network shown in Fig. 2a, for which the desired likelihood can be expressed as,

$$P(c_1|a_i) = \sum_j P(c_1|a_i, b_j)P(b_j). \tag{5}$$

Thus, we must marginalize out B, which is simple in this case because B has no parents and is marginally independent of A. Scenario two, shown in Fig. 2b, is only slightly more complicated since A and B have one parent each. The desired likelihood for this network can be expressed as,

$$P(c_1|a_i) = \sum_{jk} P(c_1|a_i, b_j)P(b_j|d_k)P(d_k). \tag{6}$$

B is marginally independent of A as in the first scenario, but this time B has a parent D which must also be marginalized out. Note that E is not involved

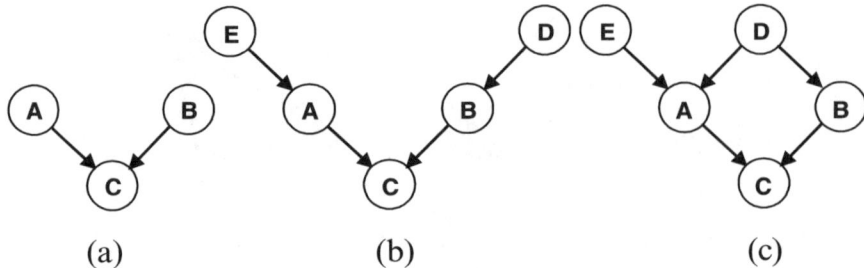

Fig. 2. Three scenarios in which the target set is $\{C\}$ and the explanation set is $\{A\}$

in the calculation. Note also that marginalizing out D gives the probability of B and so once this has been done the right hand side of (6) becomes the same as the right hand side of (5). Thus, if the marginal probabilities for each node have already been obtained scenario two becomes equivalent to scenario one.

Scenario three involves the multiply connected network shown in Fig. 2c, which has an additional directed edge from node D to node A. In this case the desired likelihood can be expressed as,

$$P(c_1|a_i) = \frac{P(c_1, a_i)}{P(a_i)}$$

$$= \frac{\sum_{jkl} P(e_l)P(d_k)P(a_i|e_l, d_k)P(b_j|d_k)P(c_1|a_i, b_j)}{\sum_{kl} P(e_l)P(d_k)P(a_i|e_l, d_k)}. \tag{7}$$

Clearly, the solution for scenario three is much more complex since B is no longer marginally independent of A, but only conditionally independent of A given D. In scenario two E could be excluded because it was conditionally independent of C given A, or to put this in different terminology, A d-separated E and C. But this does not hold in scenario three since although the chain E-A-C is blocked by A, the chain E-A-D-B-C is not. More generally, for singly connected networks any ancestors of nodes in the explanation set can be excluded from the calculation since they are d-separated from the target nodes by the nodes in the explanation set, while no such result holds for multiply connected networks.

It is also clear that marginalization only needs to be carried out for nodes which are ancestors of the nodes in the target set. (This can be seen by noting that the likelihood can be expressed as the joint probability of the target and explanation nodes divided by the joint probability of the explanation nodes, which involve only ancestors of the nodes in question.) In effect this means that calculating likelihoods when only parents of target nodes are included in the explanation set can be carried out on a subset of the initial network. This is because only ancestors of target nodes need to be taken into account and, furthermore, any of these ancestors which are d-separated from the target set by the explanation set can also be excluded.

3.2 Interventions

The approach described Sect. 3.1.1 applies to the case where the target set is found to be in a particular configuration and we wish to know what configuration of the explanation set would have the highest likelihood. If the Bayesian network represents casual relationships between variables, it also applies to the case where we wish to know what intervention on the variables in the explanation set would make the target configuration most probable. Furthermore, the restriction that no node in the target set should have a parent which is a descendent of another node in the target set can be lifted in this case. The reason for this is that intervention is different from observation and in effect involves removing from the DAG arrows coming into the node where the intervention occurs. This is the basic idea underlying Pearl's do-calculus [2].

Example 2. Consider the Bayesian network whose DAG is given in Fig. 3a and consider the target set $T = \{B, D\}$ with configuration (b_1, d_1). If this configuration is observed and we wish to know what configuration of the explanation set $\{A, C\}$ has the highest likelihood, the approach described in Sect. 3.1.1 will fail since the value of C will have an effect on B as well as D. However, if instead of observing the target configuration, we wish to know what intervention on the explanation set will make the target configuration most probable, the method will work since intervening on C will have no effect on B. In effect, the intervention amounts to replacing the DAG in Fig. 3a by that in Fig. 3b. From Fig. 3b it is clear that the approach described in Sect. 3.1.1 is now applicable.

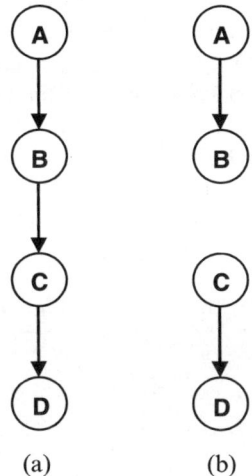

(a)   (b)

Fig. 3. DAGs considered in example 2

4 Likelihoods in Trees

4.1 Explanations

Consider again the DAG shown in Fig. 3a. Suppose now that we wish to find the likelihood of an ancestor, but not necessarily a parent, of a given node. Suppose, for example, that we wish to find $P(d_1|a_i)$ for each value of i. This can be obtained as follows,

$$P(d_1|a_i) = \sum_{j,k} P(b_j|a_i)P(c_k|b_j)P(d_1|c_k). \qquad (8)$$

To illustrate how this can be calculated efficiently consider the junction tree representation of the DAG, as shown in Fig. 4. (It is not necessary to use the junction tree approach for this problem and, in fact, an approach such as Pearl's message-passing [1] might be more appropriate in this case. Junction trees are used as they will also be considered later in the paper. For a full discussion of this approach see [13, 14]). The potential functions will be initialized as follows,

$$\begin{aligned}
\phi_{AB} &= P(A)P(B|A), \\
\phi_{BC} &= P(C|B), \qquad\qquad (9) \\
\phi_{CD} &= P(D|C).
\end{aligned}$$

$P(d_1, a_i)$ can be obtained for each i by setting the value of D to d_1, passing sum-flows from clique C_{CD} to C_{AB}, and then marginalizing out B. $P(d_1|a_i)$ can then be obtained by dividing $P(d_1, a_i)$ by $P(a_i)$ for each i. If A is not a root node in the network, this last step is not required.

It is worth noting that this approach is more efficient than an alternative approach, which involves setting the value of A to a_i and then finding $P(d_1|a_i)$ by propagating sum-flows from C_{AB} to C_{CD}. This is just the standard way of updating in a Bayesian network, but in this case it needs to be repeated for each value of i. In general, to perform this type of operation along a chain requires $O(n^3)$ operations, where n is the number of possible values of the variables, compared to $O(n^2)$ for the method described above.

Suppose we now consider the tree-structured DAG shown in Fig. 5 and wish to obtain $P(c_1, e_1|a_i)$ for each value of i. This can be written as,

$$P(c_1, e_1|a_i) = \sum_{j} P(c_1|b_j)P(b_j|a_i) \sum_{k} P(e_1|d_k)P(d_k|a_i) \qquad (10)$$

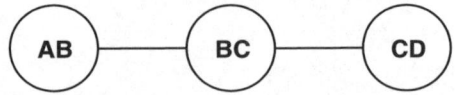

Fig. 4. The Junction Tree for the DAG in Fig. 3a

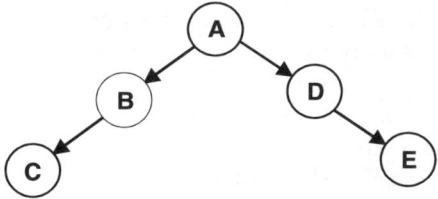

Fig. 5. A tree-structured DAG

and so, as before, the relevant summations can be carried out on the corresponding branches of the tree independently and collected at A.

It is now clear how this approach can be generalized to treat the case where the target set contains nodes in separate sub-branches of a tree and the explanation set contains an ancestor of each node in the target set and no descendents of nodes in the target set. The following steps should be taken in order to obtain the probability of the specified target configuration given each configuration of the explanation set:

Step 1: Any node in the explanation set that is blocked from all nodes in the target set by other nodes in the explanation set is irrelevant to the likelihood and can be removed from the explanation set.

Step 2: The junction tree is created and initialized and evidence is entered for each clique containing a node in the target set and its parent, i.e. the values of the clique potential that do not correspond to the value in the target configuration are set to zero.

Step 3: From each clique identified in step 2 sum-flows are passed up the tree until they reach a clique containing a node in the explanation set.

Step 4: For each clique that receives a sum-flow in step 3 the variable not contained in the explanation set is marginalized out of the clique potential.

Step 5: The likelihood can then be obtained for each configuration of the explanation set by multiplying the corresponding values for each clique potential receiving a sum-flow in step 3. The result should be divided by the prior probability of the root node if it is contained in the explanation set (and has not been ruled out in step 1).

4.2 Interventions

Returning to Fig. 5, suppose we wish to obtain $P(c_1, e_1 | b_i)$ for each value of i. Although the explanation set $\{B\}$ only contains ancestors of nodes in the target set $\{C, E\}$, we cannot obtain likelihoods (and in particular the maximum likelihood) by considering the influence of B on its child node C. This is because B and E are not independent of each other. However, as discussed in the previous section, if we are considering interventions we can focus on B's influence on C since the arrow between A and B will be removed and B and E will be rendered independent.

The procedure outlined above for calculating likelihoods applies to interventions as well as explanations, but in the case of interventions the restriction that the explanation set contain an ancestor of each node in the target set can be lifted. In this case the likelihood can be obtained in the same way except that in step 2, cliques should only be identified where the node in the target set has an ancestor in the explanation set. (Strictly speaking the result will not be the likelihood but the likelihood divided by the probability of the projection of the target configuration on variables without ancestors in the explanation set. This is just a constant factor, however, and will still permit an ordering of configurations in terms of their likelihoods.)

4.3 The Influence of Descendents

So far we have considered ancestor nodes in the explanation set, but here we consider how descendents in the explanation set can be treated as a prelude to tackling the general case in the next section. Consider the Bayesian network possessing the DAG shown in Fig. 5, but now consider the explanation set $\{B, D\}$ and the target set $\{A\}$ with target configuration a_1. Thus, we wish to obtain $P(a_1|b_i, d_j)$ for each (i, j) pair. The general expression can be written as,

$$P(a_1|b_i, d_j) = \frac{P(a_1)P(b_i|a_1)P(d_j|a_1)}{\sum_k P(a_k)P(b_i|a_k)P(d_j|a_k)}. \tag{11}$$

Note that the denominator in (11) cannot be factorized as was possible in (10) and so the variables A, B and D must all be considered together in this case. The problem only gets worse if $\{C, E\}$ is the explanation set. This case highlights a significant problem for calculating likelihoods and will be considered again in the next section.

Suppose that instead we consider the explanation set $\{B\}$ so that we now wish to obtain $P(a_1|b_i)$ for each value of i. This can be written as,

$$P(a_1|b_i) = \frac{P(a_1)P(b_i|a_1)}{\sum_k P(a_k)P(b_i|a_k)}. \tag{12}$$

Although this is a simple expression it does highlight the important point that if B is the only node in the explanation set, the other children of A do not need to be taken into account. Of course it does require prior probabilities to be calculated, a point which will be considered again in the next section.

The general point being made here is well-known, but it is worth emphasizing in this context because of its relevance for calculating likelihoods in Bayesian networks. An example illustrates the point. Suppose the relationship between the presence or absence of a disease and the results of various diagnostic tests can be represented by a Bayesian network such as that in Fig. 5 where A is the disease and B, D and perhaps other child nodes of A represent the result of various tests. A relevant consideration in the choice of which test should be performed first is the probability that the person has the disease given that the test result is positive and of course this can be calculated for each test independently. The relevance for calculation of likelihoods is that this extends to tree structures more generally where the probability of a configuration of a target node given the configurations of one or more of its descendents is required.

5 Likelihoods in a General Bayesian Network

So far we have considered various restrictions either in terms of the network being considered or the nodes to be included in the explanation set (e.g. parents of target nodes). In this section we remove these restrictions to consider Bayesian networks in general. First, we consider the most general case in which any set of non-target nodes can be included in the explanation set. We then look briefly at the case where all non-target nodes are included in the explanation set, which we refer to as a complete explanation.

5.1 The General Case

In Sect. 4.3 when considering likelihoods in the case where the explanation set contains descendents of the target set, we were effectively using the following expression to obtain the likelihood,

$$P(x_T|x_E) = \frac{P(x_T, x_E)}{P(x_E)}. \tag{13}$$

Furthermore, the procedure discussed in Sect. 4.1 in the calculation of $P(d_1|a_i)$ as found in Fig. 3a involved first of all calculating $P(d_1, a_i)$ and then dividing by $P(a_i)$. This is very similar to what normally happens in inference problems in Bayesian networks. A typical problem would be to calculate the marginal distribution for a node given some evidence elsewhere in the network. For example, suppose that given the DAG in Fig. 3a we wished to calculate the marginal distribution for A given evidence d_1 for node D. The procedure outlined in Sect. 4.1 could be used to obtain $P(d_1, a_i)$ and then normalization used to obtain $P(a_i|d_1)$. Thus, it would seem feasible to obtain likelihoods in Bayesian networks by using (13) and so performing two inferences in the network, one to obtain the numerator and one to obtain the denominator.

The difficulty is that obtaining a marginal distribution for a subset of nodes in a Bayesian network is not straightforward in the general case. The reason for this is that in general the factorization which applies to the joint distribution for all variables does not apply to a subset of them. For example, suppose we consider the denominator in (11) again. This expression does not factorize on the junction tree representation of the DAG in Fig. 5. If we wish to compute the marginal distribution for a set of variables which are all contained within one clique of the junction tree, the problem is straightforward.

Once equilibrium has been reached by passing active flows in both directions between each pair of cliques in the junction tree, each clique contains the joint probability distribution for the variables it contains. The desired marginal probability is then obtained by marginalizing out variables that are not in the required set.

This problem has been investigated by Xu [15] and, in the context of obtaining the Maximum A Posteriori (MAP) hypothesis, by Nilsson [16] for example. The MAP problem is similar to the current problem since it involves finding the most probable configuration of a subset of nodes in a network given evidence about other nodes in the network. A special case of this problem is to find the maximum configuration of *all* nodes for which evidence has not been provided. This special case has been dealt with in [16, 17]. The more general MAP problem, however, is more relevant here since it involves a function which does not factorize over the junction tree.

Since the MAP problem involves maximizing the numerator in (13) we can adopt approaches used for it to obtain the likelihoods. The basic idea is to merge relevant cliques in the junction tree so that all the nodes in explanation set are contained within a subtree of the junction tree to ensure that the both the numerator and denominator in (13) factorize over the modified junction tree. A standard propagation algorithm can then be applied. This process should be carried out twice, once without any evidence being entered and once with the target configuration being entered as evidence, to obtain the denominator and numerator respectively.

Following [18] we summarize the Xu-Nilsson algorithm for modifying the junction tree as follows:

Step 1: Identify the smallest subtree J' of the junction tree J that contains the variables of the explanation set X_E.

Step 2: Pass sum-flows from the rest of the cliques to J'.

Step 3: Select two neighbouring cliques C_i and C_j in J' and merge them into a new clique C_{ij} by deleting variables that do not belong to X_E and updating the clique potential using the following expression,

$$\varphi(C_{ij}) = \sum_{C_i \cup C_j \setminus C_{ij}} \frac{\varphi(C_i)\varphi(C_j)}{\varphi(S_{ij})} \tag{14}$$

Repeat step 3 until J' contains only variables in X_E.

This approach can then be applied to obtain both the numerator and denominator in (13) and, by calculating the relevant ratios, the desired likelihood can be obtained.

As a simple example consider again the Bayesian network shown in Fig. 5 and for which the junction tree is shown in Fig. 6. Suppose now that we wish to calculate $P(a_1|c_i, d_j)$ for each (i, j) pair. Since the nodes C and D are in the explanation set we must merge nodes until the variables A and B can be

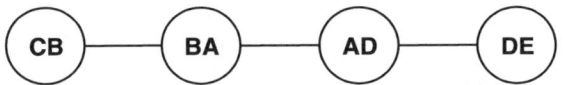

Fig. 6. A junction tree for the DAG shown in Fig. 5

removed from the subtree *CB-BA-AD*. First, merge *CB* and *BA* to obtain the
new node *CA* and then merge it with *AD* to obtain *CD*.

Most inference problems in Bayesian networks are known to be NP-hard
[19] and, in fact, the approach used here is more demanding even than MAP
since it is performing an MAP inference twice. Things are not quite this bad,
however, since savings can be made. For example, sum-flows from cliques not
involving variables in the target set do not have to be performed in the second
inference problem. Furthermore, the problem will be straightforward in some
scenarios, such as when all the explanatory nodes are in the same clique.

5.2 Complete Explanation Sets

Suppose we consider a general Bayesian network as in Sect. 5.1, but now consider a complete explanation set consisting of all the non-target nodes in the
network. From Sect. 3.1 we know that explanatory nodes can become redundant. For example, if there is just a single target node X_T and all of its parents
are in the explanation set, then all other ancestors of X_T are redundant since
they are d-separated from X_T by its parents. More generally, if we can specify
a set of nodes that d-separates the target set from the rest of the network, the
remaining nodes become redundant. Suppose T represents the set of target
nodes and M the set of nodes d-separating T from the remaining nodes R.
Since the nodes in T will be independent of those in R conditional on those
in M, changing the value of a node in R can have no effect on the value of
nodes in T once the values in M have been specified. Hence, in calculating
the likelihood we do not need to consider nodes in R.

It is well known that the set of nodes d-separating a node A from the
rest of the network is the Markov blanket of A, denoted bl(A), consisting of
the parents of A, the children of A and the parents of the children of A [13].
By taking the union of the Markov blankets of all the nodes in the target
set we can d-separate it from the rest of the network. Hence, suppose our
target set is $X_T = \{X_{t1}, \ldots X_{tm}\}$ and the explanation set X_E consists of all
the remaining nodes in the network, then all the nodes in $X_E \setminus \cap_{i=1,m} \text{bl}(X_{ti})$
will be redundant. Although finding likelihoods for complete explanations can
be obtained by the general approach outlined in Sect. 5.1, the considerations
noted here can improve the efficiency of the calculation. We will consider an
example of this in Sect. 6.

6 Results

Preliminary results are presented by considering two examples from the much discussed ASIA network as found in [14]. The network contains eight binary nodes as follows: A – Asia visit; T – Tuberculosis; L – Lung cancer; E – Either tuberculosis or lung cancer; S – Smoker; B – Bronchitis; D – Dyspnoea; X – X-ray. The Bayesian network is multiply connected containing two chains from S to D ($S \rightarrow B \rightarrow D$ and $S \rightarrow L \rightarrow E \rightarrow D$). The network is presented in Fig. 7 and a corresponding junction tree for the network, presented in [16], is given in Fig. 8.

Suppose we consider the target configuration to be a positive X-ray result ($X = x_1$) and the explanation set to be Lung cancer and Dyspnoea. In other words, we wish to obtain the probabilities $P(x_1|l_i, d_j)$ for all values of i and j. We note that L is an ancestor of X, but D is neither an ancestor nor a descendent of X. Nevertheless, it is neither marginally independent of X nor conditionally independent of X given L and so knowledge of the value of D can affect the probability of X.

First of all, we identify the subtree containing the nodes in the explanation set, i.e. (BEL – BDE). Then, we pass sum-flows from the peripheral nodes to BEL and then marginalize out B and E to obtain $P(l_i, d_j)$. We then repeat this procedure only now we incorporate the target configuration ($X = x_1$) as evidence.

The results obtained by considering the MAP and likelihood calculations reveal their complementary nature. The MAP calculation reveals that having

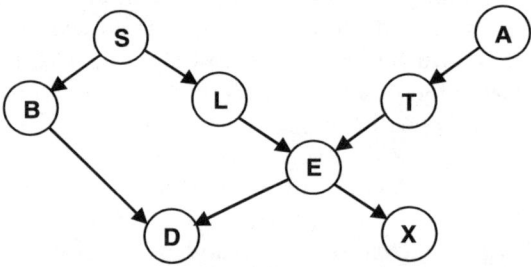

Fig. 7. The Asia network [13]

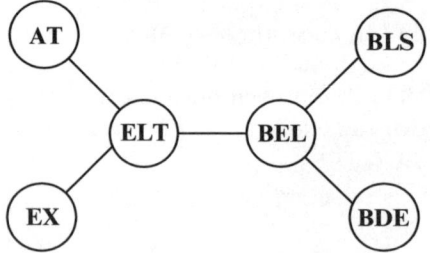

Fig. 8. A junction tree for the Asia network (adapted from [15])

neither Lung cancer nor Dyspnoea is the most probable explanation for the positive X-ray result. This is due to the fact that since Lung cancer is relatively rare the positive X-ray result can be accounted for by its false positive rate. Having both Lung cancer and Dyspnoea is the second most probable explanation. The results for the likelihood calculation are given in Table 1.

Note that in contrast to the MAP result the likelihood is equally high for Lung cancer irrespective of whether Dyspnoea is present or not. This is an artifact of the odd nature of the E variable since it is a logical variable taking on the logical OR of the L and T variables. Nevertheless, if Lung cancer is absent the presence of Dyspnoea does make the positive X-ray result more likely. This is due to the fact that if Lung cancer is absent, the presence of Dyspnoea makes the presence of Tuberculosis slightly more likely and hence also raises the probability of a positive X-ray result.

In our second calculation we consider the target configuration to be Bronchitis ($B = b_1$) and the explanation set to be the remaining nodes in the network. We note that in this network Smoking (S), Dyspnoea (D) and Either lung cancer or tuberculosis (E) form the Markov blanket of Bronchitis (B) and so all the other nodes in the network become redundant. In other words, we wish to obtain the probabilities $P(b_1|s_i, d_j, e_k)$ for all values of i, j and k. While we could proceed as in the previous case using the entire junction tree, in this case we can instead remove all the other nodes from the original Bayesian network and construct a junction tree for the remaining network. (Note that node E would now have no parents in this modified network and that there is no need to specify its prior probability as it does not enter the calculation of likelihoods.) The simplified junction tree is presented in Fig. 9.

Proceeding as before on the modified junction tree we obtain the results shown in Table 2. Note that in cases where Dyspnoea is present the probability

Table 1. Results for the Asia Network

	$P(X = x_1|L, D)$
$L = l_1\ D = d_1$	0.98
$L = l_1\ D = d_2$	0.98
$L = l_2\ D = d_1$	0.068
$L = l_2\ D = d_2$	0.056

X = x1 denotes a positive X-ray result, L = l_1(l_2) denotes the presence (absence) of Lung cancer variable and D = d_1(d_2) denotes the presence (absence) of Dyspnoea

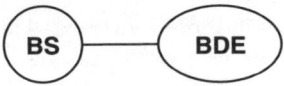

Fig. 9. The junction tree formed from the modified network consisting of just the nodes for Smoking (S), Bronchitis (B), Dyspnoea (D) and Either lung cancer or tuberculosis (E)

Table 2. Results for the Asia network when Bronchitis (B) is the only node in the target set and the only non-redundant nodes in the explanation set are Smoking (S), Dyspnoea (D) and Either lung cancer or tuberculosis (E)

			$P(B = b_1 \vert S, D, E)$
$S = s_1$	$D = d_1$	$E = e_1$	0.68
$S = s_1$	$D = d_1$	$E = e_2$	0.90
$S = s_1$	$D = d_2$	$E = e_1$	0.47
$S = s_1$	$D = d_2$	$E = e_2$	0.40
$S = s_2$	$D = d_1$	$E = e_1$	0.35
$S = s_2$	$D = d_1$	$E = e_2$	0.72
$S = s_2$	$D = d_2$	$E = e_1$	0.20
$S = s_2$	$D = d_2$	$E = e_2$	0.16

of Bronchitis is higher when both lung cancer and tuberculosis are absent than when one of them is present. This is due to the fact that either lung cancer or tuberculosis would explain away Dyspnoea. In fact, this gives rise to the fact that the second highest likelihood is obtained when the person does not smoke, but Dyspnoea is present while lung cancer and tuberculosis are not.

7 Conclusions

We have carried out a preliminary investigation into the calculation of likelihoods in a Bayesian network. A number of motivations have been proposed for this work and some discussion has been given of the relationship between likelihoods, explanation and intervention. We have looked at cases where the explanation set is restricted to parents of the nodes in the target set and also presented a procedure to obtain likelihoods in a tree-structured Bayesian network when the explanation set is restricted to ancestors of the target set. We have also explored a general procedure for obtaining the likelihood in a Bayesian network irrespective of whether the explanation set contains ancestors or descendents of the nodes in the target set.

There are numerous directions for future work. First of all, more calculations need to be performed to illustrate the merits and potential applications of the work. Some suggestions have been made as to how the procedure could be made more efficient and so these should be investigated. Furthermore, in certain cases alternative procedures will certainly be more efficient and so extension of the approaches adopted in Sect. 4 and Sect. 5.2 could also be investigated. The link between this work and algorithms for MAP emphasizes the need to investigate approximation methods. Work on the MAP problem (see for example [20, 21]) suggests how this could proceed. Thus, although this work is preliminary in nature it does open up new directions for the solution of a challenging inference problem in Bayesian networks.

References

1. Pearl J (1988) Probabilistic Reasoning in Intelligent Systems. Morgan Kaufman, San Mateo
2. Pearl J (2000) Causality: Models, Reasoning and Inference. Cambridge University Press, Cambridge
3. Shimony AE (1991) Explanation, irrelevance and statistical independence. In: Proceedings of AAAI, pp. 482–487
4. Suermondt HJ (1992) Explanation in Bayesian belief networks. Ph.D thesis, Stanford University
5. Henrion M, Druzdzel MJ (1990) Qualitative propagation and scenario-based approaches to explanation of probabilistic reasoning. In: Proceedings of the 6th Conference on Uncertainty in AI, pp. 17–32
6. Chajewska U, Halpern JY (1997) Defining explanation in probabilistic systems. In: Proceedings of the 13th Conference on Uncertainty in AI, pp. 62–71
7. Halpern JY, Pearl J (2001) Causes and Explanations: a Structural-Model Approach – Part I: Causes. In: Proceedings of the 17th Conference on Uncertainty in AI, pp. 194–202
8. Halpern JY, Pearl J (2001) Causes and Explanations: A Structural-Model Approach – Part II: Explanations. Proceedings of the 17th International Joint Conference on AI, pp. 27–34
9. Salmon WC (1998) Causality and Explanation. Oxford University Press, Oxford
10. Glass DH (2002) Coherence, explanation and Bayesian networks. In: Proceedings of the 13th Irish Conference on AI and Cognitive Science. Lecture notes in AI 2464:177–182
11. Jensen FV (1996) An Introduction to Bayesian Networks. UCL Press, London
12. Neapolitan RE (1990) Probabilistic Reasoning in Expert Systems. Wiley, New York
13. Cowell RG, Dawid AP, Lauritzen SL, Spiegelhalter DJ (1998) Probabilistic Networks and Expert Systems. Springer, Berlin Heidelberg New York
14. Lauritzen SL, Spiegelhalter DJ (1988) Local computations with probabilities on graphical structures and their application to expert systems. Journal of the Royal Statistical Society: Series B 50(2): 157–224
15. Xu H (1995) Computing marginals for arbitrary subsets from marginal representation in Markov trees. Artificial Intelligence 74:177–189
16. Nilsson D (1998) An efficient algorithm for finding the M most probable configurations in Bayesian networks. Statistics and Computing 2:159–173
17. Dawid AP (1992) Applications of a general propagation algorithm for probabilistic expert systems. Statistics and Computing 2:25–36
18. de Campos LM, Gamez JA, Moral S (2002) On the problem of performing exact partial abductive inferences in Bayesian belief networks using junction trees. Studies in Fuzziness and Soft Computing, 90:289–302
19. Cooper G (2002) Computation complexity of probabilistic inference using Bayesian belief network. SIAM Journal of Computing 42:393–405
20. de Campos LM, Gamez JA, Moral S (1999) Partial abductive inference in Bayesian belief networks using a genetic algorithm. Pattern Recognition Letter 20:1211–1217
21. Park JD, Darwiche A (2001) Approximating MAP using local search. In: Proceedings of the 17th Conference on Uncertainty in AI, pp. 403–410

Towards Elimination of Redundant and Well Known Patterns in Spatial Association Rule Mining

Vania Bogorny, João Francisco Valiati, Sandro da Silva Camargo,
Paulo Martins Engel, and Luis Otavio Alvares

Instituto de Informática – Universidade Federal do Rio Grande do Sul (UFRGS),
Caixa Postal 15.064 – 91.501-970, Porto Alegre, Brazil
vbogorny@inf.ufrgs.br, jvaliati@inf.ufrgs.br, scamargo@inf.ufrgs.br,
engel@inf.ufrgs.br, alvares@inf.ufrgs.br

Summary. In this paper we present a new method for mining spatial association rules from geographic databases. On the contrary of most existing approaches that propose syntactic constraints to reduce the number of rules, we propose to use background geographic information extracted from geographic database schemas. In a first step we remove all well known dependences explicitly represented in geographic database schemas. In a second step we remove redundant frequent sets. Experiments show a very significant reduction of the number of rules when both well known dependences and redundant frequent sets are removed.

1 Introduction

The association rule mining technique emerged with the objective to find novel, useful, and interesting associations, *hidden* among itemsets [1] and spatial predicate sets [2]. An enormous amount of algorithms with different thresholds for reducing the number of rules has been proposed. However, only the data have been considered, while the database schema, which is a rich knowledge resource, has not been used as prior knowledge to eliminate well known patterns.

In traditional association rule mining the schema might not be useful, since items and transactions can be stored in a single relation. In geographic databases, however, the number of object types to be considered for mining is large. Every different object type is normally stored in a different relation, since most geographic databases follow the relational approach [3]. Figure 1 shows an example of how geographic data are stored in relational databases. There is a different relation/table for every different object type [3] (street, water resource, gas station, and island).

V. Bogorny et al.: *Towards Elimination of Redundant and Well Known Patterns in Spatial Association Rule Mining*, Studies in Computational Intelligence (SCI) **109**, 343–360 (2008)
www.springerlink.com © Springer-Verlag Berlin Heidelberg 2008

(a) Street

Gid	Name	Length	Shape
1	BR-101	632056.03	Multiline $[(x_1, y_1), (x_2, y_2), ..]$
2	RS-226	255365.88	Multiline $[(x_1, y_1), (x_2, y_2), ..]$

(b) WaterResource

Gid	Name	Length	Shape
1	Jacui	3214328.71	Multiline $[(x_1, y_1), (x_2, y_2), ..]$
2	Guaiba	283434.23	Multiline $[(x_1, y_1), (x_2, y_2), ..]$
3	Uruguai	4523333.12	Multiline $[(x_1, y_1), (x_2, y_2), ..]$

(c) GasStation

Gid	Name	Vol_Diesel	Vol_Gas	Shape
1	Posto do Beto	20000	85000	Point$[(x_1, y_1)]$
2	Posto da Silva	30000	95000	Point$[(x_1, y_1)]$
3	Posto Ipiranga	25000	120000	Point$[(x_1, y_1)]$

(d) Island

Gid	Name	Population	Sanitary_Condition	Shape
1	Flores	5000	Yes	Point$[(x_1, y_1)]$
2	Pintada	20000	Partial	Point$[(x_1, y_1)]$
3	Da Luz	15000	No	Point$[(x_1, y_1)]$

Fig. 1. Examples of geographic data storage in relational databases

From the database design point of view, the objective of data modeling is to bring together all relevant object types of the application, their associations/relationships, and their constraints [3,4]. Many geographic object types have mandatory associations, represented in the schema by *one–one* and *one–many* cardinality constraints, which the database designer has the responsibility to warrant when the schema is conceived [4]. The representation is usually in the third normal form [4], intending to reduce anomalies and warrant integrity.

In contrast to database schema modeling, where associations between data are *explicitly* represented, association rule mining algorithms should find *implicit* and novel associations. While the former represents the data into the third normal form, the latter usually denormalizes the data in one single table or one single file. This transformation brings the associations explicitly represented in the database schema to the dataset to be mined, and by consequence, many well known associations specified in the schema, are extracted by association rule mining algorithms.

In geographic databases, the number of associations specified in the schema reflects a large number of well known geographic dependences. Figure 2 shows two layers of information of the same geographic region. On the left there is a well known pattern, i.e., a geographic dependence where gas stations do always intersect streets. If considered in association rule mining, such dependence

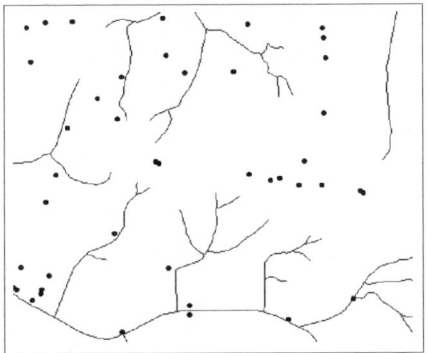 

Fig. 2. (*left*) Non-standard spatial relationships between Gas Stations (*points*) and Water Bodies (*lines*), and (*right*) well known geographic dependence between Gas Stations (*points*) and Streets (*lines*)

will produce high confidence rules (e.g. *is_a(GasStation)*→ *intersect(Street)* (100%)). On the right, however, there is no explicit pattern among gas stations and water resources which may produce well known rules. Relationships such as the example shown in Fig. 2 (right) may be interesting for association rule mining.

Users of some domains may not be interested in strong geographic domain rules such as *is_a(GasStation)*→ *intersect(Street)* (100%), but in non-obvious rules such as *is_a(GasStation) and intersect(WaterResource)* → *pollution = high* (70%).

In geographic databases, most mined rules are strongly related to geographic dependences which represent strong regularities, but do not contribute to the discovery of novel and useful knowledge. The result is the mixed presentation of thousands of interesting and uninteresting associations that can discourage users from interpreting them all in order to find interesting patterns.

We claim that well known associations, explicitly represented in geographic database schemas, should be eliminated in association rule mining to avoid their extraction and presentation to the user. Although some well known associations can be reduced in geographic data preprocessing steps [5], most dependences can only be eliminated into the data mining algorithm. Aiming to reduce the amount of both redundant and well known patterns, this paper presents a two-step approach for mining spatial association rules. In the first step geographic database schemas are used as prior knowledge to eliminate all association rules which contain obvious geographic dependences. In the second step, we eliminate all redundant frequent itemsets that generate redundant rules [6], similarly to the closed frequent pattern mining technique [7–9].

The remainder of the paper is organized as follows: Section 2 introduces geographic database schemas and how dependences can be extracted. Section 3 describes the problem of mining spatial association rules with well known geographic dependences and presents the methods to remove both redundant and well known dependences. Section 4 presents the experiments to evaluate the proposed approach, Sect. 5 presents the related works, and Sect. 6 concludes the paper and gives some directions of future work.

2 Geographic Database Schemas

Geographic database schemas are normally extended relational or object-oriented schemas [3]. There is a trend toward extending both Entity Relationship (ER) and Object-Oriented (OO) diagrams with pictograms to provide special treatment to spatial data types [3]. References [10, 11] are approaches which extend ER and OO diagrams for geographic applications. In both ER and OO approaches, relationships among entities are represented through associations with cardinality constraints. In geographic database schemas, these associations may either represent a spatial relationship or a single association, aggregation, etc.

Mandatory associations are represented by cardinality constraints *one–one* and *one–many* [3, 4]. Figure 3 shows an example of part of a conceptual geographic database schema, represented in a UML class diagram [12], and part of its respective logical schema for relational and OO databases.

The schema in Fig. 3 represents part of the data shown in Fig. 2. Notice that there are many mandatory associations (e.g. gas station and street, street and county, water resource and county, and island and water resource). These dependences, explicitly represented in the schema, produce well known patterns when considered in spatial association rule mining (see in Fig. 2 (left) that every gas station intersects one or more streets).

In the logical level, mandatory relationships expressed by cardinalities *one–one* and *one–many* normally result in foreign-keys in relational geographic databases, and in pointers to classes, in object-oriented geographic databases [3, 4].

For data mining and knowledge discovery well known geographic dependences can be either specified by the user or automatically retrieved with processes of reverse engineering [13] if the schema is not available. Many different approaches to extract dependences from relational databases using reverse engineering are available in the literature. For data mining and knowledge discovery in non-geographic databases reverse engineering has been used to understand the data model [14] in legacy systems, or to automatically extract SQL queries [15], but not as prior knowledge to reduce well known patterns.

When provided by the user, a larger set of dependences can be specified; not only associations explicitly represented in the schema, but other geographic domain dependences which produce well known patterns.

Figure 4 shows an example of a data preprocessing algorithm to extract mandatory *one–one* and *one–many* associations from geographic database schemas. If the database is relational, then the algorithm searches for all foreign keys. For each foreign key, the name of the table which it references is

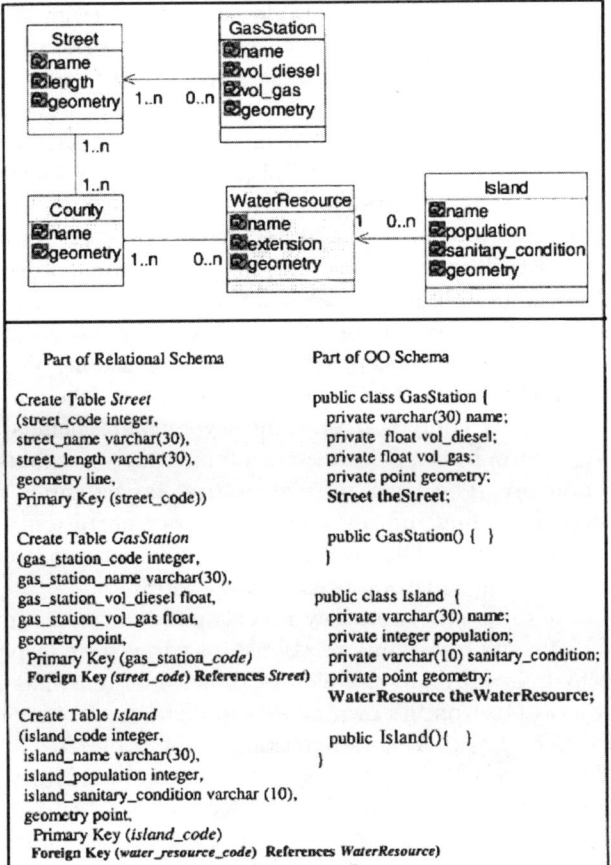

Fig. 3. Conceptual and logical geographic database schema

```
Given: a relational database schema
Find all foreign_keys
For each foreign_key
    Insert into φ the name of the table
    which the foreign_key references and the
    name of the table to which the foreign_key
    belongs


Given: an OO database schema
Find all classes
For each class in the database schema
    If there are references to classes
        Insert the class name and the
        referenced class into φ
```

Fig. 4. Algorithm to extract mandatory relationships

retrieved, as well as the name of the table where the foreign key is specified. The name of both relations is stored in a set of knowledge constraints ϕ. If the database is object-oriented, then the same steps are performed, but searching for classes with attributes which refer to other classes.

In order to evaluate the amount of mandatory associations in real geographic database schemas we analyzed the object-oriented geographic database schema developed by the Brazilian Army. As the terrain model has a large number of object types, common to different schemas, the database schema developed by the Brazilian Army includes most geographic objects abstracted from the real world as well as their associations.

On account of the large number of entities and relationships to be represented, geographic data conceptual schemas are usually designed in different layers of information. The geographic database schema developed by the Brazilian Army is composed of eight layers (subschemas): edification, infra-structure, hydrography, vegetation, administrative regions, referential, relief, and toponymy. The layer infra-structure, for example, is divided in six sub-schemas, including information about transportation, energy, economy, communication, etc. The hydrography layer, for example, represents geographic objects such as rivers, oceans, lakes, etc.

Information of different layers may be extracted for data mining, and the number of *one–one* and *one–many* relationships varies from layer to layer. For example, the hydrography layer, which is shown in Fig. 5, has a total of 24 geographic objects (16 from its own layer and eight from other layers) which share 13 mandatory 1..1 or 1..n associations.

This analysis showed that a large number of mandatory well known geographic dependences are explicitly specified in the schema, and if used as prior knowledge to avoid their extraction in association rule mining, a large amount of irrelevant patterns will be eliminated.

3 Mining Spatial Association Rules with Knowledge Constraints

We illustrate the problem of mining spatial association rules without removing explicit geographic domain dependences through an example. Considering a set of elements $\Psi = \{A, B, C, D\}$, all possible combinations of these elements produce the sets: $\{A, B\}$, $\{A, C\}$, $\{A, D\}$, $\{B, C\}$, $\{B, D\}$, $\{C, D\}$, $\{A, B, C\}$, $\{A, B, D\}$, $\{A, C, D\}$, $\{B, C, D\}$, and $\{A, B, C, D\}$. Without considering any threshold, the number of possible subsets is 11, and the maximum number of rules produced with these subsets is 50, as shown in Table 1.

Now consider that the elements C and D have a mandatory association. Notice that there are four subsets in which C and D appear together ($\{C, D\}$, $\{A, C, D\}$, $\{B, C, D\}$ and $\{A, B, C, D\}$). These four subsets will produce 28 rules, and in every rule, C and D will appear. The result is that 56% of the whole amount of rules is created with the dependence between C and D.

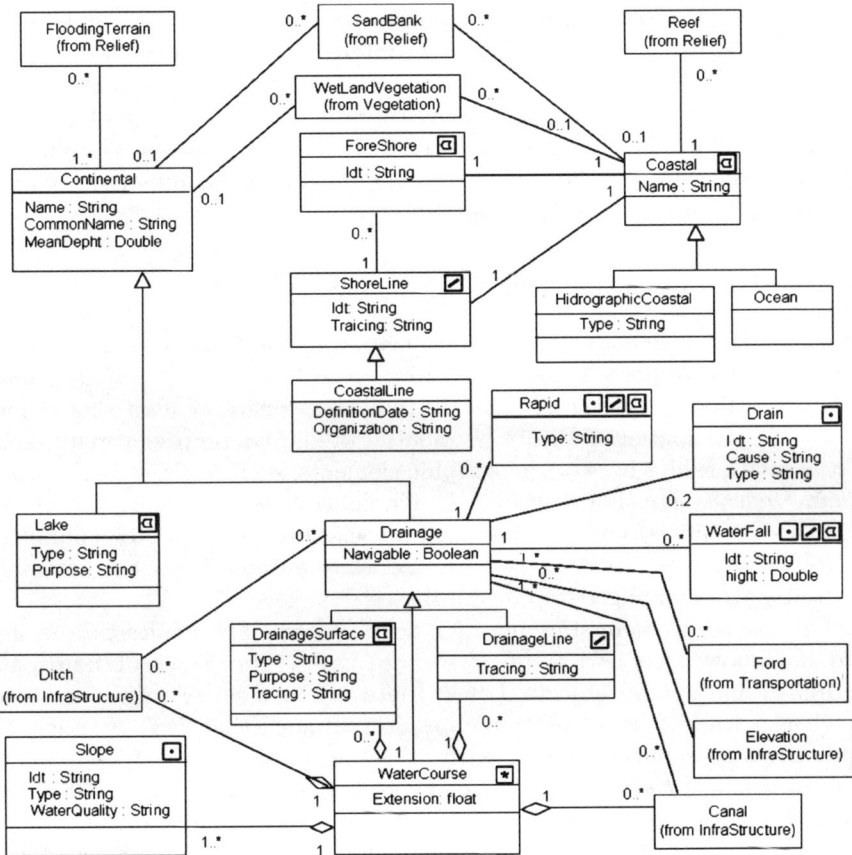

Fig. 5. Conceptual object oriented schema of the Brazilian Geographic Territory (MCOO of EBG – Brazilian Army – STI – DSG – 1°DL)

Table 1. Maximum number of sets and association rules

Sets	Possible rules	Rules
{AB}	$A \rightarrow B; B \rightarrow A$	2
{AC}	$A \rightarrow C; C \rightarrow A$	2
{AD}	$A \rightarrow D; D \rightarrow A$	2
{BC}	$B \rightarrow C; C \rightarrow B$	2
{BD}	$B \rightarrow D; D \rightarrow B$	2
{CD}	$C \rightarrow D; D \rightarrow C$	2
{ABC}	$A \rightarrow BC; B \rightarrow AC; C \rightarrow AB; BC \rightarrow A; AC \rightarrow B; AB \rightarrow C$	6
{ABD}	$A \rightarrow BD; B \rightarrow AD; D \rightarrow AB; BD \rightarrow A; AD \rightarrow B; AB \rightarrow D$	6
{ACD}	$A \rightarrow DC; D \rightarrow AC; C \rightarrow AD; DC \rightarrow A; AC \rightarrow D; AD \rightarrow C$	6
{BCD}	$D \rightarrow BC; B \rightarrow DC; C \rightarrow DB; BC \rightarrow D; DC \rightarrow B; DB \rightarrow C$	6
{ABCD}	$A \rightarrow BCD; B \rightarrow ACD; C \rightarrow ABD; D \rightarrow ABC; AB \rightarrow CD;$	14
	$AC \rightarrow BD; AD \rightarrow BC; BC \rightarrow AD; BD \rightarrow AC; CD \rightarrow AB;$	
	$BCD \rightarrow A; ACD \rightarrow B; ABD \rightarrow C; ABC \rightarrow D$	

It is important to observe that we cannot simply remove C and D from Ψ, because either C or D may have an interesting association with A or B. However, we can avoid the combination of C and D in the same set. This eliminates the possibility of generating rules including both C and D.

In the following sections we describe the formal problem of mining association rules, the method as dependent objects are eliminated, and how redundant frequent itemsets are pruned.

3.1 Spatial Association Rules

An association rule consists of an implication of the form $X \rightarrow Y$, where X and Y are sets of items co-occurring in a given tuple [1]. Spatial association rules are defined in terms of spatial predicates, where at least one element in X or Y is a spatial predicate [2]. Spatial predicates represent materialized spatial relationships between geographic elements, such as *close, far, contains, within, touches*, etc. For example, *is_a(x,slum)* ∧ *far_from(x,water_network)* $\rightarrow$ *disease(hepatitis)* (70%) is a spatial association rule with 70% confidence. In [16] we presented an intelligent framework to automatically extract spatial predicates from large geographic databases.

The formal problem statement for defining association rules can be specified as follows: Let $F = \{f_1, f_2, \ldots, f_k, \ldots, f_n\}$ be a set of non-spatial attributes and spatial objects. Let Ψ (dataset) be a set of reference objects T, where each T is a set of predicates (tuple) such that $T \subseteq F$. Each T is represented as a binary vector, with an element $t[k] = 1$, if T contains the attribute f_k, and $t[k] = 0$, otherwise. There is exactly one tuple in the dataset to be mined for each reference object. Considering X as a subset of F, T contains X if, for all f_k in X, $t[k] = 1$. Similarly, being Y a subset of F, T contains Y if, for all f_k in Y, $t[k] = 1$.

In a rule $X \rightarrow Y$, $X \subset F$, $Y \subset F$ and $X \cap Y = \phi$. The support s of a predicate set X is the number of tuples in which the predicate set X occurs as a subset. The support of the rule $X \rightarrow Y$ is given as $s(X \cup Y)$.

The rule $X \rightarrow Y$ is satisfied in Ψ with confidence factor $0 \leq c \leq 1$, if at least $c\%$ of the instances in Ψ that satisfy X also satisfy Y. The notation $X \rightarrow Y(c)$ specifies that the rule $X \rightarrow Y$ has confidence factor c. More precisely, the confidence factor is given as $s(X \cup Y)/s(X)$.

The problem of mining association rules is performed in two main steps:

(a) *Find all frequent patterns/predicates/sets*: a set of predicates is a frequent pattern if its support is at least equal to a certain threshold, called *minsup*.
(b) *Generate strong rules*: a rule is strong if it reaches minimum support and its confidence is at least equal to a certain threshold, called *minconf*.

Assertion 1. [17] If a predicate set Z is frequent, then every subset of Z will also be frequent. If the set Z is not frequent, then every set that contains Z is not frequent too. All rules derived from Z satisfy the support constraints if Z satisfies the support constraints.

Considering Assertion 1, we propose a third class of constraints, called *knowledge constraints* (ϕ). These constraints will be used to avoid the generation of frequent sets which contain the pairs of dependences specified in ϕ, as will be explained in Sect. 3.2.

3.2 Pruning Well Known Patterns

Figure 6 shows the algorithm Apriori-KC, which is based on Apriori [17], and has been the basis for dozens of association rule mining algorithms.

The algorithm shown in Fig. 6 removes from the candidate sets all pairs of elements which have geographic dependences. As in Apriori, Apriori-KC performs multiple passes over the dataset. In the first pass, the support of the individual elements is computed to determine 1-predicate sets. In the subsequent passes, given k as the number of the current pass, the large sets L_{k-1} in the previous pass ($k - 1$) are grouped into sets C_k with k elements, which are called *candidate sets*.

The support of each candidate set is computed, and if it is equal or higher than minimum support, then this set is considered frequent. This process continues until the number of frequent sets is zero.

Similarly to [18], which eliminates in the second pass candidate sets that contain both parent and child specified in concept hierarchies, we propose a method to eliminate all candidate sets which contain geographic dependences, independently of any concept hierarchy.

The dependences are eliminated in an efficient way, in one step, in the second pass, when generating candidates with two elements. Being ϕ a set of

```
Given:  φ,Ψ, minsup
L₁ = {large 1-predicate sets};
For (k = 2; Lₖ − 1 = ∅; k + +) do begin
        Cₖ = apriori_gen(Lₖ₋₁); // Generates new
                                // candidates
        Forall T ∈ Ψ do begin
                Cₜ = subset (Cₖ,T); // Candidates in t
                forall candidates c ∈ Cₜ do
                        c.count++;
        End;
        Lₖ = {c ∈ Cₖ| c.count ≥ minsup};

        If k = 2 // in the second pass
                L₂ = L₂ − φ; //removes pairs with
                            // dependences
End;
```

Fig. 6. Pseudo-code of Apriori-KC to generate large predicate sets without well known dependences

pairs of geographic objects with dependences, which can be extracted from the database schema or provided by the user, when k is 2, all pairs of elements with a dependence in ϕ are removed from C_2.

According to Assertion 1, this step *warrants* that the pairs of geographic objects in ϕ will neither appear together in the frequent sets nor in the spatial association rules. This makes our approach effective and independent of any threshold such as minimum support, minimum confidence, lift, etc.

Our dependence elimination method removes each pair of geographic objects in ϕ (e.g. $\{C, D\}$), and avoids the generation not only of the main rule $C \rightarrow D$ but of all derived rules (e.g. $D \rightarrow C$, $C \rightarrow AD$) which contain the known dependence.

Although Apriori-KC removes well known geographic dependences, many redundant frequent sets are still generated, as will be explained in the following section.

3.3 Pruning Redundant Patterns

It is known that the Apriori algorithm generates a large amount of redundant frequent sets and association rules [7]. For instance, let us consider the dataset shown in Fig. 7a and the frequent sets and the transactions where the items occur in Fig. 7b. The set $\{A, D, W\}$, for example, is a *frequent set* because it

(a) dataset	
Tid	itemset
1	A, C, D, T, W
2	C, D, W
3	A, D, T, W
4	A, C, D, W
5	A, C, D, T, W
6	C, D, T

b) *frequent* sets with minimum support 50%	
TidSet	Frequent itemsets L and closed frequent itemsets
123456	**{D}**
12456	{C}, **{C,D}**
12345	{W}, **{D,W}**
1245	{C,W}, {C,D,W}
1345	{A}, {A,D}, {A,W}, **{A,D,W}**
1356	{T}, **{D,T}**
145	{A,C}, {A,C,W}, {A,C,D}, **{A,C,D,W}**
135	{T,W}, {A,T}, {A,D,T}, {A,T,W}, {D,T,W}, **{A,D,T,W}**
156	{C,T},**{C,D,T}**

Fig. 7. (a) Example dataset and (b) the frequent itemsets with minimum support 50%

reaches minimum support (50%). It is also a *closed frequent set* [6] because in the set of transactions (1,345) where it occurs in the dataset, no set larger than $\{A, D, W\}$ (with more than three elements) in the same transactions reaches minimum support. The frequent set $\{A, D, T\}$, for example, appears in the transactions 135, but in the same transactions, a larger set $\{A, D, T, W\}$ can be generated. In this case the tidset$(A, D, T) = 135$, the tidset $(A, D, T, W) = 135$, and $\{A, D, T\} \subset \{A, D, T, W\}$, so the frequent set $\{A, D, T\}$ is not closed.

According to [7], all frequent sets L that occur in the same transactions generate rules with same support and same confidence. In Fig. 7b, for the transactions 1,345, all rules generated from the four frequent itemsets have same support and same confidence. The maximal itemset for each set of transactions contains the maximal number of elements (in transactions 1,345 $\{A, D, W\}$ is the maximal set), all other sets generate redundant rules. For instance, a rule generated from a frequent itemset $\{A, W\}$, such as A→W, is redundant in relation to a rule A→DW, generated from the closed frequent itemset $\{A, D, W\}$.

A frequent itemset L is a *closed frequent itemset* if $\Omega(L) = L$ [7]. The closure operator Ω associates with a frequent itemset L the maximal set of items common to all transactions (tidset) containing L. Rules generated from non-closed frequent itemsets are redundant to rules generated from their respective closed frequent sets. In Fig. 7b, the frequent sets in bold style are the closed frequent itemsets. All other frequent sets are redundant and will generate redundant association rules.

Redundant frequent sets and redundant association rules can be significantly reduced by generating closed frequent itemsets [7, 8]. Since redundant frequent sets generate association rules with same support and same confidence they can be eliminated in the step where frequent sets are generated [7], similarly to the dependence pruning method presented in the previous section.

After the frequent sets with well known geographic dependences have been removed, the resultant frequent sets may not be closed anymore [9]. For instance, let us consider the dependence $\{C, D\}$ and the frequent sets shown in Fig. 7b for the transactions 145. Notice that by removing the pair $\{C, D\}$, the frequent set $\{A, C, D\}$ and the closed frequent set $\{A, C, D, W\}$ will not be generated by Apriori-KC. However, the pair $\{A, C\}$ is still redundant in relation to the set $\{A, C, W\}$, while the latter is the maximal set for the transactions 145, but is not closed. Considering that after geographic dependences have been removed we may not generate closed frequent itemsets in relation to the dataset, we apply the closed frequent pattern mining technique over the resultant frequent sets generated by Apriori-KC to remove the redundant sets, but not to generate closed frequent sets.

Figure 8 shows the pruning method to eliminate redundant frequent sets that are still generated by Apriori-KC.

Given a set L of frequent sets generated with Apriori-KC presented in the previous section and the geographic dataset Ψ, the pruning of redundant frequent sets starts similarly to the closed frequent set approach. All frequent

```
Given: Lₖ; // frequent sets generated with Apriori-KC
        Ψ; // a spatial dataset
Find: Maximal M

    // find maximal generalized predicate sets
    M = L;
    For (k = 2; Mₖ != ∅; k + +) do begin
        For (j = k + 1; Mⱼ != 0; j + +) do begin
            If (tidSet (Mₖ) = tidSet (Mⱼ))
                If (Mₖ ⊂ Mⱼ) // Mⱼ is more general than Mₖ
                    Delete Mₖ from M;
        End;
    End;
    Answer = M;
```

Fig. 8. Elimination of redundant frequent sets without well known dependences

sets M, with size k are compared to the sets with size $k + 1$. When a set $M_k \subset M_{k+1}$ and the set of transactions (tidset) in which M_k appears is the same as the transactions where M_{k+1} appears, then we can say that M_k is redundant, while M_{k+1} is more general. When this occurs, M_k is removed from M. This process continues until all frequent sets in M have been tested and have no more redundant frequent sets. Association rules are then extracted from M as in Apriori.

4 Experiments and Evaluation

For the data mining user, more important than the reduction of the computational time to generate spatial association rules is the elimination of well known geographic domain rules, which will reduce the human time for analyze the rules.

In this section we present experiments performed over different datasets, extracted from a real geographic database of the city of Porto Alegre, in order to evaluate the elimination of geographic dependences and redundant frequent sets in spatial association rule mining. The first experiment was performed over a dataset with 18 spatial predicates, including the relevant feature types trees, treated water network, slums, cellular antennas, water resources, water collection points, illumination points, streets, schools, hospitals, health centers, industrial residues repositories, and artesian wells. This dataset has *two* dependences among the relevant feature types *illumination points* and *streets*, and *water resource* and *water collection points*.

In the first experiment, shown in Fig. 9, we evaluate the rule reduction when one and two pairs of well known dependences are eliminated, considering two different values of minimum support, 15 and 25%. In Fig. 9 (left)

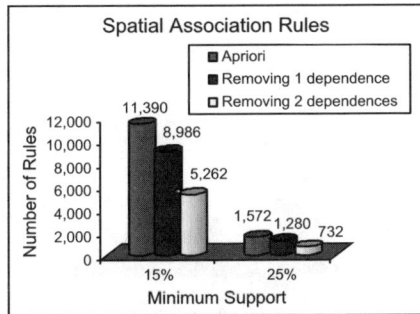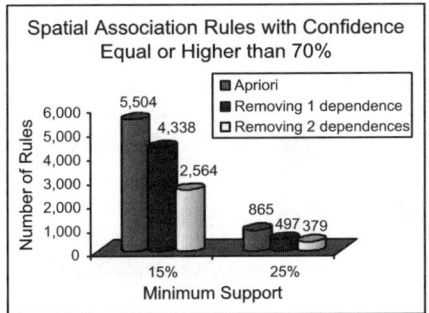

Fig. 9. Experiments with dependence elimination

we evaluate the method without considering minimum confidence. In this experiment, for different values of minimum support the elimination of one dependence reduced the number of rules in an average of 20% in relation to Apriori. The elimination of two dependences reduced the rule generation in comparison with Apriori in around 54% for the different values of minimum support.

In Fig. 9 (right) we have performed the same experiment considering 70% minimum confidence. As can be observed, the number of rules is significantly reduced by considering 70% as minimum confidence, but only considering this experiment we would not be able to say how many rules are eliminated by the confidence and how many by the dependence pruning method. For instance, for support 25% the rule reduction when one dependence is eliminated reaches 47% in relation to Apriori. In comparison with the experiment in Fig. 9 (left) we know that the dependence elimination pruned the rules in 19%, while the other 28% are reduced by the minimum confidence threshold.

In order to evaluate the pruning methods proposed in this paper we present two more experiments performed over different datasets, considering different values of minimum support, and without considering minimum confidence. The second experiment was performed over a dataset with 17 spatial predicates including feature types bus stops, streets, slums, sewer network, cellular antennas, water collection points, hydrants, hydrographic basin, etc. This dataset has a spatial dependence between the relevant feature types *bus stop* and *street*. In this experiment, which is shown in Fig. 10, we considered three different values of minimum support (5, 10, and 15%), in order to evaluate the reduction of both frequent sets and association rules. In Fig. 10 (left) we evaluate the frequent set reduction by removing both dependences and redundant frequent sets. For the three values of minimum support the elimination of one single dependence reduced the frequent sets in an average of 20%, while the elimination of the dependence and the redundant frequent sets reduced the frequent sets in an average of 86%.

By reducing the number of frequent sets, the number of association rules is by consequence reduced for all values of minimum support, as shown in

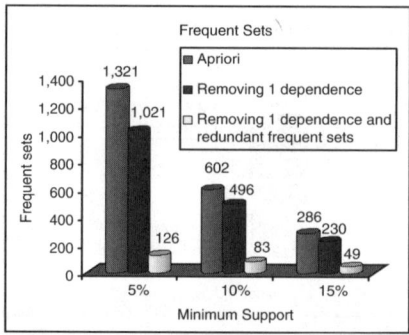
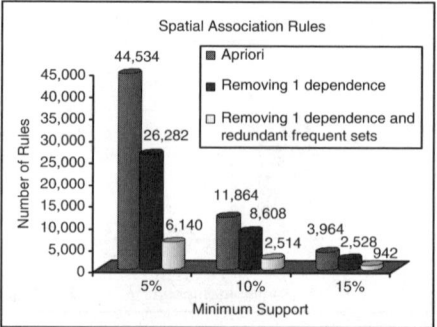

Fig. 10. Frequent set and association rule evaluation by pruning dependences and redundant frequent sets

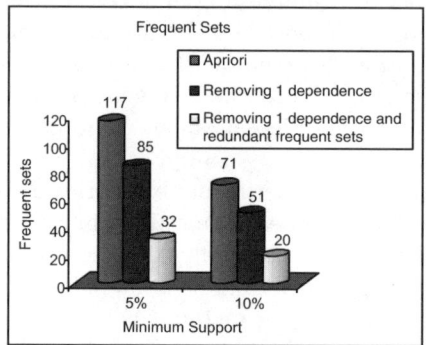
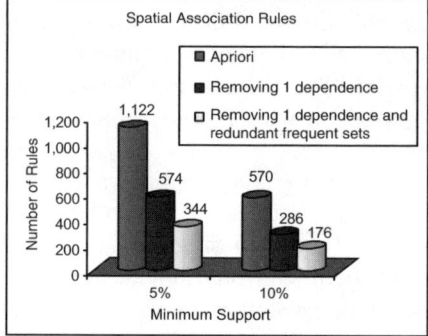

Fig. 11. Frequent sets and spatial association rule evaluation

Fig. 10 (right). The most significant reduction is for lower minimum support. The elimination of one single dependence reduced the rules in 41% for minimum support 5% in comparison to Apriori, and much more significantly when the redundant frequent sets are eliminated as well, where the rule reduction reaches 86%.

The third experiment was performed over a dataset with 15 spatial predicates including feature types streams, slums, hospitals, gas stations, streets, etc. This dataset has *one* dependence among the relevant feature types *gas station* and *streets*, and 2,159 rows, where each row corresponds to a census sector in the city of Porto Alegre. In this experiment the number of rows is much higher than in the previous experiments that had only 109 rows, but the number of generated frequent sets and association rules is much lower, since frequent sets and association rules are data dependent. Figure 11 (left) shows the number of frequent sets generated from this dataset and Fig. 11 (right) shows the respective number of association rules. Although the number of frequent sets and association rules generated from this dataset are much lower than in the previous experiments, the percentage of frequent set and rule reduction remains significant. A reduction from 1,122 rules to 344 rules as shown for minimum support 5% in Fig. 11 (right) will make a significant difference for the data mining user that has to analyze all rules, one by one, in order to find novel, interesting, and useful patterns.

5 Related Works

Existing approaches for mining spatial association rules do not make use of prior knowledge to reduce the number of well known patterns. Koperski [2] presented a top-down, progressive refinement method. In a first step spatial approximations are calculated, and in a second step, more precise spatial relationships are computed to the result of the first step. Minimum support is used in data preprocessing to extract only frequent spatial relationships. A similar method has been proposed by [19] for mining association rules among geographic objects with broad boundaries. [20] applied Apriori [17] to geographic data at different granularity levels.

In our previous work [5] we presented a data preprocessing method using prior knowledge to reduce geographic dependences between the reference object and the relevant objects. However, geographic dependences among relevant objects can only be completely eliminated during the data mining step. As a continued study in mining spatial association rules using background knowledge, in this paper we propose to remove not only well known geographic patterns among relevant feature types, but also redundant patterns.

In geographic databases, minimum support can eliminate information which may lead to novel knowledge, while geographic domain associations may still remain among the resultant set of rules.

Approaches for mining non-geographic data generate closed frequent itemsets to reduce redundant rules [6–8] and investigate the most appropriate threshold [21] or the interestingness [22] of the rules, but do not warrant the elimination of well known patterns. Our approach presented in this paper eliminates in a first step the exact pairs of dependences which produces nonnovel rules, independently of any threshold, and in a second step eliminates redundant frequent sets. In summary, our method avoids the generation of rules known a priori as non-interesting, and then removes redundant frequent sets to avoid the generation of redundant rules.

6 Conclusions and Future Work

In this paper we presented a method for mining spatial association rules using prior geographic domain knowledge. Domain knowledge refers to mandatory geographic dependences explicitly represented in geographic database schemas or which are well known by the user. We showed that explicit mandatory relationships produce irrelevant patterns, while the implicit spatial relationships may lead to more interesting rules.

Considering geographic domain dependences as prior knowledge, we proposed a frequent set pruning method that significantly reduces the number of spatial association rules. Besides pruning a large number of rules, all associations that would be created with well known geographic domain dependences are eliminated. Our method eliminates all dependences in one single step, before creating the rules. The result is that more interesting rules will be generated, independently of values of minimum support or confidence.

The main advantage of our method is the simplicity as well known dependences are eliminated. While most approaches define syntactic constraints and different thresholds to reduce the number of patterns and association rules, we consider *semantic knowledge constraints,* and eliminate the exact pairs of geographic objects that produce well known patterns.

The main contribution of our approach is for the data mining user, which does not have to analyze hundreds or thousands of rules without novel knowledge.

Traditional association rule mining algorithms that generate *frequent* sets or closed *frequent* sets eliminate redundant and non-interesting rules, but do not warrant the elimination of well known geographic dependences. In this paper we have shown the significant frequent set and rule reduction when both well known dependences and redundant frequent sets are eliminated before the generation of spatial association rules.

As future work we will evaluate the problem of mining spatial association rules with knowledge constraints at different granularity levels.

Acknowledgment

Our thanks for both CAPES and CNPq which partially provided the financial support for this research. For the Brazilian Army, which provided the geographic database schema, and for Procempa, for the geographic database.

References

1. Agrawal R, Imielinski T, Swami A (1993) Mining association rules between sets of items in large databases. In: Buneman P, Jajodia S (eds.), ACM-SIGMOD International Conference on Management of Data. ACM Press, New York, pp 207–216
2. Koperski K, Han J (1995) Discovery of spatial association rules in geographic information databases. In: Egenhofer MJ, Herring JR (eds.), SSD 4th International Symposium in Large Spatial Databases, (SSD'95). Springer, Berlin Heidelberg New York, pp 47–66
3. Shekhar S, Chawla S (2003) Spatial Databases: A Tour. Prentice Hall, Upper Saddle River.
4. Elmasri R, Navathe S (2003) Fundamentals of database systems, 4th edn. Addison Wesley, Reading

5. Bogorny V, Engel PM, Alvares LO (2006) GEOARM: an interoperable framework to improve geographic data preprocessing and spatial association rule mining. In: Zhang K, Spanoudakis G, Visaggio G (eds.), 18th International Conference on Software Engineering and Knowledge Engineering, (SEKE'06). Knowledge Systems Institute, Skokie

6. Bastide Y, Pasquier N, Taouil R, Stumme G, Lakhal L (2000) Mining minimal non-redundant association rules using frequent closed itemsets. In: Lloyd JW, Dahl V, Furbach U, Kerber M, Lau KK, Palamidessi C, Pereira LM, Sagiv Y, Stuckey PJ (eds.), First International Conference on Computational Logic. LNCS. Springer, Berlin Heidelberg New York, pp 972–986

7. Pasquier N, Bastide Y, Taouil R, Lakhal L (1999) Discovering frequent closed itemsets for association rules. In: Beeri C, Buneman P (eds.), Seventh International Conference on Database Theory (ICDT'99). LNCS 1540. Springer, Berlin Heidelberg New York, pp 398–416

8. Zaki M, Ching-Jui H (2002) CHARM: an efficient algorithm for closed itemset mining. In: Grossman RL, Han J, Kumar V, Mannila H, Notwani R (eds.), Second SIAM International Conference on Data Mining. SIAM, Philadelphia, pp 457–473

9. Pei P, Han J, Mao R (2000) CLOSET an efficient algorithm for mining frequent closed itemsets. In: Chen W, Naughton JF, Bernstein PA (eds.), ACM SIGMOD Workshop on Research Issues in Data Mining and Knowledge Discovery (DMKD'00). ACM Press, New York

10. Parent C, Spaccapietra S, Zimanyi E, Donini P, Plazanet C, Vangenot C (1998) Modeling spatial data in the MADS conceptual model, In: Poiker T, Chrisman N (eds.), 8th International Symposium on Spatial Data Handling, (SDH'98), pp 138–150

11. Borges KAV, Laender AHF, Davis JR C (2001) OMT-G: an object-oriented data model for geographic applications. GeoInformatica, 5(3): 221–260

12. Booch G, Rumbaugh J, Jacobson I (1998) The unified modeling language: user guide, Addison-Wesley, Reading

13. Chifosky EJ, Cross JH (1990) Reverse engineering and design recovery: a taxonomy. IEEE Software, 7: 13–17

14. MCKearney S, Roberts H (1996) Reverse engineering databases for knowledge discovery. In: Simoudis E, Han J, Fayyad U (eds.), Second International Conference on Knowledge Discovery and Data Mining (KDD'96). ACM Press, New York, pp 375–378

15. Shoval P, Shreiber N (1993) Database reverse engineering: from the relational to the binary relationship model. Data and Knowledge Engineering, 10: 293–315

16. Bogorny V, Engel PM, Alvares LO (2005) A reuse-based spatial data preparation framework for data mining, In: Chu WC, Juzgado NJ, Wong WE (eds.), 17th International Conference on Software Engineering and Knowledge Engineering, (SEKE'05). Knowledge Systems Institute, Skokie, pp 649–652

17. Agrawal R, Srikant R (1994) Fast algorithms for mining association rules. In: Bocca JB, Jarke M, Zaniolo C (eds.), 20th International Conference on Very Large Databases (VLDB'94). Morgan Kaufmann, San Francisco

18. Srikant R, Agrawal R (1995) Mining generalized association rules, In: Dayal U, Gray PMD, Nishio S (eds.), 21st International Conference on Very Large Databases (VLDB'1995). Morgan Kaufmann, San Francisco, pp 407–419

19. Clementini E, Di Felice P, Koperski K (2000) Mining multiple-level spatial association rules for objects with a broad boundary. Data & Knowledge Engineering, 34(3): 251–270
20. Mennis J, Liu J (2005) Mining association rules in spatio-temporal data: an analysis of urban socioeconomic and land cover change. Transactions in GIS, 9(1): 5–17
21. Tan PN, Kumar V, Srivastava J (2002) Selecting the right interestingness measure for association patterns. In: 8th International Conference on Knowledge Discovery and Data Mining, ACM Press, New York, pp 32–41
22. Silberschatz A, Tuzhilin A (1996) What makes patterns interesting in knowledge discovery systems. Knowledge and Data Engineering, IEEE Transactions on Knowledge and Data Engineering, 8(6): 970–974

Alternative Method for Incrementally Constructing the FP-Tree

Muhaimenul[1], Reda Alhajj[1,2], and Ken Barker[1]

[1] Department of Computer Science, University of Calgary, Calgary, AB, Canada
[2] Department of Computer Science, Global University, Beirut, Lebanon

Summary. The FP-tree is an effective data structure that facilitates the mining of frequent patterns from transactional databases. But, transactional databases are dynamic in general, and hence modifications on the database must be reflecting onto the FP-tree. Constructing the FP-tree from scratch and incrementally updating the FP-tree are two possible choices. However, from scratch construction turns unfeasible as the database size increases. So, this chapter addresses incremental update by extending the FP-tree concepts and manipulation process. Our new approach is capable of handling all kinds of changes, include additions, deletions and modifications. The target is achieved by constructing and incrementally dealing with the complete FP-tree, i.e., with one as the minimum support threshold. Constructing the complete FP-tree has the other advantage that it provides the freedom of mining for lower minimum support values without the need to reconstruct the tree. However, directly reflecting the changes onto the FP-tree may invalidate the basic FP-tree structure. Thus, we apply a sequence of shuffling and merging operations to validate and maintain the modified tree. The experiments conducted on synthetic and real datasets clearly highlight advantages of the proposed incremental approach over constructing the FP-tree from scratch.

1 Introduction

Data mining is the process of discovering and predicting hidden and unknown knowledge by analyzing known databases. It is different from querying in the sense that querying is a retrieval process, while mining is a discovery process. Data mining has received considerable attention over the past decades and a number of effective mining techniques already exist. They are well investigated and documented in the literature. However, existing and emerging applications of data mining motivated the development of new techniques and the extension of existing ones to adapt to the change. Data mining has several applications, including market analysis, pattern recognition, gene expression data analysis, spatial data analysis, among others.

Muhaimenul et al.: *Alternative Method for Incrementally Constructing the FP-Tree*, Studies in Computational Intelligence (SCI) **109**, 361–377 (2008)
www.springerlink.com

Association rules mining is a technique that investigates the correction between items within the transactions of a given database. Explicitly, given a database of transactions, such that each transaction contains a set of items, the association rules mining process determines correlations of the form $X \Rightarrow Y$, such that X and Y are disjoint sets of items from the investigated database. A correlation $X \Rightarrow Y$ is characterized by support and confidence. The former refers to the percentage of transactions that contain all items in $X \cup Y$ by considering all the transactions in the database; and the latter is the percentage of transactions that contain all items in Y by considering only transactions that contain all items in X. A rule is worth further investigation if it has both high support and confidence as compared to predefined minimum support and confidence values specified mostly by the user who is expected to be an expert. Basket market analysis is one of the first applications of association rules mining [3]. Organizations which deal with transactional data are more concerned to use the outcome to decide on better marketing strategies, to design better promotional activities, to make better product shelving decisions, and above all to use these as a tool to gain competitive advantages.

A formal definition of the problem of association rule mining as described in [3, 11] can be stated as follows. Let $I = \{i_1, i_2, \ldots, i_m\}$ be a set of literals, called items, and let D be a set of transactions, where each transaction T is a set of items such that $T \subseteq I$. Each transaction in D has a unique identifier, denoted TID. A transaction T is said to contain an itemset X if $X \subseteq T$. The support $\sigma_D(X)$ of an itemset X in D is the fraction of the total number of transactions in D that contain X. Let σ ($0 ¡ \sigma ¡ 1$) be a constant called minimum support, mostly user-specified. An itemset X is said to be frequent on D if $\sigma_D(X) \geq \sigma$. The set of all frequent itemsets $L(D, \sigma)$ is defined formally as, $L(D, \sigma) = \{X : X \subset I, \sigma_D(X) \geq \sigma\}$. An Association rule is a correlation of the form "$X \Rightarrow Y$," where $X \subseteq I$, $Y \subseteq I$ and $X \cup Y = \phi$. The rule $X \Rightarrow Y$ has a support σ in the transactional database D if $\sigma\%$ of the transaction in D contain $X \cup Y$. The rule $X \Rightarrow Y$ holds in the transactional dataset D with confidence c if $c\%$ of the transactions in D that contain X also contain Y.

So, the process of association rules mining starts by determining subsets of items that satisfy the minimum support criteria. Such subsets are called frequent itemsets. There are several approaches described in the literature to decide on frequent itemsets, and one of the basic rules used by Apriori [3] is the fact that all subsets of a frequent itemset must be frequent. However, most of the approaches described in the literature are capable of handling static databases; they do not consider dynamic database that are frequently updated to reflect such updates incrementally onto the outcome of the mining process without repeating the whole process from scratch. In fact, real life databases are mostly dynamic and for an approach to be widely accepted it should have the flexibility to handle dynamic databases effectively and efficiently. So, the target is having incremental mining algorithms capable of updating the underlying mining models without scanning, or with minimal scanning of, the old data depending on the mining task at hand.

This need has been realized by several research groups who successfully developed incremental algorithms, e.g, [6–9, 18]. Such algorithms can maintain mining models and can update the association rules if the database is updated without rebuilding the mining model for the whole updated database. But most of these incremental algorithms are level-wise in nature and depend on candidate generation and testing; hence require multiple database scans. On the other hand, FP-growth proposed by Han et al. [12], is an interesting approach that mines frequent patterns without explicitly re-generating the candidate itemsets. This approach uses a data structure called the FP-tree, which is a compact representation of the original database. But the problem with the FP-tree approach is that it is not incremental in nature. In other words, the FP-tree is constructed to include items that satisfy a given pre-specified minimum support threshold and is expected to be reconstructed from scratch each time the original database is modified. This reconstruction from scratch becomes unacceptable as the database size increases and the frequency of modifications becomes high. This is our main motivation for the incremental approach developed within the realm of this research project. Our initial testing reported in [2] highly encouraged us to improve the process further for better overall efficiency and performance.

The target that we successfully achieved in this research project is to incrementally update the FP-tree as the database gets updated and without scanning the old data or reconstructing the tree from scratch. To achieve this, we construct and maintain the complete FP-tree. This way, all occurrences of items are reflected from the database onto the FP-tree. So, new updates are added into the complete FP-tree without scanning the old data. This is not a simple process. Just reflecting updates into the FP-tree may turn it into a structure that contradicts its definition. To overcome this, we adjust the modified structure to fit to the FP-tree definition by applying two basic operations, namely shuffling and merging. We further improved the implementation by recoding several of the Java built-in functions and operations and hence achieved much better performance than the initial results reported in [2]. In this chapter, we demonstrate the applicability, effectiveness and efficiency of the proposed approach by comparing its performance with the performance of constructing the complete FP-tree from scratch. The achieved results are very promising and clearly reflect the novelty of the proposed incremental approach.

The rest of the chapter is organized as follows. Section 2 presents the necessary background and the related work. Section 3 presents our proposed approach with an illustrative example. Section 4 includes the experimental results. Section 5 is summary and conclusions.

2 Background and Related Work

2.1 Construction of the FP-Tree

FP-tree: An FP-tree is a prefixed tree of the frequent items by considering all transactions in the database. It is called prefixed tree because it is constructed in such a way that common prefixes of the sorted transactions are merged to be shared as tree paths or sub-paths. Items in each transaction are sorted in descending order according to the overall frequency in the database. So, it is most likely that the most frequent items will be shared the most. As a result, we have a very compact representation of the original transactional database. Each node of the FP-tree contains the following fields: *item-name*, *item-count*, and *node-link*, where *item-name* denotes the represented item, *item-count* designates the number of transactions sharing a prefix path up to the node, and *node-link* refers to the next node in the FP-tree with the same item-name. The root is labeled as "null" and links to a set of item prefix sub-trees as its children. A frequent item header table is associated with each FP-tree; each entry in the table contains two fields: *item-name* and *head of node-link* which points in the FP-tree to the first occurrence of the node carrying the same *item-name*.

To construct the FP-tree, first scan the whole transactional database to collect the frequency of each item in the database. Then, sort the items according to their frequency in descending order to build the list of frequent items in the database. In the next phase, create the root of the FP-tree. Then, scan the whole transaction database once again, and for each transaction in the database first sort its items in descending order according to the overall frequency. Items that do not meet the support criteria are removed from the transaction. Finally, add the sorted and truncated transaction to the root of the FP-tree in the following path.

Let the sorted and truncated item-list in the transaction be $[p|P]^*$; this notation was adopted by Han et al. [12], where p is the first element of the list and P is the remaining list. If the root has a child corresponding to item p, then just increment the count of the child by one. But if the root does not contain a child corresponding to the same item, then create a new child of the root corresponding to item p, and set its count to one. To add the remaining list P, consider the child as the root for list P and add items in P recursively; proceed by the same way until list P is empty.

2.2 Related Work

FUP (Fast Update) [6] is one of the earliest algorithms proposed for maintaining association rules discovered in large databases. The approaches described in [6–8] all belong to the same class of candidate generation and testing; they all require k scans to discover large itemsets, with k as the length of the maximal large itemset. There are other incremental algorithms, which are based on the FUP algorithm, e.g., MAAP [20] and PELICAN [19]. These FUP based algorithms not only require multiple database scans, but also have to generate two sets of candidate itemsets (one for the original data and one for the incremental data), resulting in sharp increase in the number of candidate itemsets.

On the other hand, the work described in [9,18] proposes an incremental algorithm which requires at most one scan of the whole database to find the set of frequent patterns for the updated database. However, it is shown in the work described in [1] that the main bottleneck of the work described in [9,18] is that it has to generate the *negative border closure*. The negative border consists of all itemsets that were candidates, but did not have the minimum support to become frequent during the kth pass of the Apriori algorithm, i.e., NBd(L_k)= $C_k - L_k$, where L_k is the set of frequent itemsets and C_k is the set of candidate itemsets that were generated during the kth iteration. Negative border closure generation induces the possibility of rapid increase in the candidate generation, and the time required to count the support for these explosive number of candidates outweighs the time saving achieved by evading repeated database scans.

Recently, Ganti et al. [11] proposed an improved version of the approach described in [9], known as $ECUT/ECUT^+$, where they keep TID-lists: $\theta(i_1)$, ..., $\theta(i_k)$ of all the items in an itemset $X = \{i_1, \ldots, i_n\}$, to count the support of X. The intuition behind keeping the TID-list is that it reduces the amount of data to scan when counting the support of an itemset X, by only scanning the relative portion of the database. In order to count the support of the candidate itemsets efficiently, they keep the candidate itemsets in a prefix-tree data structure proposed by Mueller [17]. These approaches work based on the assumption that when database updates occur, only a small number of new candidate itemsets need to be considered. This assumption is very unrealistic, and so these approaches also suffer from the same candidate explosion problem.

The sliding window filtering (SWF) approach [14] is another incremental association rule mining approach that relies on database partitioning. SWF partitions the database into several non-overlapping partitions, and employs two filtering thresholds in each partition to deal with candidate itemset generation. Each partition is independently scanned, once to gather the support count of the itemsets of length 1, and then the first filtering threshold is used to generate what they call type β candidate itemsets C_2 of length 2 (i.e., the newly selected candidate itemsets for that particular partition). By using the prior knowledge gained from the previous partitions, it also employs another filtering threshold known as *cumulative filter* (or *CF*) to generate type α candidate itemsets, to progressively update C_2. These are length 2 candidate itemsets that were carried over from the previous progressive candidate sets and remain as candidate itemsets after the current partition is taken into consideration. Using two different support thresholds in this way helps in reducing candidate generation by a considerable amount.

After generating C_2 from the first database, the scan reduction technique is used to generate candidate itemsets C_k of length k, for all $k > 2$. But, the performance of the SWF approach is dependent on the partition size, and deletion can only be done at the partition level, which is an unrealistic assumption. Moreover, the scan reduction technique usually generates more candidates when the length of the candidates is greater than 2.

The work of Amir et al. [4] involves projecting all the transactions in the database in a trie structure (a trie is an ordered prefix-tree like data structure to store an associative array or map; in the case of [4], the key is the set of itemsets, which maps to the corresponding frequency). It takes transactions of the database one by one, extracts the powerset of the transaction, and projects each member of the powerset into the trie structure for updating the support count. Once built, the trie structure can be traversed in a depth first manner to find all the frequent patterns.

The advantage of the trie structure is that it is inherently incremental – transactions can be easily added or deleted from the trie structure. But, when the support threshold to mine is low and the database size is large, it is very likely that the trie would become closer to the powerset $2^{|I|}$ of the set of items I present in the database D, and hence would exhaust the main memory very soon. Moreover, this approach requires generating the powerset of each transaction while constructing the trie, which can be very time consuming operation when the average transaction length is large.

AFPIM [13] finds new frequent itemsets with minimum re-computation when transactions are added to, deleted from, or modified in the transactional database. In AFPIM, the FP-tree structure of the original database is maintained in addition to the frequent itemsets. The authors claim that in most cases, without needing to re-scan the whole database, the FP-tree structure of the updated database is obtained by adjusting the preserved FP-tree according to the inserted and deleted transactions. The frequent itemsets of the updated database are mined from the new FP-tree structure. However, this claim is not supported by the approach as described in their paper [13]. Actually, AFPIM is not a true incremental approach because it does not keep the complete FP-tree, rather it is based on two values of support for constructing the FP-tree. It keeps in the tree only items with support larger than the minimum of the two maintained support values, i.e., potential candidates to become frequent in the near future, this is not realistic and mostly requires full scan of the whole database almost every time the database updates are beyond the anticipated change.

CanTree (CANonical-order TREE) is another incremental tree construction approach proposed by Leung et al. [15]. It is basically a derivative of the FP-tree, where the items are ordered according to some canonical ordering established prior to or during the mining process. All the other properties of CanTree are similar to those of the FP-tree. Based on the canonical ordering, the authors are able to eliminate the initial database scan required to gather the frequency ordering of items. The use of this canonical ordering also makes their approach incremental. But they provide no realistic assumptions on how to determine this canonical ordering in the first place. We know that it is very common for the real world databases to have trends or to have data skew – i.e., the itemsets that makes the transactions – changes over time. For example, if we consider a retail store transactional database – the itemsets that are commonly purchased in the winter may be different from the itemsets that are purchased during the summer; or even this may vary based on the time period of the day or week and the financial status of the customers. Such a CanTree will grow exponentially as we keep adding the transactional data, and as it does not follow the global frequency in descending ordering – the end result being the exhaustion of the main memory at a very early stage. Moreover, it takes more time to traverse a larger tree compared to a smaller one, this may actually increase the tree construction time once the tree becomes very large.

3 Incremental Construction of the FP-Tree

In this section, we describe the proposed method for constructing the FP-tree incrementally. This method avoids full database scan, and reflects onto the FP-tree that was valid at time t all database updates done between t and $t + 1$. Thereby, the proposed method saves substantial amount of processor time.

When new transactions are added to the database, some items that were previously infrequent may become frequent, and some items that were previously frequent may become infrequent. As we do not know which items are going to be frequent and which items are going to be infrequent, and to be able to construct the FP-tree incrementally, we start with a complete FP-tree having minimum support threshold of 1. In other words, for our method to work, we always keep a complete FP-tree based on transactions that are valid at the current time. This approach provides the additional advantage that it does not require building new FP-trees for mining frequent patterns with different support thresholds.

The incremental method starts from the initial FP-tree constructed at time t. When a new set of transactions d_{t+1} arrive at time $t + 1$, the method first scans the incremental database d_{t+1} to collect the frequency of items in the incremental database d_{t+1}. For each item, its frequency in the initial database D_t and the incremental database d_{t+1} (if exists) are combined to get the cumulative frequency list of all the items in the updated database D_{t+1}. Then, the items are sorted in descending ordering according to their already computed cumulative frequencies to get what we call the $F_{t+1}-$List. Corresponding entries in the old frequency list F_t-List and the new frequency list $F_{t+1}-$List are compared to identify items which have changed their relative order in the sorted frequency ordering. Such items are stored in a separate list, which we call the S-List (shuffle list). So in essence, the S-List contains the set of items for which the order of the nodes in the FP-tree need to be changed to reflect the addition of new transactions and to keep the FP-tree valid.

Before describing the incremental method, we need to present the necessary definitions and the basic terminology required to understand the material in the rest of this section.

$Freq_t(a_{t,i})$: Let D_t be the database at time t, $Freq_t(a_{t,i})$ is defined as the fraction of transactions that appear in D_t and which contain item $a_{t,i}$.

F_t-List: The Frequency List at time t, denoted F_t-List, contains the list of items sorted in descending ordering according to the frequency of items that appear in the transactions at time t. It is defined as the set $\{a_{t,1}, a_{t,2}, \ldots, a_{t,n}\}$ where n is the number of items present in the database at time t, $a_{t,i}$ corresponds to an item in the database and $Freq_t(a_{t,i}) \geq Freq_t(a_{t,j})$ for $1 \leq i < j \leq n$.

$Pos_t(a_{t,i})$: $Pos_t(a_{t,i})$ is defined as the ordinal position of item $a_{t,i}$ in the F_t-List at time t.

S-List: The S-List is constructed from the F_{t+1}-List by repeating the following steps until the F_{t+1}-List is empty:

1. Take the first item, $a_{t+1,1}$ from the F_{t+1}-List.
2. Insert item, $a_{t+1,1}$ into the S-List if $Pos_{t+1}(a_{t+1,1}) < Pos_t(a_{t,i})$, where there is an item $a_{t,i}$ in the F_t-List and $a_{t+1,1} = a_{t,i}$ for $1 \leq i \leq n$.
3. Remove item $a_{t+1,1}$ from the F_{t+1}-List and item $a_{t,i}$ from the F_t-List, while maintaining the ordering of the items.
4. If there are more items in the F_{t+1}-List, go back to step 1.

The incremental update process assumes two main instances as input, namely, the current complete FP-tree and the incremental database to be reflected onto the current complete FP-tree. Then, a sequence of steps are performed to produce the most up-to-date complete FP-tree, which incorporates all the updates. First, the frequencies of the items in the incremental database are produced by scanning the incremental database. These frequencies are added to the corresponding frequencies of the current FP-tree. The new total frequencies should be sorted in descending order to be able to check whether the order of items has changed. Items that have changed their positions produce the S-list, which is used to shuffle the current FP-tree to move each node to its desired position in the new version of the FP-tree, the version to be obtained after adding the items in the incremental database. Such arrangement will easily facilitate updating the FP-tree to include the incremental database, and hence produce the desired most up-to-date FP-tree.

The shuffling operation takes FP-tree T_{D_t} at time t and the ordered S-List as input to produce FP-tree $T_{D'_{t+1}}$ in which nodes are ordered according to the cumulative frequency of items. For each item in the S-List starting from items that have the highest frequent: its corresponding node is *pulled* to its *maximum attainable frequency peak* in the FP-tree T_{D_t}; and then nodes that correspond to the same item and have same parent are recursively *Merged*.

The insertion operation takes FP-tree $T_{D'_{t+1}}$ and the incremental database d_{t+1} as input to produce a modified FP-tree $T_{D_{t+1}}$. for each transaction *Tran* in the database d_{t+1}: the items in *Tran* are sorted in descending order according to their total frequency (from the original and the incremental database), and then *Tran* is added to tree $T_{D'_{t+1}}$ while maintaining the prefix path property.

To get a better understanding of the shuffling process, we describe the *Pull* and *Merge* operations next.

The Pull Operation: We need to *pull* each node in the S-List to its *maximum attainable frequency peak* in the FP-tree. By *maximum attainable frequency peak* we mean the position in the FP-tree for which node frequency is higher than any node of the sub-trees below the peak. We start from the maximum frequent node in the S-List and pull subsequent frequent nodes in the S-List one by one. Depending on the *item-count* C_n of the node to be *Pulled* and the *item-count* C_p of its parent, the following two cases are possible.

Case-I : $C_n = C_p$

In this case, the *item-count* of the node to be *pulled* is the same as the *item-count* of its parent. We can safely assume that the parent has no children other than the node to be *pulled*. Figure 1 shows one example of *case I* and how the node is *pulled* in such case. In this case, we can just swap the two nodes in question by swapping their *item-name*, *item-count* and *node-link* pointers. In Fig. 1, the dotted blue links represent the *node-link* pointers.

Case-II: $C_n < C_p$

In this case, however, *item-count* of the parent node is greater than *item-count* of the node to be pulled. So, there is a strong possibility that either the node to be *pulled* has more that one sibling or the parent node was the last item for some transactions. In both cases, the *pulling* process starts by first *splitting* the parent node into two nodes. Then, *case-II* simply reduces to *case-I*, and we can easily swap the two nodes to *pull* the desired node upwards.

The Split Operation: This is achieved by creating one new node, which corresponds to the same item as the parent node. We make this node a sibling

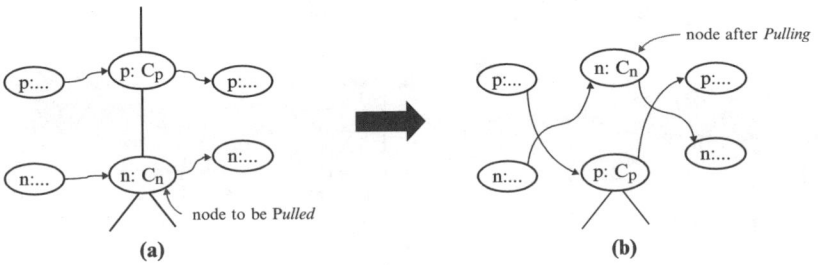

Fig. 1. Case-I ($C_n = C_p$)

of the parent. We then adjust the *item-count* for the parent as C_n (the same count as for the node to be *pulled*), and make the *item-count* for the new node as $C_p - C_n$. Then, we have to adjust the children of the parent node and the new node in question. We can do this by making all siblings of the node to be *pulled* as children of the new node. This way, the parent node will contain only one child, which is the node to be *pulled*. This successfully reduces *case-II* to a situation similar to *case-I*. Figure 2 shows one example of *case-II* and steps of the *pulling* operation.

We continue *pulling* recursively as long as the item frequency of the node to be pulled is greater than the item frequency of its parent. This condition is broken when we reach the *maximum attainable frequency peak corresponding* to the node and we do not need to *pull* the node any farther; this is in fact the stopping criterion of the *pulling* operation.

Merge Operation: Once we have *pulled* the node to its *maximum attainable frequency peak* and because of the shuffling of nodes, there can be situations where the parent of the *pulled* node has two child nodes that correspond to the same item. In this case, we have to *merge* the two nodes to remain consistent with the definition of the FP-tree. We have to carry out this *merging* operation recursively, as there can be situations where two children of the *merged* node correspond to the same item. After the final *Merging* process, we get a tree where the ordering of nodes corresponds to the cumulative frequency of items in descending ordering (Fig. 3).

We can handle the deletion of transactions in a way similar to the addition of transactions. We start by shuffling/rearranging the tree by taking into account the new frequency ordering. Once all the nodes of the tree are arranged

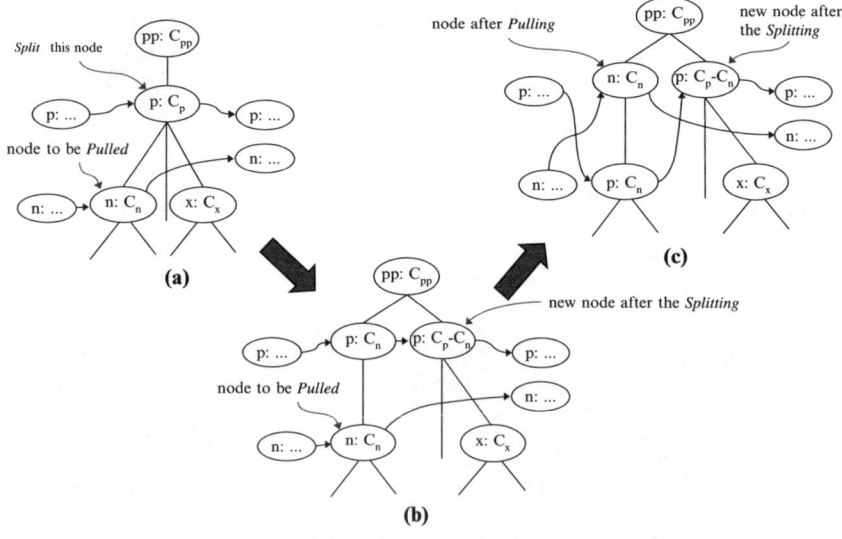

Fig. 2. Case-II (C_n ¡ C_p)

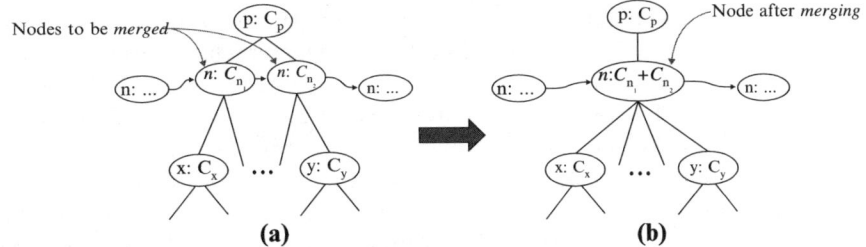

Fig. 3. *Merge* operation

properly, we can delete existing transactions from the FP-tree by just only decrementing the item-count of the nodes that participate in the transactions paths.

After updating the FP-tree for the modified database, we are currently mining the FP-tree by using using the FP-growth approach [12]. We are still working on our own mining algorithm which combines the advantages of several of the algorithms already described in the literature.

4 Experimental Results

In this section, we describe the performance of the proposed incremental update on different real and synthetic datasets. The method has been implemented in JAVA. We decided on minimizing the usage of JAVA build-in functions and classes because we realized that our first implementation did not produce good results even for the classical FP-tree, mainly because it heavily used such predefined JAVA functions.

All the experiments have been conducted on an IBM Pentium IV machine with 2.0 GHz CPU and 512 MB main memory; running Windows-XP. We used five synthetic and one real datasets in the experiments; the synthetic datasets are: D_1(T40I10D100K) with 1,000 items and 100K transactions, D_2(T10I4D1M) with 1,000 items and 1M transactions, D_3(T10I4D100K) and D_4(T25I10D100K) with 10,000 items and 100K transactions; and D_5(Connect) with 129 items and 67,557 transactions. The real dataset is D_6(Retail) with 16,469 items and 88,162 transactions. The datasets D_1, D_5, and D_6 are taken from the FIM dataset repository [10], and dataset D_2-D_4 is generated using Paolo Palmirini's synthetic data generator [16]. Dataset D_6 is real data from some anonymous retail store and was donated by Tom Brijs [5]. These datasets have different number of items (129~16,469) and different number of transactions (67,557~1M); the aim is to demonstrate well the different aspects of the proposed approach.

We have tested the performance of the incremental update against constructing the FP-tree from scratch for each of the six datasets while varying the size of the incremental dataset. To construct the incremental dataset, we have randomly chosen $x\%$ of the dataset to split the dataset into two parts. The part with $x\%$ of the transactions forms the incremental dataset and the remaining part with $(100\text{-}x)\%$ of the transactions form the initial dataset for which the initial FP-tree is built. This initial FP-tree and the incremental dataset is the input of the incremental update process. We are measuring how much time is required to add $x\%$ of the transactions to the FP-tree representing $(100\text{-}x)\%$ of the transactions. In Figs. 4–9, we have presented the time comparison of the incremental update against constructing the FP-tree from scratch for the six different datasets and for five different splits, namely 1, 5, 10, 20 and 50%. For each of these datasets, we have found that by adopting the incremental approach we can save a substantial amount of time by not constructing the FP-tree from scratch whenever a database update occurs. Even when the split size is 50% of the original database, adopting an incremental approach is considerably better that constructing the FP-tree from scratch.

In Fig. 10, we present a different view of the results presented in Figs. 4–9. Here, the x-axis like before represents the split percentage of the database, while the y-axis represents the percentage of time saving achieved by taking the incremental approach and not constructing the FP-tree from scratch. In Fig. 10, we observe that the shape of the performance lines for datasets D_1 and D_2 are identical because of their similarity in the number of different items ($1K$); but they have a performance gap because of the difference in the average size of the transactions. Similar situations are observed for datasets

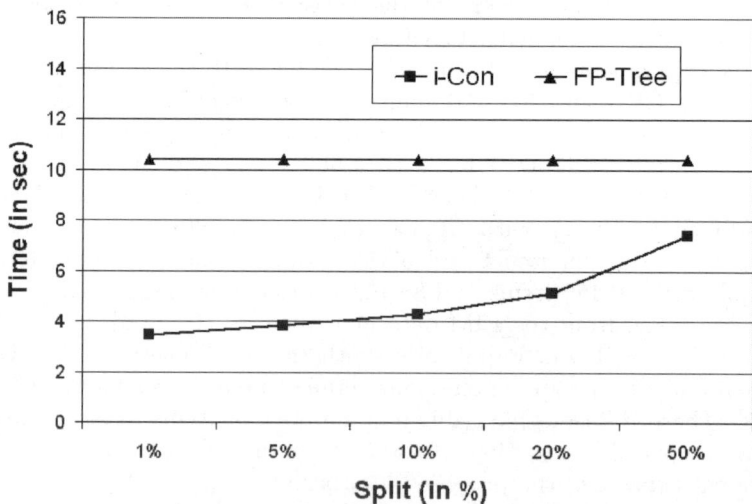

Fig. 4. Scalability of the incremental update with various sizes of incremental database D_1, T40I10D100K (N=1K)

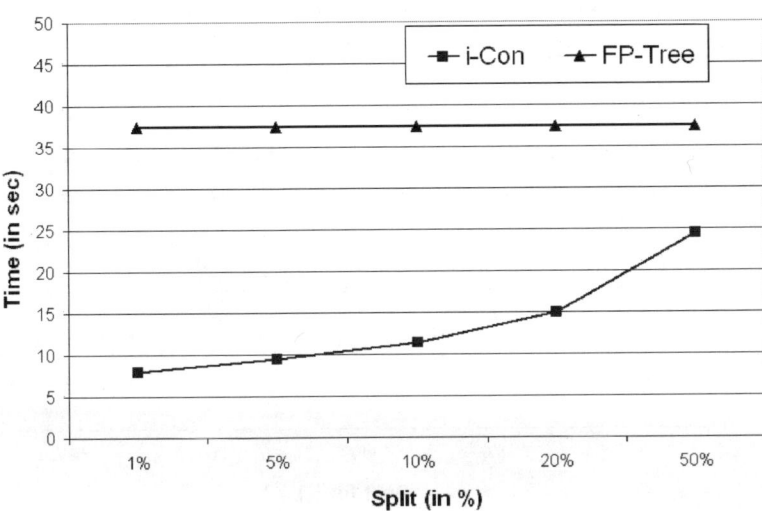

Fig. 5. Scalability of the incremental update with various sizes of incremental database D_2, T10I4D1M (N=1K)

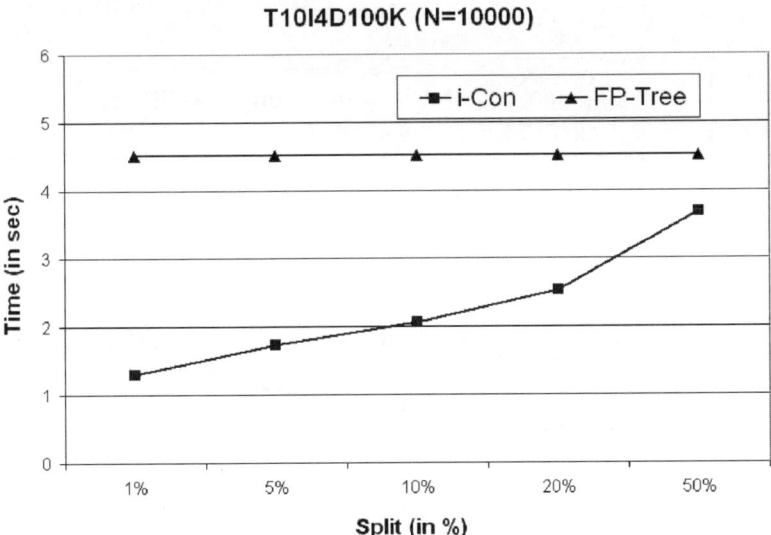

Fig. 6. Scalability of the incremental update with various sizes of incremental database D_3, T10I4D100K (N=10K)

D_3, D_4. In general, the incremental update takes more time when the average size of the transactions are higher. We also observe another performance gap based on the number of items present in the dataset. The trend line for dataset D_6, however, is somewhat different because of its excessively large

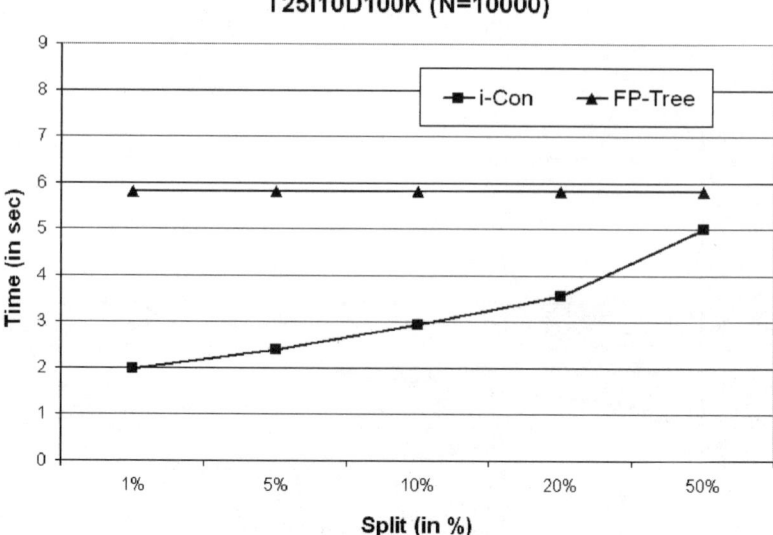

Fig. 7. Scalability of the incremental update with various sizes of incremental database D_4, T25I10D100K (N=10K)

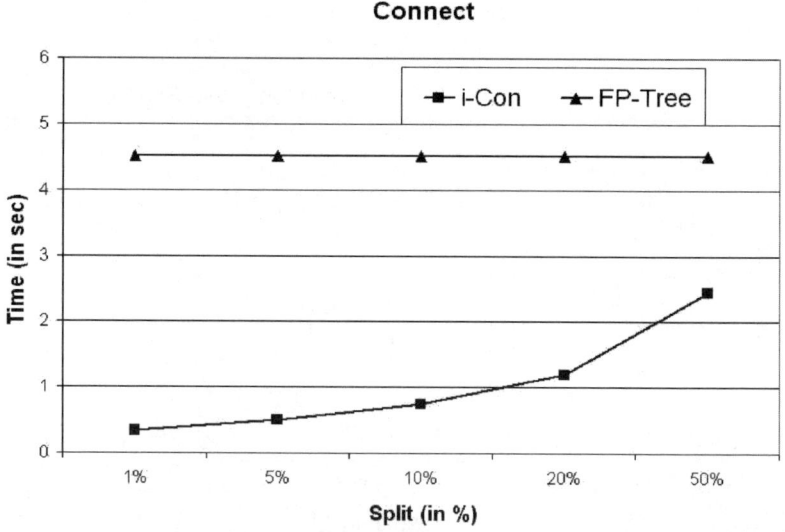

Fig. 8. Scalability of the incremental update with various sizes of incremental database connect

number of items. The best performance is achieved for dataset D_5, where the number of different items is only 129. These results demonstrate the power of the proposed method regardless of the number of transactions and attributes present in the database.

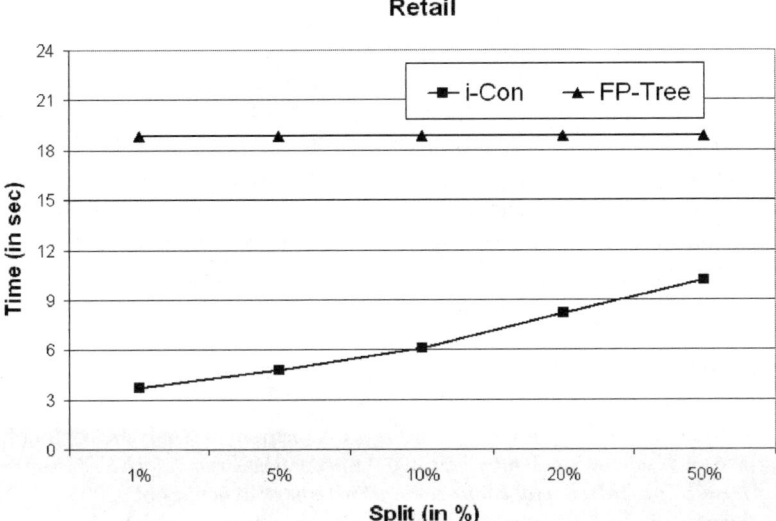

Fig. 9. Scalability of the incremental update with various sizes of incremental database retail

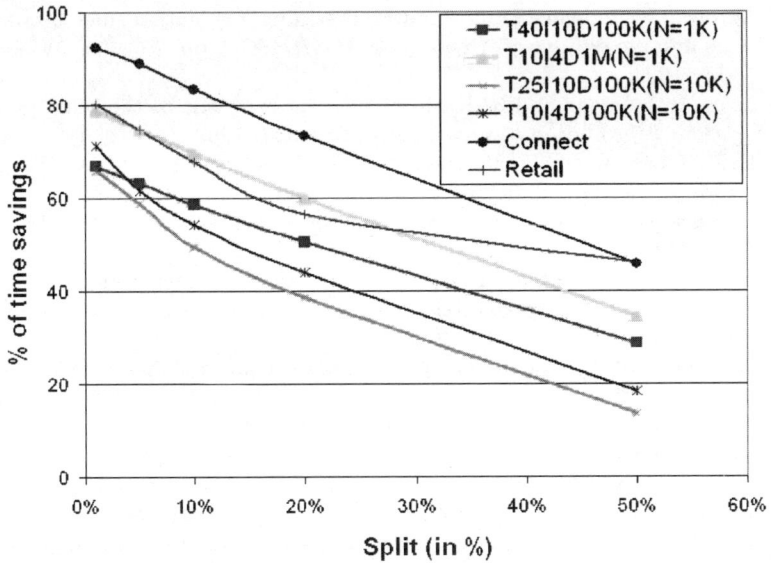

Fig. 10. Time savings by the incremental update process

5 Summary and Conclusions

In this chapter, we described an incremental update method that can update the FP-tree incrementally without scanning the old dataset and with a minimal scanning (two scans) of the incremental database. Experimental results show that our approach performs considerably better compared to building the tree from scratch, and thereby achieves a huge amount of time saving. Following our approach, we can also mine for frequent patterns for any support threshold because we build the complete FP-tree, with minimum support specified as one. Our approach works for both addition and deletion of transactions. We have highly improved the performance of the system by reimplementing several of the JAVA built-in constructs used in our first implementation. We have also developed a disk based version of the complete FP-tree; the target is to be able to scale well for huge databases.

References

1. M. Adnan, R. Alhajj and K. Barker, Performance Analysis of Incremental Update of Association Rules Mining Approaches *Proc. of IEEE International Conference on Intelligent Engineering Systems*, Greece, Sept. 2005.
2. M. Adnan, R. Alhajj and K. Barker, Constructing complete FP-tree for incremental mining of frequent patterns in dynamic databases, *Proc. of the International Conference on Industrial & Engineering Applications of Artificial Intelligence & Expert Systems*, Springer LNCS, Annecy, France, June 2006.
3. R. Agrawal, T. Imielinski and A. Swami, Mining association rules between sets of items in large databases, *Proc. of ACM-SIGMOD,* pp. 207–216, Washington, DC, May 1993.
4. A. Amir, R. Feldman and R. Kashi, A new and versatile method for association generation, *Information Systems*, 22(6), 333–347, 1999.
5. T. Brijs, G. Swinnen, K. Vanhoof and G. Wets, The use of association rules for product assortment decisions: a case study, *Proc. of ACM International Conference on Knowledge Discovery and Data Mining*, pp. 254–260, San Diego, August 1999.
6. D. W. Cheung, J. Han, V. T. Ng and C. Y. Wong, Maintenance of discovered association rules in large databases: an incremental updating technique, *Proc. of IEEE-ICDE*, pp. 106–114 , 1996.
7. D. W. Cheung, V. T. Ng and B. W. Tam, Maintenance of discovered knowledge: a case in multi-level association rules, *Proc. of ACM International Conference on Knowledge Discovery and Data Mining*, pp. 307–310, 1996.
8. D. W. Cheung, S. D. Lee and B. Kao, A general incremental technique for mining discovered association rules, *Proc. of the International Conference on Database System for Advanced Applications*, pp. 185–194, 1997.
9. R. Feldman, Y. Aumann, A. Amir and H. Mannila, Efficient algorithms for discovering frequent sets in incremental databases, *Proc. of the Workshop on Research Issues on Data Mining and Knowledge Discovery*, 1997.
10. http://fimi.cs.helsinki.fi/data/
11. V. Ganti, J. E. Gehrke and R. Ramakrishnan, DEMON: mining and monitoring evolving data, *IEEE Transactions on Knowledge and Data Engineering*, 13(1), 50–63, 2001.
12. J. Han, J. Pei and Y. Yin, Mining frequent patterns without candidate generation, *Proc. of ACM-SIGMOD*, pp. 1–12, Dallas, TX, May 2000.

13. J.-L. Koh and S.-F. Shieh, An efficient approach for maintaining association rules based on adjusting FP-Tree structures, *Proc. of DASFAA*, Korea, March 2004.
14. C.-H. Lee and C.-R. Lin and M.-S. Chen, Sliding-window filtering: an efficient algorithm for incremental mining, *Proc. of ACM International Conference on Information and Knowledge Management*, pp. 263–270, New York, NY, 2001.
15. C. K.-S. Leung, Q. I. Khan and T. Hoque, CanTree: a tree structure for efficient incremental mining of frequent patterns, *Proc. of IEEE International Conference on Data Mining*, pp. 274–281, Los Alamitos, CA, 2005.
16. http://miles.cnuce.cnr.it/~palmeri/datam/DCI/datasets.php
17. A. Mueller, Fast sequential and parallel algorithms for association rule mining: a comparison, *University of Maryland Institute for Advanced Computer Studies Report*, University of Maryland at College Park, College Park, MD, 1995.
18. S. Thomas, S. Bodagala, K. Alsabti and S. Ranka, An efficient algorithm for the incremental updation of association rules in large databases, *Proc. of the International Conference on Knowledge Discovery in Databases*, pp. 263–266, 1997.
19. A. Veloso, B. Possas, W. Meira Jr., M. B. de Carvalh, Knowledge management in association rule mining, *Proc. of the International Workshop on Integrating Data Mining and Knowledge Management*, 2001.
20. Z. Zhou and C. I. Ezeife, A Low-scan incremental association rule maintenance method based on the apriori property, *Proc. of the Biennial Conference of the Canadian Society on Computational Studies of Intelligence*, pp. 26–35, London, ON, 2001.

Part VI

Intuitonistic Fuzzy Sets and Systems

On the Intuitionistic Fuzzy Implications and Negations

Krassimir T. Atanassov

CLBME-Bulgarian Academy of Sciences, P.O. Box 12, Sofia-1113, Bulgaria
krat@bas.bg

Summary. Definitions of 174 intuitionistic fuzzy logic implications are introduced. Some of their properties are studied and some open problems, related to them are formulated.[1]

1 Introduction: On Some Previous Results

The concept of the intuitionistic fuzzy set (IFS, see [2]) was introduced in 1983 as an extension of Zadeh's fuzzy set. During the last twenty years intuitionistic fuzzy logics (IFL) was developed. In it two implications were discussed. In the last two years a new direction of the research in the IFL started with [3], where ten variants of intuitionistic fuzzy implications are discussed, using as a basis the book [12] by Georg Klir and Bo Yuan. Other five implications, defined by the author and his colleagues Boyan Kolev and Trifon Trifonov, are introduced in [1,2,4–6,8–10]. These fifteen implications generate five different negations, that are a basis for new eight implications introduced in [9]. The latest ones give rise to negations that coincide with the respective negations generated by the corresponding implications, i.e., the generating process finishes. Hence, using this scheme, we obtain 23 implications. Two completely different implications are introduced in [10], that generated (by the above meen) four other implications. Here, for the first time we introduce 145 new implications generated by the first 29 ones. Therefore, we obtain 174 implications. As we shall show, some of them coincide and finally, we will obtain 99 different implications, a large part of which – with essentially intuitionistic behaviour.

In IFL if x is a variable (in more general case – propositional form), then its truth-value is represented by the ordered couple

$$V(x) = <a, b>,$$

---

[1] Originally published in Proceedings of the 3rd IEEE Conference on Intelligent Systems, London, 2006 (see [6]).

K.T. Atanassov: *On the Intuitionistic Fuzzy Implications and Negations*, Studies in Computational Intelligence (SCI) **109**, 381–394 (2008)
www.springerlink.com

so that $a, b, a + b \in [0, 1]$, where a and b are degrees of validity and of non-validity of x.

Everywhere below we shall assume that for the three variables x and y there hold the equalities:

$$V(x) = <a, b>$$
$$V(y) = <c, d>$$
$$(a, b, c, d, e, f, a + b, c + d, e + f \in [0, 1]).$$

For the needs of the discussion below we shall define the notion of Intuitionistic Fuzzy Tautology (IFT, [1, 2]) by:

$$x \text{ is an IFT if and only if } a \geq b,$$

while x will be a tautology if and only if $a = 1$ and $b = 0$.

We shall define the following relation:

$$V(x) \geq V(y) \text{ if and only if } a \geq c \text{ and } b \leq d.$$

In some definitions we shall use functions sg and $\overline{sg}$:

$$sg(x) = \begin{cases} 1 & \text{if } x > 0 \\ 0 & \text{if } x \leq 0 \end{cases},$$

$$\overline{sg}(x) = \begin{cases} 0 & \text{if } x > 0 \\ 1 & \text{if } x \leq 0 \end{cases}.$$

For two variables x and y operation "conjunction" (&) is defined (see [1, 2] by:

$$V(x \& y) = <\min(a, c), \max(b, d)>.$$

In Table 1 we include the implications from [3], but also the implications, introduced by the author in [1, 2, 4–6, 8–10] with coauthors Boyan Kolev (for [8]) and Trifon Trifonov (for [9, 10]).

2 Main Results

Following and extending [4, 7], we shall define operations negation on the basis of the above implications. Here, we shall note ith negation that corresponds to ith implication by $\neg^i$ so that

$$\neg^i x = x \rightarrow_i 0^*,$$

where

$$V(0^*) = <0, 1>.$$

Table 1. List of the first 29 intuitionistic fuzzy implications

Notation	Name	Form of implication
$\rightarrow_1$	Zadeh	$< \max(b, \min(a, c)), \min(a, d)>$
$\rightarrow_2$	Gaines-Rescher	$<1-\text{sg}(a - c), d.\text{sg}(a - c)>$
$\rightarrow_3$	Gödel	$<1-(1-c).\text{sg}(a - c), d.\text{sg}(a - c)>$
$\rightarrow_4$	Kleene-Dienes	$< \max(b, c), \min(a, d)>$
$\rightarrow_5$	Lukasiewicz	$< \min(1, b + c), \max(0, a + d - 1)>$
$\rightarrow_6$	Reichenbach	$<b + ac, ad>$
$\rightarrow_7$	Willmott	$< \min(\max(b, c), \max(a, b), \max(c, d)), \max(\min(a, d),$ $\min(a, b), \min(c, d))>$
$\rightarrow_8$	Wu	$<1-(1-\min(b, c)).\text{sg}(a - c),$ $\max(a, d).\text{sg}(a - c).\text{sg}(d - b)>$
$\rightarrow_9$	Klir and Yuan 1	$<b + a^2 c, ab + a^2 d>$
$\rightarrow_{10}$	Klir and Yuan 2	$<c.\overline{sg}(1-a) + \text{sg}(1-a).(\overline{sg}(1-c) +$ $b.\text{sg}(1-c)), d.\overline{sg}(1-a) + a.\text{sg}(1-a).\text{sg}(1-c)>$
$\rightarrow_{11}$	Atanassov 1	$<1-(1-c).\text{sg}(a - c), d.\text{sg}(a - c).\text{sg}(d - b)>$
$\rightarrow_{12}$	Atanassov 2	$< \max(b, c), 1-\max(b, c)>$
$\rightarrow_{13}$	Atanassov and Kolev	$<b + c - b.c, a.d>$
$\rightarrow_{14}$	Atanassov and Trifonov	$<1-(1-c).\text{sg}(a - c)-d.\overline{sg}(a - c).\text{sg}(d - b), d.\text{sg}(d - b)>$
$\rightarrow_{15}$	Atanassov 3	$<1-(1-\min(b, c)).\text{sg}(\text{sg}(a - c) + \text{sg}(d - b))$ $- \min(b, c).\text{sg}(a - c).\text{sg}(d - b),$ $1-(1 - \max(a, d)).\text{sg}(\overline{sg}(a - c) + \overline{sg}(d - b)) - \max(a, d).$ $\overline{sg}(a - c).\overline{sg}(d - b)>$
$\rightarrow_{16}$		$< \max(1-\text{sg}(a), c), \min(\text{sg}(a), d)>$
$\rightarrow_{17}$		$< \max(b, c), \min(a.b + a^2, d)>$
$\rightarrow_{18}$		$< \max(b, c), \min(1-b, d)>$
$\rightarrow_{19}$		$< \max(1-\text{sg}(\text{sg}(a) + \text{sg}(1-b)), c), \min(\text{sg}(1-b), d)>$
$\rightarrow_{20}$		$< \max(1-\text{sg}(a), 1-\text{sg}(1-\text{sg}(c))),$ $\min(\text{sg}(a), \text{sg}(1-\text{sg}(c)))>$
$\rightarrow_{21}$		$< \max(b, c(c + d)), \min(a(a + b), d(c^2 + d + cd))>$
$\rightarrow_{22}$		$< \max(b, 1-d), \min(1-b, d)>$
$\rightarrow_{23}$		$<1- \min(\text{sg}(1-b), \text{sg}(1-\text{sg}(1-d))),$ $\min(\text{sg}(1-b), \text{sg}(1-\text{sg}(1-d)))>$
$\rightarrow_{24}$		$<\overline{sg}(a - c).\overline{sg}(d - b), \text{sg}(a - c).\text{sg}(d-b)>$
$\rightarrow_{25}$		$< \max(b.\overline{sg}(a).\overline{sg}(1-b), c.\overline{sg}(d).\overline{sg}(1-c)),$ $\min(a.\text{sg}(1-b), d.\text{sg}(1-c))>$
$\rightarrow_{26}$		$< \max(\overline{sg}(1-b), c), \min(\text{sg}(a), d)>$
$\rightarrow_{27}$		$< \max(\overline{sg}(1-b), \text{sg}(c)), \min(\text{sg}(a), \overline{sg}(1-d))>$
$\rightarrow_{28}$		$< \max(\overline{sg}(1-b), c), \min(a, d)>$
$\rightarrow_{29}$		$< \max(\overline{sg}(1-b), \overline{sg}(1-c)), \min(a, \overline{sg}(1-d))>$

Then we obtain the following negations:

$$\neg^1 x = <b, a>,$$
$$\neg^2 x = <1-\text{sg}(a), \text{sg}(a)>,$$

$\neg^3 x = <1-\mathrm{sg}(a),\ \mathrm{sg}(a)>,$

$\neg^4 x = <b, a>,$

$\neg^5 x = <b, a>,$

$\neg^6 x = <b, a>,$

$\neg^7 x = <b, a>,$

$\neg^8 x = \{<\mathrm{x}, 1-\mathrm{sg}(a),\ \mathrm{sg}(a).\mathrm{sg}(1-b)>,$

$\neg^9 x = <b, ab + a^2>,$

$\neg^{10} x = <\mathrm{sg}(1-a).b,\ \overline{sg}(1-a) + a.\mathrm{sg}(1-a)>,$

$\neg^{11} x = <1-\mathrm{sg}(a),\ \mathrm{sg}(a).\mathrm{sg}(1-b)>,$

$\neg^{12} x = <b, 1-b>,$

$\neg^{13} x = <b, a>,$

$\neg^{14} x = <1-\mathrm{sg}(a)-\overline{sg}(a).\mathrm{sg}(1-b),\ \mathrm{sg}(1-b)>,$

$\neg^{15} x = <1-\mathrm{sg}(\mathrm{sg}(a) + \mathrm{sg}(1-b)),\ 1-\overline{sg}(1-b)-\overline{sg}(a).\overline{sg}(1-b)>,$

$\neg^{16} x = <1-\mathrm{sg}(a),\ \mathrm{sg}(a)>,$

$\neg^{17} x = <b, a.b + a^2>,$

$\neg^{18} x = <b, 1-b>,$

$\neg^{19} x = <1-\mathrm{sg}(\mathrm{sg}(a) + \mathrm{sg}(1-b)),\ \mathrm{sg}(1-b)>,$

$\neg^{20} x = <1-\mathrm{sg}(a)),\ \mathrm{sg}(a))>,$

$\neg^{21} x = <b, a.(a + b)>,$

$\neg^{22} x = <b, 1-b>,$

$\neg^{23} x = <1-\mathrm{sg}(1-b),\ \mathrm{sg}(1-b)>,$

$\neg^{24} x = <\overline{sg}(1-b),\ \mathrm{sg}(a)>,$

$\neg^{25} x = <b.\overline{sg}(1-b),\ a.\mathrm{sg}(1-b)>,$

$\neg^{26} x = <\overline{sg}(a),\ \mathrm{sg}(a)>,$

$\neg^{27} x = <\overline{sg}(a),\ \mathrm{sg}(a)>,$

$\neg^{28} x = <\overline{sg}(1-b),\ \mathrm{sg}(a)>,$

$\neg^{29} x = <\overline{sg}(1-b),\ \mathrm{sg}(a)>,$

Therefore, the following coincidences are valid for the above negations:

$\neg_1 x = \neg^1 x = \neg^4 x = \neg^5 x = \neg^6 x = \neg^7 x = \neg^{10} x = \neg^{13} x'$

$\neg_2 x = \neg^2 x = \neg^3 x = \neg^8 x = \neg^{11} x = \neg^{13} x = \neg^{16} x = \neg^{20} x,$

$\neg_3 x = \neg^9 x = \neg^{17} x = \neg^{21} x,$

$\neg_4 x = \neg^{12} x = \neg^{18} x = \neg^{22} x,$

$\neg_5 x = \neg^{14} x = \neg^{15} x = \neg^{19} x = \neg^{23} x,$

$\neg_6 x = \neg^{24} x = \neg^{26} x = \neg^{27} x = \neg^{28} x = \neg^{29} x,$

$\neg_7 x = \neg^{25} x.$

The intuitionistic fuzzy modal operators can be defined (see, e.g., [2]):

$$V(\Box x) = <a, 1-a>,$$
$$V(\Diamond x) = <1-b, b>.$$

By analogy to [11] we can introduce two new types of implications:

$$X_{2,\mathrm{i}} = V(<\mathrm{a}, b> \to_{2,\mathrm{i}} <\mathrm{c}, d>) = V(\Box <\mathrm{a}, b> \to_{\mathrm{i}} \Diamond <\mathrm{c}, d>)$$

and
$$X_{3,i} = V(<a, b> \rightarrow_{3,i} <c, d>) = V(\Diamond <a, b> \rightarrow_i \Box <c, d>)$$
for $1 \leq i \leq 29$.

Therefore, from the above 29 implications we can introduce the following two groups of new implications, every one of which – with 29 implications:

$X_{2,1} = <1- \min(a, \max(1-a, d)), \min(a, d)>$,
$X_{2,2} = <1-sg(a + c-1), d.sg(a + d-1)>$,
$X_{2,3} = <1-d.sg(a + c-1), d.sg\ a(x) + d-1)>$,
$X_{2,4} = <1- \min(a, d), \min(a, d)>$,
$X_{2,5} = < \min(1, 2-a-d), \max(0, a + d-1)>$,
$X_{2,6} = <1-a.d, a.d>$,
$X_{2,7} = < \min(1- \min(a, d), \max(a, 1-a), \max(1-d, d)), \max(\min(a, d),$
$\qquad \min(a, 1-a), \min(1-d, d))>$,
$X_{2,8} = <1- \max(a, d).sg(a(x) + d-1), \max(a, d).sg(a + d-1)>$,
$X_{2,9} = <1-a + a^2 -a^2.c, a-a^2 + a^2.d>$,
$X_{2,10} = <(1-d).\overline{sg}(1-a) + sg(1-a).(\overline{sg}(d + (1-a).sg(d), d.\overline{sg}(1-a)$
$\qquad +a.sg(1-a).sg(d)>$,
$X_{2,11} = <1-d.sg(a + d-1), d.sg(a + d-1)>$,
$X_{2,12} = <1- \min(a, d), \min(a, d)>$,
$X_{2,13} = <1-a.d, a.d>$,
$X_{2,14} = <(1-d).sg(a + d--1), d.sg(a + d-1)>$,
$X_{2,15} = <1-sg(a + d-1), 1-\overline{sg}(a + d-1)>$,
$X_{2,16} = <1- \min(sg(a), 1-d), \min(sg(a), d)>$,
$X_{2,17} = <1- \min(a, d), \min(a, d)>$,
$X_{2,18} = <1- \min(a, d), \min(a, d)>$,
$X_{2,19} = <1- \min(sg(a), d), \min(sg(a), d)>$,
$X_{2,20} = <1- \min(sg(a), \overline{sg}(1-d)), \min(sg(a), \overline{sg}(1-d))>$,
$X_{2,21} = <1- \min(a, d), \min(a, d)>$,
$X_{2,22} = <1- \min(a, d), \min(a, d)>$,
$X_{2,23} = <1- \min(sg(a), \overline{sg}(1-d)), \min(sg(a), \overline{sg}(1-d))>$,
$X_{2,24} = <\overline{sg}(a + d-1), sg(a + d-1)>$,
$X_{2,25} = < \max(\overline{sg}(a), 1-d), \min(sg(a), d)>$,
$X_{2,26} = < \max(\overline{sg}(a), d), \min(sg(a), d)>$,
$X_{2,27} = < \max(\overline{sg}(a), sg(1-d)), \min(sg(a), \overline{sg}(1-d))>$,
$X_{2,28} = < \max(\overline{sg}(a), 1-d), \min(a, d)>$,
$X_{2,29} = < \max(\overline{sg}(a), \overline{sg}(d)), \min(\overline{sg}(a), \overline{sg}(1-d))>$.

$X_{3,1} = < \max(b, \min(1-b, c)), 1- \max(b, c))>$,
$X_{3,2} = <1-sg(1-b-c), (1-c).sg(1-b-c)>$,
$X_{3,3} = <1-(1-c).sg(1-b-c), (1-c).sg(1-b-c)>$,
$X_{3,4} = < \max(b, c), 1- \max(b, c)>$,
$X_{3,5} = < \min(1, b + c), \max(0, 1-b-c)>$,
$X_{3,6} = <b + c-b.c, (1-b).(1-c)>$,

$X_{3,7} = <\min(\max(b,c), \max(1-b,b), \max(c,1-c)), \max(1-\max(b,c),$
$\qquad \min(1-b,b), \min(c,1-c))>,$

$X_{3,8} = <1-(1-\min(b,c)).\mathrm{sg}(1-b-d),\ (1-\min(b,c).\mathrm{sg}(1-b-c)>,$

$X_{3,9} = <b+(1-b)^2.c, 1-b-(1-b)^2.d>,$

$X_{3,10} = <c.\overline{sg}(b) + \mathrm{sg}(b).(\overline{sg}(1-c)+b.\mathrm{sg}(1-c)),$
$\qquad (1-c).\overline{sg}(b) + (1-b).\mathrm{sg}(b).\mathrm{sg}(1-c)>,$

$X_{3,11} = <1-(1-c).\mathrm{sg}(1-b-c),\ (1-c).\mathrm{sg}(1-b-c)>,$

$X_{3,12} = <\max(b,c), 1-\max(b,c)>,$

$X_{3,13} = <b+c-b.c, 1-b-c+b.c>,$

$X_{3,14} = <1-(1-c).\mathrm{sg}(1-b-c),\ (1-c).\mathrm{sg}(1-b-c)>,$

$X_{3,15} = <1-\mathrm{sg}(1-b-c), 1-\overline{sg}(1-b-c)>,$

$X_{3,16} = <1-\min(\mathrm{sg}(1-d),c), \min(\mathrm{sg}(1-d),1-c)>,$

$X_{3,17} = <\max(b,c), 1-\max(b,c)>,$

$X_{3,18} = <\max(b,c), 1-\max(b,c)>,$

$X_{3,19} = <\max(1-\mathrm{sg}(1-d),c), 1-\max(1-\mathrm{sg}(1-b),c)>,$

$X_{3,20} = <1-\min(\mathrm{sg}(1-d), \overline{sg}(c)), \min(\mathrm{sg}(1-d), \overline{sg}(c))>,$

$X_{3,21} = <\max(b,c), 1-\max(b,c)>,$

$X_{3,22} = <\max(b,c), 1-\max(b,c)>,$

$X_{3,23} = <1-\min(\mathrm{sg}(1-b), \overline{sg}(c)), \min(\mathrm{sg}(1-b),\overline{sg}(c))>,$

$X_{3,24} = <\overline{sg}(1-b-c), \mathrm{sg}(1-b-c)>,$

$X_{3,25} = <\max(\overline{sg}(1-b),\overline{sg}(1-c)),\ 1-\max(b,c)>,$

$X_{3,26} = <\max(\overline{sg}(1-c),\mathrm{sg}(d),\ \min(\mathrm{sg}(d), \overline{sg}(1-a))>,$

$X_{3,27} = <\max(\overline{sg}(1-b),\mathrm{sg}(c)),\ \min(\mathrm{sg}(1-b), \overline{sg}(c))>,$

$X_{3,28} = <\max(\overline{sg}(1-b),c),\ 1-\max(b,c)>,$

$X_{3,29} = <\max(\overline{sg}(1-b), \overline{sg}(1-c)),\ \min(\overline{sg}(1-b), \overline{sg}(c))>.$

Now, we shall discuss another way for generating of new implications. We shall use operation "substitution" in the following form for a given propositional form $f(a,\ldots,b,x,c,\ldots,d)$:

$$g(a,\ldots,b,y,c,\ldots,d) = \frac{x}{y}f(a,\ldots,b,x,c,\ldots,d),$$

i.e., $g(a,\ldots,b,y,c,\ldots,d)$ coincide with $f(a,\ldots,b,y,c,\ldots,d)$ with exception of the participations of variable x that is changed to variable y. For example

$$\frac{x}{y}(a+x) = a+y.$$

It is very important to note that the changes are made simultaneously, i.e., they change all participations of y to x. On the other hand, if we have to change variable x to variable y, and variable y to variable x, then we will also do it simultaneously. For example

$$\frac{x}{y}\frac{y}{x}(a+x-y) = a+y-x.$$

All above implications generate new implications by operation "substitution" with the formulae:

$$X_{4,i} = V(<a,b> \rightarrow_{4,i} <c,d>) = V\left(\frac{a\ d\ b\ c}{d\ a\ c\ b}(<a,b> \rightarrow_i <c,d>)\right),$$

$$X_{5,i} = V(<a,b> \rightarrow_{5,i} <c,d>) = V\left(\frac{a\ d\ b\ c}{d\ a\ c\ b}(<a,b> \rightarrow_{2,i} <c,d>)\right),$$

$$X_{6,i} = V(<a,b> \rightarrow_{6,i} <c,d>) = V\left(\frac{a\ d\ b\ c}{d\ a\ c\ b}(<a,b> \rightarrow_{3,i} <c,d>)\right)$$

for $1 \leq i \leq 29$.

So, we construct the following 87 new implications:

$X_{4,1} = <\max(c, \min(d,b)),\ \min(d,a)) >,$

$X_{4,2} = <1-sg(d-b),\ a.sg(d-b)>,$

$X_{4,3} = <1-(1-b).sg(d-b),\ a.sg(d-b)>,$

$X_{4,4} = <\max(c,b),\ \min(d,a)>,$

$X_{4,5} = <\min(1, c+b),\ \max(0, d+a-1)>,$

$X_{4,6} = <c + d.b,\ d.a>,$

$X_{4,7} = <\min(\max(c,b), \max(d,c), \max(b,a)), \max(\min(d,a), \min(d,c),$
$\qquad \min(b,a))>,$

$X_{4,8} = <1-(1-\min(c,b)).sg(d-b),\ \max(d,a).sg(d-b).sg(a-c)>,$

$X_{4,9} = <c + d^2b,\ dc + d^2a>,$

$X_{4,10} = <b.\overline{sg}(1-d) + sg(1-d).(\overline{sg}(1-b) + c.sg(1-b)),\ a.\overline{sg}(1-d)$
$\qquad + d.sg(1-d).sg(1-b)>,$

$X_{4,11} = <1-(1-b).sg(d-b),\ a.sg(d-b).sg(a-c)>,$

$X_{4,12} = <\max(c,b),\ 1-\max(c,b)>,$

$X_{4,13} = <c + b - c.b,\ d.a>,$

$X_{4,14} = <1-(1-b).sg(d-b)-a.\ \overline{sg}(d-b).sg(a-c),\ a.sg(a-c)>,$

$X_{4,15} = <1-(1-\min(c,b)).sg(sg(d-b) + sg(a-c)) - \min(c,b).sg(d-b)$
$\qquad .sg(d-c),\ 1-(1-\max(d,a)).sg(\overline{sg}(d-b)+\overline{sg}(a-c)) - \max(d,a)$
$\qquad .\overline{sg}(d-b).\overline{sg}(a-c)>,$

$X_{4,16} = <\max(1-sg(d),\ b),\ \min(sg(d),a)>,$

$X_{4,17} = <\max(c,b),\ \min(d.c + d^2, a)>,$

$X_{4,18} = <\max(c,b),\ \min(1-c,a)>,$

$X_{4,19} = <\max(1-sg(sg(d) + sg(1-c)), b),\ \min(sg(1-c), a)>,$

$X_{4,20} = <\max(1-sg(d), 1-sg(1-sg(b))),\ \min(sg(d), sg(1-sg(b)))>,$

$X_{4,21} = <\max(c, b(b+a)),\ \min(a(d+c), a(b^2 + d + ba))>,$

$X_{4,22} = <\max(c, 1-a),\ \min(1-c, a)>,$

$X_{4,23} = <1-\min(sg(1-c), sg(1-sg(1-a))),\ \min(sg(1-c),$
$\qquad sg(1-sg(1-a)))>,$

$X_{4,24} = <\overline{sg}(d-b).\overline{sg}(a-c),\ sg(d-b).sg(a-c)>,$

$X_{4,25} = <\max(c.\overline{sg}(d).\overline{sg}(1-c), b.\overline{sg}(a).\overline{sg}(1-b)),\ \min(d.sg(1-c),$
$\qquad a.sg(1-b))>,$

$X_{4,26} = <\max(\overline{sg}(1-c), b),\ \min(sg(d), a)>,$

$X_{4,27} = <\max(\overline{sg}(1-c), sg(b)), \min(sg(d), \overline{sg}(1-a))>,$

$X_{4,28} = <\max(\overline{sg}(1-c), \nu_A(x), \min(a, d)>,$

$X_{4,29} = <\max(\overline{sg}(1-c), \overline{sg}(1-b)), \min(\overline{sg}(1-a), \overline{sg}(1-c))>.$

$X_{5,1} = <1 - \min(d, \max(1-d, a)), \min(d, a))>,$

$X_{5,2} = <1 - sg(d+b-1), a.sg(d+a-1)>,$

$X_{5,3} = <1 - a.sg(d+b-1), a.sg(d+a-1)>,$

$X_{5,4} = <1 - \min(d, a), \min(d, a)>,$

$X_{5,5} = <\min(1, 2-d-a), \max(0, d+a-1)>,$

$X_{5,6} = <1 - d.a, d.a>,$

$X_{5,7} = <\min(1 - \min(d, a), \max(d, 1-d), \max(1-a, a)),$
$\qquad \max(\min(d, a), \min(d, 1-d), \min(1-a, a))>,$

$X_{5,8} = <1 - \max(d, a).sg(d+a-1), \max(d, a).sg(d+a-1)>,$

$X_{5,9} = <1 - d + d^2 - d^2.b, d - d^2 + d^2.a>,$

$X_{5,10} = <(1-a).\overline{sg}(1-d) + sg(1-d).(\overline{sg}(a + (1-d).sg(a), a.\overline{sg}(1-d)$
$\qquad + d.sg(1-d).sg(a)>,$

$X_{5,11} = <1 - a.sg(d+a-1), a.sg(d+a-1)>,$

$X_{5,12} = <1 - \min(d, a), \min(d, a)>,$

$X_{5,13} = <1 - d.a, d.a>,$

$X_{5,14} = <(1-a).sg(d+a-1), a.sg(d+a-1)>,$

$X_{5,15} = <1 - sg(d+a-1), 1 - \overline{sg}(d+a-1)>,$

$X_{5,16} = <1 - \min(sg(d), 1-a), \min(sg(d), a)>,$

$X_{5,17} = <1 - \min(d, a), \min(d, a)>,$

$X_{5,18} = <1 - \min(d, a), \min(d, a)>,$

$X_{5,19} = <1 - \min(sg(d), a), \min(sg(d), a)>,$

$X_{5,20} = <1 - \min(sg(d), \overline{sg}(1-a)), \min(sg(d), \overline{sg}(1-a))>,$

$X_{5,21} = <1 - \min(d, a), \min(d, a)>,$

$X_{5,22} = <1 - \min(d, a), \min(d, a)>,$

$X_{5,23} = <1 - \min(sg(d), \overline{sg}(1-a)), \min(sg(d), \overline{sg}(1-a))>,$

$X_{5,24} = <\overline{sg}(d+a-1), sg(d+a-1)>,$

$X_{5,25} = <\max(\overline{sg}(d), a), \min(d, a)>,$

$X_{5,26} = <\max(\overline{sg}(d), 1-a), \min(sg(d, a)>,$

$X_{5,27} = <\max(\overline{sg}(d), sg(1-a)), \min(sg(d), \overline{sg}(1-a))>,$

$X_{5,28} = <\max(\overline{sg}(d), 1-a), \min(a, d)>,$

$X_{5,29} = <\max(\overline{sg}(a), \overline{sg}(d)), \min(\overline{sg}(\nu_B(x)), \overline{sg}(1-a))>.$

$X_{6,1} = <\max(c, \min(1-c, b)), 1 - \max(c, b))>,$

$X_{6,2} = <1 - sg(1-c-b), (1-b).sg(1-c-b)>,$

$X_{6,3} = <1 - (1-b).sg(1-c-b), (1-b).sg(1-c-b)>,$

$X_{6,4} = <\max(c, b), 1 - \max(c, b)>,$

$X_{6,5} = <\min(1, c+b), \max(0, 1-c-b)>,$

$X_{6,6} = <c + b - c.b, (1-c).(1-b)>,$

$X_{6,7} = <\min(\max(c, b), \max(1-c, c), \max(b, 1-b)),$
$\qquad \max(1 - \max(c, b), \min(1-c, c), \min(b, 1-b))>,$

$X_{6,8} = <1 - (1 - \min(c, b)).sg(1-c-a), (1 - \min(c, b).sg(1-c-b)>,$

$X_{6,9} = <c + (1-c)^2.b, 1 - c - (1-c)^2.a>,$

$X_{6,10} = <b.\overline{sg}(c) + sg(c).(\overline{sg}(1-b) + c.sg(1-b)),$
$\qquad (1-b).\overline{sg}(c) + (1-c).sg(c).sg(1-b)>,$

$$X_{6,11} = <1 - (1-b).\mathrm{sg}(1-c-b),\ (1-b).\mathrm{sg}(1-c-b)>,$$
$$X_{6,12} = <\max(c,b),\ 1 - \max(c,b)>,$$
$$X_{6,13} = <c + b - c.b,\ 1 - c - b + c.b>,$$
$$X_{6,14} = <1 - (1-b).\mathrm{sg}(1-c-b),\ (1-b).\mathrm{sg}(1-c-b)>,$$
$$X_{6,15} = <1 - \mathrm{sg}(1-c-b),\ 1 - \overline{sg}(1-c-b)>,$$
$$X_{6,16} = <1 - \min(\mathrm{sg}(1-a),b),\ \min(\mathrm{sg}(1-a),1-b)>,$$
$$X_{6,17} = <\max(c,b),\ 1 - \max(c,b)>,$$
$$X_{6,18} = <\max(c,b),\ 1 - \max(c,b)>,$$
$$X_{6,19} = <\max(1 - \mathrm{sg}(1-a),b),\ 1 - \max(1 - \mathrm{sg}(1-c),b)>,$$
$$X_{6,20} = <1 - \min(\mathrm{sg}(1-a),\overline{sg}(b)),\ \min(\mathrm{sg}(1-a),\overline{sg}(b))>,$$
$$X_{6,21} = <\max(c,b),\ 1 - \max(c,b)>,$$
$$X_{6,22} = <\max(c,b),\ 1 - \max(c,b)>,$$
$$X_{6,23} = <1 - \min(\mathrm{sg}(1-c),\overline{sg}(b)),\ \min(\mathrm{sg}(1-c),\overline{sg}(b))>,$$
$$X_{6,24} = <\overline{sg}(1-c-b),\ \mathrm{sg}(1-c-b)>,$$
$$X_{6,25} = <\max(\overline{sg}(1-c),\overline{sg}(1-b)),\ 1 - \max(c,b)>\}.$$
$$X_{6,26} = <\max(\overline{sg}(1-b),c),\ \min(\mathrm{sg}(a,\overline{sg}(1-d))>,$$
$$X_{6,27} = <\max(\overline{sg}(1-c),\mathrm{sg}(b)),\ \min(\mathrm{sg}(1-c),\overline{sg}(b))>.$$
$$X_{6,28} = <\max(\overline{sg}(1-c),\nu_A(x)),\ 1 - \max(b,c)>,$$
$$X_{6,29} = <\max(\overline{sg}(1-b),\overline{sg}(1-c)),\ \min(\overline{sg}(1-c),\overline{sg}(b))>.$$

Let

$$V(E^*) = <1,0>.$$

Theorem 1. *All above implications are extensions of ordinary first-order logic implication.*

The validity of this theorem follows from the fact that for every $i(1 \leq i \leq 6)$ and for every $j(1 \leq j \leq 29)$ it can be checked that

(a) $V(O^* \rightarrow_{i,j} O^*) = V(E^*),$
(b) $V(O^* \rightarrow_{i,j} E^*) = V(E^*),$
(c) $V(E^* \rightarrow_{i,j} O^*) = V(O^*),$
(d) $V(E^* \rightarrow_{i,j} E^*) = V(E^*),$

where $\rightarrow_{1,j} = \rightarrow_j$.

Theorem 2. *The following implications coincide (we shall denote this fact by "="):*

(1) $\rightarrow_4 = \rightarrow_{4,4}$
(2) $\rightarrow_5 = \rightarrow_{4,5}$
(3) $\rightarrow_7 = \rightarrow_{4,7}$
(4) $\rightarrow_{12} = \rightarrow_{3,4} = \rightarrow_{3,12} = \rightarrow_{3,17} = \rightarrow_{3,18} = \rightarrow_{3,21} = \rightarrow_{3,22} = \rightarrow_{4,12} = \rightarrow_{6,4} = \rightarrow_{6,12}$
$\quad = \rightarrow_{6,17} = \rightarrow_{6,18} = \rightarrow_{6,21} = \rightarrow_{6,22}$
(5) $\rightarrow_{13} = \rightarrow_{4,13}$
(6) $\rightarrow_{15} = \rightarrow_{4,15}$
(7) $\rightarrow_{24} = \rightarrow_{4,24}$
(8) $\rightarrow_{25} = \rightarrow_{4,25}$

(9) $\rightarrow_{2,3}=\rightarrow_{2,11}=\rightarrow_{5,14}$

(10) $\rightarrow_{2,4}=\rightarrow_{2,12}=\rightarrow_{2,17}=\rightarrow_{2,18}=\rightarrow_{2,21}=\rightarrow_{2,22}=\rightarrow_{5,4}=\rightarrow_{5,12}=\rightarrow_{5,17}=\rightarrow_{5,18}$
$=\rightarrow_{5,21}=\rightarrow_{5,22}$

(11) $\rightarrow_{2,5}=\rightarrow_{5,5}$

(12) $\rightarrow_{2,6}=\rightarrow_{2,13}=\rightarrow_{5,6}=\rightarrow_{5,13}$

(13) $\rightarrow_{2,7}=\rightarrow_{5,7}$

(14) $\rightarrow_{2,8}=\rightarrow_{5,8}$

(15) $\rightarrow_{2,15}=\rightarrow_{2,24}=\rightarrow_{5,15}=\rightarrow_{5,24}$

(16) $\rightarrow_{2,16}=\rightarrow_{2,19}=\rightarrow_{2,26}$

(17) $\rightarrow_{2,20}=\rightarrow_{2,23}=\rightarrow_{2,27}$

(18) $\rightarrow_{2,25}=\rightarrow_{5,25}$

(19) $\rightarrow_{3,3}=\rightarrow_{3,11}=\rightarrow_{3,14}$

(20) $\rightarrow_{3,5}=\rightarrow_{6,5}$

(21) $\rightarrow_{3,6}=\rightarrow_{3,13}=\rightarrow_{6,6}=\rightarrow_{6,13}$

(22) $\rightarrow_{3,7}=\rightarrow_{6,7}$

(23) $\rightarrow_{3,8}=\rightarrow_{6,8}$

(24) $\rightarrow_{3,15}=\rightarrow_{3,24}=\rightarrow_{6,15}=\rightarrow_{6,24}$

(25) $\rightarrow_{3,16}=\rightarrow_{3,19}=\rightarrow_{3,26}$

(26) $\rightarrow_{3,20}=\rightarrow_{3,23}=\rightarrow_{3,27}$

(27) $\rightarrow_{3,25}=\rightarrow_{6,25}$

(28) $\rightarrow_{5,3}=\rightarrow_{5,11}=\rightarrow_{5,14}$

(29) $\rightarrow_{5,16}=\rightarrow_{5,19}=\rightarrow_{5,26}$

(30) $\rightarrow_{5,20}=\rightarrow_{5,23}=\rightarrow_{5,27}$

(31) $\rightarrow_{6,3}=\rightarrow_{6,11}=\rightarrow_{6,14}$

(32) $\rightarrow_{6,16}=\rightarrow_{6,19}=\rightarrow_{6,26}$

(33) $\rightarrow_{6,20}=\rightarrow_{6,23}=\rightarrow_{6,27}$

The check of these equalities is direct.

This Theorem permits us to construct a new table with all the different implications with notations between 30 and 99.

Table 2. List of the new 70 intuitionistic fuzzy implications

Notation	Form of implication
$\rightarrow_{30}$	$<1-\min(a,\max(1-a,d)),\ \min(a,d)>$
$\rightarrow_{31}$	$<1-\mathrm{sg}(a+c-1),\ d.\mathrm{sg}(a+d-1)>$
$\rightarrow_{32}$	$<1-d.\mathrm{sg}(a+c-1),\ d.\mathrm{sg}\ a(x)+d-1)>$
$\rightarrow_{33}$	$<1-\min(a,d),\ \min(a,d)>$
$\rightarrow_{34}$	$<\min(1,2-a-d),\ \max(0,a+d-1)>$
$\rightarrow_{35}$	$<1-a.d,\ a.d>$
$\rightarrow_{36}$	$<\min(1-\min(a,d),\ \max(a,1-a),\ \max(1-d,d)),\ \max(\min(a,d),$ $\min(a,1-a),\ \min(1-d,d))>$
$\rightarrow_{37}$	$<1-\max(a,d).\mathrm{sg}(a(x)+d-1),\ \max(a,d).\mathrm{sg}(a+d-1)>$
$\rightarrow_{38}$	$<1-a+a^2-a^2.c,\ a-a^2+a^2.d>$
$\rightarrow_{39}$	$<(1-d).\overline{sg}(1-a)+\mathrm{sg}(1-a).(\overline{sg}(d+(1-a).\mathrm{sg}(d),\ d.\overline{sg}(1-a)$ $+a.\mathrm{sg}(1-a).\mathrm{sg}(d)>$
$\rightarrow_{40}$	$<1-\mathrm{sg}(a+d-1),\ 1-\overline{sg}(a+d-1)>$

Table 2. (*Continued*)

Notation	Form of implication
$\rightarrow_{41}$	$<1 - \min(\text{sg}(a), 1 - d), \min(\text{sg}(a), d)>$
$\rightarrow_{42}$	$<1 - \min(\text{sg}(a), \overline{sg}(1 - d)), \min(\text{sg}(a), \overline{sg}(1 - d))>$
$\rightarrow_{43}$	$< \max(\overline{sg}(a), 1 - d), \min(\text{sg}(a), d)>$
$\rightarrow_{44}$	$< \max(\overline{sg}(a), 1 - d), \min(a, d)>$
$\rightarrow_{45}$	$< \max(\overline{sg}(a), \overline{sg}(d)), \min(\overline{sg}(a), \overline{sg}(1 - d))>$
$\rightarrow_{46}$	$< \max(b, \min(1 - b, c)), 1 - \max(b, c))>$
$\rightarrow_{47}$	$<1 - \text{sg}(1 - b - c), (1 - c).\text{sg}(1 - b - c)>$
$\rightarrow_{48}$	$<1 - (1 - c).\text{sg}(1 - b - c), (1 - c).\text{sg}(1 - b - c)>$
$\rightarrow_{49}$	$< \min(1, b + c), \max(0, 1 - b - c)>$
$\rightarrow_{50}$	$<b + c - b.c, (1 - b).(1 - c)>$
$\rightarrow_{51}$	$< \min(\max(b, c), \max(1 - b, b), \max(c, 1 - c)),$ $\max(1 - \max(b, c), \min(1 - b, b), \min(c, 1 - c)))>$
$\rightarrow_{52}$	$<1 - (1 - \min(b, c)).\text{sg}(1 - b - d), (1 - \min(b, c).\text{sg}(1 - b - c)>$
$\rightarrow_{53}$	$<b + (1 - b)^2.c, 1 - b - (1 - b)^2.d>$
$\rightarrow_{54}$	$<c.\overline{sg}(b) + \text{sg}(b).(\overline{sg}(1 - c) + b.\text{sg}(1 - c)), (1 - c).\overline{sg}(b) + (1 - b).\text{sg}(b)$ $.\text{sg}(1 - c)>$
$\rightarrow_{55}$	$<1 - \text{sg}(1 - b - c), 1 - \overline{sg}(1 - b - c)>$
$\rightarrow_{56}$	$<1 - \min(\text{sg}(1 - d), c), \min(\text{sg}(1 - d), 1 - c)>$
$\rightarrow_{57}$	$<1 - \min(\text{sg}(1 - d), \overline{sg}(c)), \min(\text{sg}(1 - d), \overline{sg}(c))>$
$\rightarrow_{58}$	$< \max(\overline{sg}(1 - b), \overline{sg}(1 - c)), 1 - \max(b, c)>$
$\rightarrow_{59}$	$< \max(\overline{sg}(1 - b), c), 1 - \max(b, c)>$
$\rightarrow_{60}$	$< \max(\overline{sg}(1 - b), \overline{sg}(1 - c)), \min(\overline{sg}(1 - b), \overline{sg}(c))>$
$\rightarrow_{61}$	$< \max(c, \min(d, b)), \min(d, a))>$
$\rightarrow_{62}$	$<1 - \text{sg}(d - b), a.\text{sg}(d - b)>$
$\rightarrow_{63}$	$<1 - (1 - b).\text{sg}(d - b), a.\text{sg}(d - b)>$
$\rightarrow_{64}$	$<c + d.b, d.a>$
$\rightarrow_{65}$	$<1 - (1 - \min(c, b)).\text{sg}(d - b), \max(d, a).\text{sg}(d - b).\text{sg}(a - c)$
$\rightarrow_{66}$	$<c + d^2b, dc + d^2a$
$\rightarrow_{67}$	$<b.\overline{sg}(1 - d) + \text{sg}(1 - d).(\overline{sg}(1 - b) + c.\text{sg}(1 - b)), a.\overline{sg}(1 - d) + d.\text{sg}(1 - $ $d).\text{sg}(1 - b)>$
$\rightarrow_{68}$	$<1 - (1 - b).\text{sg}(d - b), a.\text{sg}(d - b).\text{sg}(a - c)>$
$\rightarrow_{69}$	$<1 - (1 - b).\text{sg}(d - b) - a.\overline{sg}(d - b).\text{sg}(a - c), a.\text{sg}(a - c)>$
$\rightarrow_{70}$	$< \max(1 - \text{sg}(d), b), \min(\text{sg}(d), a)>$
$\rightarrow_{71}$	$< \max(c, b), \min(d.c + d^2, a)>$
$\rightarrow_{72}$	$< \max(c, b), \min(1 - c, a)>$
$\rightarrow_{73}$	$< \max(1 - \text{sg}(\text{sg}(d) + \text{sg}(1 - c)), b), \min(\text{sg}(1 - c), a)$
$\rightarrow_{74}$	$< \max(1 - \text{sg}(d), 1 - \text{sg}(1 - \text{sg}(b))), \min(\text{sg}(d), \text{sg}(1 - \text{sg}(b)))>$
$\rightarrow_{75}$	$< \max(c, b(b + a)), \min(a(d + c), a(b^2 + d + ba))$
$\rightarrow_{76}$	$< \max(c, 1 - a), \min(1 - c, a)>$
$\rightarrow_{77}$	$<1 - \min(\text{sg}(1 - c), \text{sg}(1 - \text{sg}(1 - a))), \min(\text{sg}(1 - c), \text{sg}(1 - \text{sg}(1 - a)))>$
$\rightarrow_{78}$	$< \max(\overline{sg}(1 - c), b), \min(\text{sg}(d, a)>$
$\rightarrow_{79}$	$< \max(\overline{sg}(1 - c), \text{sg}(b)), \min(\text{sg}(d), \overline{sg}(1 - a))>$
$\rightarrow_{80}$	$< \max(\overline{sg}(1 - c), \nu_A(x), \min(a, d)>$
$\rightarrow_{81}$	$< \max(\overline{sg}(1 - c), \overline{sg}(1 - b)), \min(\overline{sg}(1 - a), \overline{sg}(1 - c))>$
$\rightarrow_{82}$	$<1 - \min(d, \max(1 - d, a)), \min(d, a))>$
$\rightarrow_{83}$	$<1 - \text{sg}(d + b - 1), a.\text{sg}(d + a - 1)>$
$\rightarrow_{84}$	$<1 - a.\text{sg}(d + b - 1), a.\text{sg}(d + a - 1)>$

Table 2. (*Continued*)

Notation	Form of implication
$\to_{85}$	$<1 - d + d^2 - d^2.b,\ d - d^2 + d^2.a>$
$\to_{86}$	$<(1-a).\overline{sg}(1-d) + sg(1-d).(\overline{sg}(a + (1-d).sg(a),\ a.\overline{sg}(1-d) + d.sg(1-d).sg(a)>$
$\to_{87}$	$<1 - \min(sg(d), 1 - a),\ \min(sg(d), a)>$
$\to_{88}$	$<1 - \min(sg(d), \overline{sg}(1 - a)),\ \min(sg(d), \overline{sg}(1 - a))>$
$\to_{89}$	$< \max(\overline{sg}(d), 1 - a),\ \min(a, d)>$
$\to_{90}$	$< \max(\overline{sg}(a), \overline{sg}(d)),\ \min(\overline{sg}(\nu_B(x)), \overline{sg}(1 - a))>$
$\to_{91}$	$< \max(c, \min(1 - c, b)),\ 1 - \max(c, b))>$
$\to_{92}$	$<1 - sg(1 - c - b),\ (1 - b).sg(1 - c - b)>$
$\to_{93}$	$<1 - (1 - b).sg(1 - c - b),\ (1 - b).sg(1 - c - b)>$
$\to_{94}$	$<c + (1 - c)^2.b,\ 1 - c - (1 - c)^2.a>$
$\to_{95}$	$<b.\overline{sg}(c) + sg(c).(\overline{sg}(1 - b) + c.sg(1 - b)),\ (1 - b).\overline{sg}(c) + (1 - c).sg(c).sg(1 - b)>$
$\to_{96}$	$<1 - \min(sg(1 - a), b),\ \min(sg(1 - a), 1 - b)>$
$\to_{97}$	$<1 - \min(sg(1 - a), \overline{sg}(b)),\ \min(sg(1 - a), \overline{sg}(b))>$
$\to_{98}$	$< \max(\overline{sg}(1 - c), \nu_A(x)),\ 1 - \max(b, c)>$
$\to_{99}$	$< \max(\overline{sg}(1 - b), \overline{sg}(1 - c)),\ \min(\overline{sg}1 - c), \overline{sg}(b))>$

3 Conclusion and Open Problems

At the moment the following problems are open:

1. To construct other implications.
2. To study properties of all implications and negations, and especially, to determine the implications and negations that satisfy axioms of intuitionistic logic (see, e.g., [13]).
3. To compare the different implications, to construct a graph of their orders and to study its properties.
4. To study validity of Klir and Yuan's axioms for all implications.

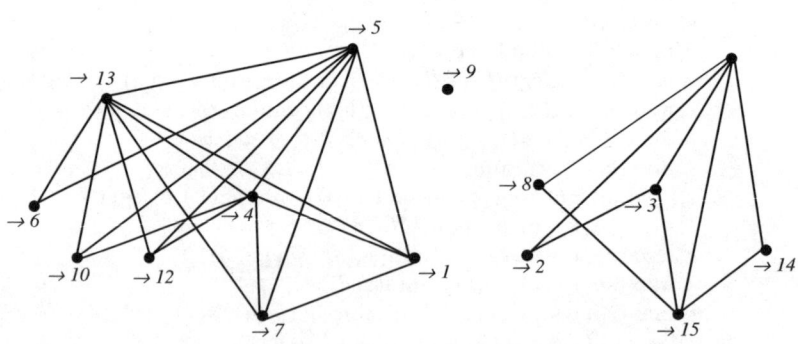

For example, following [5] we can show the relations between the first 15 implications from Table 1. If the implication $\rightarrow_i$ is located on a higher position than implication $\rightarrow_j$ this means that for all $a, b, c, d \in [0, 1]$ so that $a + b, c + d \in [0, 1]$:

$$V(<a, b> \rightarrow_i <c, d>) \geq V(<a, b> \rightarrow_j <c, d>).$$

The author hopes that the answers of these problems will be known in near future. This will give possibility to be determined some the most suitable implications. Of course, one of them will be the classical (Kleene-Dienes) implication. But the future will show the other ones, that will have intuitionistic behaviour.

When the concept of IFS was introduced, Georgi Gargov (1947–1996), who gave the name of this fuzzy set extension and the author, thought that the intuitionistic idea is best embodied in the form of elements of the IFSs with their two degrees – degree of membership or of validity, or of correctness, etc. and degree of non-membership or of non-validity, or of incorrectness, etc. . . Because of this, for a long period of time the author used the operations for "classical" negation() and "classical" implication (), only. In the last couple of years it became clear that the operations over intiutuinostic fuzzy objects can be of intuitionistic nature themselves and that was the reason for the appearance of the new results. Now it is known for which of the implications from Table 1, and the negations the following formulas are valid:

$$x \rightarrow \neg\neg x,$$
$$\neg\neg x \rightarrow x.$$

And that would be extended to all the implications from Table 2.

References

1. K. Atanassov. Two variants of intuitonistc fuzzy propositional calculus. Preprint IM-MFAIS-5-88, Sofia, 1988.
2. K. Atanassov. Intuitionistic Fuzzy Sets. Springer Physica, Heidelberg, 1999.
3. K. Atanassov. Intuitionistic fuzzy implications and Modus Ponens. Notes on Intuitionistic Fuzzy Sets, 11(1), 2005, 1–5.
4. K. Atanassov. A new intuitionistic fuzzy implication from a modal type. Advanced Studies in Contemporary Mathematics, 12(1), 2006, 117–122.
5. K. Atanassov. On some intuitionistic fuzzy implications. Comptes Rendus de l'Academie bulgare des Sciences, Tome 59(1), 2006, 19–24.
6. K. Atanassov. On eight new intuitionistic fuzzy implications. Proceedings of IEEE Conference "Intelligent Systems'06", London, 5–7 Sept. 2006, 741–746.
7. K. Atanassov. On the implications and negations over intuitionistic fuzzy sets. Proceedings of the Scientific Conf. of Burgas Free University, Burgas, 9–11 June 2006, Vol. III, 374–384.

8. K. Atanassov and B. Kolev. On an intuitionistic fuzzy implication from a possibilistic type. Advanced Studies in Contemporary Mathematics, 12(1), 2006, 111–116.

9. K. Atanassov and T. Trifonov, On a new intuitionistic fuzzy implication from Gödel's type. Proceedings of the Jangjeon Mathematical Society, 8(2), 2005, 147–152.

10. K. Atanassov and T. Trifonov, Two new intuitionistic fuzzy implications. Advanced Studies in Contemporary Mathematics, 13(1), 2006, 69–74.

11. R. Feys. Modal Logics. Gauthier-Villars, Paris, 1965.

12. G. Klir and Bo Yuan. Fuzzy Sets and Fuzzy Logic. Prentice Hall, New Jersey, 1995.

13. H. Rasiova and R. Sikorski. The Mathematics of Metamathematics, Warszawa, Polish Academy of Science, 1963.

On the Probability Theory on the Atanassov Sets

Beloslav Riečan

Matej Bel University, Tajovského 40, SK–97401 Banská Bystrica, Slovakia
Mathematical Institute of Slovak Academy of Sciences, Štefánikova 49,
SK–81473 Bratislava, Slovakia
riecan@fpv.umb.sk

Summary. We present an axiomatic approach to the probability theory on IF-sets = intuitionistic fuzzy sets = Atanassov sets (Intuitionistic Fuzzy Sets: Theory and Applications. Physica, New York, 1999). Starting with two constructive definitions (*Issues in Intelligent Systems: Paradigms*, EXIT, Warszawa 2005, pp. 63–58; *Soft Methods in Probability, Statistics and Data Analysis*, Physica, New York 2002, pp. 105–115) we consider a theory including not only the special cases but also the general form of the probabilities on IF-sets. Moreover an embedding of the theory to the probability theory on MV-algebras is given. This fact enables to use the well developed probability theory on MV-algebras for constructing probability theory on IF-events.

1 Introduction

Although there are different opinions about IF-events, the following definitions are accepted generally [1, 3, 4]. Let $(\Omega, \mathcal{S})$ be a measurable space. By an IF-event [4] we mean any pair

$$A = (\mu_A, \nu_A)$$

of $\mathcal{S}$-measurable functions, such that $\mu_A \geq 0$, $\nu_A \geq 0$, and

$$\mu_A + \nu_A \leq 1.$$

The function μ_A is the membership function and the function ν_A the non-membership function. The family $\mathcal{F}$ of all IF-events is ordered by the following way:

$$A \leq B \iff \mu_A \leq \mu_B, \ \nu_A \geq \nu_B.$$

Evidently the notion of an IF-event is a natural generalization of the notion of the fuzzy events. Given a fuzzy event μ_A, the pair $(\mu_A, 1 - \mu_A)$ is an IF-event, so IF-events can be seen as generalizations of fuzzy events. We hence want to define probability on IF-events generalizing probability on fuzzy events. And

B. Riečan: *On the Probability Theory on the Atanassov Sets*, Studies in Computational Intelligence (SCI) **109**, 395–413 (2008)
www.springerlink.com

actually, two constructions were proposed independently [3, 4]. It is interesting that both definitions can be regarded as a special case of a descriptive definition [12, 13] given axiomatically (see Definition 2.3). This axiomatic definition has been based on the Lukasiewicz connectives

$$a \oplus b = \min(a + b, 1),$$
$$a \odot b = \max(a + b - 1, 0).$$

The operations can be naturally extended to IF-events by the following way. If $A = (\mu_A, \nu_A)$ and $B = (\mu_B, \nu_B)$ then

$$A \oplus B = (\mu_A \oplus \mu_B, \nu_A \odot \nu_B)$$
$$A \odot B = (\mu_A \odot \mu_B, \nu_A \oplus \nu_B)$$

It is easy to see that in the case of a fuzzy set $A = (\mu_A, 1 - \mu_A)$ the Lukasiewicz operations are obtained. Similarly as in the classical case and in the fuzzy case, probability is a mapping (in our case from $\mathcal{F}$ to the unit interval) which is continuous, additive and satisfies some boundary conditions. Here the main difference is in additivity. There are infinitely many possibilities how to define additivity

$$m(A) + m(B) = m(S(A, B)) + m(T(A, B))$$

where

$$S(A, B) = (S(\mu_A, \mu_B), T(\nu_A, \nu_B))$$
$$S, T : [0, 1]^2 \to [0, 1]$$

being such binary operations that

$$S(u, v) + T(1 - u, 1 - v) \leq 1.$$

In the present contribution we have chosen only two possibilities: the Lukasiewicz $S(a, b) = a \oplus b, T(a, b) = a \odot b$, and the Zadeh

$$S(a, b) = a \vee b = \max(a, b),$$
$$T(a, b) = a \wedge b = \min(a, b).$$

Namely, in these two choices we are able to formulate a meaningful theory including such fundamental assertions as the law of large numbers (a bridge between frequency and probability) or central limit theorem (as a possible starting point to statistical inference). The main idea of the chapter is in the embedding of the family $\mathcal{F}$ into an appropriate MV-algebra. Although the MV-algebra considered in our IF-case is very simple, it would be not very economic to work only with the special case and do not use some known results contained in the general MV-algebra probability theory. On the other hand the simple formulations in the IF-events case could lead to a larger variety of possible applications.

The Kolmogorov probability theory has three fundamental notions: probability, random variable, and expectation. In our fuzzy case an analogous situation occurs. The existence of the joint observable plays the crucial role. It is interesting that the corresponding existence theorems (for the Lukasiewicz connectives in Theorem 5.5, for the Zadeh connectives in [17]) have been proved by some thoroughly different methods. Therefore it is hardly to expect that there exists a general methods working for a larger set of pairs (S, T). In this moment this is an open problem.

Some ideas of the chapter has been used separately in previous author papers (e.g. the embedding theorem in the chapter [15] about entropy of dynamical systems, the representation theorem in [16]) and here they are presented in a simple and clear way as a source for possible applications.

2 Probability on IF-Events

First recall some basic definitions. By an IF-set we consider a pair $A = (\mu_A, \nu_A)$ of functions $\mu_A, \nu_A : \Omega \to [0, 1]$ such that

$$\mu_A + \nu_A \leq 1.$$

μ_A is called a membership function of A, ν_A a nonmembership function of A. If $(\Omega, \mathcal{S}, p)$ is a probability space and μ_A, ν_A are $\mathcal{S}$-measurable, then A is called and IF-event and the probability of A has been defined by the following two definitions.

Definition 2.1 (Grzegorzewski, Mrowka [4]). *The probability* $\mathcal{P}(A)$ *of an IF-event* A *is defined as the interval*

$$\mathcal{P}(A) = \left[\int_\Omega \mu_A \, dp, 1 - \int_\Omega \nu_A \, dp \right].$$

Definition 2.2 (Gersternkorn, Manko [3]). *The probability* $\mathcal{P}(A)$ *of an IF-event* A *is defined as the number*

$$\mathcal{P}(A) = \frac{1}{2} \left(\int_\Omega \mu_A \, dp + 1 - \int_\Omega \nu_A \, dp \right).$$

Both definitions are special cases of the following one [10, 12, 13]. Denote by $\mathcal{F}$ the family of all IF-events, and by $\mathcal{J}$ the family of all compact intervals. In the following definition we shall assume that $[a, b] + [c, d] = [a + c, b + d]$, and $[a_n, b_n] \nearrow [a, b]$, if $a_n \nearrow a$, $b_n \nearrow b$. On the other hand we define

$$(\mu_A, \nu_A) \oplus (\mu_B, \nu_B) = (\mu_A \oplus \mu_B, \nu_A \odot \nu_B)$$
$$(\mu_A, \nu_B) \odot (\mu_B, \nu_B) = (\mu_A \odot \mu_B, \nu_A \oplus \nu_B)$$

where

$$f \oplus g = \min(f+g, 1) \quad \text{and} \quad f \odot g = \max(f+g-1, 0).$$

Moreover

$$(\mu_{A_n}, \nu_{A_n}) \nearrow (\mu_A, \nu_A)$$

means

$$\mu_{A_n} \nearrow \mu_A, \nu_{A_n} \searrow \nu_A.$$

Definition 2.3 ([10]). *An IF-probability on $\mathcal{F}$ is a mapping $\mathcal{P} : \mathcal{F} \to \mathcal{J}$ satisfying the following conditions:*

(i) $\mathcal{P}((0,1)) = [0,0]$, $\mathcal{P}((1,0)) = [1,1]$
(ii) $\mathcal{P}((\mu_A, \nu_A)) + \mathcal{P}((\mu_B, \nu_B)) = \mathcal{P}((\mu_A, \nu_A)) \oplus (\mu_B, \nu_B)) + \mathcal{P}((\mu_A, \nu_A) \odot (\mu_B, \nu_B))$
    for any $(\mu_A, \nu_A), (\mu_B, \nu_B) \in \mathcal{F}$
(iii) $(\mu_{A_n}, \nu_{A_n}) \nearrow (\mu_A, \nu_A) \Longrightarrow \mathcal{P}((\mu_{A_n}, \nu_{A_n})) \nearrow \mathcal{P}((\mu_A, \nu_A))$

Example 2.4. The function from Definition 2.1 satisfies the conditions stated in Definition 2.3. It was proved in [4, 12]. Moreover, in [12] an axiomatic characterization of the example was given. On the other hand the general definition represents a larger variety as the singular example [8].

Example 2.5. If we put $\mathcal{P}((\mu_A, \nu_A)) = \{\frac{1}{2}(\int_\Omega \mu_A \, dp + 1 - \int_\Omega \nu_A \, dp)\}$, then $\mathcal{P}$ satisfies all conditions stated in Definition 2.3. We shall demonstrate it on the property (ii):

$$\mathcal{P}((\mu_A, \nu_A) \oplus (\mu_B, \nu_B)) + \mathcal{P}((\mu_A, \nu_A) \odot (\mu_B, \nu_B))$$
$$= \mathcal{P}(\mu_A \oplus \mu_B, \nu_A \odot \nu_B) + \mathcal{P}(\mu_A \odot \mu_B, \nu_A \oplus \nu_B)$$
$$= \frac{1}{2}\left(\int_\Omega \mu_A \oplus \mu_B \, dp + 1 - \int_\Omega \nu_A \odot \nu_B\right) dp$$
$$+ \frac{1}{2}\left(\int_\Omega \mu_A \odot \mu_B \, dp + 1 - \int_\Omega \nu_A \oplus \nu_B \, dp\right).$$

If we use the identity $x \oplus y + x \odot y = x + y$, we obtain that

$$\mathcal{P}((\mu_A, \nu_A) \oplus (\mu_B, \nu_B)) + \mathcal{P}((\mu_A, \nu_A) \odot (\mu_B, \nu_B))$$
$$= \frac{1}{2}\int_\Omega (\mu_A \oplus \mu_B + \mu_A \odot \mu_B + 2 - (\nu_A \odot \nu_B + \nu_A \oplus \nu_B)) \, dp$$
$$= \frac{1}{2}\int_\Omega (\mu_A + \mu_B + 2 - (\nu_A + \nu_B)) \, dp$$
$$= \frac{1}{2}\left(\int_\Omega \mu_A \, dp + 1 - \int_\Omega \nu_B \, dp\right) + \frac{1}{2}\left(\int_\Omega \mu_B \, dp + 1 - \int_\Omega \nu_B \, dp\right)$$
$$= \mathcal{P}((\mu_A, \nu_A)) + \mathcal{P}((\mu_B, \nu_B)).$$

Anyway, we have here two points of view: probability as a real function (Definition 2.2) and probability as an interval-valued function (Definition 2.1). Therefore in the first case a special terminology will be introduced (at least in the chapter).

Definition 2.6. *A function $p : \mathcal{F} \to [0,1]$ will be called a state if the following conditions are satisfied:*

(i) $p((0,1)) = 0$, $p((1,0)) = 1$
(ii) $p((\mu_A,\nu_A)\oplus(\mu_B,\nu_B))+p((\mu_A,\nu_A)\odot(\mu_B,\nu_B)) = p((\mu_A,\nu_A))+p((\mu_B,\nu_B))$
  for any $(\mu_A,\nu_A), (\mu_B,\nu_B) \in \mathcal{F}$
(iii) $(\mu_{A_n},\nu_{A_n}) \nearrow (\mu_A,\nu_A) \Longrightarrow p((\mu_{A_n},\nu_{A_n})) \nearrow p((\mu_A,\nu_A))$

It is easy to see that the preceding definitions are equivalent in the following sense.

Theorem 2.7. *Let $\mathcal{P} : \mathcal{F} \to \mathcal{J}$ be a mapping. Denote $\mathcal{P}(A) = [\mathcal{P}^\flat(A), \mathcal{P}^\sharp(A)]$ for any $A \in \mathcal{F}$. Then $\mathcal{P}$ is a probability if and only if $\mathcal{P}^\flat$, $\mathcal{P}^\sharp$ are states.*

3 MV-Algebras

MV-algebra is an algebraic system $(M, \oplus, \odot, \neg, 0, 1)$, where $\oplus, \odot$ are binary operations, $\neg$ an unary operation, and 0, 1 fixed elements such that the following properties are satisfied: $\neg 0 = 1$, $\neg 1 = 0$, $x \oplus 1 = 1$, $x \odot y = \neg(\neg x \oplus \neg y)$ and $y \oplus \neg(y \oplus \neg x) = x \oplus \neg(x \oplus \neg y)$ for all $x, y \in M$.

Example 3.1. Put $M = [0,1]$, $a \oplus b = \min(a+b, 1)$, $a \odot b = \max(a+b-1, 0)$, $\neg a = 1 - a$. This structure is a typical example of an MV-algebra.

In every MV-algebra M the binary relation $\leq$ given by $a \leq b$ if and only if $a \odot \neg b = 0$, is a partial order such that M is a distributive lattice.

By the Mundici theorem [2] any MV-algebra can be obtained by a similar way as it was shown in Example 3.1, of course, instead of the set R of reals an l-group must be considered. Recall that an l-group is a system $(G, +, \leq)$ such that $(G, +)$ is an Abelian group, $(G, \leq)$ is a lattice and $a \leq b$ implies $a + c \leq b + c$.

Up to isomorphism, every MV-algebra $(M, \oplus, \odot, \neg, 0, 1)$ can be identified with the unit interval $[0, u]$(u is a strong unit in G, i.e. to any $a \in M$ there exists $n \in N$ such that $nu \geq a$) such that

$$(a \oplus b) = (a + b) \wedge u$$
$$a \odot b = (a + b - u) \vee 0$$
$$\neg a = u - a.$$

Consider now the family $\mathcal{F}$ of all IF-events on a measurable space $(\Omega, \mathcal{S})$, i.e. such pairs $A = (\mu_A, \nu_A)$ of $\mathcal{S}$-measurable functions that $0 \leq \mu_A$, $0 \leq \nu_A$, $\mu_A + \nu_A \leq 1$.

We shall construct an MV-algebra $\mathcal{M}$ and then embed $\mathcal{F}$ to $\mathcal{M}$ and show that there exists one-to-one correspondence between probabilities on $\mathcal{F}$ and probabilities on $\mathcal{M}$.

Definition 3.2. *Define* $\mathcal{M} = \{(\mu_A, \nu_A); \mu_A, \nu_A$ *are* $\mathcal{S}$-*measurable,* $\mu_A, \nu_A : \Omega \to [0,1]\}$ *together with operations*

$$(\mu_A, \nu_A) \oplus (\mu_B, \nu_B) = (\mu_A \oplus \mu_B, \nu_A \odot \nu_B),$$
$$(\mu_A, \nu_A) \odot (\mu_B, \nu_B) = (\mu_A \odot \mu_B, \nu_A \oplus \nu_B),$$
$$\neg(\mu_A, \nu_A) = (1 - \mu_A, 1 - \nu_A).$$

Theorem 3.3. *The system* $(\mathcal{M}, \oplus, \odot, \neg, 0, 1)$ *is an MV-algebra.*

Proof. Consider the set $\mathcal{G} = \{(f, g); f, g : \Omega \to R, f, g$ are measurable$\}$. The ordering $\leq$ is induced by the IF-ordering, hence $(f, g) \leq (h, k) \iff f \leq h, g \geq k$. Evidently $(\mathcal{G}, \leq)$ is a lattice, $(f, g) \vee (h, k) = (f \vee h, g \wedge k)$, $(f, g) \wedge (h, k) = (f \wedge h, g \vee k)$. Now we shall define the group operation $+$ by the following formula:

$$(f, g) + (h, k) = (f + h, g + k - 1).$$

It is not difficult to see that $+$ is commutative and associative, and $(0, 1)$ is the neutral element. The inverse element to (f, g) is the pair $(-f, 2 - g)$, since

$$(f, g) + (-f, 2 - g) = (f - f, g + 2 - g - 1) = (0, 1),$$

therefore

$$(f, g) - (h, k) = (f, g) + (-h, 2 - k) = (f - h, g - k + 1).$$

If we put $u = (1, 0)$, then $\mathcal{M} = \{(f, g) \in \mathcal{G}; (0, 1) \leq (f, g) \leq (1, 0)\} = \{(f, g) \in \mathcal{G}; 0 \leq f \leq 1, 0 \leq g \leq 1\}$ with the MV-algebra operations, i.e.

$$(f, g) \oplus (h, k) = ((f, g) + (h, k)) \wedge (1, 0) = (f + h, g + k - 1) \wedge (1, 0)$$
$$= ((f + h) \wedge 1, (g + g - 1) \vee 0) = (f \oplus h, g \odot k),$$

and similarly

$$(f, g) \odot (h, k) = (f \odot h, g \oplus k).$$

We see that $\mathcal{M}$ is an MV-algebra with respect to the Mundici theorem. $\square$

Definition 3.4. *A function* $p : \mathcal{M} \to [0,1]$ *is a state if the following conditions are satisfied:*

(i) $p((0,1)) = 0$, $p((1,0)) = 1$
(ii) $p((\mu_A, \nu_A)) + p((\mu_B, \nu_B)) = p((\mu_A, \nu_A)) \oplus (\mu_B, \nu_B)) + p((\mu_A, \nu_A) \odot (\mu_B, \nu_B))$ *for any* $(\mu_A, \nu_A), (\mu_B, \nu_B) \in \mathcal{F}$
(iii) $(\mu_{A_n}, \nu_{A_n}) \nearrow (\mu_A, \nu_A) \implies p((\mu_{A_n}, \nu_{A_n})) \nearrow p((\mu_A, \nu_A))$

Theorem 3.5. *To any state* $p : \mathcal{F} \to [0,1]$ *there exists exactly one state* $\bar{p} : \mathcal{M} \to [0,1]$ *such that* $\bar{p}|\mathcal{F} = p$.

Proof. The function $\bar{p}$ can be defined by the equality $\bar{p}((\mu_A, \nu_A)) = p((\mu_A, 0)) - p((0, 1 - \nu_A))$. First we prove that $\bar{p}$ is an *extension* of p.

Let $(\mu_A, \nu_A) \in \mathcal{F}$, i.e. $\mu_A + \nu_A \leq 1$. Then

$$(\mu_A, \nu_A) \odot (0, 1 - \nu_A) = (\mu_A \odot 0, \nu_A \oplus (1 - \nu_A))$$
$$= ((\mu_A + 0 - 1) \vee 0, (\nu_A + 1 - \nu_A) \wedge 1) = (0, 1)$$

On the other hand

$$(\mu_A, \nu_A) \oplus (0, 1 - \nu_A) = (\mu_A \oplus 0, \nu_A \odot (1 - \nu_A))$$
$$= ((\mu_A + 0) \wedge 1, (\nu_A + 1 - \nu_A - 1) \vee 0) = (\mu_A, 0).$$

Therefore by (ii)

$$p((\mu_A, \nu_A)) + p((0, 1 - \nu_A)) = p((\mu_A, 0)) + p((0, 1)) = p((\mu_A, 0)),$$

hence

$$p((\mu_A, \nu_A)) = p((\mu_A, 0)) - p((0, 1 - \nu_A)) = \bar{p}((\mu_A, \nu_A)).$$

It follows that $\bar{p}$ satisfies (i).

We are going to prove *additivity* (ii).

By the definition

$$\bar{p}((\mu_A, \nu_A)) + \bar{p}((\mu_B, \nu_B)) = p((\mu_A, 0)) - p((0, 1 - \nu_A)) + p((\mu_B, 0))$$
$$- p((0, 1 - \nu_B)). \tag{3.5.1}$$

On the other hand

$$\bar{p}((\mu_A, \nu_A) \oplus (\mu_B, \nu_B)) = \bar{p}((\mu_A \oplus \mu_B, \nu_A \odot \nu_B))$$
$$= p((\mu_A \oplus \mu_B, 0)) - p((0, 1 - \nu_A \odot \nu_B), \tag{3.5.2}$$

$$\bar{p}((\mu_A, \nu_A) \odot (\mu_B, \nu_B)) = \bar{p}((\mu_A \odot \mu_B, \nu_A \oplus \nu_B))$$
$$= p((\mu_A \odot \mu_B, 0)) - p((0, 1 - (\nu_A \oplus \nu_B))).$$

Of course,

$$p((\mu_A, 0)) + p((\mu_B, 0)) = p((\mu_A, 0) \oplus (\mu_B, 0)) + p((\mu_A, 0) \odot (\mu_B, 0))$$
$$= p((\mu_A \oplus \mu_B, 0 \odot 0)) + p((\mu_A \odot \mu_B, 0 \oplus 0))$$
$$= p((\mu_A \oplus \mu_B, 0)) + p((\mu_A \odot \mu_B, 0)),$$

hence

$$p((\mu_A, 0)) + p((\mu_B, 0)) = p((\mu_A \oplus \mu_B. 0)) + p((\mu_A \odot \mu_B, 0)). \tag{3.5.3}$$

Further

$$p((0, 1 - \nu_A)) + p((0, 1 - \nu_B))$$
$$= p((0, 1 - \nu_A) \oplus (0, 1 - \nu_B)) + p((0, 1 - \nu_A) \odot (0, 1 - \nu_B))$$
$$= p((0 \oplus 0, (1 - \nu_A) \odot (1 - \nu_B))) + p((0 \odot 0, (1 - \nu_A) \oplus (1 - \nu_B)))$$
$$= p((0, 1 - (\nu_A \oplus \nu_B))) + p((0, 1 - (\nu_A \odot \nu_B))),$$

hence

$$p((0, 1 - \nu_A)) + p((0, 1 - \nu_B)) = p((0, 1 - (\nu_A \oplus \nu_B)))$$
$$+ p((0, 1 - (\nu_A \odot \nu_B))). \qquad (3.5.4)$$

By (3.5.1), (3.5.2), (3.5.3) and (3.5.4) it follows

$$\bar{p}((\mu_A, \nu_A)) + \bar{p}((\mu_B, \nu_B)) = \bar{p}((\mu_A, \nu_A) \oplus (\mu_B, \nu_B)) + \bar{p}((\mu_A, \nu_A) \odot (\mu_B, \nu_B)).$$

The next step of our proof is proving the *continuity* of $\bar{p}$.

Let $(\mu_{A_n}, \nu_{A_n}) \nearrow (\mu_A, \nu_A)$, i.e. $\mu_{A_n} \nearrow \mu_A, \nu_{A_n} \searrow \nu_A$. Then $(\mu_{A_n}, 0) \nearrow (\mu_A, 0)$, hence

$$p((\mu_{A_n}, 0)) \nearrow p((\mu_A, 0)). \qquad (3.5.5)$$

On the other hand the relation $\nu_{A_n} \searrow \nu_A$ implies $1 - \nu_{A_n} \nearrow (0, 1 - \nu_A)$. Since $\nu_{A_n} \searrow \nu_A$, we have $(0, \nu_{A_n}) \nearrow (0, \nu_A)$, hence $p((0, \nu_{A_n})) \nearrow p((0, \nu_A))$. Of course

$$(0, \nu_B) \odot (0, 1 - \nu_B) = (0 \odot 0, \nu_B \oplus (1 - \nu_B)) = (0, 1),$$
$$(0, \nu_B) \oplus (0, 1 - \nu_B) = (0 \oplus 0, \nu_B \odot (1 - \nu_B)) = (0, 0).$$

Therefore

$$p((0, \nu_B)) + p((0, 1 - \nu_B)) = p((0, 0)) = \alpha.$$
$$p((0, 1 - \nu_{A_n})) = \alpha - p((0, \nu_{A_n})) \nearrow \alpha - p((0, \nu_A)) = p((0, 1 - \nu_A)),$$

hence

$$-p((0, 1 - \nu_{A_n})) \nearrow -p((0, 1 - \nu_A)). \qquad (3.5.6)$$

The relations (3.5.5) and (3.5.6) implies

$$\bar{p}((\mu_{A_n}, \nu_{A_n})) = p((\mu_{A_n}, 0)) - p((0, 1 - \nu_{A_n})) \nearrow p((\mu_A, 0)) - p(0, 1 - \nu_A))$$
$$= \bar{p}((\mu_A, \nu_A)).$$

and the continuity of $\bar{p}$ is proved.

Now we shall show the *uniqueness* of the extension.

Let q be any state on $\mathcal{M}$ such that $p \mid \mathcal{F} = p$. Since

$$(0, 1 - \nu_A) \odot (\mu_A, \nu_A) = (0, 1),$$
$$(0, 1 - \nu_A) \oplus (\mu_A, \nu_A) = (\mu_A, 0),$$

we obtain

$$p((\mu_A, 0)) = q((\mu_A, 0)) = q((0, 1 - \nu_A)) + q((\mu_A, \nu_A))$$
$$= p((0, 1 - \nu_A)) + q((\mu_A, \nu_A)),$$

hence $q((\mu_A, \nu_A)) = p((\mu_A, 0)) - p((0, 1 - \nu_A)) = \bar{p}((\mu_A, \nu_A))$.   □

Theorem 3.6. *To any probability* $\mathcal{P} : \mathcal{F} \to \mathcal{J}$ *there exists a probability* $\bar{\mathcal{P}} :$ $\mathcal{M} \to \mathcal{J}$ *such that* $\bar{\mathcal{P}}|\mathcal{F} = \mathcal{P}$.

Proof. Put $\mathcal{P}(A) = [\mathcal{P}^\flat(A), \mathcal{P}^\sharp(A)]$. By Theorem 2.7 $\mathcal{P}^\flat$, $\mathcal{P}^\sharp$ are states. By Theorem 3.5 there exist states $\bar{\mathcal{P}}^\flat, \bar{\mathcal{P}}^\sharp : \mathcal{M} \to [0, 1]$ such that $\bar{\mathcal{P}}^\flat|\mathcal{F} = \mathcal{P}^\flat$, $\bar{\mathcal{P}}^\sharp|\mathcal{F} = \mathcal{P}^\sharp$. For $A = (\mu_A, \nu_A) \in \mathcal{M}$ put

$$\bar{\mathcal{P}}(A) = [\bar{\mathcal{P}}^\flat(A), \bar{\mathcal{P}}^\sharp(A)].$$

If $A \in \mathcal{F}$, then

$$\bar{\mathcal{P}}(A) = [\bar{\mathcal{P}}^\flat(A), \bar{\mathcal{P}}^\sharp(A)] = [\mathcal{P}^\flat(A), \mathcal{P}^\sharp(A)] = \mathcal{P}(A),$$

hence $\bar{\mathcal{P}}|\mathcal{F} = \mathcal{P}$. If $A, B \in \mathcal{M}$, then

$$\bar{\mathcal{P}}^\flat(A \oplus B) + \bar{\mathcal{P}}^\flat(A \odot B) = \bar{\mathcal{P}}^\flat(A) + \bar{\mathcal{P}}^\flat(B),$$
$$\bar{\mathcal{P}}^\sharp(A \oplus b) + \bar{\mathcal{P}}^\sharp(A \odot B) = \bar{\mathcal{P}}^\sharp(A) + \bar{\mathcal{P}}^\sharp(B),$$

hence

$$\bar{\mathcal{P}}(A \oplus B) + \bar{\mathcal{P}}(A \odot B) = [\bar{\mathcal{P}}^\flat(A \oplus B), \bar{\mathcal{P}}^\sharp(A \oplus B)] + [\bar{\mathcal{P}}^\flat(A \odot B), \bar{\mathcal{P}}^\sharp(A \odot B)]$$
$$= [\bar{\mathcal{P}}^\flat(A) + \bar{\mathcal{P}}^\flat(B), \bar{\mathcal{P}}^\sharp(A) + \bar{\mathcal{P}}^\sharp(B)]$$
$$= [\bar{\mathcal{P}}^\flat(A), \bar{\mathcal{P}}^\sharp(A)] + [\bar{\mathcal{P}}^\flat(B), \bar{\mathcal{P}}^\sharp(B)] = \bar{\mathcal{P}}(A) + \bar{\mathcal{P}}(B).$$

Finally, let $A_n \nearrow A$. Then $\bar{\mathcal{P}}^\flat(A_n) \nearrow \bar{\mathcal{P}}^\flat(A), \bar{\mathcal{P}}^\sharp(A_n) \nearrow \bar{\mathcal{P}}^\sharp(A)$, hence

$$\bar{\mathcal{P}}(A_n) = [\bar{\mathcal{P}}^\flat(A_n), \bar{\mathcal{P}}^\sharp(A_n)] \nearrow [\bar{\mathcal{P}}^\flat(A), \bar{\mathcal{P}}^\sharp(A)] = \bar{\mathcal{P}}(A).$$

□

4 Weak Probability

Definition 4.1. *The weak probability* $\mathcal{P} : \mathcal{F} \to \mathcal{J}$ *is defined here by the following axioms:*

(i) $\mathcal{P}((1, 0)) = [1, 1], \mathcal{P}((0, 1)) = [0, 0]$
(ii) if $A \odot B = (0, 1)$, *then* $\mathcal{P}(A \oplus B) = \mathcal{P}(A) + \mathcal{P}(B)$
(iii) if $A_n \nearrow A$, *then* $\mathcal{P}(A_n) \nearrow \mathcal{P}(A)$

The weak probability can be represent in the form $\mathcal{P} : \mathcal{F} \to \mathcal{J}$

$$\mathcal{P}(A) = \left[f\left(\int \mu_A \, dP, \int \nu_A \, dP \right), \ g\left(\int \mu_A \, dP, \int \nu_A \, dP \right) \right].$$

The main result is obtained in the following theorem.

Theorem 4.2. *To any weak probability $\mathcal{P} : \mathcal{F} \to \mathcal{J}$ there exist $\alpha, \beta \in [0, 1]$, $\alpha \leq \beta$ such that*

$$\mathcal{P}((\mu_A, \nu_A))$$
$$= \left[(1 - \alpha) \int \mu_A dP + \alpha \left(1 - \int \nu_A dP \right), (1 - \beta) \int \mu_A dP + \beta \left(1 - \int \nu_A dP \right) \right].$$

Proof. See [16]. □

Denote as before $\mathcal{P}(A) = [\mathcal{P}^\flat(A), \mathcal{P}^\sharp(A)]$. As a consequence of Theorem 4.2 one can obtain that any weak probability is probability.

Theorem 4.3. *Any weak probability (in the sense of Definition 4.1) is a probability (in the sense of Definition 2.3).*

Proof. Consider $(\mu_A, \nu_A) \in \mathcal{F}, (\mu_B, \nu_B) \in \mathcal{F}$. Then there exists such $\alpha \in [0, 1]$ that

$$\mathcal{P}^\flat(\mu_A, \nu_A) = (1 - \alpha) \int_\Omega \mu_A \, dP + \alpha \left(1 - \int_\Omega \nu_A \, dP \right)$$

$$\mathcal{P}^\flat(\mu_B, \nu_B) = (1 - \alpha) \int_\Omega \mu_B \, dP + \alpha \left(1 - \int_\Omega \nu_B \, dP \right)$$

$$\mathcal{P}^\flat(\mu_{A \oplus B}, \nu_A) = (1 - \alpha) \int_\Omega \mu_{A \oplus B} \, dP + \alpha \left(1 - \int_\Omega \nu_{A \oplus B} \, dP \right)$$

$$\mathcal{P}^\flat(\mu_{A \odot B}, \nu_A) = (1 - \alpha) \int_\Omega \mu_{A \odot B} \, dP + \alpha \left(1 - \int_\Omega \nu_{A \odot B} \, dP \right)$$

Since

$$\mu_A + \mu_B = \mu_{A \oplus B} + \mu_{A \odot B}$$
$$\nu_A + \nu_B = \nu_{A \oplus B} + \nu_{A \odot B}$$

we obtain

$$\mathcal{P}^\flat(\mu_A, \nu_A) + \mathcal{P}^\flat(\mu_B, \nu_B) = \mathcal{P}^\flat(\mu_{A \oplus B}, \nu_{A \oplus B}) + \mathcal{P}^\flat(\mu_{A \odot B}, \nu_{A \odot B}).$$

Similarly

$$\mathcal{P}^\sharp(\mu_A, \nu_A) + \mathcal{P}^\sharp(\mu_B, \nu_B) = \mathcal{P}^\sharp(\mu_{A \oplus B}, \nu_{A \oplus B}) + \mathcal{P}^\sharp(\mu_{A \odot B}, \nu_{A \odot B}).$$

Therefore

$$\mathcal{P}(\mu_A, \nu_A) + \mathcal{P}(\mu_B, \nu_B) = \mathcal{P}(\mu_{A \oplus B}, \nu_{A \oplus B}) + \mathcal{P}(\mu_{A \odot B}, \nu_{A \odot B})$$

We have proved that $\mathcal{P}$ is additive in the strong sense. □

Recently M. Krachounov introduced the notion of the probability on $\mathcal{F}$ by the following way (here $f \vee g = \max(f, g)$, $f \wedge g = \min(f, g)$):

Definition 4.4. *A mapping $p : \mathcal{F} \to [0, 1]$ is an M-state if the following properties are satisfied:*

(i) $p((0, 1)) = 0, p((1, 0)) = 1$;

(ii) $p((\mu_A, \nu_A)) + p((\mu_B, \nu_B)) = p((\mu_A \vee \mu_B, \nu_A \wedge \nu_B)) + p((\mu_A \wedge \mu_B, \nu_A \vee \nu_B))$.

(iii) If $(\mu_{A_n}, \nu_{A_n}) \searrow (\mu_A, \nu_A)$ then $p((\mu_{A_n}, \nu_{A_n})) \searrow p((\mu_A, \nu_A))$.

Theorem 4.5. *Any state is an M-state. There exists M-state that is not a state.*

Proof. Again by Theorem 4.2 there exists $\alpha \in [0, 1]$ and a probability P such that

$$p((\mu_A, \nu_A)) = (1 - \alpha) \int_\Omega \mu_A \, dP + \alpha \left(1 - \int_\Omega \nu_A \, dP \right)$$

$$p((\mu_B, \nu_B)) = (1 - \alpha) \int_\Omega \mu_B \, dP + \alpha \left(1 - \int_\Omega \nu_B \, dP \right)$$

$$p((\mu_{A \vee B}, \nu_{A \vee B})) = (1 - \alpha) \int_\Omega \mu_{A \vee B} \, dP + \alpha \left(1 - \int_\Omega \nu_{A \vee B} \, dP \right)$$

$$p((\mu_{A \wedge B} A, \nu_A)) = (1 - \alpha) \int_\Omega \mu_{A \wedge B} \, dP + \alpha \left(1 - \int_\Omega \nu_{A \wedge B} \, dP \right)$$

Since

$$\mu_A + \mu_B = \mu_{A \vee B} + \mu_{A \wedge B},$$
$$\nu_A + \nu_B = \nu_{A \vee B} + \nu_{A \wedge B},$$

we see that

$$p((\mu_A, \nu_A)) + p((\mu_B, \nu_B)) = p((\mu_{A \vee B}, \nu_{A \vee B})) + p((\mu_{A \wedge B}, \nu_{A \wedge B})).$$

Wa have seen that any state is an M-state. Now we shall present an example of an M-state that is not a state. Fix $x_0 \in \Omega$ and put

$$m(A) = \frac{1}{2}(\mu_A^2(x_0) + 1 - \nu_A^2(x_0)).$$

Since

$$(\mu_A \vee \mu_B)^2 + (\mu_A \wedge \mu_B)^2 = \mu_A^2 + \mu_B^2,$$
$$(\nu_A \vee \nu_B)^2 + (\nu_A \wedge \nu_B)^2 = \nu_A^2 + \nu_B^2,$$

it is not difficult to see that m is an M-state. Put

$$\mu_A(x) = \mu_B(x) = \frac{1}{4}, \quad \nu_A(x) = \nu_B(x) = \frac{1}{2}$$

for any $x \in \Omega$. Then $m(A) = m(B) = \frac{13}{32}$. On the other hand

$$A \oplus B = \left(\left(\frac{1}{2} \right)_\Omega, 0_\Omega \right), \quad A \odot B = (0_\Omega, 1_\Omega),$$

hence

$$m(A \oplus B) + m(A \odot B) = \frac{5}{8} + 0 \neq \frac{13}{32} + \frac{13}{32} = m(A) + m(B).$$

$\square$

5 Observables

As we have mentioned yet, the Kolmogorov probability theory has two important notions: the first one is probability, the second one the notion of a random variable. Since a random variable ξ is a measurable mapping $\xi : \Omega \to R$, to any random variable a mapping $x : \mathcal{B}(R) \to \mathcal{S}$ can be defined by the formula $x(A) = \xi^{-1}(A)$.

Such a homomorphism is called in quantum structures an observable. We shall use the terminology also in our IF probability theory.

Definition 5.1. *An IF-observable is a mapping $x : \mathcal{B}(R) \to \mathcal{F}$ satisfying the following conditions:*

(i) $x(R) = (1_\Omega, 0_\Omega)$, $x(\emptyset) = (0_\Omega, 1_\Omega)$
(ii) $A \cap B = \emptyset \implies x(A) \odot x(B) = (0_\Omega, 1_\Omega)$, $x(A \cup B) = x(A) \oplus x(B)$;
(iii) $A_n \nearrow A \implies x(A_n) \nearrow x(A)$.

Since $\mathcal{F} \subset \mathcal{M}$, any observable $x : \mathcal{B}(R) \to \mathcal{F}$ is an observable in the sense of the MV-algebra probability theory [19, 20].

Proposition 5.2. *If $\mathcal{P} = (\mathcal{P}^\flat, \mathcal{P}^\sharp) : \mathcal{F} \to \mathcal{J}$ is an IF-probability, and x is an IF-observable, then the mappings $\mathcal{P}^\flat \circ x, \mathcal{P}^\sharp \circ x : \mathcal{B}(R) \to [0, 1]$ are probability measures.*

Proof. We prove only additivity. If $A \cap B = \emptyset$, then $x(A) \odot x(B) = (0_\Omega, 1_\Omega)$, hence

$$\mathcal{P}^\flat(x(A \cup B)) = \mathcal{P}^\flat(x(A) \oplus x(B)) + \mathcal{P}^\flat(x(A) \odot x(B)) = \mathcal{P}^\flat(x(A)) + \mathcal{P}^\flat(x(B)).$$

$\square$

Definition 5.3. *The product $A.B$ of two IF-events A, B is defined by the equality*

$$A.B = (\mu_A \mu_B, 1 - (1 - \nu_A)(1 - \nu_B)).$$

Definition 5.4. *If x, y are IF-observables, then their joint IF-observable is a mapping $h : \mathcal{B}(R^2) \to \mathcal{F}$ satisfying the following conditions:*

(i) $h(R^2) = (1_\Omega, 0_\Omega)$, $h(\emptyset) = (0_\Omega, 1_\Omega)$

(ii) $A \cap B = \emptyset \implies h(A) \wedge h(B) = (0_\Omega, 1_\Omega)$, $\quad h(A \cup B) = h(A) + h(B)$,
$A, B \in \mathcal{B}(R^2)$

(iii) $A_n \nearrow A \implies h(A_n \nearrow h(A))$

(iv) $h(C \times D) = x(C).y(D)$, $\quad C, D \in \mathcal{B}(R)$

A motivation for the notion of joint observable is the random vector $T = (\xi, \eta) : \Omega \to R^2$. Any random vector $T : \Omega \to R^2$ induces a homomorphism $h : A \mapsto T^{-1}(A)$, $\mathcal{B}(R^2) \to \mathcal{S}$ such that $T^{-1}(C \times D) = \xi^{-1}(C) \cap \eta^{-1}(D)$.

Theorem 5.5. *For any IF-observables* $x, y : \mathcal{B}(R) \to \mathcal{F}$ *there exists their joint IF-observable.*

Proof. Put $x(A) = (x^\flat(A), 1 - x^\sharp(A))$, $y(B) = (y^\flat(B), 1 - y^\sharp(B))$. We want to construct $h(C) = (h^\flat(C), 1 - h^\sharp(C))$. Fix $\omega \in \Omega$ and put

$$\mu(A) = x^\flat(A)(\omega), \quad \nu(B) = y^\flat(B)(\omega).$$

It is not difficult to prove that $\mu, \nu : \mathcal{B}(R) \to [0, 1]$ are probability measures. Let

$$\mu \times \nu : \mathcal{B}(R^2) \to [0, 1]$$

be the product of measures and define

$$h^\flat(A)(\omega) = \mu \times \nu(A).$$

Then $h^\flat : \mathcal{B}(R^2) \to \mathcal{T}$, where $\mathcal{T}$ is the family of all $\mathcal{S}$-measurable functions from Ω to $[0,1]$. If $C, D \in \mathcal{B}(R)$, then

$$h^\flat(C \times D)(\omega) = \mu \times \nu(C \times D) = \mu(C).\mu(D) = x^\flat(C)(\omega).y^\flat(D)(\omega),$$

hence

$$h^\flat(C \times D) = x^\flat(C).y^\flat(D).$$

Similarly $h^\sharp : \mathcal{B}(R^2) \to \mathcal{T}$ can be constructed such that

$$h^\sharp(C \times D) = x^\sharp(C).y^\sharp(D).$$

Put

$$h(A) = (h^\flat(A), 1 - h^\sharp(A)), \quad A \in \mathcal{B}(R^2).$$

By Definition 5.4 we have for $C, D \in \mathcal{B}(R)$

$$
\begin{aligned}
x(C).y(D) &= (x^\flat(C), 1 - x^\sharp(C)).(y^\flat(D), 1 - y^\sharp(D)) \\
&= (x^\flat(C).y^\flat(D), 1 - (1 - (1 - x^\sharp(C))).(1 - (1 - y^\sharp(D)))) \\
&= (x^\flat(C).y^\flat(D), 1 - x^\sharp(C).y^\sharp(D)) \\
&= (h^\flat(C \times D), 1 - h^\sharp(C \times D)) = h(C \times D).
\end{aligned}
$$

$\square$

Definition 5.6. *For any probability* $\mathcal{P} = [\mathcal{P}^\flat, \mathcal{P}^\sharp] : \mathcal{F} \to \mathcal{J}$ *and any IF-observable* $x : \mathcal{B}(R) \to \mathcal{F}$ *we define*

$$E_\flat(x) = \int_R t\, d\mathcal{P}_x^\flat(t), \qquad\qquad E_\sharp(x) = \int_R t\, d\mathcal{P}_x^\sharp(t),$$

$$\sigma_\flat^2 = \int_R (t - E_\flat(x))^2 d\mathcal{P}_x^\flat(t), \qquad \sigma_\sharp^2(x) = \int_R (t - E_\sharp(x))^2 d\mathcal{P}_x^\sharp(t)$$

assuming that these integrals exist.

Definition 5.7. *Let* $g_n : R^n \to R$ *be a Borel function,* $x_1, \ldots, x_n : \mathcal{B}(R) \to \mathcal{F}$ *be IF-observables,* $h_n : \mathcal{B}(R^n) \to \mathcal{F}$ *their joint IF-observable. Then we define the IF-observable* $y_n = g_n(x_1, \ldots, x_n) : \mathcal{B}(R) \to \mathcal{F}$ *by the prescription*

$$y_n(A) = h_n(g_n^{-1}(A)).$$

Again here a motivation is a function of random variables. Let $T = (\xi, \eta)$ be a random vector. Then $\zeta = g(\xi, \eta) = g \circ T$, hence $\zeta^{-1}(A) = (g \circ T)^{-1}(A) = T^{-1}(g^{-1}(A)) = h \circ g^{-1}(A)$, where $h = T^{-1} : \mathcal{B}(R^2) \to \mathcal{S}$ is the joint observable corresponding to the random vector T.

Definition 5.8. *A sequence* (x_n) *of IF-observables is independent if for any* n

$$\mathcal{P}^\flat(h_n(A_1 \times \cdots \times A_n)) = \mathcal{P}_{x_1}^\flat(A_1). \ldots . \mathcal{P}_{x_n}^\flat(A_n),$$

$$\mathcal{P}^\sharp(h_n(A_1 \times \cdots \times A_n)) = \mathcal{P}_{x_1}^\sharp(A_1). \ldots . \mathcal{P}_{x_n}^\sharp(A_n),$$

where $h_n : \mathcal{B}(R^n) \to \mathcal{F}$ *is the joint observable of* $x_1, \ldots, x_n$.

6 Central Limit Theorem

Theorem 6.1. *Let* (x_n) *be a sequence of independent equally distributed, square integrable IF-observables,*

$$E_\flat(x_n) = E_\sharp(x_n) = a, \quad \sigma_\flat^2(x_n) = \sigma_\sharp^2(x_n) = \sigma^2 \qquad (n = 1, 2, \ldots)$$

Then for any $t \in R$ *there holds*

$$\lim_{n \to \infty} \mathcal{P}^\flat \left(\frac{x_1 + \cdots + x_n - na}{\sigma \sqrt{n}} ((-\infty, t)) \right) = \frac{1}{\sqrt{2\pi}} \int_{-\infty}^t e^{-\frac{u^2}{2}} \, du$$

$$\lim_{n \to \infty} \mathcal{P}^\sharp \left(\frac{x_1 + \cdots + x_n - na}{\sigma \sqrt{n}} ((-\infty, t)) \right) = \frac{1}{\sqrt{2\pi}} \int_{-\infty}^t e^{-\frac{u^2}{2}} \, du$$

Proof. We have seen that $\mathcal{F} \subset \mathcal{M}$ where $\mathcal{M}$ is an MV-algebra and there exists states $\overline{\mathcal{P}}^\flat, \overline{\mathcal{P}}^\sharp : \mathcal{M} \to [0,1]$ such that $\overline{\mathcal{P}}^\flat | \mathcal{F} = \mathcal{P}^\flat$, $\overline{\mathcal{P}}^\sharp | \mathcal{F} = \mathcal{P}^\sharp$.

Moreover, x_n are IF-observables, $x_n : \mathcal{B}(R) \to \mathcal{F} \subset \mathcal{M}$, hence also observables in the sense of MV-algebra probability theory.

Therefore by Theorem 2.12 of [19] (see also Theorem 9.2.6. in [20]) and Theorem 5.5 we have

$$\lim_{n\to\infty} \overline{\mathcal{P}}^{\flat}\left(\frac{x_1 + \cdots + x_n - na}{\sigma\sqrt{n}}((-\infty, t))\right) = \frac{1}{\sqrt{2\pi}} \int_{-\infty}^{t} e^{-\frac{u^2}{2}}\, du$$

and the analogous assertion holds for the mapping $\overline{\mathcal{P}}^{\sharp}$.   □

7 Weak Law of Large Numbers

Theorem 7.1. *Let* (x_n) *be a sequence of independent equally distributed IF-observables, (i.e.* $\mathcal{P}^{\flat} \circ x_1 = \mathcal{P}^{\flat} \circ x_n$, $\mathcal{P}^{\sharp} \circ x_1 = \mathcal{P}^{\sharp} x_n$ *for any* n), $E_{\flat}(x_n) = E_{\sharp}(x_n) = a$, $(n = 1, 2, \dots)$ *Then for any* $\varepsilon > 0$ *there holds*

$$\lim_{n\to\infty} \mathcal{P}^{\flat}\left(\left(\frac{x_1 + \cdots + x_n}{n} - a\right)((-\varepsilon, \varepsilon))\right) = 1$$

$$\lim_{n\to\infty} \mathcal{P}^{\sharp}\left(\left(\frac{x_1 + \cdots + x_n}{n} - a\right)((-\varepsilon, \varepsilon))\right) = 1$$

Proof. We have seen that $\mathcal{F} \subset \mathcal{M}$ where $\mathcal{M}$ is an MV-algebra and there exists states $\overline{\mathcal{P}}^{\flat}, \overline{\mathcal{P}}^{\sharp} : \mathcal{M} \to [0,1]$ such that $\overline{\mathcal{P}}^{\flat}|\mathcal{F} = \mathcal{P}^{\flat}$, $\overline{\mathcal{P}}^{\sharp}|\mathcal{F} = \mathcal{P}^{\sharp}$.

Moreover, x_n are IF-observables, $x_n : \mathcal{B}(R) \to \mathcal{F} \subset \mathcal{M}$, hence also observables in the sense of MV-algebra probability theory. Therefore by Theorem 2.15 of [19] and Theorem 5.5 we have

$$\lim_{n\to\infty} \overline{\mathcal{P}}^{\flat}\left(\left(\frac{x_1 + \cdots + x_n}{n} - a\right)((-\varepsilon, \varepsilon))\right) = 1$$

and the analogous assertion holds for the mapping $\overline{\mathcal{P}}^{\sharp}$.   □

8 Strong Law of Large Numbers

The strong law of large numbers is concerned with the P-almost everywhere convergence. If (η_n) is a sequence of random variable on a probability space $\Omega, \mathcal{S}, P)$, then (η_n) converges to 0 P-almost everywhere, if

$$\lim_{p\to\infty} \lim_{k\to\infty} \lim_{i\to\infty} P\left(\bigcap_{n=k}^{k+i} \eta_n^{-1}\left(\left(\frac{-1}{p}, \frac{1}{p}\right)\right)\right) = P\left(\bigcap_{p=1}^{\infty} \bigcup_{k=1}^{\infty} \bigcap_{n=k}^{\infty} \eta_n^{-1}\left(\left(\frac{-1}{p}, \frac{1}{p}\right)\right)\right) = 1.$$

This fact leads to the following definition:

Definition 8.1. *A sequence* y_n *of observables converges m-almost everywhere to* 0 *if and only if*

$$\lim_{p\to\infty} \lim_{k\to\infty} \lim_{i\to\infty} P\left(\bigcap_{n=k}^{k+i} y_n^{-1}\left(\left(\frac{-1}{p}, \frac{1}{p}\right)\right)\right) = P\left(\bigcap_{p=1}^{\infty} \bigcup_{k=1}^{\infty} \bigcap_{n=k}^{\infty} y_n^{-1}\left(\left(\frac{-1}{p}, \frac{1}{p}\right)\right)\right) = 1.$$

Theorem 8.2. *Let (x_n) be a sequence of independent, square integrable IF-observables, such that $\sum_{n=1}^{\infty} \sigma^2(x_n)/n^2 < \infty$. Then the sequence*

$$\left(\frac{x_1 - E(x_1) + \cdots + x_n - E(x_n)}{n}\right)_{n=1}^{\infty}$$

converges to 0 $\mathcal{P}^b$-almost everywhere as well as $\mathcal{P}^\sharp$-almost everywhere.

Proof. We have seen that $\mathcal{F} \subset \mathcal{M}$ where $\mathcal{M}$ is an MV-algebra and there exists states $\overline{\mathcal{P}}^b, \overline{\mathcal{P}}^\sharp : \mathcal{M} \to [0,1]$ such that $\overline{\mathcal{P}}^b | \mathcal{F} = \mathcal{P}^b$, $\overline{\mathcal{P}}^\sharp | \mathcal{F} = \mathcal{P}^\sharp$.

Moreover, x_n are IF-observables, $x_n : \mathcal{B}(R) \to \mathcal{F} \subset \mathcal{M}$, hence also observables in the sense of MV-algebra probability theory.

Therefore by Theorem 2.16 of [19] (see also Theorem 9.3.4 in [20]) and Theorem 5.5 the sequence

$$\left(\frac{x_1 - E(x_1) + \cdots + x_n - E(x_n)}{n}\right)_{n=1}^{\infty}$$

converges to 0 $\mathcal{P}^b$-almost everywhere as well as $\mathcal{P}^\sharp$-almost everywhere.   □

9 Conditional Probability

In the Kolmogorov theory, given a probability space $(\Omega, \mathcal{S}, P)$, by the conditional expectation of two random variables ξ and η we understand a Borel function $E(\xi|\eta) : R \to R$ such that, for all $B \in \mathcal{B}(R)$,

$$\int_B E(\xi|\eta) \, dP_\eta = \int_{\eta^{-1}(B)} \xi \, dP.$$

This construction has been adapted for MV-algebras with product. This notion has been introduced independently by F. Montagna [9] and the author [11]. One of possible definitions is the following.

Definition 9.1. *A product in an MV-algebra M is an associative and commutative binary operation $*$ satisfying the following conditions, for all $a, b, c, a_n, b_n \in M$:*

(i) $1 * a = a$.
(ii) If $a \odot b = 0$, then $(c * a) \odot (c * b) = 0$ and $c * (a \oplus b) = (c * a) \oplus (c * b)$.
(iii) If $a_n \searrow 0$ and $b_n \searrow 0$, then $a_n * b_n \searrow 0$.

An MV-algebra with product is a pair $(M, *)$, where M is an MV-algebra, and $*$ is a product.

Theorem 9.2. *Define on the family $\mathcal{M}$ the binary operation $*$ by the formula*

$$A * B = (\mu_A, \nu_A) * (\mu_B, \nu_B) = (\mu_A \mu_B, 1 - (1 - \nu_A)(1 - \nu_B))$$
$$= (\mu_A \mu_B, \nu_A + \nu_B - \nu_A \nu_B).$$

Then $*$ is a product on $\mathcal{M}$.

Proof. First we prove (i). Indeed, $(1,0) * (\mu_A, \nu_A) = (1.\mu_A, 0 + \nu_A - 0.\nu_A) = (\mu_A, \nu_A)$.

Let now $A \odot B = (\mu_A, \nu_A) \odot (\mu_B, \nu_B) = (0, 1)$. It means that $0 = \mu_A \odot \mu_B = (\mu_A + \mu_B - 1) \vee 0$, hence $\mu_A + \mu_B \leq 1$, and $\mu_A \oplus \mu_B = \mu_A + \mu_B$.

On the other hand $1 = \nu_A \oplus \nu_B = (\nu_A + \nu_B) \wedge 1$, hence $\nu_A + \nu_B \geq 1$, and $\nu_A \odot \nu_B = \nu_A + \nu_B - 1$. Now

$$(C * A) \odot (C * B) = ((\mu_C, \nu_C) * (\mu_A, \nu_A)) \odot ((\mu_C, \nu_C) * (\mu_B, \nu_B))$$
$$= (\mu_A \mu_C, \nu_A + \nu_C - \nu_A \nu_C) \odot (\mu_B \mu_C, \nu_B + \nu_C - \nu_B \nu_C)$$
$$= (\mu_A \mu_C \odot \mu_C \mu_B, (\nu_C + \nu_A - \nu_A \nu_C) \oplus (\nu_C + \nu_B - \nu_C \nu_B)).$$

Of course, $(\mu_A \mu_C + \mu_B \mu_C - 1) \vee 0 \leq (\mu_A + \mu_B - 1) \vee 0 = 0$, hence $\mu_A \mu_C \odot \mu_C \mu_B = 0$. On the other hand

$$(\nu_C + \nu_A - \nu_C \nu_A) \oplus (\nu_C + \nu_B - \nu_C \nu_B)$$
$$= (\nu_C + \nu_A - \nu_A \nu_C + \nu_C + \nu_B - \nu_B - \nu_C \nu_B) \wedge 1$$
$$\geq (1 + \nu_C(1 - \nu_A) + \nu_C(1 - \nu_B)) \wedge 1 \geq 1 \wedge 1 = 1.$$

Therefore $(C * A) \odot (C * B) = (0, 1)$. Count now

$$C * (A \oplus B) = (\mu_C, \nu_C) * ((\mu_A, \nu_A) \oplus (\mu_B, \nu_B))$$
$$= (\mu_C, \nu_C)(\mu_A \oplus \mu_B, \nu_A \odot \nu_B)$$
$$= (\mu_C(\mu_A \oplus \mu_B), \nu_C + \nu_A \odot \nu_B - \nu_C(\nu_A \odot \nu_B))$$
$$= (\mu_C \mu_A + \mu_C \mu_B, \nu_C + \nu_A + \nu_B - 1 - \nu_C \nu_A - \nu_C \nu_B + \nu_C).$$

On the other hand

$$(C * A) \oplus (C * B) = ((\mu_C, \nu_C) * (\mu_A, \nu_A)) \oplus ((\mu_C, \nu_C) * (\mu_B, \nu_B))$$
$$= (\mu_C \mu_A, \nu_C + \nu_A - \nu_C \nu_A) \oplus (\mu_C \mu_B, \nu_C + \nu_B - \nu_C \nu_B)$$
$$= (\mu_C \mu_A + \mu_C \mu_B, \nu_C + \nu_A - \nu_C \nu_A + \nu_C + \nu_B - \nu_C \nu_B - 1).$$

We see that $C * (A \oplus B) = (A * C) \oplus (A * C)$.

Finally let

$$A_n = (\mu_{A_n}, \nu_{A_n}) \searrow (0, 1), \qquad B_n = (\mu_{B_n}, \mu_{B_n}) \searrow (0, 1)$$

hence $\mu_{A_n} \searrow 0$, $\mu_{B_n} \searrow 0$, $\nu_{A_n} \nearrow 1$, $\nu_{B_n} \nearrow 1$. First $\mu_{A_n * B_n} = \mu_{A_n} \mu_{B_n} \searrow 0$. On the other hand $1 - \nu_{A_n} \searrow 0, 1 - \nu_{B_n} \searrow 0$. Therefore

$$\nu_{A_n * B_n} = \nu_{A_n} + \nu_{B_n} - \nu_{A_n} \nu_{B_n} = 1 - (1 - \nu_{A_n})(1 - \nu_{B_n}) \nearrow 1.$$

Therefore

$$A_n * B_n = (\mu_{A_n} \mu_{B_n}, \nu_{A_n} + \nu_{B_n} - \nu_{A_n} \nu_{B_n}) \searrow (0, 1).$$

We have proved that $\mathcal{M}$ is an MV-algebra with product. $\square$

Again we see that the results of the probability theory on MV-algebras may be used. The details published in Lendelová see [7].

10 Conclusion

In the chapter some results are presented concerning the probability on IF-events. Probably more important is the method how to obtain these results and their proofs; an embedding the family of all IF-events to an appropriate MV-algebra. This fact is important at least from two points of view.

First, MV-algebra is a very important algebraic model for many valued logic [2, 9]. It plays the same role as Boolean algebra in two-valued logic. Moreover, MV-algebras present a very simple model, because of the Mundici theorem: they can be represent by lattice ordered groups.

Secondly, there exists well developed probability theory on MV-algebras [19, 20]. It is based on a local representation of any sequence [18] of observables by a sequence of random variables in a suitable Kolmogorov probability space. Of course, the construction is quite abstract, and from the point of view of applications it is more effective to use directly the results as to study the construction.

Of course, the communication channel between MV-algebras and IF-events should continue in both directions. there are some results of MV-algebra measure theory still not used in the IF-events theory, e.g. some important investigations concerning ergodicity and entropy of dynamical systems; some experiences are contained in [14, 15].

On the other hand some problems of the probability on IF-events could be possibly solved by simpler methods in the more concrete situation. Namely in the IF-probability theory a general form is known [16, 8], Theorem 4.2).

Although the method presented in the chapter is very effective, it is not unique possible (see e.g. [6, 7] has not been published). Moreover there is another concept [5] using the Zadeh connectives instead of the Lukasiewicz ones. We have mentioned it in Theorem 4.2, in a more complete form it will published in [17].

At the end let us mention that the probability theory on IF-events cannot be reduced to the probability theory on fuzzy sets. It follows from the general form of the probability on IF-events (Theorem 4.2).

Acknowledgement

The paper was supported by Grant VEGA 1/2002/05.

References

1. Atanassov, K.: Intuitionistic Fuzzy Sets: Theory and Applications. Physica, New York 1999.
2. Cignoli, R. L. O., D'Ottaviano I. M. L., Mundici, D.: Algebraic Foundations of Many-valued Reasoning. Kluwer, Dordrecht 2000.

3. Gerstenkorn, T., Manko, J.: Probabilities of intuitionistic fuzzy events. In: Issues in Intelligent Systems: Paradigms (O. Hryniewicz et al. eds.). EXIT, Warszawa 2005, pp. 63–58.

4. Grzegorzewski, P., Mrowka, E.: Probability of intuitionistic fuzzy events. In: Soft Methods in Probability, Statistics and Data Analysis (P. Grzegorzewski et al. eds.). Physica, New York 2002, pp. 105–115.

5. Krachounov, M.: Intuitionostic probability and intuitionistic fuzzy sets. In: First Intern. Workshop on IFS, Generalized Nets and Knowledge Engineering (E. El-Darzi, K. Atanassov, P. Chountas eds.). Univ. of Westminster, London 2006, pp. 18–24, 714–717.

6. Lendelová, K.: Measure theory on multivalued logics and its applications, Ph.D. thesis. M.Bel University Banská Bystrica 2005.

7. Lendelová, K.: Conditional probability on L-poset with product. Proc. Eleventh Int. Conf. IPMU, Paris 2006, pp. 946–951.

8. Lendelová, K., Michalíková, A., Riečan, B.: Representation of probability on triangle. Issues in Soft Computing – Decision and Operation research (O. Hryniewicz, J. Kaczprzyk, D. Kuchta eds.) Warszawa, EXIT 2005, pp. 235–242.

9. Montagna, F.: An algebraic approach to propositional fuzzy logic. J. Logic, Lang. Inf. 9, 2000, 91–124.

10. Renčová, M., Riečan, B.: Probability on IF-sets: an elementary approach. In: First Intern. Workshop on IFS, Generalized Nets and Knowledge Engineering (E. El-Darzi, K. Atanassov, P. Chountas eds.). Univ. of Westminster, London 2006, pp. 8–17.

11. Riečan, B.: On the product MV-algebras. Tatra Mt. Math. Publ. 16, 1999, 143–149.

12. Riečan, B.: A descriptive definition of the probability on intuitionistic fuzzy sets. In: Proc. EUSFLAT'2003 (Wagenecht, M. and Hampet, R. eds.), Zittau - Goerlitz Univ. Appl. Sci, Dordrecht 2003, pp. 263–266.

13. Riečan, B.: Representation of probabilities on IFS events. Advances in Soft Computing, Soft Methodology and Random Information Systems (M. Lopez - Diaz et al. eds.) Springer, Berlin Heidelberg New York 2004, pp. 243–246.

14. Riečan, B.: On the entropy on the Lukasiewicz square. Proc. Joint EUSFLAT - LFA 2005, Barcelona, pp. 330–333.

15. Riečan, B.: On the entropy of IF dynamical systems. In: Proceedings of the Fifth International workshop on IFS and Generalized Nets, Warsaw, Poland 2005, pp. 328–336.

16. Riečan, B.: On a problem of Radko Mesiar: general from of IF probabilities. Fuzzy Sets and Systems 152, 2006, 1485–1490.

17. Riečan, B.: Probability theory on IF-events. In: Algebraic and Proof-theoretic spects of Non-classical Logics. Papers in honour of daniele Mundici on the occasion of his 60th birthday. Lecture Notes in Computer Science, Vol. 4460, Springer, Berlin Heidelberg New York 2007.

18. Riečan, B.: On some contributions to quantum structures inspired by fuzzy sets. Kybernetika 2007, 43(4), pp. 481–490.

19. Riečan, B., Mundici, D.: Probability on MV-algebras. In: Handbook of measure Theory (E. Pap ed.) Elsevier, Amsterdam 2002, pp. 869–909.

20. Riečan, B., Neubrunn, T.: Integral, Measure, and Ordering. Kluwer, Dordrecht 1997.

Dilemmas with Distances Between Intuitionistic Fuzzy Sets: Straightforward Approaches May Not Work

Eulalia Szmidt and Janusz Kacprzyk

Systems Research Institute, Polish Academy of Sciences, ul. Newelska 6, 01–447 Warsaw, Poland
szmidt@ibspan.waw.pl, kacprzyk@ibspan.waw.pl

Summary. We show and justify how to calculate distances for intuitionistic fuzzy sets (A-IFSs, for short)[1]. We show a proper way of calculations not only from a mathematical point of view but also of an intuitive appeal making use of all the relevant information.

1 Introduction

The concept of a distance in the context of fuzzy sets [26], or some generalization – intuitionistic fuzzy sets, or A-IFSs for short [1, 3] is of utmost importance, for theory and applications, notably in similarity related issues in pattern recognition, classificatons, group decisions, soft consensus measures, etc.

There are well-known formulas for measuring distances between fuzzy sets using, e.g. the *Minkowski r-metrics* (e.g. the Hamming distances for $r = 1$, the Euclidean distances for $r = 2$, the dominance metric for $r = \infty$), or the Hausdorff metric.

The situation is quite different for A-IFSs for which there are two ways of measuring distances. Some researchers use two parameters only (the memberships and non-memberships) in the formulas whereas the others Szmidt and Kacprzyk [13], Deng-Feng [5], Tan and Zhang [24], Narukawa and Torra [6]) use all three parameters (the membership, non-membership and hesitation margin) characterizing A-IFSs. Both methods are correct in the *Minkowski r-metrics* as all necessary and sufficient conditions are fulfilled for distances in spite of the formulas used (with two or with three parameters). The situation

---

[1] There is currently a discussion on the appropriateness of the name *intuitionistic fuzzy set* (cf. Dubois, Gottwald, Hajek, Kacprzyk and Prade (Fuzzy Sets Syst. 156:485–491, 2005)) introduced by Atanassov. However, this is beyond the scope of this chapter which is just concerned with an application of the concept.

E. Szmidt and J. Kacprzyk: *Dilemmas with Distances Between Intuitionistic Fuzzy Sets: Straightforward Approaches May Not Work*, Studies in Computational Intelligence (SCI) **109**, 415–430 (2008)
www.springerlink.com   © Springer-Verlag Berlin Heidelberg 2008

is quite different (to the disadvantage of two-parameter formulas) in the Hausdorff metric but it will be discussed in details later (Sect. 5.2).

One could say that if both methods follow (in the *Minkowski r-metrics*) all mathematical assumptions, the problem does not exist – both methods are correct and can be used interchangeably. Unfortunately, the fact which method we use does influence the final results – the results of calculations differ not only in the values (what is obvious) but also give quite different, qualitively!, answers (Sect. 5.3).

The advocates of the method with two parameters (memberships and non-memberships) claim that it is useless to take into account the third parameter (hesitation margin) which value is always known as the result of the two used parameters. However, the fact that we know the values does not mean that we can discard them from the formulas on distances provided it gives more information and/or insight. It is particularly clear for the Hausdorff metric (two-parameter formulas do not work – Sect. 5.2). For the *Minkowski r-metrics*, from a mathematical point of view, we can calculate distances using two parameters only but we lose a lot of important information. While calculating distances for A-IFSs we should not only observe the mathematical correctness but should look for formulas that use all possible information important from the point of view of a particular application in mind. Therefore, one can argue that by neglecting the hesitation margins we make no use of an important part of the knowledge available. The hesitation margins inform us, generally speaking, about an amount of lacking information (Sect. 2). Obviously if we know a lot, a little or maybe nothing is relevant, and when this not used, we can draw improper conclusions (Sect. 5.2).

It seems that the advocates of two-parameter methods just generalize the following reasoning known for fuzzy sets:

1. For fuzzy sets one parameter (the memberships) is enough
2. The sum of membership and non-membership is "automatically" equal to one
3. The non-memberships as a direct result of memberships are redundant

So, for the A-IFSs the similar reasoning seems to be automatically adopted:

1. We know that the sum of membership, non-membership and hesitation margin is equal to one.
2. The hesitation margins are a direct result of the memberships and non-memberships.
3. Hesitation margins are redundant in the formulas.

Although the reasoning in the same, we show that there are differences when we omit the non-memberships for fuzzy sets (Sect. 3), and the hesitation margins for A-IFSs (Sect. 4). Both analytical and geometric considerations justify this claims.

The material in the chapter is organized as follows. In Sect. 2 we briefly remind the concept of A-IFSs and discuss their geometrical representations

(Sect. 2.1). In Sect. 3 we remind the distances for fuzzy sets showing why non-memberships play no role in the formulas. In Sect. 4 we discuss the Hamming and the Euclidean distances between A-IFSs using three-parameter formulas. We gives arguments (both analytical and geometrical) why the third parameters should not be excluded from the formulas. In Sect. 5 we show the decisive drawbacks and errors when one insists on calculating distances using two-parameter formulas. First, two-parameter formulas contradict the obvious facts from intuitionistic fuzzy set theory (Sect. 5.1). Second, calculating the Hausdorff distance with two-parameter formulas is incorrect in the sense of breaking mathematical rules – we show it in Sect. 5.2, and propose the proper way of calculations (using all three parameters, of course). Third, discarding hesitation margins from the formulas leads to quality differences in conclusions – it is discussed in Sect. 5.3. In Sect. 6 we finish with some conclusions.

2 Intuitionistic Fuzzy Sets

One of the possible generalizations of a fuzzy set in X [26], given by

$$A^{'} = \{<x, \mu_{A'}(x)>|x \in X\} \tag{1}$$

where $\mu_{A'}(x) \in [0, 1]$ is the membership function of the fuzzy set $A^{'}$, is an Atanassov's intuitionistic fuzzy set [1–3] A given by

$$A = \{<x, \mu_A(x), \nu_A(x)>|x \in X\} \tag{2}$$

where: $\mu_A : X \to [0, 1]$ and $\nu_A : X \to [0, 1]$ such that

$$0 \le \mu_A(x) + \nu_A(x) \le 1 \tag{3}$$

and $\mu_A(x)$, $\nu_A(x) \in [0, 1]$ denote a degree of membership and a degree of non-membership of $x \in A$, respectively.

Obviously, each fuzzy set may be represented by the following A-IFS

$$A = \{<x, \mu_{A'}(x), 1 - \mu_{A'}(x)>|x \in X\} \tag{4}$$

For each A-IFS in X, we will call

$$\pi_A(x) = 1 - \mu_A(x) - \nu_A(x) \tag{5}$$

an *intuitionistic fuzzy index* (or a *hesitation margin*) of $x \in A$ which expresses a lack of knowledge of whether x belongs to A or not (cf. [3]). $0 \le \pi_A(x) \le 1$ for each $x \in X$.

The use of A-IFSs means the introduction of another degree of freedom that gives us an additional possibility to represent imperfect knowledge and more adequately describe many real problems. Applications of A-IFSs to group decision making, negotiations and other situations are presented in Szmidt and Kacprzyk [9–12, 14–21].

2.1 A Geometrical Representation

Having in mind that for each element x belonging to an A-IFS A, the values
of membership, non-membership and the intuitionistic fuzzy index add up
to one:

$$\mu_A(x) + \nu_A(x) + \pi_A(x) = 1$$

and that each: membership, non-membership, and the intuitionistic fuzzy index are from the interval $[0,1]$, we can imagine a unit cube (Fig. 1) inside
which there is an ABD triangle where the above equation is fulfilled. In other
words, the ABD triangle represents a surface where coordinates of any element belonging to an A-IFS can be represented. Each point belonging to the
ABD triangle is described via three coordinates: (μ, ν, π). Points A and B
represent crisp elements. Point $A(1,0,0)$ represents elements fully belonging
to an A-IFS as $\mu = 1$. Point $B(0,1,0)$ represents elements fully not belonging
to an A-IFS as $\nu = 1$. Point $D(0,0,1)$ represents elements about which we are
not able to say if they belong or not belong to an A-IFS (intuitionistic fuzzy
index $\pi = 1$). Such an interpretation is intuitively appealing and provides
means for the representation of many aspects of imperfect information. Segment AB (where $\pi = 0$) represents elements belonging to a classical fuzzy set
$(\mu + \nu = 1)$. Any other combination of values characterizing an A-IFS can be
represented inside the triangle ABD. In other words, each element belonging
to an A-IFS can be represented as a point (μ, ν, π) belonging to the triangle
ABD (Fig. 1).

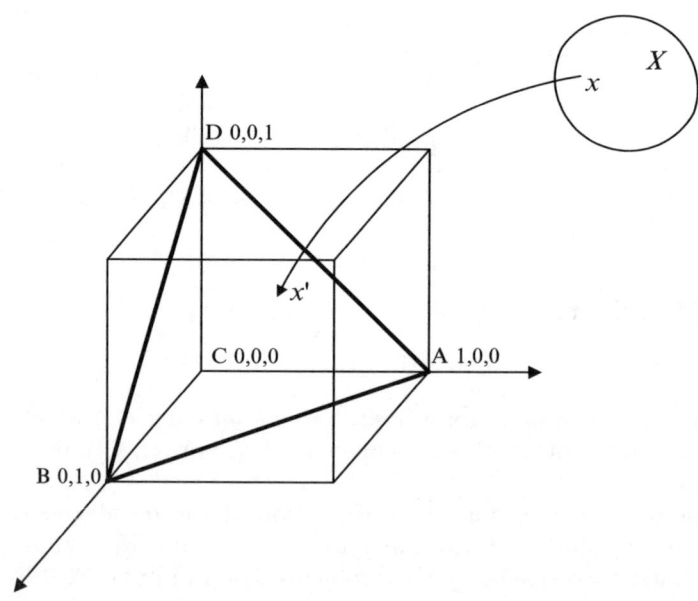

Fig. 1. Geometrical representation

It is worth mentioning that the geometrical interpretation is directly related to the definition of an A-IFS introduced by Atanassov [1,3]. By employing the above geometrical representation, we can calculate distances between any two A-IFSs A and B containing n elements. To start discussing distances for A-IFSs, we first recall distances for fuzzy sets as in some papers the formulas which are correct for fuzzy sets are automatically(!) transformed for A-IFSs leading: at least – to using only a part of information we have, at worst – to incorrect results. We will discuss the problem in details in Sect. 4.

3 Distances for Fuzzy Sets

Let us briefly recall some basic concepts concerning distances for fuzzy sets. The most widely used distances for fuzzy sets A', B' in $X = \{x_1, x_2, \ldots, x_n\}$ are:

– The Hamming distance $d'(A', B')$

$$d'(A', B) = \sum_{i=1}^{n} |\mu_{A'}(x_i) - \mu_{B'}(x_i)| \tag{6}$$

– The normalized Hamming distance $l'(A', B')$:

$$l'(A', B') = \frac{1}{n} \sum_{i=1}^{n} |\mu_{A'}(x_i) - \mu_{B'}(x_i)| \tag{7}$$

– The Euclidean distance $e'(A', B')$:

$$e'(A', B') = \left\{ \sum_{i=1}^{n} (\mu_{A'}(x_i) - \mu_{B'}(x_i))^2 \right\}^{0.5} \tag{8}$$

– The normalized Euclidean distance $q'(A', B')$:

$$q'(A', B') = \left\{ \frac{1}{n} \sum_{i=1}^{n} (\mu_{A'}(x_i) - \mu_{B'}(x_i))^2 \right\}^{0.5} \tag{9}$$

In all the above formulas (6)–(9), the membership functions are only present which is because for a fuzzy set $\mu(x_i) + \nu(x_i) = 1$.

As we can represent a fuzzy set A' in X in an equivalent intuitionistic-type representation (4), we will employ such a representation while rewriting the distances (6)–(9).

So, first, taking into account an intuitionistic-type representation of a fuzzy set, we can express the very essence of the Hamming distance by putting

$$d'(A',B') = \sum_{i=1}^{n}(|\mu_{A'}(x_i) - \mu_{B'}(x_i)| + |\nu_{A'}(x_i) - \nu_{B'}(x_i)|)$$

$$= \sum_{i=1}^{n}(|\mu_{A'}(x_i) - \mu_{B'}(x_i)| + |1 - \mu_{A'}(x_i) - 1 + \mu_{B'}(x_i)|)$$

$$= 2\sum_{i=1}^{n}|\mu_{A'}(x_i) - \mu_{B'}(x_i)| = 2d'(A',B') \tag{10}$$

i.e. it is twice as large as the Hamming distance of a fuzzy set (6).

And similarly, the normalized Hamming distance $l'(A',B')$ taking into account an intuitionistic-type representation of a fuzzy set is in turn equal to

$$l'(A',B') = \frac{1}{n} \cdot d'(A',B') = \frac{2}{n}\sum_{i=1}^{n}|\mu_{A'}(x_i) - \mu_{B'}(x_i)| \tag{11}$$

i.e. the result of (11) is two times multiplied as compared to (7).

Then, by the same line of reasoning, the Euclidean distance, taking into account an intuitionistic-type representation of a fuzzy set, is

$$e'(A',B') = \left(\sum_{i=1}^{n}(\mu_{A'}(x_i) - \mu_{B'}(x_i))^2 + (\nu_{A'}(x_i) - \nu_{B'}(x_i))^2\right)^{\frac{1}{2}}$$

$$= \left(\sum_{i=1}^{n}(\mu_{A'}(x_i) - \mu_{B'}(x_i))^2 + (1 - \mu_{A'}(x_i) - 1 + \mu_{B'}(x_i))^2\right)^{\frac{1}{2}}$$

$$= \left(2\sum_{i=1}^{n}(\mu_{A'}(x_i) - \mu_{B'}(x_i))^2\right)^{\frac{1}{2}} \tag{12}$$

i.e. it is only multiplied by $\sqrt{2}$ as compared to the Euclidean distance for the usual representation of fuzzy sets given by (8). Multiplication by a constant value does not bring any additional information connected to non-memberships.

The normalized Euclidean distance $q'(A',B')$ taking into account an intuitionistic-type representation of a fuzzy set is then

$$q'(A',B') = \sqrt{\frac{1}{n} \cdot e'(A',B')} = \sqrt{\frac{2}{n}\sum_{i=1}^{n}(\mu_{A'}(x_i) - \mu_{B'}(x_i))^2} \tag{13}$$

and again the result of (13) is multiplied by $\sqrt{2}$ as compared to (9).

The above results confirm the very well-known fact that for fuzzy sets taking into account the memberships only is enough while calculating distances.

An additional use of the non-membership into the formulas is redundant as the results obtained are only multiplied by a constant.

One could extend the conclusions (and unfortunately, it is done not so rarely) to A-IFSs – asserting that as for A-IFSs we have $\mu_A(x) + \nu_A(x) + \pi_A(x) = 1$, it means that just two of the three parameters are quite enough to calculate distances because when knowing two parameters we immediately know the third as well. Of course, we do know the third parameter but it need not imply that it should be neglected in the formulas for calculating distances for A-IFSs We will show a justification in Sect. 4.

4 Distances Between A-IFSs

Distances between A-IFSs are calculated in the literature in two ways, using two parameters only or all three parameters describing elements belonging to the sets. Both ways are proper from the point of view of pure mathematical conditions concerning distances (all properties are fulfilled in both cases). Unfortunately one cannot say that both ways are equal when assessing the results obtained by the two approaches. Now we will present arguments why in our opinion all three parameters should be used in the respective formulas, and what additional qualities their inclusion can give.

In Szmidt and Kacprzyk [13], Szmidt and Baldwin [7,8], it is shown why in the calculation of distances between A-IFSs one should use all three parameters describing A-IFSs The considerations in [7,8,13] are illustrated via a geometrical representation of A-IFSs (Sect. 2.1).

Employing the above geometrical representation, we can calculate distances between any two A-IFSs A and B in $X = \{x_1, x_2, \ldots, x_n\}$ [7,8,13], e.g.

- The normalized Hamming distance:

$$l_{IFS}(A, B) = \frac{1}{2n} \sum_{i=1}^{n} (|\mu_A(x_i) - \mu_B(x_i)|$$
$$+ |\nu_A(x_i) - \nu_B(x_i)| + |\pi_A(x_i) - \pi_B(x_i)|) \qquad (14)$$

- The normalized Euclidean distance:

$$e_{IFS}(A, B) = \left(\frac{1}{2n} \sum_{i=1}^{n} (\mu_A(x_i) - \mu_B(x_i))^2 \right.$$
$$\left. + (\nu_A(x_i) - \nu_B(x_i))^2 + (\pi_A(x_i) - \pi_B(x_i))^2 \right)^{\frac{1}{2}} \qquad (15)$$

Both distances are from the interval $[0, 1]$.

It is easy to notice why all three parameters should be used when calculating distances. As the geometrical representation shows (Fig. 1), each side of the considered triangle is of the same length, i.e. $AB = BD = AD$. But while using two parameters only, i.e.

$$l_2(A, B) = \frac{1}{2n} \sum_{i=1}^{n} (|\mu_A(x_i) - \mu_B(x_i)| + |\nu_A(x_i) - \nu_B(x_i)|) \tag{16}$$

we obtain

$$l_2(A, B) = 0.5(|1 - 0| + |0 - 1|) = 1 \tag{17}$$

$$l_2(A, D) = 0.5(|1 - 0| + |0 - 0|) = 0.5 \tag{18}$$

$$l_2(B, D) = 0.5(|0 - 0| + |1 - 0|) = 0.5 \tag{19}$$

so that

$$l_2(A, B) \neq l_2(A, D) \text{ and } l_2(A, B) \neq l_2(B, D) \tag{20}$$

i.e. the use of formula (16) means the use of two different scales: one for measuring distances for fuzzy sets (segment AB), and the other one for "pure" A-IFSs (for which the hesitation margin is greater than zero – the whole area of the triangle ABD over the segment AB). Only while using (14) with all the three parameters we obtain

$$l_{IFS}(A, B) = 0.5(|1 - 0| + |0 - 1| + |0 - 0|) = 1 \tag{21}$$

$$l_{IFS}(A, D) = 0.5(|1 - 0| + |0 - 0| + |0 - 1|) = 1 \tag{22}$$

$$l_{IFS}(B, D) = 0.5(|0 - 0| + |1 - 0| + |0 - 1|) = 1 \tag{23}$$

which means that the condition $l_{IFS}(A, D) = l_{IFS}(A, B) = l_{IFS}(B, D)$ is fulfilled, i.e. the distances for fuzzy sets and A-IFSs are measured using the same scale.

In other words, when taking into account two parameters only, for elements from the classical fuzzy sets (which are a special case of A-IFSs – segment AB in Fig. 1) we obtain distances from a different interval than for elements belonging to A-IFSs. This may be a serious deficiency in practice as for the fuzzy and intuitionistic fuzzy sets two different measurement scales are used.

The same conclusions – that all three parameters should be taken into account in the formulas for distances can be drawn in an analytical way as well.

Let us verify if we can discard the values π from the formula (14). Taking into account (5) we have

$$|\pi_A(x_i) - \pi_B(x_i)| = |1 - \mu_A(x_i) - \nu_A(x_i) - 1 + \mu_B(x_i) + \nu_B(x_i)|$$
$$\leq |\mu_B(x_i) - \mu_A(x_i)| + |\nu_B(x_i) - \nu_A(x_i)| \tag{24}$$

Inequality (24) means that the third parameter in (14) should not be omitted as it was in the case of fuzzy sets for which taking into account the second parameter would only result in the multiplication by a constant value. For A-IFSs omitting the third parameter has an influence on the results.

A similar situation occurs for the Euclidean distance. Let us verify the effect of omitting the third parameter (π) in (15). Taking into account (5), we have

$$
\begin{aligned}
(\pi_A(x_i) - \pi_B(x_i))^2 &= (1 - \mu_A(x_i) - \nu_A(x_i) - 1 + \mu_B(x_i) + \nu_B(x_i))^2 \\
&= (\mu_A(x_i) - \mu_B(x_i))^2 + (\nu_A(x_i) - \nu_B(x_i))^2 \\
&\quad + 2(\mu_A(x_i) - \mu_B(x_i))(\nu_A(x_i) - \nu_B(x_i))
\end{aligned} \tag{25}
$$

what means that taking into account the third parameter π while calculating the Euclidean distance for the A-IFSs does have an influence on the final result. It was also obvious because a two-dimensional geometrical representation (being a foundation for using two parameters only) is an orthogonal projection of the three-dimension geometrical representation. For a deeper discussion of the problem of connections between geometrical representations of A-IFSs we refer an interested reader to Szmidt and Kacprzyk [13], Tasseva et al. [25], Atanassov et al. [4].

So far we have presented both analytical and geometrical arguments why the formulas with all three parameters should be used when calculating distances for the A-IFSs. Now we will show some negative effects of using two parameters only.

5 Some Negative Effects of Using Two Parameters

5.1 Distances Built on "Personal Opinions"

Some researchers claim that the calculation of distances for the A-IFSs using two parameters only – like in (16), and the obtained results (17)–(19) are proper. In their opinion the distance of the element representing full lack of knowledge $(0,0,1)$ from a crisp element, e.g. $(1,0,0)$ should be less than the distance between two crisp elements what is guaranteed by two parameter representation.

Albeit it is rather difficult to fully stick to the construction of general models on personal opinions, we agree for a (short) moment for such an explanation. But we immediately see a contradiction. On the one hand, $(0,0,1)$, due to the theory of A-IFSs, represents any (!) element as $(0,0,1)$ means that we have no knowledge about this element whatsoever, so that it could be a crisp element, any fuzzy element $(\mu, \nu, 0)$ (where: $\mu + \nu = 1$), or any intuitionistic fuzzy element (μ, ν, π) (where: $\mu + \nu + \pi = 1$). Let assume that in particular it could be a crisp element. But it can not be a crisp element (for the case of two parameters only) as the distance between $(0,0,1)$ and a crisp element is less than the distance between two crisp elements (for the Hamming distance it is equal to $1/2$). This fact seems to best subsume this procedure.

Of course, we can agree that – from a theoretical point of view – one could apply ANY formula to calculate distances (due to some mathematical rules) and rendering ANY personal convictions concerning the ratio of the distances: between the crisp elements $(1, 0, 0)$ and $(0, 1, 0)$ on the one hand, and between $(0, 0, 1)$ and a crisp element on the other hand. However, it should be stressed that a particular solution adopted in a particular work may be proper in a particular specific case but not generally.

5.2 The Hausdorff Distances

The calculation of distances using two parameters only leads to completely wrong results when a Hausdorff distance is introduced on the basis of such formulas. We will show it following the results presented by Grzegorzewski in his paper: "Distances between intuitionistic fuzzy sets and/or interval-valued fuzzy sets based on Hausdorff metric" in Fuzzy Sets and Systems, 148 (2004) 319–328.

Grzegorzewski claims in Abstract: "The proposed new distances are straightforward generalizations of the well known Hamming distance, the Euclidean distance and their normalized counterparts", (p. 324): "Our definitions are natural counterparts of the Hamming distance, the Euclidean distance and their normalized versions", and in Conclusions (p. 327) it is repeated again. He has in mind the Hamming and the Euclidean distances taking into account two parameters only, and the following:
the "Hausdorff" distance

$$H_2(A, B) = \frac{1}{2n} \sum_{i=1}^{n} max\{|\mu_A(x_i) - \mu_B(x_i)|, |\nu_A(x_i) - \nu_B(x_i)|\} \quad (26)$$

which in fact is not a Hausdorff distance.

It is important to assess properly the results obtained when using (26). The best explanations are given by Grzegorzewski in Example 2 (the paper cited, p. 324) where the Hausdorff distances for separate elements are calculated. As we know, for separate elements the Hausdorff distances reduce just to "normal" distances, i.e. Hamming or Euclidean, etc. So let us look at the results in Example 2 and let us compare them to the results obtained from (16), i.e. the Hamming distance (using two parameters only) which – according to Grzegorzewski – are proper.

In Example 2 the following A-IFSs: A, B, D, G, $E \in X = \{x\}$ are considered

$$A = \{<x, 1, 0>\}, B = \{<x, 0, 1>\}, D = \{<x, 0, 0>\},$$

$$G = \{<x, \frac{1}{2}, \frac{1}{2}>\}, E = \{<x, \frac{1}{4}, \frac{1}{4}>\} \quad (27)$$

It is easy to notice that in each of the above sets only one element is present. So in fact while calculating distances between the above sets, we calculate distances between (two) elements. Now we will verify if the proposed formula (26) and its counterpart, the Hamming distance (16), give the same results as it should be the case (by the well-known definition of the Hausdorff distance) for distances between separate elements.

Here are the results obtained from (26):

$$H_2(A, B) = \max\{|1 - 0|, |0 - 1|\} = 1$$
$$H_2(A, D) = \max\{|1 - 0|, |0 - 0|\} = 1$$
$$H_2(B, D) = \max\{|0 - 0|, |1 - 0|\} = 1$$
$$H_2(A, G) = \max\{|1 - 1/2|, |0 - 1/2|\} = 0.5$$
$$H_2(A, E) = \max\{|1 - 1/4|, |0 - 1/4|\} = 0.75$$
$$H_2(B, G) = \max\{|0 - 1/2|, |1 - 1/2|\} = 0.5$$
$$H_2(B, E) = \max\{|0 - 1/4|, |1 - 1/4|\} = 0.75$$
$$H_2(D, G) = \max\{|0 - 1/2|, |0 - 1/2|\} = 0.5$$
$$H_2(D, E) = \max\{|0 - 1/4|, |1 - 1/4|\} = 0.25$$
$$H_2(G, E) = \max\{|1/2 - 1/4|, |1/2 - 1/4|\} = 0.25$$

Their counterpart Hamming distances calculated from (16) (without the hesitation margins) are:

$$l_2(A, B) = 0.5(|1 - 0| + |0 - 1|) = 1$$
$$l_2(A, D) = 0.5(|1 - 0| + |0 - 0||) = 0.5$$
$$l_2(B, D) = 0.5(|0 - 0| + |1 - 0||) = 0.5$$
$$l_2(A, G) = 0.5(|0 - 1/2| + |0 - 1/2|) = 0.5$$
$$l_2(A, E) = 0.5(|1 - 1/4| + |0 - 1/4||) = 0.5$$
$$l_2(B, G) = 0.5(|1 - 1/4| + |0 - 1/4|) = 0.5$$
$$l_2(B, E) = 0.5(|1 - 1/4| + |0 - 1/4|) = 0.5$$
$$l_2(D, G) = 0.5(|0 - 1/2| + |0 - 1/2|) = 0.5$$
$$l_2(D, E) = 0.5(|0 - 1/4| + |0 - 1/4|) = 0.25$$
$$l_2(G, E) = 0.5(|1/2 - 1/4| + |1/2 - 1/4|) = 0.25$$

i.e. the values of the Hamming distances obtained from (16) and used for the derivation of the Hausdorff distances (26) are not consistent. The differences:

$$H_2(A, D) \neq l_2(A, D)$$
$$H_2(B, D) \neq l_2(B, D)$$
$$H_2(A, E) \neq l_2(A, E)$$
$$H_2(B, E) \neq l_2(B, E)$$

Conclusion: the way of calculating the Hausdorff distance for the A-IFSs while using two parameters only leads to wrong results – the mathematical assumptions for the Hausdorff distance are broken.

Now let us calculate the Hausdorff distance in a proper way, i.e. using the formula proposed here:

$$H_3(A, B) = \frac{1}{2n} \sum_{i=1}^{n} \max\{(|\mu_A(x_i) - \mu_B(x_i)|,$$

$$|\nu_A(x_i) - \nu_B(x_i)|, |\pi_A(x_i) - \pi_B(x_i)|)\} \qquad (28)$$

We use (28) for data from the Example 2. But now, as we also take into account the hesitation margins, instead of (27) we use the "full description" of the data:

$$A = \{<x, 1, 0, 0>\}, \quad B = \{<x, 0, 1, 0>\},$$

$$D = \{<x, 0, 0, 1>\}, \quad G = \{<x, \frac{1}{2}, \frac{1}{2}, 0>\},$$

$$E = \{<x, \frac{1}{4}, \frac{1}{4}, \frac{1}{2}>\}, \qquad (29)$$

and obtain:

$$H_3(A, B) = \max(|1 - 0|, |0 - 1|, |0 - 0|) = 1$$
$$H_3(A, D) = \max(|1 - 0|, |0 - 0|, |0 - 1|) = 1$$
$$H_3(B, D) = \max(|0 - 0|, |1 - 0|, |0 - 1|) = 1$$
$$H_3(A, G) = \max(|0 - 1/2|, |0 - 1/2|, |0 - 0|) = 1/2$$
$$H_3(A, E) = \max(|1 - 1/4|, |0 - 1/4|, |0 - 1/2|) = 3/4$$
$$H_3(B, G) = \max(|1 - 1/4|, |0 - 1/4|, |0 - 1/2|) = 3/4$$
$$H_3(B, E) = \max(|1 - 1/4|, |0 - 1/4|, |0 - 1/2|) = 3/4$$
$$H_3(D, G) = \max(|0 - 1/2|, |0 - 1/2|, |1 - 0|) = 1$$
$$H_3(D, E) = \max(|0 - 1/4|, |0 - 1/4|, |1 - 1/2|) = 1/2$$
$$H_3(G, E) = \max(|1/2 - 1/4|, |1/2 - 1/4|, |0 - 1/2|) = 1/2$$

Now we calculate the counterpart Hamming distances using (14) (with all three parameters). The results are:

$$l_{IFS}(A, B) = 0.5(|1 - 0| + |0 - 1| + |0 - 0|) = 1$$
$$l_{IFS}(A, D) = 0.5(|1 - 0| + |0 - 0| + |0 - 1|) = 1$$
$$l_{IFS}(B, D) = 0.5(|0 - 0| + |1 - 0| + |0 - 1|) = 1$$
$$l_{IFS}(A, G) = 0.5(|0 - 1/2| + |0 - 1/2| + |0 - 0|) = 0.5$$
$$l_{IFS}(A, E) = 0.5(|1 - 1/4| + |0 - 1/4| + |0 - 1/2|) = 0.75$$
$$l_{IFS}(B, G) = 0.5(|1 - 1/4| + |0 - 1/4| + |0 - 1/2|) = 0.75$$
$$l_{IFS}(B, E) = 0.5(|1 - 1/4| + |0 - 1/4| + |0 - 1/2|) = 0.75$$
$$l_{IFS}(D, G) = 0.5(|0 - 1/2| + |0 - 1/2| + |1 - 0|) = 1$$
$$l_{IFS}(D, E) = 0.5(|0 - 1/4| + |0 - 1/4| + |1 - 1/2|) = 0.5$$
$$l_{IFS}(G, E) = 0.5(|1/2 - 1/4| + |1/2 - 1/4| + |0 - 1/2|) = 0.5$$

i.e. the Hausdorff distance proposed by us (28) (using the membership, non-membership and hesitation margin) and the Hamming distance (14) give fully consistent results.

On the other hand, the Hamming distance (16) (with two parameters) does not give the results consistent with the results of the counterpart Hausdorff distance (26).

It is worth repeating again: although the values of the hesitation margins are well known as the consequence of the memberships and non-memberships, these values should not be discarded from the formulas while calculating the Hamming or the Euclidean or the Hausdorff distances.

5.3 Qualitative Differences When Coming to Conclusions

The argument stated in the end of the previous section is of utmost importance for virtually all similarity related and based approaches to pattern recognition, classifications, decision making, and a whole array of other areas. In all these cases final results do depend on which formula we use to calculate a distance. The answers can be quite different when neglecting the information about hesitation margins as shown below.

Example 1. Let $A\{<x, 0.1, 0.6, 0.3>\}$, $B\{<x, 0.4, 0.2, 0.4>\}$, and $C\{<0.3, 0.1, 0.6>\}$.

The Hamming distances among these (one-element) sets calculated while using formula (16) with two parameters only are:

$$l_2(A, B) = 0.5(|0.1 - 0.4| + |0.6 - 0.2|) = 0.35 \tag{30}$$

$$l_2(A, C) = 0.5(|0.1 - 0.3| + |0.6 - 0.1|) = 0.35 \tag{31}$$

i.e. the distance from A to B is equal to the distance from A to C.

Now let us calculate the Hamming distances using formula (14) with all the three parameters. We obtain:

$$l_{IFS}(A, B) = 0.5(|0.1 - 0.4| + |0.6 - 0.2| + |0.3 - 0.4|) = 0.4 \tag{32}$$

$$l_{IFS}(A, C) = 0.5(|0.1 - 0.3| + |0.6 - 0.1| + |0.3 - 0.6|) = 0.5 \tag{33}$$

The answer is now qualitatively different as the distance from A to B is not equal to the distance from A to C.

In fact there is nothing strange in the obtained qualitative differences. They are obvious in the light of (24), and (25) showing in an analytical way why the third parameters should not be discarded from the formulas. Neglecting a part of information available (representing by hesitation margins) may be rather strange from a point of view of, for instance, decision making. The use of two-parameter formulas means just such a voluntary resignation from an important part of information.

The situation is the same while the distances are used to construct other measures like entropy or similarity. One should have in mind that the obvious fact, namely, that the possibility to calculate a hesitation margin from (5) is quite different thing than neglecting this values (hesitation margins) in some non-linear (!) measures (e.g. similarity measures). The problem is discusses in detail in Szmidt and Kacprzyk [22, 23] as far as some similarity measures and entropy measures are concerned, respectively.

6 Conclusions

We discussed two ways of calculating distances between A-IFSs: using two-parameter and three-parameter formulas. Our considerations can be summarized as follows:

- Both kind of formulas are correct from a pure mathematical point of view in the *Minkowski r-metrics* – all conditions are fulfilled for distances.
- In the Hausdorff metric only the three-parameters formula is correct. The two-parameter formula is incorrect.

Although from a pure mathematical point of view we could use the two-parameter formulas in the *Minkowski r-metrics*, there are important reasons why we should not do this:

- The mathematical correctness of the two-parameter formulas in *Minkowski r-metrics* does not coincide with the meaning of some concepts essential for A-IFSs.
- Both the analytical and geometrical considerations show that the third parameter should be present in the formulas.
- By using the two-parameter formulas we loose important information and in effect the results obtained may be inconsistent with intuition, common-sense, rational arguments, etc.

Therefore the three-parameter formulas are not only formally correct but since they use all the available information, they give in effect results that are correct and consistent with the essence of A-IFSs. Unfortunately, the same can not be said about two-parameter formulas. This can often be a decisive factor in practice.

References

1. Atanassov K. (1983) Intuitionistic Fuzzy Sets. VII ITKR Session. Sofia (Deposed in Central Sci. Technol. Library of Bulg. Acad. of Sci., 1697/84) (in Bulgarian).
2. Atanassov K. (1986) Intuitionistic fuzzy sets. Fuzzy Sets and Systems, 20, 87–96.
3. Atanassov K. (1999) Intuitionistic Fuzzy Sets: Theory and Applications. Springer, Berlin Heidelberg New York.

4. Atanassov K., Taseva V., Szmidt E., Kacprzyk J. (2005) On the geometrical interpretations of the intuitionistic fuzzy sets. In: K. T. Atanassov, J. Kacprzyk, M. Krawczak, E. Szmidt (Eds.): Issues in the Representation and Processing of Uncertain and Imprecise Information. Series: Problems of the Contemporary Science. EXIT, Warszawa, pp. 11–24.

5. Deng-Feng Li (2005) Multiattribute decision making models and methods using intuitionistic fuzzy sets. Journal of Computer and System Sciences, 70, 73–85.

6. Narukawa Y. and Torra V. (in press) Non-monotonic fuzzy measure and intuitionistic fuzzy set. Accepted for MDAI'06.

7. Szmidt E. and Baldwin J. (2003) New similarity measure for intuitionistic fuzzy set theory and mass assignment theory. Notes on IFSs, 9(3), 60–76.

8. Szmidt E. and Baldwin J. (2004) Entropy for intuitionistic fuzzy set theory and mass assignment theory. Notes on IFSs, 10(3), 15–28.

9. Szmidt E. and Kacprzyk J. (1996) Intuitionistic fuzzy sets in group decision making. Notes on IFS, 2, 15–32.

10. Szmidt E. and Kacprzyk J. (1996) Remarks on some applications of intuitionistic fuzzy sets in decision making. Notes on IFS, 2(3), 22–31.

11. Szmidt E. and Kacprzyk J. (1998) Group Decision Making under Intuitionistic Fuzzy Preference Relations. IPMU'98, pp. 172–178.

12. Szmidt E. and Kacprzyk J. (1998) Applications of Intuitionistic Fuzzy Sets in Decision Making. EUSFLAT'99, pp. 150–158.

13. Szmidt E. and Kacprzyk J. (2000) Distances between intuitionistic fuzzy sets. Fuzzy Sets and Systems, 114(3), 505–518.

14. Szmidt E. and Kacprzyk J. (2000) On Measures on Consensus Under Intuitionistic Fuzzy Relations. IPMU 2000, pp. 1454–1461.

15. Szmidt E. and Kacprzyk J. (2001) Distance from consensus under intuitionistic fuzzy preferences. Proc. EUROFUSE Workshop on Preference Modelling and Applications, Granada, pp. 73–78.

16. Szmidt E. and Kacprzyk J. (2001) Analysis of consensus under intuitionistic fuzzy preferences. International Conference in Fuzzy Logic and Technology. Leicester, UK, pp. 79–82.

17. Szmidt E. and Kacprzyk J. (2002) An intuitionistic fuzzy set based approach to intelligent data analysis (an application to medical diagnosis). In A. Abraham, L. Jain, J. Kacprzyk (Eds.): Recent Advances in Intelligent Paradigms and and Applications. Springer, Berlin Heidelberg New York, pp. 57–70.

18. Szmidt E. and Kacprzyk J. (2002) Analysis of Agreement in a Group of Experts via Distances Between Intuitionistic Fuzzy Preferences. IPMU 2002, pp. 1859–1865.

19. Szmidt E. and Kacprzyk J. (2002) An intuitionistic fuzzy set based approach to intelligent data analysis (an application to medical diagnosis). In A. Abraham, L. Jain, J. Kacprzyk (Eds.): Recent Advances in Intelligent Paradigms and Applications. Springer, Berlin Heidelberg New York, pp. 57–70.

20. Szmidt E. and Kacprzyk J. (2002) Evaluation of agreement in a group of experts via distances between intuitionistic fuzzy sets. Proc. IS'2002 – Int. IEEE Symposium: Intelligent Systems, Varna, Bulgaria, IEEE Catalog Number 02EX499, pp. 166–170.

21. Szmidt E. and Kacprzyk J. (2005) A new concept of a similarity measure for intuitionistic fuzzy sets and its use in group decision making. In V. Torra, Y. Narukawa, S. Miyamoto (Eds.): Modelling Decisions for AI. LNAI 3558, Springer, Berlin Heidelberg New York, pp. 272–282.

22. Szmidt E. and Kacprzyk J. (2007) A New Similarity Measure for Intuitionistic Fuzzy Sets: Straightforward Approaches may not work. Proc. FUZZ-IEEE 2007, pp. 1–6.
23. Szmidt E. and Kacprzyk J. (2007) Some Problems with Entropy Measures for the Atanassov Intuitionistic Fuzzy Sets. WILF, pp. 291–297.
24. Tan Ch. and Zhang Q. (2005) Fuzzy Multiple Attribute Topsis Decision Making Method Based on Intuitionistic Fuzzy Set Theory. Proc. IFSA 2005, pp. 1602–1605.
25. Tasseva V., Szmidt E. and Kacprzyk J. (2005) On one of the geometrical interpretations of the intuitionistic fuzzy sets. Notes on IFS, 11(3), 21–27.
26. Zadeh L.A. (1965) Fuzzy sets. Information and Control, 8, 338–353.

Fuzzy-Rational Betting on Sport Games with Interval Probabilities

Kiril I. Tenekedjiev, Natalia D. Nikolova, Carlos A. Kobashikawa, and Kaoru Hirota

Technical University-Varna, 9010 Varna, 1 Studentska Str.
Bulgaria, +359 52 383 670
kiril@dilogos.com
Technical University-Varna, 9010 Varna, 1 Studentska Str.
Bulgaria, +359 52 383 670
natalia@dilogos.com
Tokyo Institute of Technology, Interdisciplinary Graduate School of Science and Engineering, Dept. of Computational Intelligence and Systems Science, G3-49, 4259 Nagatsuta, Midori-ku, Yokohama-city, 226-8502, Japan, +81 45 924 5682
carlosk@hrt.dis.titech.ac.jp
Tokyo Institute of Technology, Interdisciplinary Graduate School of Science and Engineering, Dept. of Computational Intelligence and Systems Science, G3-49, 4259 Nagatsuta, Midori-ku, Yokohama-city, 226-8502, Japan, +81 45 924 5682
hirota@dis.titech.ac.jp

Summary. The paper discusses betting on sport events by a fuzzy-rational decision maker, who elicits interval subjective probabilities, which may be conveniently described by intuitionistic fuzzy sets. Finding the optimal bet for this decision maker is modeled and solved using fuzzy-rational generalized lotteries of II type. Approximation of interval probabilities is performed with the use of four criteria under strict uncertainty. Four expected utility criteria are formulated on that basis. The scheme accounts for the interval character of probability elicitation results. *Index Terms.* – generalized lotteries of II type, intuitionistic fuzzy sets, fuzzy rationality, interval probabilities.

1 Introduction

Decision theory [11] supports the decision maker (DM) in situations of choice by balancing her/his preferences, beliefs and risk attitude. The utility theory [19] analyzes preferences over risky alternatives with quantified uncertainty, modeled as lotteries. The set of lotteries within a situation of choice forms the lottery set L. A lottery is a prize set X with a probability function over it, defining the chance to win some of X elements.

Let's compare according to preference countless alternatives $l_{\vec{c}}^{cr}$, defined by a z-dimensional parameter $\vec{c}$ that belongs to a continuous set C. Then

K.I. Tenekedjiev et al.: *Fuzzy-Rational Betting on Sport Games with Interval Probabilities*, Studies in Computational Intelligence (SCI) **109**, 431–454 (2008)
www.springerlink.com   © Springer-Verlag Berlin Heidelberg 2008

$L = \{l_{\vec{c}}^{cr} | \vec{c} \in C\}$. Let the alternative $l_{\vec{c}}^{cr}$ gives $t_{\vec{c}}$ number of one-dimensional prizes $x_{\vec{c},r}$ (for $r = 1, 2, \ldots, t_{\vec{c}}$). A continuous one-dimensional set of prizes X results from the union of all $x_{\vec{c},r}$: $X = \{x_{\vec{c},r} | \vec{c} \in C, r = 1, 2, \ldots, t_{\vec{c}}\}$. Such alternatives may be modeled with one-dimensional generalized lotteries of II type (1D GL-II with z-dimensional parameter) [21].

Let the probability of the event $\theta_{\vec{c},r}$ "to receive the prize $x_{\vec{c},r}$ from alternative $l_{\vec{c}}^{cr}$" is $P_{\vec{c}}(\theta_{\vec{c},r})$. Then the group of hypotheses $\theta_{\vec{c},1}, \theta_{\vec{c},2}, \ldots, \theta_{\vec{c},r}, \ldots, \theta_{\vec{c},t_{\vec{c}}}$ is described in terms of probabilities by a one-dimensional classical discrete probability function (DPF) – $f_{d,\vec{c}}(.)$:

$$<\theta_{\vec{c},r}, \ P_{\vec{c}}(\theta_{\vec{c},r})>, \text{ for } r = 1, 2, \ldots, t_{\vec{c}} \tag{1}$$

Let 1D GL-II with z-dimensional parameter, where uncertainty is described by classical DPF, is called classical-risky:

$$l_{\vec{c}}^{cr} = << \theta_{\vec{c},1}, \ P_{\vec{c}}(\theta_{\vec{c},1}) >, x_{\vec{c},1}; < \theta_{\vec{c},2}, \ P_{\vec{c}}(\theta_{\vec{c},2}) >, \ x_{\vec{c},2}; \ldots; < \theta_{\vec{c},t_{\vec{c}}}, \tag{2}$$
$$P_{\vec{c}}(\theta_{\vec{c},t_{\vec{c}}}) >, x_{\vec{c},t_{\vec{c}}} >, \text{ for } \vec{c} \in C.$$

Here

$$P_{\vec{c}}(\theta_{\vec{c},r}) \geq 0, \text{ for } r = 1, 2, \ldots, t_{\vec{c}}, \text{ at } \vec{c} \in C, \tag{3}$$

$$\sum_{r=1}^{t_{\vec{c}}} P_{\vec{c}}(\theta_{\vec{c},r}) = 1, \text{ for } \vec{c} \in C. \tag{4}$$

An equivalent more simplified representation of classical-risky 1D GL-II is:

$$l_{\vec{c}}^{cr} = <P_{\vec{c}}(\theta_{\vec{c},1}), x_{\vec{c},1}; \ P_{\vec{c}}(\theta_{\vec{c},2}), x_{\vec{c},2}; \ \ldots; P_{\vec{c}}(\theta_{\vec{c},t_{\vec{c}}}), x_{\vec{c},t_{\vec{c}}} >. \tag{5}$$

A utility function $u(.)$ typical for the DM is defined over the elements of X, such that

$$\sum_{j=1}^{r} P_{\vec{c}_i}(\theta_{\vec{c}_i,j}) u(x_{\vec{c}_i,j}) \geq \sum_{j=1}^{r} P_{\vec{c}_k}(\theta_{\vec{c}_k,j}) u(x_{\vec{c}_k,j}) \Leftrightarrow l_{\vec{c}_i}^{cr} \succsim l_{\vec{c}_k}^{cr}, \forall l_{\vec{c}_i}^{cr} \in L, \forall \ l_{\vec{c}_k}^{cr} \in L.$$
$$\tag{6}$$

Here, $\succ$, $\sim$, and $\succsim$ denote the binary relations of the DM "preferred to", "indifferent with" and "at least as preferred as" called respectively "strict preference", "indifference", and "weak preference". The validity of (6) is based on rationality axioms and theorems [12]. It follows from (6) that the alternatives should be ranked according to a real-valued index called expected utility:

$$E_{\vec{c}}(u/f_{d,\vec{c}}) = \sum_{j=1}^{r} P_{\vec{c}}(\theta_{\vec{c},j}) u(x_{\vec{c},j}). \tag{7}$$

Thus, a problem of multi-dimensional optimization on $\vec{c}$ of the expected utility (7) over the set C arises:

$$\vec{c}_{\mathbf{opt}} = ? \text{ such that } E_{\vec{c}_{\mathbf{opt}}}(u/f_{d,\vec{c}_{\mathbf{opt}}}) \geq E_{\vec{c}}(u/f_{d,\vec{c}}), \vec{c} \in C. \tag{8}$$

The most preferred lottery is $l^{cr}_{\vec{c}_{opt}}$. This task may be approached numerically as the only alternative in most practical cases.

The solution of the optimization task (8) at $z = 1$ may be found using different numerical techniques (e.g., golden section search, Keefer-Johnson, parabolic interpolation, etc.) for maximization of one-dimensional function in a closed interval [10]. The Brendt method for one-dimensional optimization [24] is used in [34]. The scanning method is to be preferred [14] since information for the number of extremums of the expected utility is not present in the general case.

The optimization task (8) at $z > 1$ is numerically solved, which is the only alternative in most practical cases. There are many numerical methods (e.g. gradient method, random search, Nelder-Meed method, Gauss-Newton method, Levenberg-Marquardt, etc.) for optimization of multi-dimensional function. If $\vec{c}$ is of low dimension, it is best to use multi-dimensional scaling [14] due to the same reasons emphasized for the one-dimensional case at $z = 1$.

Betting problems over the results of sport games may be modeled via classical-risky 1D GL-II. However, subjective probabilities are always elicited in the form of uncertainty intervals. Then the expected utility is also calculated in the form of uncertainty interval. This leads to non-transitivity of $\sim$ and mutual non-transitivity of $\succ$ and $\sim$. Since transitivity is a key aspect of rationality [11], and the declared preferences may be modeled by fuzzy sets [30], then the DM is called fuzzy-ratinal [23]. There are different approaches to rank alternatives, where uncertainty is described by interval probabilities. One possible approach is to rank the elements of L according to a criterion, resulting from scalarization of the expected utility uncertainty interval Examples of such techniques are the $\Gamma_{max\ i\ min}$ and $\Gamma_{max\ i\ max}$ criteria (where lotteries are ranked in descending order respectively of the lower and upper margin of the expected utility) [15, 26, 35]. The second possible approach requires comparing the elements of L by the length of real intervals [13, 18], e.g. interval dominance method [37]. The third approach does not take into account the expected utility uncertainty interval, but emphasizes on the dependencies of expected utilty for different alternatives at different probability values. Maximality [16, 36] and E-dominance fall within this group of techniques [27]. Only the first approach gives equal results. In this paper the former is realized by a two-step procedure, which employs criteria under strict uncertainty Q for approximation of interval probabilities into point estimates and subsequent ranking of the approximating alternatives according to expected utility. The resulting criterion is called Q-expected utility.

In what follows, Sect. 2 analyzes the elicitation of subjective probabilities in the form of uncertainty intervals and the subsequent partial non-transitivity. Section 3 introduces ribbon DPFs, and fuzzy-rational GL-II are introduced on that basis. In Sect. 4 an algorithm to rank fuzzy-rational 1D GL-II is presented, where the approximation of interval probabilities is performed by the Laplace, Wald, Hurwicz$_\alpha$ and maximax criteria under strict uncertainty. Modelling and solving problems with fuzzy-rational 1D GL-II is demonstrated in Sect. 5 and discusses how to bet on the results of sport games.

2 Subjective Elicitation of Probabilities

Whereas utilities measure DM's preferences, probabilities are the measure of uncertainty in risky choices. Of the three approaches to probabilities – classical, frequentist and subjective – the latter is most adequate for the purpose of quantitative decision analysis. Subjective probability reflects the degree of belief of the observer of the system (OS) that a random event shall occur [6]. Thus, unlike the other two cases, here probability is a function of the subject, and is unique for each OS, depending on her/his knowledge of the world, beliefs and expectations. De Finetti's coherence approach [7] gives the connection between beliefs and Kolmogorov's probabilities [9], whereas the axiomatic approach [8, 25] defines formal conditions that guarantee the rationality of the OS.

To elicit the subjective probability of an event $\theta_{\tilde{e},r}$ two bets are defined. The first bet $l_1(\theta_{\tilde{e},r})$ gives a huge prize at the moment T if the event $\theta_{\tilde{e},r}$ occurs, whereas the second bet $l_2(m, n)$ gives the same huge prize at the same moment, if a white ball is drawn from an urn of n balls, of which m are white.

The preferential equation $l_1(\theta_{\tilde{e},r}) \sim l_2(m, n)$ is solved according to m and the root, divided by n, is $P(\theta_{\tilde{e},r})$. For an ideal DM, i.e., one with infinite discriminating abilities, a unique m may be identified using dichotomy, which is a solution of the equation. The real DM does not have infinite discriminating abilities and his preferences deviate from rational preferences of the ideal DM. As a result, the solution of the preferential equation is an interval of values, which are a solution of the preferential equation $l_1(\theta_{\tilde{e},r}) \sim l_2(m, n)$. That is why, the following is required: (1) the greatest possible $m = m_{down}$, where the DM holds $l_1(\theta_{\tilde{e},r}) \succ l_2(m_{down}, n)$; (2) the smallest possible $m = m_{up}$, where the DM holds $l_2(m_{up}, n) \succ l_1(\theta_{\tilde{e},r})$. Then the root $m(\theta_{\tilde{e},r}) \in [m_{down}; m_{up}]$, and the subjective probability uncertainty interval is $P(\theta_{\tilde{e},r}) \in [m_{up}/n; m_{down}/n]$. It may be elicited using different techniques, e.g. structured belief in event trees [28], qualitative statements [5], infinite random sets [1], etc. A quick and easy-to-use algorithm is proposed in [22] that uses triple dichotomy [31]. The latter is a modification of dichotomy [24], which is used three times – first, to find a point from the uncertainty interval of the root, and two more times to find the margins of the interval.

It is obvious that for the real DM, $\sim$ is not transitive, whereas $\succ$ and $\sim$ are not mutually transitive. Let $m_{up} \geq m_2 > m_1 \geq m_{down}$. Then for the real DM: a) $l_2(m_1, n) \sim l_1(\theta_{\tilde{e},r})$, $l_1(\theta_{\tilde{e},r}) \sim l_2(m_2, n)$ and $l_2(m_2, n) \succ l_2(m_1, n)$, although transitivity of $\sim$ assumes $l_2(m_2, n) \sim l_2(m_1, n)$; b) $l_2(m_2, n) \succ l_2(m_1, n)$, $l_2(m_1, n) \sim l_1(\theta_{\tilde{e},r})$ and $l_2(m_2, n) \sim l_1(\theta_{\tilde{e},r})$, although transitivity of $\succ$ and $\sim$ assume $l_2(m_2, n) \succ l_1(\theta_{\tilde{e},r})$; c) $l_1(\theta_{\tilde{e},r}) \sim l_2(m_2, n)$, $l_2(m_2, n) \succ l_2(m_1, n)$ and $l_1(\theta_{\tilde{e},r}) \sim l_2(m_1, n)$, although transitivity of $\sim$ and $\succ$ assume $l_1(\theta_{\tilde{e},r}) \succ l_2(m_1, n)$.

3 Ribbon DPF and Fuzzy-Rational GL-II

Let the uncertainty in a one-dimensional discrete random variable X is partially measured by one-dimensional DPF, which entirely lies within a lower and an upper one-dimensional border functions. Such a one-dimensional DPF will be called ribbon. If $f_d^R(.)$ is a one-dimensional ribbon DPF of X, which may take only one of the fixed values $x_{\tilde{c},1}, x_{\tilde{c},2}, \ldots, x_{\tilde{c},t_{\tilde{c}}}$, whereas $P^d(.)$ and $P^u(.)$ are lower and upper border functions of $f_{d,\tilde{c}}(.)$, then:

$$P^d(x_{\tilde{c},r}) \le f_{d,\tilde{c}}^R(x_{\tilde{c},r}) \le P^u(x_{\tilde{c},r}), \text{ for } r = 1, 2, \ldots, t_{\tilde{c}}, \tag{9}$$

$$0 \le P^d(x_{\tilde{c},r}) \le P^u(x_{\tilde{c},r}) \le 1, \text{ for } r = 1, 2, \ldots, t_{\tilde{c}}, \tag{10}$$

$$\sum_{r=1}^{t_{\tilde{c}}} P^d(x_{\tilde{c},r}) \le 1 \le \sum_{r=1}^{t_{\tilde{c}}} P^u(x_{\tilde{c},r}). $$

Each full group of disjoint events (hypotheses) may be partially probabilistically described by ribbon DPF. Let $\theta_{\tilde{c},1}, \theta_{\tilde{c},2}, \ldots, \theta_{\tilde{c},r}, \ldots, \theta_{\tilde{c},t_{\tilde{c}}}$ are a group of hypotheses, and the index of the hypotheses is a one-dimensional discrete random variable I with $t_{\tilde{c}}$ possible values. Then:

$$P^d(\theta_{\tilde{c},r}) = P^d(r) \le f_{d,\tilde{c}}^R(r) = P(I = r) = P(\theta_{\tilde{c},r}) \le P^u(r) = P^u(\theta_{\tilde{c},r}),$$
$$\text{for } r = 1, 2, \ldots, t_{\tilde{c}}. \tag{11}$$

The equation (11) allows representing the ribbon DPF as:

$$<\theta_{\tilde{c},r}, \; P^d(\theta_{\tilde{c},r}), 1 - P^u(\theta_{\tilde{c},r})>, \text{ for } r = 1, 2, \ldots, t_{\tilde{c}}. \tag{12}$$

The border functions here depend on the realization r of I, and are defined over the random variables $\theta_{\tilde{c},r}$. The quantities $P^d(\theta_{\tilde{c},r})$ and $P^u(\theta_{\tilde{c},r})$ physically represent the lower and upper margin of the probability uncertainty interval of $\theta_{\tilde{c},r}$.

The form of the ribbon DPF (12) use the representation of intuitionistic fuzzy sets [2]. This is based on the analogies between subjective probabilities and degrees of membership to intuitionistic fuzzy sets discussed in [30], as well as on the idea that any intuitionistic fuzzy event can be assigned interval probability [29].

1D GL-II with a one-dimensional DPF is called fuzzy-rational. Then the elements in L are denoted $l_{\tilde{c}}^{fr} : L = \{l_{\tilde{c}}^{fr} | \tilde{c} \in C\}$.

Let the event $\theta_{\tilde{c},r}$ "to receive the prize $x_{\tilde{c},r}$ from alternative $l_{\tilde{c}}^{fr}$" has probability belonging to the closed interval $[P_{\tilde{c}}^d(\theta_{\tilde{c},r}); P_{\tilde{c}}^u(\theta_{\tilde{c},r})]$. Then the group of hypotheses $\theta_{\tilde{c},1}, \theta_{\tilde{c},2}, \ldots, \theta_{\tilde{c},r}, \ldots, \theta_{\tilde{c},t_{\tilde{c}}}$ is partially probabilistically described by a one-dimensional ribbon DPF $- f_{d,\tilde{c}}^R(.)$, with lower and upper border functions $P_{\tilde{c}}^d(.)$ and $P_{\tilde{c}}^u(.)$ of the kind (12).

Now the alternatives may be represented as fuzzy-rational 1D GL-II:

$$l_{\vec{c}}^{fr} = << \theta_{\vec{c},1}, P_{\vec{c}}^d(\theta_{\vec{c},1}), 1 - P_{\vec{c}}^u(\theta_{\vec{c},1}) >, x_{\vec{c},1}; < \theta_{\vec{c},2}, P_{\vec{c}}^d(\theta_{\vec{c},2}), 1 - P_{\vec{c}}^u(\theta_{\vec{c},2}) >,$$
$$x_{\vec{c},2}; \dots; < \theta_{\vec{c},t_{\vec{c}}}, P_{\vec{c}}^d(\theta_{\vec{c},t_{\vec{c}}}), 1 - P_{\vec{c}}^u(\theta_{\vec{c},t_{\vec{c}}}) >, x_{\vec{c},t_{\vec{c}}} >, \text{ for } \vec{c} \in C. \qquad (13)$$

The conditions (9) hold here for each $\vec{c} \in C$.

4 Ranking Fuzzy-Rational 1D GL-II

Expected utility criterion (7) can be applied to rank fuzzy-rational 1D GL-II only if $P_{\vec{c}}^d(\theta_{\vec{c},r}) = P_{\vec{c}}^u(\theta_{\vec{c},r})$. As long as $P_{\vec{c}}^d(\theta_{\vec{c},r}) < P_{\vec{c}}^u(\theta_{\vec{c},r})$ for at least one p they need to be approximated by classical-risky ones, which is a problem under strict uncertainty [4]. Thus, ranking fuzzy-rational 1D GL-II is a problem of mixed "strict uncertainty-risk" type. The main idea is to use a criterion Q under strict uncertainty for the transformation at this stage, which despite their disadvantages are well known techniques that reflect the degree of pessimism/optimism of the fuzzy-rational DM. The resulting approximating lotteries are called Q-lotteries. At the second stage, the Q-lotteries must be ranked. This is a problem under risk and is approached by expected utility. This two-stage procedure to rank fuzzy-rational lotteries is equivalent to the introduction of the a ranking criterion, called Q-expected utility.

Since the alternatives in L are countless then it is impossible to rank them all. The work [33] proposes that the most preferred fuzzy-rational GL-II from L is defined on three stages:

(1) Through a chosen criterion under strict uncertainty Q, each ribbon DPF – $f_{d,\vec{c}}^R(.)$, is approximated by a classical DPF – $f_{d,\vec{c}}^Q(.)$:

$$P_{\vec{c}}^d(\theta_{\vec{c},r}) \leq f_{d,\vec{c}}^Q(r) = P_{\vec{c}}^Q(\theta_{\vec{c},r}) \leq P_{\vec{c}}^u(\theta_{\vec{c},r}), \text{ for } \vec{c} \in C. \qquad (14)$$

In that way each fuzzy-rational GL-II is approximated by a classical-risky GL-II, called Q-GL-II:

$$l_{\vec{c}}^Q = << \theta_{\vec{c},1}, P_{\vec{c}}^Q(\theta_{\vec{c},1}) >, x_{\vec{c},1}; < \theta_{\vec{c},2}, P_{\vec{c}}^Q(\theta_{\vec{c},2}) >, x_{\vec{c},2}; \dots \qquad (15)$$
$$\dots; < \theta_{\vec{c},t_{\vec{c}}}, P_{\vec{c}}^Q(\theta_{\vec{c},t_{\vec{c}}}) >, x_{\vec{c},t_{\vec{c}}} >, \text{ for } \tilde{c} \in C.$$

(2) The alternatives may be compared according to preference according to the expected utility of the Q-GL-II with a multi-dimensional parameter:

$$E_{\vec{c}}^Q\left(u/f_{d,\vec{c}}^R\right) = \sum_{r=1}^{t_{\vec{c}}} P_{\vec{c}}^Q(\theta_{\vec{c},r}) u(x_{\vec{c},r}). \qquad (16)$$

(3a) At $z = 1$, the most preferred fuzzy-rational GL-II with a one-dimensional parameter, $l_{c_{opt}}^{fr}$, is defined by the one-dimensional optimization task:

$$c_{opt} = ? \text{ such that } E^Q_{c_{opt}} \left(u/f^R_{d,c_{opt}} \right) \geq E^Q_c \left(u/f^R_{d,c} \right), \text{ for } c \in C. \qquad (17)$$

(3b) At $z > 1$ the most preferred fuzzy-rational GL-II with a multi-dimensional parameter, $l^{fr}_{\vec{c}_{opt}}$, is defined by the z-dimensional optimization task:

$$\vec{c}_{opt} = ? \text{ such that } E^Q_{\vec{c}_{opt}}(u/f^R_{d,\vec{c}_{opt}}) \geq E^Q_{\vec{c}}(u/f^R_{d,\vec{c}}), \text{ for } \vec{c} \in C. \qquad (18)$$

The resulting criterion to rank fuzzy-rational GL-II is called Q-expected utility. If certain criteria under strict uncertainty are applied, then the discrete utility function $u(.)$is used in the approximation of $f^R_{d,\vec{c}}$ by $f^Q_{d,\vec{c}}$. If the chosen criterion is Laplace, Wald, maximax or Hurwicz$_\alpha$, then the approximation may be performed using the procedures for ordinary lotteries in [32].

4.1 Approximation Using Laplace ($Q = L$)

The probabilities $P^L_{\vec{c}}(\theta_{\vec{c},r})$, for $r = 1, 2, \ldots, t_{\vec{c}}$, do not depend on the utility function. According to the Laplace principle of insufficient reason, if no information is available for a group of hypotheses, then each hypothesis is assumed equally probable. Then probability estimates are weighed average of the lower and upper margin, such that $\sum_{r=1}^{t_{\vec{c}}} P^L_{\vec{c}}(\theta_{\vec{c},r}) = 1$. Let $\alpha^{(\vec{c})}_L$ be the Laplace weight coefficient, defined in [20]:

$$\alpha^{(\vec{c})}_L = \begin{cases} \dfrac{1 - \sum\limits_{r=1}^{t_{\vec{c}}} P^d_{\vec{c}}(\theta_{\vec{c},r})}{\sum\limits_{r=1}^{t_{\vec{c}}} P^u_{\vec{c}}(\theta_{\vec{c},r}) - \sum\limits_{r=1}^{t_i} P^d_{\vec{c}}(\theta_{\vec{c},r})}, & \text{for } \sum\limits_{r=1}^{t_{\vec{c}}} P^u_{\vec{c}}(\theta_{\vec{c},r}) > \sum\limits_{r=1}^{t_i} P^d_{\vec{c}}(\theta_{\vec{c},r}) \\[4ex] 0.5, & \text{for } \sum\limits_{r=1}^{t_{\vec{c}}} P^u_{\vec{c}}(\theta_{\vec{c},r}) = \sum\limits_{r=1}^{t_i} P^d_{\vec{c}}(\theta_{\vec{c},r}) \end{cases} \qquad (19)$$

Then:

$$P^L_{\vec{c}}(\theta_{\vec{c},r}) = [1 - \alpha^{(\vec{c})}_L]P^d_{\vec{c}}(\theta_{\vec{c},r}) + \alpha^{(\vec{c})}_L P^u_{\vec{c}}(\theta_{\vec{c},r}), \text{ for } r = 1, 2, \ldots, t_{\vec{c}}. \qquad (20)$$

4.2 Approximation Using Wald ($Q = W$)

The Wald criterion for coincides with the criterion $\Gamma_{max\ i\ min}$ (see Sect. 1). Then the probabilities $P^W_{\vec{c}}(\theta_{\vec{c},r})$, for $r = 1, 2, \ldots, t_{\vec{c}}$ are defined so that the W-expected utility of the lottery is minimal and $\sum_{r=1}^{t_{\vec{c}}} P^W_{\vec{c}}(\theta_{\vec{c},r}) = 1$. Let $\rho(1), \rho(2), \ldots, \rho(t_i)$ is the permutation of $1, 2, \ldots, t_{\vec{c}}$, such that

$$u(x_{\vec{c},\rho(1)}) \geq u(x_{\vec{c},\rho(2)}) \geq \ldots \geq u(x_{\vec{c},\rho(t_{\vec{c}})}). \qquad (21)$$

The index of the critical Wald prize $r_W^{(\bar{c})}$ and the Wald weight coefficient $\beta^{(\bar{c})}$ [20] must be defined for each alternative. For that purpose the required Wald weight coefficient $\beta_r^{(\bar{c})}$, for $r = 1, 2, \ldots, t_{\bar{c}}$ for each prize is defined:

$$\beta_{\rho(r)}^{(\bar{c})} = \begin{cases} 1 - \dfrac{\sum\limits_{k=r+1}^{t_{\bar{c}}} P_{\bar{c}}^u(\theta_{\bar{c},\rho(k)}) - \sum\limits_{k=1}^{r} P_{\bar{c}}^d(\theta_{\bar{c},\rho(k)})}{P_{\bar{c}}^u(\theta_{\bar{c},\rho(r)}) - P_{\bar{c}}^d(\theta_{\bar{c},\rho(r)})}, & \text{for } P_{\bar{c}}^u(\theta_{\bar{c},\rho(r)}) > P_{\bar{c}}^d(\theta_{\bar{c},\rho(r)}) \\[2mm] 0, & \text{for } P_{\bar{c}}^u(\theta_{\bar{c},\rho(r)}) = P_{\bar{c}}^d(\theta_{\bar{c},\rho(r)}) \\ & \text{and } \left(\sum\limits_{k=1}^{t_{\bar{c}}} P_{\bar{c}}^u(\theta_{\bar{c},\rho(k)}) \right. \\ & \qquad > \sum\limits_{k=1}^{t_{\bar{c}}} P_{\bar{c}}^d(\theta_{\bar{c},\rho(k)}) \text{ or } r < t_{\bar{c}} \Big) \\[2mm] 1, & \text{for } \sum\limits_{k=1}^{t_{\bar{c}}} P_{\bar{c}}^u(\theta_{\bar{c}}, \rho(k)) \\ & \quad = \sum\limits_{k=1}^{t_{\bar{c}}} P_{\bar{c}}^d(\theta_{\bar{c}}, \rho(k)) \text{ and } r = t_{\bar{c}} \end{cases} \quad (22)$$

$$r_W^{(\bar{c})} = arg\{\beta_r^{(\bar{c})} \in (0; 1]\}, \tag{23}$$

$$\beta^{(\bar{c})} = \beta_{r_W^{(\bar{c})}}^{(\bar{c})}. \tag{24}$$

On the basis of (22)–(24) it follows that

$$P_{\bar{c}}^W(\theta_{\bar{c},r}) = \begin{cases} P_{\bar{c}}^d(\theta_{\bar{c},r}), & \text{for } \rho(r) < \rho(r_W^{(\bar{c})}) \\ [1 - \beta^{(i)}] P_{\bar{c}}^d(\theta_{\bar{c},r}) + \beta^{(i)} P_{\bar{c}}^u(\theta_{\bar{c},r}), & \text{for } \rho(r) = \rho(r_W^{(\bar{c})}), \quad r = 1, 2, \ldots, t_{\bar{c}} \\ P_{\bar{c}}^u(\theta_{\bar{c},r}), & \text{for } \rho(r) > \rho(r_W^{(\bar{c})}) \end{cases}$$

$$(25)$$

4.3 Approximation Using Maximax Criterion $(Q = \neg W)$

The maximax criterion coincides with the $\Gamma_{max\ i\ max}$ criterion. The probabilities $P_{\bar{c}}^{\neg W}(\theta_{\bar{c},r})$, for $r = 1, 2, \ldots, t_{\bar{c}}$ may be found by the dependencies in Sect. 4.2 using the dependence:

$$u(x_{\bar{c},r}) = -u(x_{\bar{c},r}), \text{for } r = 1, 2, \ldots, t_{\bar{c}}. \tag{26}$$

4.4 Approximation Using Hurwicz$_\alpha(Q = H_\alpha)$

The Hurwicz$_\alpha$ criterion under strict uncertainty assumes that the choice may be made by a numerical index, which is a sum of the worst and the best that may occur, weighted by the pessimistic index $\alpha \in [0; 1]$. The application of this idea means to define the probabilities $P_{\bar{c}}^{H_\alpha}(\theta_{\bar{c},r})$, for $r = 1, 2, \ldots, t_{\bar{c}}$, as a weighed value of the probabilities $P_{\bar{c}}^W(\theta_{\bar{c},r})$ and $P_{\bar{c}}^{\neg W}(\theta_{\bar{c},r})$ in Sects. 4.2 and 4.3 [32]:

$$P_{\vec{c}}^{H\alpha}(\theta_{\vec{c},r}) = \alpha P_{\vec{c}}^{W}(\theta_{\vec{c},r}) + (1 - \alpha)P_{\vec{c}}^{-W}(\theta_{\vec{c},r}), \text{ for } r = 1, 2, \dots, t_{\vec{c}}. \quad (27)$$

Here, $\alpha \in [0; 1]$ is a pessimistic index, measuring the pessimism of the DM. Each of the transformations may be also described by intuitionistic operators necessity (?), possibility ($\Diamond$) and their fuzzy generalization D_α [3].

5 Example for Ranking Fuzzy-Rational GL-II

The example in this section is solved in [33] with a specialized software, available upon request.

5.1 (A) Setup

A bookmakers' house offers bets on the results of a football match between the teams A (host) and B (guest). The DM may bet non-negative sums (in dollars) c_1, c_2 and c_3 of each result – home win (event θ_1), draw (event θ_2) and away win (event θ_3). The coefficients over the three results are $coef_1 = 4/3, coef_2 = 3.5$ and $coef_3 = 6$. The betting rules from the general setup in Appendix apply here.

According to the fuzzy-rational DM, the uncertainty interval of the events θ_1, θ_2 and θ_3 are:

$$P(\theta_1) = [0.18; 0.22], P(\theta_2) = [0.45; 0.55], P(\theta_3) = [0.21; 0.29]. \quad (28)$$

The DM must decide how to bet \$100 on the results from the competition. At the given coefficients, the DM may have maximal profit of \$500 (if she/he successfully bets \$100 on θ_1) and maximal loss of \$100 (if she/he unsuccessfully bets \$100 on an arbitrary outcome). The utility function of the DM over the net profit in US dollars from the bet x is approximated in the interval $[-\$100; \$500]$ using the analytical form (29) and is depicted on Fig. 1:

$$u(x) = \frac{\arctg(0.01x + 2) - \arctg(1)}{\arctg(7) - \arctg(1)}. \quad (29)$$

5.2 (B) Modelling Using Fuzzy-Rational GL-II

The defined problem may be restated using one-dimensional fuzzy-rational GL-II with a three-dimensional parameter $\vec{c} = (c_1, c_2, c_3)$, belonging to a continuous three-dimensional set C:

$$C = \{\vec{c} = (c_1, c_2, c_3) | c_1 \geq 0 \wedge c_2 \geq 0 \wedge c_3 \geq 0 \wedge c_1 + c_2 + c_3 \leq 100\}. \quad (30)$$

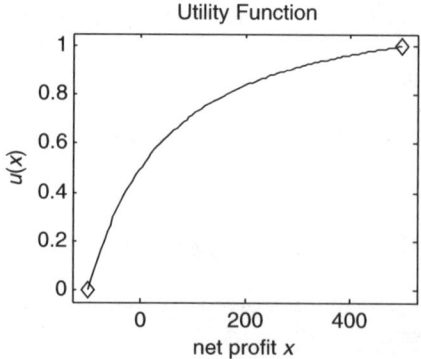

Fig. 1. Graphics of the utility function over net profits in the interval $[-\$100; \$500]$

Alternatives may be represented by fuzzy-rational GL-II, according to (13):

$$l_{\vec{c}}^{fr} = << \theta_1, 0.18, 0.78 >, x_{\vec{c},1}; < \theta_2, 0.45, 0.45 >, x_{\vec{c},2}; \qquad (31)$$
$$< \theta_3, 0.21, 0.71 >, x_{\vec{c},3} >, \text{for } \vec{c} \in C.$$

In (31), the events and their probabilities do not depend on $\vec{c}$ unlike one-dimensional prizes, which at the given coefficients are

$$x_{\vec{c},1} = c_1/3 - c_2 - c_3, x_{\vec{c},2} = 5c_2/2 - c_1 - c_3, x_{\vec{c},3} = 5c_3 - c_1 - c_2. \qquad (32)$$

At a chosen criterion Q, the Q-expected utility $E_{\vec{c}}^Q(u/p)$ of each alternative may be calculated. The optimal bet may be defined by the three-dimensional optimization task:

$$\vec{c}_{\mathbf{opt}} = ?, \text{че} E_{\vec{c}_{\mathbf{opt}}}^Q(u/p) \geq E_{\vec{c}}^Q(u/p), \text{for } \vec{c} \in C. \qquad (33)$$

Solving (33) is not required for the reasons listed below.
Obviously if

$$(1/coef_j) > 1, \qquad (34)$$

Then the only meaningful bet is $c_j = 0$, because all positive bets are Dutch books leading to losses for all results. Bookmakers' houses never set coefficients lower than one, since no one bets on them.

Although not that obvious, the first statement may be summarized for the case of bets over two results from a football match. If

$$(1/coef_j) + (1/coef_i) > 1, \qquad (35)$$

then for each bet with positive c_i and c_j there exists a bet c_i' and c_j', always resulting in higher profits, where $c_i' < c_i, c_j' < c_j$ and at least one c_i' and c_j' is 0. This means that if (35) holds, there will never be an optimal bet with positive c_i and c_j, because this can be represented as a simultaneous betting on c'_i and c'_j and a Dutch book.

Similar statement may be formulated for the case of bets over all possible results from the football match. If

$$(1/coef_1) + (1/coef_2) + (1/coef_3) > 1, \tag{36}$$

then for each bet with positive c_1, c_2 and c_3, there exists a bet with c'_1, c'_2 and c'_3, always resulting in higher profits, where $c'_1 < c_1, c'_2 < c_2, c'_3 < c_3$ and at least one of c'_1, c'_2 and c'_3 is 0. This means that if (36) holds, no bet with positive c_1, c_2 и c_3 will be optimal, since it can be represented as a simultaneous betting of c'_1, c'_2 and c'_3 and a Dutch book. The condition (36) I always holds, otherwise bookmakers' houses will be in a Dutch book themselves. That is why gamblers never bet on all three outcomes from a football match.

All stated assumptions are a special cases of the theorem for the dominating bet, defined and proven in Appendix, at $n = 3$ and cardinality of J^p respectively 1, 2 and 3.

In the analyzed problem,

$$(1/coef_1) + (1/coef_2) + (1/coef_3) = 3/4 + 2/7 + 1/6 = 101/84 > 1, \tag{37}$$
$$(1/coef_1) + (1/coef_2) = 3/4 + 2/7 = 29/28 > 1, \tag{38}$$
$$(1/coef_1) + (1/coef_3) = 3/4 + 1/6 = 11/12 < 1, \tag{39}$$
$$(1/coef_2) + (1/coef_3) = 2/7 + 1/6 = 19/42 < 1, \tag{40}$$

The aforementioned statements allow to decompose the three-dimensional optimization task (33) to three one-dimensional and two two-dimensional optimization tasks.

(B1) First (One-Dimensional) Task

The three-dimensional parameter $\vec{c}_1$ belongs to a continuous one-dimensional set C_1:

$$C_1 = \{\vec{c}_1 = (c_1, c_2, c_3) | c_1 \geq 0 \wedge c_1 \leq 100 \wedge c_2 = 0 \wedge c_3 = 0\}. \tag{41}$$

Since $c_{2,opt}$ and $c_{3,opt}$ in C_1 are zero, then the optimal bet can be defined by the one-dimensional optimization task on c_1:

$$c_{1,opt} = ?, \text{for } E^Q_{\vec{c}_{1,opt}}(u/p) = E^Q_{c_{1,opt}}(u/p) \geq E^Q_{c_1}(u/p) = E^Q_{\vec{c}_1}(u/p), \text{for } \vec{c}_1 \in C_1. \tag{42}$$

(B2) Second (One-Dimensional) Task

The three-dimensional parameter $\vec{c}_2$ belongs to a continuous one-dimensional set C_2:

$$C_2 = \{\vec{c}_2 = (c_1, c_2, c_3) | c_2 \geq 0 \wedge c_2 \leq 100 \wedge c_1 = 0 \wedge c_3 = 0\}. \tag{43}$$

Since $c_{1,opt}$ and $c_{3,opt}$ in C_2 are zero, then the optimal bet may be defined by the one-dimensional optimization task on c_2:

$$c_{2,opt} = ?, \text{for } E^Q_{\vec{c}_{2,opt}}(u/p) = E^Q_{c_{2,opt}}(u/p) \geq E^Q_{c_2}(u/p) = E^Q_{\vec{c}_2}(u/p), \text{ for } \vec{c}_2 \in C_2. \tag{44}$$

(B3) Third (One-Dimensional) Task

The three-dimensional parameter $\vec{c}_3$ belongs to a continuous one-dimensional set C_3:

$$C_3 = \{\vec{c}_3 = (c_1, c_2, c_3)|c_3 \geq 0 \wedge c_3 \leq 100 \wedge c_1 = 0 \wedge c_2 = 0\}. \tag{45}$$

Since $c_{1,opt}$ and $c_{2,opt}$ in C_3 are zero, then the optimal bet may be defined by the one-dimensional optimization task on c_3:

$$c_{3,opt} = ?, \text{че} E^Q_{\vec{c}_{3,opt}}(u/p) = E^Q_{c_{3,opt}}(u/p) \geq E^Q_{c_3}(u/p)$$
$$= E^Q_{\vec{c}_3}(u/p), \text{за } \vec{c}_3 \in C_3. \tag{46}$$

(B4) Fourth (Two-Dimensional) Task

The three-dimensional parameter $\vec{c}_4$ belongs to a continuous two-dimensional set C_4:

$$C_4 = \{\vec{c}_4 = (c_1, c_2, c_3)|c_1 > 0 \wedge c_3 > 0 \wedge c_1 + c_3 \leq 100 \wedge c_2 = 0\}. \tag{47}$$

Since $c_{2,opt}$ in C_4 is zero, then the optimal bet may be defined by the two-dimensional optimization task on c_1 and c_3:

$$\vec{c}_{4,opt} = ?, \text{where } E^Q_{\vec{c}_{4,opt}}(u/p) = E^Q_{c_{1,opt},c_{3,opt}}(u/p) \geq E^Q_{c_1,c_3}(u/p)$$
$$= E^Q_{\vec{c}_4}(u/p), \vec{c}_4 \in C_4. \tag{48}$$

(B5) Fifth (Two-Dimensional) Task

The three-dimensional parameter $\vec{c}_5$ belongs to a continuous two-dimensional set C_5:

$$C_5 = \{\vec{c}_5 = (c_1, c_2, c_3)|c_2 > 0 \wedge c_3 > 0 \wedge c_2 + c_3 \leq 100 \wedge c_1 = 0\}. \tag{49}$$

Since $c_{1,opt}$ in C_5 is zero, then the optimal bet may be defined by the two-dimensional optimization task on c_2 and c_3:

$$\vec{c}_{5,opt} = ?, \text{where } E^Q_{\vec{c}_{5,opt}}(u/p) = E^Q_{c_{2,opt},c_{3,opt}}(u/p) \geq E^Q_{c_2,c_3}(u/p)$$
$$= E^Q_{\vec{c}_5}(u/p), \vec{c}_5 \in C_5. \tag{50}$$

After solving tasks B1 from B5, the solution of (33) is the optimal bet with the highest Q-expected utility:

$$\vec{c}_{opt} = \vec{c}_{i_{opt},opt} \tag{51}$$

where

$$i_{opt} = arg\ \underset{i}{max}\left\{E^Q_{\vec{c}_{i,opt}}(u/p)|i = 1,2,\ldots,5\right\} \tag{52}$$

5.3 (C) Solution Using Hurwicz$_\alpha$

The solution according to the Wald expected utility is a special case of the solution according to Hurwicz$_\alpha$ expected utility for $\alpha = 1$, which shall be discussed later. The five optimization tasks are solved using Hurwicz$_\alpha$ expected utility for 51 values of $\alpha = 0, 0.02, 0.04, \ldots, 1$. It turns out that at all values of α, $i_{opt} = 5$ and according to (51), $\vec{c}_{opt} = \vec{c}_{5,opt}$. Figure 2 presents the optimal bets $c_{2,opt}$ and $c_{3,opt}$, as well as the Hurwicz$_\alpha$ expected utility $E^{H_\alpha}_{c_{2,opt},c_{3,opt}}(u/p)$ as functions of α. For example, at $\alpha = 0$ the solution degenerates to the one using the maximax expected utility. The optimal bet of $c_{2,opt} = \$65$ and $c_{3,opt} = \$33$ (corresponding to $\vec{c}_{5,opt} = (0,65,33)$) has maximal maximax expected utility $E^{\neg W}_{\vec{c}_{5,opt}}(u/p) = E^{\neg W}_{c_{2,opt},c_{3,opt}}(u/p) = 0.62$.

5.4 (C) Solution According to Laplace

Figure 3 presents the graphics of the Laplace expected utility depending on the parameter in the three one-dimensional tasks, which are solved using the scanning method with a step of \$1. The optimal single bets $c_{1,opt} = \$0$, $c_{2,opt} = \$39$ or $c_{3,opt} = \$14$ (corresponding to $\tilde{c}_{1,opt} = (0,0,0)$, $\vec{c}_{2,opt} = (0,39,0)$

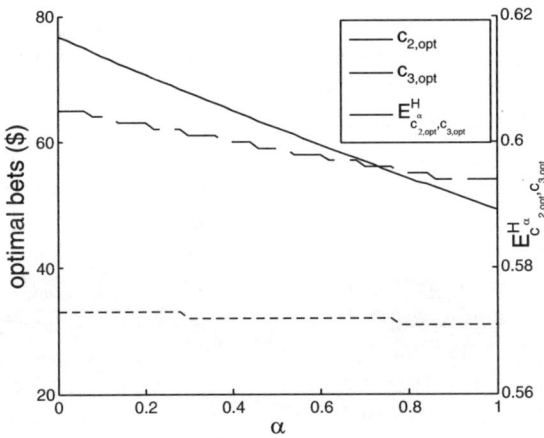

Fig. 2. Optimal bets and maximal Hurwicz$_\alpha$ expected utility as a function of α

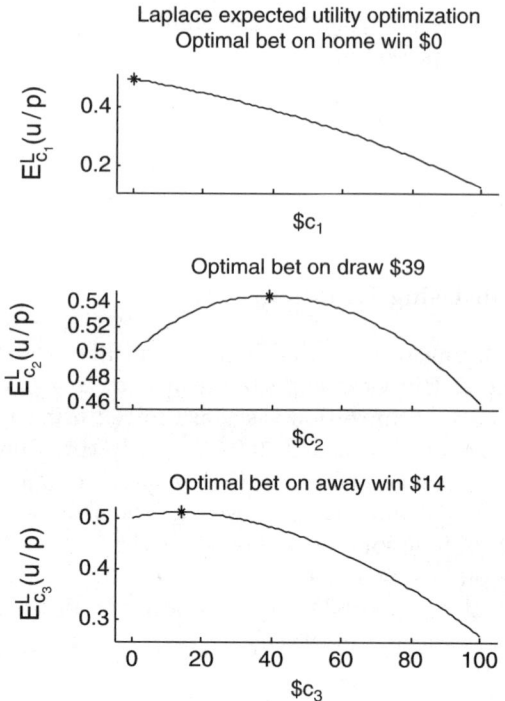

Fig. 3. Laplace expected utility of possible single bets

and $\vec{c}_{3,opt} = (0, 0, 14))$ have maximal Laplace expected utilities respectively $E^L_{\vec{c}_{1,opt}}(u/p) = E^L_{c_{1,opt}}(u/p) = 0.50$, $E^L_{\vec{c}_{2,opt}}(u/p) = E^L_{c_{2,opt}}(u/p) = 0.545$ and $E^L_{\vec{c}_{3,opt}}(u/p) = E^L_{c_{3,opt}}(u/p) = 0.512$.

When solving the fourth (two-dimensional) optimization task it turns out that there is no maximum in C_4, but a supremum $E^L_{\vec{c}_{4,sup}}(u/p)$, coinciding with the maximum in C_3. Since this result coincides with the solution of the third (optimization) task, then the solution is not presented, and the optimal value of the Laplace expected utility is $E^L_{\vec{c}_{4,opt}}(u/p) = E^L_{c_{1,opt},c_{3,opt}}(u/p) = -\infty$.

The fifth (two-dimensional) task is solved using two-dimensional scanning, with a step of \$1 on both coordinates. Figure 4 (lower graph) presents the optimal bet $c_{3,opt}(c_2)$ depending on c_2, which has the maximal Laplace expected utility at a draw bet c_2. Fig. 4 (upper graph) represents the maximal Laplace expected utility depending on c_2, calculated at c_2 and $c_{3,opt}(c_2)$. The optimal bet of $c_{2,opt} = \$58$ and $c_{3,opt} = \$30$ (corresponding to $\tilde{c}_{5,opt} = (0, 58, 30)$) has a maximal Laplace expected utility $E^L_{\vec{c}_{5,opt}}(u/p) = E^L_{c_{2,opt},c_{3,opt}}(u/p) = 0.96$.

Form (52) it follows that $i_{opt} = 5$. According to (51), $\vec{c}_{opt} = \vec{c}_{5,opt} = (0, 58, 30)$ and the best option for the fuzzy-rational DM that uses the Laplace expected utility criterion, at the given coefficients and the elicited subjective probabilities, is to put \$0 on home win, \$58 on draw and \$30 on away win.

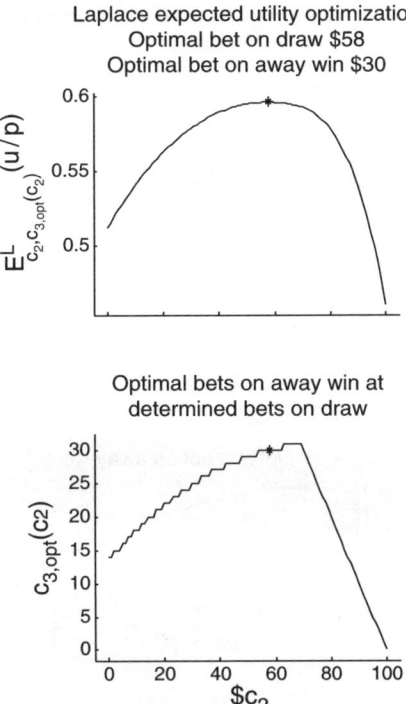

Fig. 4. Optimization according to Laplace and expected utility of the double bet on draw and away win

The fifth (two-dimensional) task is solved using two-dimensional scanning, with a step of \$1 on both coordinates. Figure 4 (lower graph) shows the optimal bet $c_{3,opt}(c_2)$ depending on c_2, which gives maximal Laplace expected utility at a draw bet c_2. Figure 4 (upper graph) shows the Laplace expected utility depending on c_2, calculated at c_2 and $c_{3,opt}(c_2)$. The optimal bet of $c_{2,opt} = \$58$ and $c_{3,opt} = \$30$ (corresponding to $\vec{c}_{5,\mathbf{opt}} = (0, 58, 30)$) has a maximal Laplace expected utility $E^L_{\vec{c}_{5,\mathbf{opt}}}(u/p) = E^L_{c_{2,opt},c_{3,opt}}(u/p) = 0.96$.

From (52) it follows that $i_{opt} = 5$. According to (51), $\vec{c}_{\mathbf{opt}} = \vec{c}_{5,\mathbf{opt}} = (0, 58, 30)$ and the bets option for the fuzzy-rational DM that uses the Laplace expected utility criterion, at the given coefficients and the elicited probabilities, is to bet \$0 on home win, \$58 on draw, and \$30 on away win.

5.5 (D) Solution According to Wald

Figure 5 shows the graphics of the Wald expected utility depending on the corresponding parameter in the three one-dimensional tasks, which are solved using the scanning method with a step of \$1. The optimal single

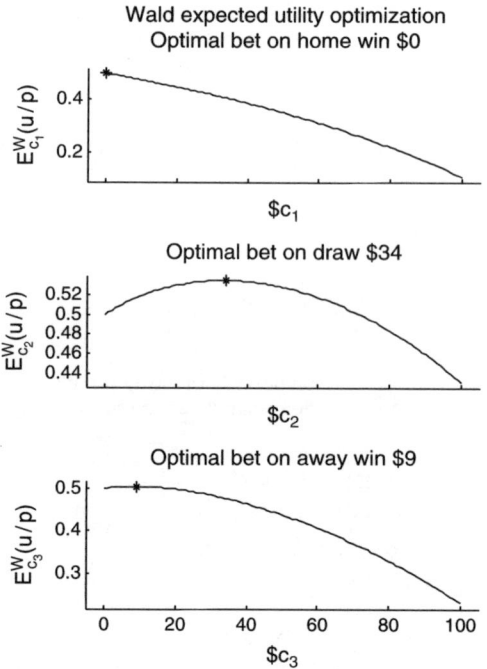

Fig. 5. Wald expected utility of the possible single bets

bets $c_{1,\text{opt}} = \$0, c_{2,\text{opt}} = \$34$ or $c_{3,\text{opt}} = \$9$ (corresponding respectively to $\vec{c}_{1,\text{opt}} = (0,0,0), \vec{c}_{2,\text{opt}} = (0,34,0)$ and $\vec{c}_{3,\text{opt}} = (0,0,9)$) give maximal Wald expected utilities respectively $E^W_{\vec{c}_{1,\text{opt}}}(u/p) = E^W_{c_{1,opt}}(u/p) = 0.5,$ $E^W_{\vec{c}_{2,\text{opt}}}(u/p) = E^W_{c_{2,opt}}(u/p) = 0.534$ and $E^W_{\vec{c}_{3,\text{opt}}}(u/p) = E^W_{c_{3,opt}}(u/p) = 0.505.$

When solving the fourth (two-dimensional) task it turns out that there is no maximum in C_4, but a supremum $E^W_{\vec{c}_{4,\text{sup}}}(u/p)$ that coincides with the maximum in C_3. Since this result coincides with the result of the third (one-dimensional) task, the solution is not presented, and the optimal value of the Wald expected utility is $E^W_{\vec{c}_{4,\text{opt}}}(u/p) = E^W_{c_{1,opt},c_{3,opt}}(u/p) = -\infty.$

The fifth (two-dimensional) task is solved using two-dimensional scanning with a step of \$1 on both coordinates. Figure 6 (lower graph) shows the optimal bet $c_{3,\text{opt}}(c_2)$ depending on c_2, which gives the maximal Wald expected utility at a draw bet c_2. Figure 6 (upper graph) shows the Wald expected utility depending on c_2, calculated at c_2 and $c_{3,\text{opt}}(c_2)$. The optimal bet of $c_{2,\text{opt}} = \$54$ and $c_{3,opt} = \$31$ (corresponding to $\vec{c}_{5,\text{opt}} = (0,54,31)$) has a maximal Wald expected utility $E^W_{\vec{c}_{5,\text{opt}}}(u/p) = E^W_{c_{2,opt},c_{3,opt}}(u/p) = 0.59.$

Form (52) it follows that $i_{opt} = 5$. According to (51), $\vec{c}_{\text{opt}} = \vec{c}_{5,\text{opt}} = (0,54,31)$ and the best option for the fuzzy-rational DM using the Wald

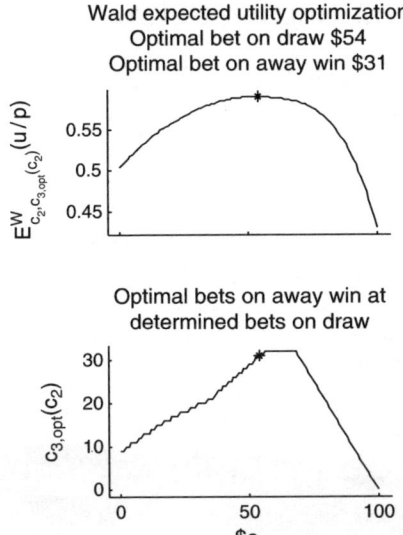

Fig. 6. Optimization according to Wald and expected utility of the double bets over draw and away win

expected utility criterion, at the given coefficients and the elicited probabilities, is to put \$0 on home win, \$54 on draw and \$31 on away win.

6 Conclusions

In the optimization according to Hurwicz$_\alpha$, the elicitation of α is performed using the classical scheme from [11]. Further studies in the application of this criterion should be based on more sophisticated ways of measuring the pessimistic criterion. It is important to mention that in the use of some criteria under strict uncertainty, the probabilities depend on the preferences of the DM, which violated expected utility. In such cases the DM is called "probabilistically sophisticated non-expected utility maximizer" according to [17].

The presented scheme in the paper is a universal method to bet on the results of sport games, which allows choosing the optimal bet depending on the interval subjective probabilities of the DM and her/his optimism/pessimism. In the revised example, the subjective probabilities do not depend on the sum, but at large bets it is possible to illegally manipulate the result of the game. Such situation may be modeled with GL-II, where not only prizes, but also states together with the DPF that describe them shall be functions of the z-dimensional parameter $\vec{c}$. All calculations in the example were performed using original software, which is accessible free of charge upon request from the authors.

Appendix: General Betting Setup

Let us accept non-negative bets at coefficients $\alpha_1, \alpha_2, \ldots, \alpha_n$ for the occurrence respectively of one out of $n > 1$ number of hypotheses $\theta_1, \theta_2, \ldots, \theta_n$, for which

$$\theta_j \cap \theta_k = \phi, \text{ at } j = 1, 2, \ldots, n \text{ and at } k = 1, 2, \ldots, j-1, j+1, \ldots, n, \quad \text{(A1)}$$
$$\theta_1 \cup \theta_2 \cup \ldots \cup \theta_n = \Omega.$$

In (A1), $\cap$ stands for "intersection", $\cup$ stands for "union", ϕ is the null event, and Ω is the certain event. Let

$$J^f = \{1, 2, \ldots, n\}. \quad \text{(A2)}$$

All coefficients are positive:

$$\alpha_j > 0, \text{за } j \in J^f. \quad \text{(A3)}$$

In practice, the condition (A3) is much stronger, since coefficients equal to at least one. It is possible to bet $\vec{c} = (c_1, c_2, \ldots, c_n)$ on each outcome, where

$$c_j \geq 0, \text{за } j \in J^f \quad \text{(A4)}$$

$$\sum_{j \in J^f} c_j > 0. \quad \text{(A5)}$$

If θ_j occurs, the sum of $\alpha_j c_j$ shall be paid to the DM, and the net profit will be

$$Pr_j = \alpha_j c_j - \sum_{j \in J^f} c_j. \quad \text{(A6)}$$

Definition for a prize of the bet:
Let the prize of a given bet be the investment that must be made in advance:

$$Inv(\vec{c}) = \sum_{j \in J^f} c_j. \quad \text{(A7)}$$

Definition of dominance:
The bet $\tilde{c}' = (c'_1, c'_2, \ldots, c'_n)$ dominates the bet $\vec{c} = (c_1, c_2, \ldots, c_n)$ if at all θ_j the profit from the second bet is not smaller than that of the first bet, and for at least one θ_j the profit is higher:

$$Pr'_j - Pr_j \geq 0, \text{ for } j \in J^f, \quad \text{(A8)}$$

$$\sum_{j \in J^f} (Pr'_j - Pr_j) > 0. \quad \text{(A9)}$$

In (A8) and (A9), Pr'_j is the profit from $\tilde{c}'$ at the occurrence of θ_j

$$Pr'_j = \alpha_j c'_j - \sum_{j \in J^f} c'_j. \tag{A10}$$

Theorem for the dominated bet

Let the bet $\vec{c}$ be made under the general betting setup, which defines the set J^p:

$$J^p = \{j/j \in J^f \wedge c_j > 0\}. \tag{A11}$$

where $\wedge$ means "conunction". The cardinality of (A11) is at least 1, because according to (A5) there is at least one event, over which a positive sum has been placed. If the following condition holds:

$$\sum_{j \in J^p} \frac{1}{\alpha_j} > 1, \tag{A12}$$

then there exists a cheaper bet $\tilde{c}' = (c'_1, c'_2, \ldots, c'_n)$, which dominates $\vec{c} = (c_1, c_2, \ldots, c_n)$.

Proof. Let J^z be the set of indices of the events $\theta_1, \theta_2, \ldots, \theta_n$, over which no positive sum has been placed:

$$J^z = \{j/j \in J^f \wedge j \notin J^p\}. \tag{A13}$$

The set J^z may as well be empty.

Let i be the index of the hypothesis that gives minimal positive income:

$$i = \arg\min_{j \in J^p} \{\alpha_j c_j\}. \tag{A14}$$

The set J contains all indices of events that are different from i, over which a positive sum has been placed:

$$J = \{j/j \in J^p \wedge j \neq i\}. \tag{A15}$$

The set J may be empty.

Let the bet $\tilde{c}' = (c'_1, c'_2, \ldots, c'_n)$ be defined as follows:

$$c'_j = \begin{cases} c_j & \text{at } j \in J^z \\ c_j - \alpha_i c_i / \alpha_j & \text{at } j \in J^p \end{cases}. \tag{A16}$$

It will be proven in three steps that the bet $\tilde{c}'$ is cheaper than $\vec{c}$ and dominates $\vec{c}$.

Step 1. The bet $\tilde{c}' = (c'_1, c'_2, \ldots, c'_n)$ is possible, i.e. $c'_j \geq 0$, for $j \in J^f$. $\square$

Proof. Two cases shall be revised to prove the statement.

Case 1: Let $j \in J^z$.

Then according to (A16),

$$c'_j = c_j. \tag{A17}$$

By definition, $j \in J^z$, thus

$$c_j = 0. \tag{A18}$$

From (A17) and (A18) it follows that

$$c'_j = 0 \text{ at } j \in J^z. \tag{A19}$$

Case 2: Let $j \in J^p$.
Then according to (A16),

$$c'_j = c_j - \frac{\alpha_i c_i}{\alpha_j} = \frac{\alpha_j c_j - \alpha_i c_i}{\alpha_j}. \tag{A20}$$

According to (A4) the numerator of (A20) is non-negative, and the denominator is positive according to (A3). Then it follows that the fraction (A20) is non-negative:

$$c'_j \geq 0 \text{ at } j \in J^p. \tag{A21}$$

Summary of step 1: Since $J^p \cup J^z = J^f$, then from (A19) and (A20) it follows that the bet $\tilde{\mathbf{c}}' = (c'_1, c'_2, \ldots, c'_n)$ is possible, because $c'_j \geq 0$ holds for $j \in J^f$.   □

Step 2. The price $Inv(\tilde{\mathbf{c}})$ of $\tilde{\mathbf{c}}$ is higher than the price $Inv(\tilde{\mathbf{c}}')$ of $\tilde{\mathbf{c}}'$:

$$Inv(\tilde{\mathbf{c}}) > Inv(\tilde{\mathbf{c}}'). \tag{A22}$$

Proof. The price $Inv(\tilde{\mathbf{c}}')$ of $\tilde{\mathbf{c}}'$ is:

$$Inv(\tilde{\mathbf{c}}') = \sum_{j \in J^f} c'_j = \sum_{j \in J^p} c'_j + \sum_{j \in J^z} c'_j. \tag{A23}$$

From (A23), according to (A16) and (A13) it follows that

$$Inv(\tilde{\mathbf{c}}') = \sum_{j \in J^p} \left(c_j - \frac{\alpha_i c_i}{\alpha_j} \right) + \sum_{j \in J^z} c_j = \sum_{j \in J^p} c_j - \alpha_i c_i \sum_{j \in J^p} \frac{1}{\alpha_j} + \sum_{j \in J^z} 0. \tag{A24}$$

From (A24), taking into account the definition (A7) it follows that

$$Inv(\tilde{\mathbf{c}}') = Inv(\tilde{\mathbf{c}}) - \alpha_i c_i \sum_{j \in J^p} \frac{1}{\alpha_j}. \tag{A25}$$

From (A14) it follows that $\alpha_i c_i > 0$ and then according to (A3) it follows that:

$$\alpha_i c_i \sum_{j \in J^p} \frac{1}{\alpha_j} > 0. \tag{A26}$$

From (A25) and (A26) it follows that $\tilde{\mathbf{c}}' = (c'_1, c'_2, \ldots, c'_n)$ is cheaper than $\tilde{\mathbf{c}} = (c_1, c_2, \ldots, c_n)$:

$$Inv(\tilde{\mathbf{c}}) = \sum_{j \in J^f} c_j > \sum_{j \in J^f} c'_j = Inv(\tilde{\mathbf{c}}'). \tag{A27}$$

□

Step 3. At the occurrence of an arbitrary event $\theta_1, \theta_2, \ldots, \theta_n$, the profit from $\tilde{c}'$ is higher than that from $\vec{c}$:

$$Pr'_j > Pr_j, \text{ for } j \in J^f. \tag{A28}$$

Proof. Three cases shall be revised to prove the statement.

Case 1: Let $j = i$.

Then from (A6), (A11) and (A13), the following holds for the bet $\vec{c}$:

$$Pr_j = Pr_i = \alpha_i c_i - \sum_{j \in J^f} c_j = \alpha_i c_i - \sum_{j \in J^p} c_j - \sum_{j \in J^z} c_j$$

$$= \alpha_i c_i - \sum_{j \in J^p} c_j - \sum_{j \in J^z} 0 = \alpha_i c_i - \sum_{j \in J^p} c_j \tag{A29}$$

From (A10), (A16), (A11) and (A13), the following holds for the bet $\tilde{c}'$:

$$Pr'_j = Pr'_i = \alpha_i c'_i - \sum_{j \in J^f} c' = \alpha_i \left(c_i - \frac{\alpha_i c_i}{\alpha_i} \right) - \sum_{j \in J^p} c'_j - \sum_{j \in J^z} c'_j$$

$$= \alpha_i \times 0 - \sum_{j \in J^p} c'_j - \sum_{j \in J^z} c_j = -\sum_{j \in J^p} \left(c_j - \frac{\alpha_i c_i}{\alpha_j} \right) - \sum_{j \in J^z} 0$$

$$= \alpha_i c_i \sum_{j \in J^p} \frac{1}{\alpha_j} - \sum_{j \in J^p} c_j. \tag{A30}$$

From (A29) and (A30) it follows that the difference in the profits from both bets is

$$Pr'_j - Pr_j = \alpha_i c_i \sum_{j \in J^p} \frac{1}{\alpha_j} - \alpha_i c_i = \alpha_i c_i \left(\sum_{j \in J^p} \frac{1}{\alpha_j} - 1 \right). \tag{A31}$$

From (A12) it follows that

$$\sum_{j \in J^p} \frac{1}{\alpha_j} - 1 > 0. \tag{A32}$$

From (A14) it follows that $\alpha_i c_i > 0$ and then according to (A32) and (A31) it follows that:

$$Pr'_j > Pr_j, \text{ for } j = i. \tag{A33}$$

Case 2: Let $j \in J$.

Then from (A6), (A11) and (A13), the following holds for the bet $\vec{c}$:

$$Pr_j = \alpha_j c_j - \sum_{j \in J^f} c_j = \alpha_j c_j - \sum_{j \in J^p} c_j - \sum_{j \in J^z} c_j = \alpha_j c_j - \sum_{j \in J^p} c_j - 0$$

$$= \alpha_j c_j - \sum_{j \in J^p} c_j \tag{A34}$$

From (A10), (A16), (A11) and (A13), the following holds for the bet $\tilde{\mathbf{c}}'$:

$$Pr'_j = \alpha_j \times c'_j - \sum_{j \in J^f} c'_j = \alpha_j \left(c_j - \frac{\alpha_i c_i}{\alpha_j} \right) - \sum_{j \in J^p} c'_j - \sum_{j \in J^z} c'_j \quad \text{(A35)}$$

$$= \alpha_j c_j - \alpha_i c_i - \sum_{j \in J^p} \left(c_j - \frac{\alpha_i c_i}{\alpha_j} \right) - \sum_{j \in J^z} c_j$$

$$= \alpha_j c_j - \alpha_i c_i + \alpha_i c_i \sum_{j \in J^p} \frac{1}{\alpha_j} - \sum_{j \in J^p} c_j - \sum_{j \in J^z} 0$$

$$= \alpha_j c_j - \alpha_i c_i + \alpha_i c_i \sum_{j \in J^p} \frac{1}{\alpha_j} - \sum_{j \in J^p} c_j.$$

From (A34) and (A35) it follows that the difference in the profits from both bets is

$$Pr'_j - Pr_j = \alpha_j c_j - \alpha_i c_i + \alpha_i c_i \sum_{j \in J^p} \frac{1}{\alpha_j} - \sum_{j \in J^p} c_j - \alpha_j c_j + \sum_{j \in J^p} c_j$$

$$= \alpha_i c_i \sum_{j \in J^p} \frac{1}{\alpha_j} - \alpha_i c_i = \alpha_i c_i \left(\sum_{j \in J^p} \frac{1}{\alpha_j} - 1 \right). \quad \text{(A36)}$$

This is the same difference as in (A31), which was proven to be positive. Thus

$$Pr'_j > Pr_j, \text{for } j \in J. \quad \text{(A37)}$$

Case 3: Let $j \in J^z$

Then from (A6), (A13) and (A7), the following holds for the bet $\tilde{\mathbf{c}}$

$$Pr_j = \alpha_j c_j - \sum_{j \in J^f} c_j = \alpha_j \times 0 - Inv(\tilde{\mathbf{c}}) = -Inv(\tilde{\mathbf{c}}). \quad \text{(A38)}$$

From (A10), (A16), (A13) and (A23), the following holds for the bet $\tilde{\mathbf{c}}'$:

$$Pr'_j = \alpha_j \times c'_j - \sum_{j \in J^f} c'_j = \alpha_j c_j - Inv(\tilde{\mathbf{c}}') = \alpha_j \times 0 - Inv(\tilde{\mathbf{c}}') = -Inv(\tilde{\mathbf{c}}'). \quad \text{(A39)}$$

From (A39) and (A40) it follows that the difference in the profits from both bets is

$$Pr'_j - Pr_j = -Inv(\tilde{\mathbf{c}}') + Inv(\tilde{\mathbf{c}}) = Inv(\tilde{\mathbf{c}}) - Inv(\tilde{\mathbf{c}}'). \quad \text{(A40)}$$

In step 2 it has been proven that $Inv(\tilde{\mathbf{c}}) > Inv(\tilde{\mathbf{c}}')$, according to (A27). Then from (A40) it follows that

$$Pr'_j > Pr_j, \text{ for } j \in J^z. \quad \text{(A41)}$$

$\square$

Summary of step 3: Since $\{i\} \cup J \cup J^z = J^f$, then from (A33), (A37) and (A41) it follows that the bet $\tilde{\mathbf{c}}' = (c'_1, c'_2, \ldots, c'_n)$ dominates $\tilde{\mathbf{c}} = (c_1, c_2, \ldots, c_n)$, because (A28) holds.

References

1. Alvarez DA (2006) On the calculation of the bounds of probability of events using infinite random sets, International Journal of Approximate Reasoning 43: 241–267
2. Atanassov K (1999) Intuitionistic fuzzy sets. Berlin Heidelberg New York: Springer
3. Atanassov K (2002) Elements of intuitionistic fuzzy logics. Part II: Intuitionistic fuzzy modal logics. Advanced Studies on Contemporary Mathematics, 5(1): 1–13.
4. Augustin Th (2001) On decision making under ambiguous prior and sampling information. In: de Cooman G, Fine T, Moral S, Seidenfeld T. (eds.): ISIPTA '01: Proceedings of the Second International Symposium on Imprecise Probabilities and their Applications. Cornell University, Ithaca (N.Y.), Shaker, Maastricht, 9–16
5. Augustin Th (2005) Generalized basic probability assignments. International Journal of General Systems, 34(4): 451–463
6. Bernstein L (1996) Against the gods – the remarkable story of risk. Wiley, USA
7. De Finetti B (1937) La prevision: ses lois logiques, ses sorces subjectives. In: Annales de l'Institut Henri Poincare, 7: 1–68. Translated in Kyburg HE Jr., Smokler HE (eds.) (1964) Studies in subjective probability. Wiley, New York, 93–158
8. De Groot MH (1970) Optimal statistical decisions. McGraw-Hill
9. Fine TL (1973) Theories of probability. Academic Press
10. Forsythe GE, Malcolm A, Moler CB (1977) Computer methods for mathematical computations. Prentice Hall
11. French S. (1993) Decision theory: an introduction to the mathematics of rationality. Ellis Horwood.
12. French S, Insua DR (2000) Statistical decision theory. Arnold, London
13. Gatev G (1995) Interval analysis approach and algorithms for solving network and dynamic programming problems under parametric uncertainty. Proc. of the Technical University-Sofia, 48(4): 343–350, Electronics, Communication, Informatics, Automatics
14. Gould NIM, Leyffer S (2003) An introduction to algorithms for nonlinear optimization. In Blowey JF, Craig AW, Shardlow, Frontiers in numerical analysis. Springer, Berlin Heidelberg New York, 109–197
15. Grunwald PD, Dawid AP (2002) Game theory, maximum entropy, minimum discrepancy, and robust bayesian decision theory. Tech. Rep. 223, University College London
16. Levi I (1999) Imprecise and indeterminate probabilities. Proc. First International Symposium on Imprecise Probabilities and Their Applications, University of Ghent, Ghent, Belgium, 65–74.
17. Machina M, Schmeidler D (1992) A more rrobust definition of subjective probability. Econometrica, 60(4): 745–780
18. Moore R (1979) Methods and applications of interval analysis. SIAM, Philadelphia
19. von Neumann J, Morgenstern O (1947) Theory of games and economic behavior. Second Edition. Princeton University Press, Princeton, NJ

20. Nikolova ND (2006) Two criteria to rank fuzzy rational alternatives. Proc. International Conference on Automatics and Informatics, 3–6 October, Sofia, Bulgaria, 283–286
21. Nikolova ND (2007) Quantitative fuzzy-rational decision analysis, PhD Dissertation, Technical University-Varna (in Bulgarian)
22. Nikolova ND, Dimitrakiev D, Tenekedjiev K (2004) Fuzzy rationality in the elicitation of subjective probabilities. Proc. Second International IEEE Conference on Intelligent Systems IS'2004, Varna, Bulgaria, III: 27–31
23. Nikolova ND, Shulus A, Toneva D, Tenekedjiev K (2005) Fuzzy rationality in quantitative decision analysis. Journal of Advanced Computational Intelligence and Intelligent Informatics, 9(1): 65–69
24. Press WH, Teukolski SA, Vetterling WT, Flannery BP (1992) Numerical recipes – the art of scientific computing, Cambridge University Press
25. Raiffa H (1968) Decision analysis. Addison Wesley
26. Satia JK, Lave RE, Lave Jr (1973) Markovian decision processes with uncertain transition probabilities. Operations Research, 21(3): 728–740
27. Schervish M, Seidenfeld T, Kadane J, Levi JI (2003) Extensions of expected utility theory and some limitations of pairwise comparisons. In Bernard JM, Seidenfeld T, Zaffalon M (eds.): ISIPTA 03: Proc. of the Third International Symposium on Imprecise Probabilities and their Applications, Lugano. Carleton Scientific, Waterloo, 496–510.
28. Shafer G, Gillett P, Scherl R (2003) A new understanding of subjective probability and its generalization to lower and upper prevision. International Journal of Approximate Reasoning, (33): 1–49
29. Szmidt E, Kacprzyk J (1999) Probability of intuitionistic fuzzy events and their application in decision making, Proc. EUSFLAT-ESTYLF Joint Conference, September 22–25, Palma de Majorka, Spain, 457–460
30. Tenekedjiev K (2006) Hurwicz$_\alpha$ expected utility criterion for decisions with partially quantified uncertainty, In First International Workshop on Intuitionistic Fuzzy Sets, Generalized Nets and Knowledge Engineering, 2–4 September, London, UK, 2006, pp. 56–75.
31. Tenekedjiev K (2007) Triple bisection algorithms for 1-D utility elicitation in the case of monotonic preferences. Advanced Studies in Contemporary Mathematics, 14(2): 259–281.
32. Tenekedjiev K, Nikolova ND (2007) Ranking discrete outcome alternatives with partially quantified uncertainty. International Journal of General Systems, 37(2): 249–274
33. Tenekedjiev K, Nikolova ND, Kobashikawa C, Hirota K (2006) Conservative betting on sport games with intuitionistic fuzzy described uncertainty. Proc. Third International IEEE Conference on Intelligent Systems IS'2006, Westminster, UK, 747–754
34. The MathWorks (2007) MATLAB optimization toolbox 3 – user's guide. The MathWorks Inc.
35. Vidakovic B (2000) Γ-minimax: a paradigm for conservative robust Bayesians. In Insua DR, Ruggeri F (eds.) Robust bayesian analysis, 241–259
36. Walley P (1991) Statistical reasoning with imprecise probabilities. Chapman and Hall, London.
37. Zaffalon M, Wesnes K, Petrini O (2003) Reliable diagnoses of dementia by the naive credal classifier inferred from incomplete cognitive data, Artificial Intelligence in Medicine, 29(1–2): 61–79

Atanassov's Intuitionistic Fuzzy Sets in Classification of Imbalanced and Overlapping Classes

Eulalia Szmidt and Marta Kukier

Systems Research Institute, Polish Academy of Sciences, ul. Newelska 6, 01–447 Warsaw, Poland
szmidt@ibspan.waw.pl, kukier@ibspan.waw.pl

Summary. We discuss the problem of classification of imbalanced and overlapping classes. A fuzzy set approach is presented first – the classes are recognized using a fuzzy classifier. Next, we use intuitionistic fuzzy sets (A-IFSs, for short)[1] to represent and deal with the same data. We show that the proposed intuitionistic fuzzy classifier has an inherent tendency to deal efficiently with imbalanced and overlapping data. We explore in detail the evaluation of the classifier results (especially from the point of view of recognizing the smaller class). We show on a simple example the advantages of the intuitionistic fuzzy classifier. Next, we illustrate its desirable behavior on a benchmark example (from UCI repository).

1 Introduction

The problem of imbalanced classes arises for a two-class classification problem, when the training data for one class greatly outnumbers the other class. Imbalanced and overlapping classes hinder the performance of the standard classifiers which are heavily biased in recognizing mostly the bigger class since there are built to achieve overall accuracy to which the smaller class contributes very little.

The problem is not only a theoretical challenge but it concerns many different types of real tasks. Examples are given by Kubat et al. [14], Fawcett and Provost [12], Japkowicz [13], Lewis and Catlett [15], Mladenic and Grobelnik [16]. To deal with the imbalance problems usually up-sampling and down-sampling are used. Alas, both methods interfere in the structure of the data, and in a case of overlapping classes even the artificially obtained balance does not solve the problem (some data points may appear as valid examples in

---

[1] Recently there is a debate on the suitability of the name *intuitionistic fuzzy set* but this is beyond the scope of this paper and we will not deal with this. To avoid any confusion we will call the sets: "Atanassov's intuitionistic fuzzy sets" (A-IFSs for short).

E. Szmidt and M. Kukier: *Atanassov's Intuitionistic Fuzzy Sets in Classification of Imbalanced and Overlapping Classes*, Studies in Computational Intelligence (SCI) **109**, 455–471 (2008)
www.springerlink.com   © Springer-Verlag Berlin Heidelberg 2008

both classes). More, Provost [17] claims that up-sampling does not add any information whereas down-sampling results in removing information. These facts give a motivation for looking for new approaches dealing more efficiently with recognition of the imbalanced data sets.

In this paper we propose an intuitionistic fuzzy approach to the problem of classification of imbalanced and overlapping classes. We consider a two–class classification problem (*legal* and *illegal* class).

The proposed method using A-IFSs has its roots in the fuzzy set approach given by Baldwin et al. [9]. In that approach the classes are represented by fuzzy sets. The fuzzy sets are generated from the relative frequency distributions representing the data points used as examples of the classes [9]. In the process of generating fuzzy sets a mass assignment based approach is adopted (Baldwin et al. [6,9]). For the obtained model (fuzzy sets describing the classes), using a chosen classification rule, a testing phase is performed to assess the performance of the proposed method.

The approach proposed in this paper is similar to the above one in the sense of the same steps we perform. The main difference lies in using A-IFSs for the representation of classes, and next – in exploiting the structure of A-IFSs to obtain a classifier better recognizing a smaller class.

The crucial point of the method is in representing the classes by A-IFSs (first, training phase). The A-IFSs are generated from the relative frequency distributions representing the data points used as examples of the classes. The A-IFSs are obtained according to the procedure given by Szmidt and Baldwin [21]. Having in mind recognition of the smaller class as good as possible we use the information about the hesitation margins making it possible to improve the results of data classification in the (second) testing phase (cf. Sects. 2 and 5.2). The obtained results in the testing phase were examined not only in the sense of general error/accuracy but also with using confusion matrices making possible to explore detailed behaviour of the classifiers.

The material in this paper is organized as follows: In Sect. 2 a brief introduction to A-IFSs is given. In Sect. 3 the mechanism converting relative frequency distributions into A-IFSs is presented. In Sect. 4 the models of the classifier errors are reminded. In Sect. 5 a simple classification problem is considered – in Sect. 5.1: using a fuzzy classifier, in Sect. 5.2: using an intuitionistic fuzzy classifier. We compare the performance of both classifiers analyzing the errors. In Sect. 6 a benchmark problem is analysed. In Sect. 7 we end with some conclusions.

2 Atanassov's Intuitionistic Fuzzy Sets

One of the possible generalizations of a fuzzy set in X [27], given by

$$A' = \{<x, \mu_{A'}(x)>|x \in X\} \tag{1}$$

where $\mu_{A'}(x) \in [0,1]$ is the membership function of the fuzzy set A', is an Atanassov's intuitionistic fuzzy set [1–3] A given by

$$A = \{<x, \mu_A(x), \nu_A(x)> | x \in X\} \tag{2}$$

where: $\mu_A : X \to [0,1]$ and $\nu_A : X \to [0,1]$ such that

$$0 \le \mu_A(x) + \nu_A(x) \le 1 \tag{3}$$

and $\mu_A(x)$, $\nu_A(x) \in [0,1]$ denote a degree of membership and a degree of non-membership of $x \in A$, respectively.

Obviously, each fuzzy set may be represented by the following A-IFS

$$A = \{<x, \mu_{A'}(x), 1 - \mu_{A'}(x)> | x \in X\} \tag{4}$$

For each A-IFS in X, we will call

$$\pi_A(x) = 1 - \mu_A(x) - \nu_A(x) \tag{5}$$

an *intuitionistic fuzzy index* (or a *hesitation margin*) of $x \in A$. It expresses a lack of knowledge of whether x belongs to A or not (cf. [3]). For each $x \in X$ $0 \le \pi_A(x) \le 1$.

The application of A-IFSs instead of fuzzy sets means the introduction of another degree of freedom into a set description. Such a generalization of fuzzy sets gives us an additional possibility to represent imperfect knowledge what leads to describing many real problems in a more adequate way. We refer an interested reader to Szmidt and Kacprzyk [22, 23] where the applications of A-IFSs to group decision making, negotiations and other situations are presented.

3 Converting Relative Frequency Distributions into Atanassov's Intuitionistic Fuzzy Sets

The mechanism of converting a relative frequency distribution into an A-IFS is mediated by the relation of an A-IFS to the mass assignment theory. Detailed description is given by Szmidt and Baldwin [19–21].

The theory of mass assignment has been developed by Baldwin [5–7] to provide a formal framework for manipulating both probabilistic and fuzzy uncertainty.

A fuzzy set can be converted into a mass assignment [4]. This mass assignment represents a family of probability distributions.

Definition 1. (Mass Assignment) *Let A' be a fuzzy subset of a finite universe Ω such that the range of the membership function of A', is $\{\mu_1, ..., \mu_n\}$ where $\mu_i > \mu_{i+1}$. Then the mass assignment of A' denoted $m_{A'}$, is a probability distribution on 2^{Ω} satisfying*

$$m_{A'}(F_i) = \mu_i - \mu_{i+1} \tag{6}$$

where: $\tag{7}$

$$F_i = \{x \in \Omega | \mu(x) \ge \mu_i\} \text{ for } i = 1, ..., n$$

Table 1. Equality of the parameters for Baldwin's voting model and A-IFS voting model

	Baldwin's voting model	A-IFS voting model
Voting in favour	n	μ
Voting against	$1 - p$	ν
Abstaining	$p - n$	π

The sets $F_1, ..., F_n$ are called the focal elements of $m_{A'}$. The detailed introduction to mass assignment theory is given by Baldwin et al. [6].

In Table 1 equality of parameters from Baldwin's voting model and from A-IFS voting model is presented [19, 20]. The equivalence occurs under the condition that each value of membership/non-membership of A-IFS occurs with the same probability for each x_i (for a deeper discussion of the problem we refer an interested reader to [19, 20]). In other words both Support Pairs (mass assignment theory) and A-IFS models give the same intervals containing the probability of the fact being true, and the difference between the upper and lower values of intervals is a measure of the uncertainty associated with the fact [19, 20].

The mass assignment structure is best used to represent knowledge that is statistically based such that the values can be measured, even if the measurements themselves are approximate or uncertain [8].

Definition 2. (Least Prejudiced Distribution) [6]
For A' a fuzzy subset of a finite universe Ω such that A' is normalized, the least prejudiced distribution of A', denoted $lp_{A'}$, is a probability distribution on Ω given by

$$lp_{A'}(x) = \sum_{F_i : x \in F_i} \frac{m_{A'}(F_i)}{|F_i|} \tag{8}$$

Theorem 1. *[9] Let P be a probability distribution on a finite universe Ω taking as a range of values $\{p_1, ..., p_n\}$ where $0 \leq p_{i+1} < p_i \leq 1$ and $\sum_{i=1}^{n} p_i = 1$. Then P is the least prejudiced distribution of a fuzzy set A' if and only if A' has a mass assignment given by*

$$m_{A'}(F_i) = \mu_i - \mu_{i+1} \ for \ i = 1, ..., n-1$$
$$m_{A'}(F_n) = \mu_n$$
$$where$$
$$F_i = \{x \in \Omega | P(x) \geq p_i\}$$
$$\mu_i = |F_i|p_i + \sum_{j=i+1}^{n} (|F_j| - |F_{j+1}|)p_j$$

Proof (see [9])   □

It is worth mentioning that the above algorithm is identical to the bijection method proposed by Dubois and Prade [10] although the motivation in [9] is quite different. Also Yager [25] considered a similar approach to mapping between probability and possibility. A further justification for the transformation was given by Yamada [26].

In other words, Theorem 1 gives a general procedure converting a relative frequency distribution into a fuzzy set, i.e. gives us means for generating fuzzy sets from data.

But Theorem 1 gives also an idea how to convert the relative frequency distributions into A-IFSs.

When discussing A-IFSs we consider memberships and independently given non-memberships so Theorem 1 gives only a part of the description we look for. To receive the full description of an A-IFS (with independently given memberships and non-memberships), it is necessary to repeat the procedure as in Theorem 1 two times. In result we obtain two fuzzy sets. To interpret them properly in terms of A-IFSs we recall first a semantic for membership functions.

Dubois and Prade [11] have explored three main semantics for membership functions – depending on the particular applications. Here we apply the interpretation proposed by Zadeh [28] when he introduced the possibility theory. Membership $\mu(x)$ is there the degree of possibility that a parameter x has value μ.

In effect of repeating the procedure as in Theorem 1 two times (first – for data representing memberships, second – for data representing non-memberships), and taking into account interpretation that the obtained values are the degrees of possibility we receive the following results.

- First time we perform the steps from Theorem 1 for the relative frequencies connected to memberships. In effect we obtain (fuzzy) possibilities $Pos^+(x) = \mu(x) + \pi(x)$ that x has value Pos^+.
  $Pos^+(x)$ (left side of the above equation) mean the values of a membership function for a fuzzy set (possibilities). In terms of A-IFSs (right side of the above equation) these possibilities are equal to possible (maximal) memberships of an A-IFS, i.e. $\mu(x) + \pi(x)$, where $\mu(x)$ – the values of the membership function for an A-IFS, and $\mu(x) \in [\mu(x), \mu(x) + \pi(x)]$.
- Second time we perform the steps from Theorem 1 for the (independently given) relative frequencies connected to non-memberships. In effect we obtain (fuzzy) possibilities $Pos^-(x) = \nu(x) + \pi(x)$ that x has not value Pos^-.
  $Pos^-(x)$ (left side of the above equation) mean the values of a membership function for another (than in the previous step) fuzzy set (possibilities). In terms of A-IFSs (right side of the above equation) these possibilities are equal to possible (maximal) non-memberships, i.e. $\nu(x) + \pi(x)$, where $\nu(x)$ – the values of the non-membership function for an A-IFS, and $\nu(x) \in [\nu(x), \nu(x) + \pi(x)]$.

The algorithm of assigning the parameters of A-IFSs:

1. From Theorem 1 we calculate the values of the left sides of the equations:

$$Pos^+(x) = \mu(x) + \pi(x) \tag{9}$$

$$Pos^-(x) = \nu(x) + \pi(x) \tag{10}$$

2. From (9)–(10), and taking into account that $\mu(x) + \nu(x) + \pi(x) = 1$, we obtain the values $\pi(x)$

$$Pos^+(x) + Pos^-(x) = \mu(x) + \pi(x) + \nu(x) + \pi(x) = 1 + \pi(x) \tag{11}$$

$$\pi(x) = Pos^+(x) + Pos^-(x) - 1 \tag{12}$$

3. Having the values $\pi(x)$, from (9) and (10) we obtain for each x: $\mu(x)$, and $\nu(x)$.

This way, starting from relative frequency distributions, and using Theorem 1, we receive full description of an A-IFS.

4 The Models of a Classifier Error

Traditionally *accuracy* of a classifier is measured as the percentage of instances that are correctly classified, and *error* is measured as the percentage of incorrectly classified instances (unseen data). *Accuracy* and *error* are simple to calculate and understand but it is well known (e.g., [18]) that in a case of imbalanced classes (typically having highly non-uniform error cost eg., medical diagnosis, fraud detection), a smaller class, being a class of primary interest is poorly recognized. To see this, consider a case with 96% of the instances belonging to the bigger (*illegal*) class, and 4% of the instances belonging to the smaller (*legal*) class – the class we are interested in. *Accuracy* of a classifier which recognizes all instances as *illegal* is equal to 96%. Although it looks high, the classifier would be useless because it totally fails to recognize the smaller class.

To avoid such situations, other measures, like TPR and FPR are also considered (cf. Table 2) while assessing a classifier dealing with imbalanced classes.

4.1 Confusion Matrix

The confusion matrix (Table 2) is often used to assess a two–class classifier. The meaning of the symbols is
a – the number of correctly classified legal points,
b – the number of correctly classified illegal points,
c – the number of incorrectly classified legal points,

Table 2. The confusion matrix

	Tested legal	Tested illegal
Actual legal	a	b
Actual illegal	c	d

d – the number of incorrectly classified illegal points,

$$TPR = \frac{legalls\ correctly\ classified}{total\ legalls} = \frac{a}{a+b} \qquad (13)$$

$$FPR = \frac{illegals\ incorrectly\ classified}{total\ illegals} = \frac{d}{c+d} \qquad (14)$$

5 Classifiers for the Imbalanced Classes

The data set D used to demonstrate the problems with imbalanced data, and to present and compare two classifiers (a fuzzy classifier and an intuitionistic fuzzy classifier) is presented in Fig. 1. The task lies in classification of the points belonging to the ellipse (Fig. 1).

The data set D consists of 288 data points from a regular grid with universes Ω_X and Ω_Y being $[-1.5, 1.5]$. Legal points lie within the ellipse

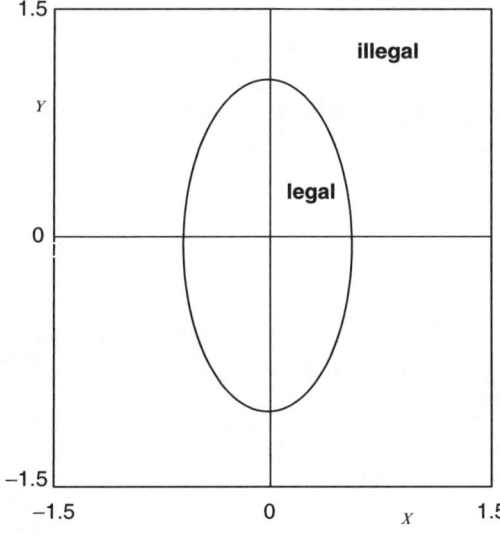

Fig. 1. Ellipse inequality in Cartesian space. Points inside the ellipse are classified as legal, points outside the ellipse are classified as illegal

$y^2 + 2x^2 \leq 1$, illegal points outside the ellipse. We divide D into two equal parts: D_1 – the training set, and D_2 – the testing set. Each data sample consists of a triple $<X, Y, CLASS>$; class is LEGAL when the point (X, Y) satisfies the ellipse inequality and ILLEGAL otherwise.

5.1 Classification via Fuzzy Sets

In the ellipse problem we have two single attribute input features: X and Y. We formed Cartesian granule fuzzy sets corresponding to the legal and illegal classes over the Cartesian product space of the partitions P_X and P_Y. The number of fuzzy sets to form the fuzzy partition of each universe is a separate problem. We verified several possibilities and decided for ten fuzzy sets over each universe – for more fuzzy sets no significant gain in terms of model prediction was made. We divided the training database D_1 into two smaller databases according to the output classification. Then we took the points corresponding to the LEGAL/ILLEGAL class and formed a Cartesian granule fuzzy set for the LEGAL/ILLEGAL class.

In order to generate the body fuzzy sets we partitioned X and Y universe with the following fuzzy sets

$$p_{X_1} = p_{Y_1} = [-1.5 : 1, -1.167 : 0]$$
$$p_{X_2} = p_{Y_2} = [-1.5 : 0, -1.167 : 1, -0.833 : 0]$$
$$p_{X_3} = p_{Y_3} = [-1.167 : 0, -0.833 : 1, -0.5 : 0]$$
$$p_{X_4} = p_{Y_4} = [-0.833 : 0, -0.5 : 1, -0.167 : 0]$$
$$p_{X_5} = p_{Y_5} = [-0.5 : 0, -0.167 : 1, 0.167 : 0]$$
$$p_{X_6} = p_{Y_6} = [-0.167 : 0, 0.167 : 1, 0.5 : 0]$$
$$p_{X_7} = p_{Y_7} = [0.167 : 0, 0.5 : 1, 0.833 : 0]$$
$$p_{X_8} = p_{Y_8} = [0.5 : 0, 0.833 : 1, 1.167 : 0]$$
$$p_{X_9} = p_{Y_9} = [0.833 : 0, 1.167 : 1, 1.5 : 0]$$
$$p_{X_{10}} = p_{Y_{10}} = [1.167 : 0, 1.5 : 1]$$

Next, from D_1 (training 144 triples) we evaluated the probability distributions on the above fuzzy partition, taking

$$P_X(p_i|legal) = \sum_{x \in D_1 : CLASS=legal} \mu_{p_i}(X)/|x \in D_1|CLASS = legal|$$

$$P_X(p_i|illegal) = \sum_{x \in D_1 : CLASS=illegal} \mu_{p_i}(X)/|x \in D_1|CLASS = illegal|$$

$$P_Y(p_i|legal) = \sum_{y \in D_1 : CLASS=legal} \mu_{p_i}(Y)/|y \in D_1|CLASS = legal|$$

$$P_Y(p_i|illegal) = \sum_{y \in D_1 : CLASS=illegal} \mu_{p_i}(Y)/|y \in D_1|CLASS = illegal|$$

Table 3. Model of the data – Probability distributions on the fuzzy partition

	Interval									
	1	2	3	4	5	6	7	8	9	10
$P_X(p_i\|legal)$	0	0	0.0341	0.1648	0.3011	0.3011	0.1648	0.0341	0	0
$P_X(p_i\|illegal)$	0.1266	0.1266	0.1169	0.0796	0.0503	0.0503	0.0796	0.1169	0.1266	0.1266
$P_Y(p_i\|legal)$	0	0.0227	0.1080	0.1477	0.2216	0.2216	0.1477	0.1080	0.0227	0
$P_Y(p_i\|illegal)$	0.1266	0.1201	0.0958	0.0844	0.0731	0.0731	0.0844	0.0958	0.1201	0.1266

The results (the probability distributions on Ω_X and Ω_Y) are given in Table 3. Then Theorem 1 was used to find the following approximation of the fuzzy sets (for data D_1)

– Legal data Ω_X

$$
\begin{aligned}
CLASS(legal_X) \: : \: & F_{X_1}/0 + F_{X_2}/0 + F_{X_3}/0.204545455 \\
& + F_{X_4}/0.727272727 + F_{X_5}/1 + F_{X_6}/1 \\
& + F_{X_7}/0.727272727 + F_{X_8}/0.204545455 \\
& + F_{X_9}/0 + F_{X_{10}}/0
\end{aligned}
\tag{15}
$$

– Illegal data Ω_X

$$
\begin{aligned}
CLASS(illegal_X) \: : \: & F_{X_1}/1 + F_{X_2}/1 + F_{X_3}/0.961038961 \\
& + F_{X_4}/0.737012987 + F_{X_5}/0.503246753 \\
& + F_{X_6}/0.503246753 + F_{X_7}/0.737012987 \\
& + F_{X_8}/0.961038961 + F_{X_9}/1 + F_{X_{10}}/1
\end{aligned}
\tag{16}
$$

– Legal data Ω_Y

$$
\begin{aligned}
CLASS(legal_Y) \: : \: & F_{Y_1}/0 + F_{Y_2}/0.181818182 + F_{Y_3}/0.693181818 \\
& + F_{Y_4}/0.852272727 + F_{Y_5}/1 + F_{X_6}/1 \\
& + F_{Y_7}/0.852272727 + F_{Y_8}/0.693181818 \\
& + F_{Y_9}/0.181818182 + F_{Y_{10}}/0
\end{aligned}
\tag{17}
$$

– Illegal data Ω_Y

$$
\begin{aligned}
CLASS(illegal_Y) \: : \: & F_{Y_1}/1 + F_{Y_2}/0.987012987 + F_{Y_3}/0.88961039 \\
& + F_{Y_4}/0.821428571 + F_{Y_5}/0.730519481 \\
& + F_{X_6}/0.730519481 + F_{Y_7}/0.821428571 \\
& + F_{Y_8}/0.88961039 + F_{Y_9}/0.987012987 + F_{Y_{10}}/1
\end{aligned}
\tag{18}
$$

Having the above fuzzy sets describing legal and illegal data in X and in Y we tested the model (on D_2 – another 144 triplets). We used a very simple classification rule – assigning a data point (X, Y) to the class to which it belongs most (highest membership values for both universes).

Table 4. The confusion matrix for fuzzy set classifier

	Tested legal	Tested illegal
Actual legal	16	16
Actual illegal	112	0

Table 5. Fuzzy classifier: tested results – confusion matrices for each of ten regions

Actual classes ↓	Tested results																			
	1		2		3		4		5		6		7		8		9		10	
	+	−	+	−	+	−	+	−	+	−	+	−	+	−	+	−	+	−	+	−
X: +	0	0	0	0	0	2	4	4	12	10	12	10	4	4	0	2	0	0	0	0
X: −	24	0	36	0	22	0	16	0	14	0	14	0	16	0	22	0	36	0	24	0
Y: +	0	0	0	2	0	6	4	4	12	4	12	4	4	4	0	6	0	2	0	0
Y: −	24	0	34	0	18	0	16	0	20	0	20	0	16	0	18	0	34	0	24	0

+ Means: legal; − means: illegal

Table 6. Fuzzy classifier: errors for tested results in each of ten regions

	Tested results									
	1	2	3	4	5	6	7	8	9	10
X: accuracy	1	1	0.92	0.83	0.72	0.72	0.83	0.92	1	1
X: TPR	0	0	0	0.5	0.55	0.55	0.5	0	0	0
X: FPR	0	0	0	0	0	0	0	0	0	0
Y: accuracy	1	0.94	0.75	0.83	0.89	0.89	0.83	0.75	0.94	1
Y: TPR	0	0	0	0.5	0.75	0.75	0.5	0	0	0
Y: FPR	0	0	0	0	0	0	0	0	0	0

$(X, Y) \in legal$

$$\Leftrightarrow legal = arg \max[\mu_{CLASS_X}(X, Y); CLASS_X \in \{legal, illegal\}]$$

and

$$\Leftrightarrow legal = arg \max[\mu_{CLASS_Y}(X, Y); CLASS_Y \in \{legal, illegal\}] \quad (19)$$

The accuracy on the test data D_2 using (19) was 88.9%. We also assessed the results using the confusion matrix – Table 4. It turned out that the classifier had difficulties with recognition of the legal (smaller) class. Only 16, i.e. the half of the tested points belonging to the legal class were correctly classified ($TPR = 0.5$). On the other hand, all 112 points belonging to the bigger – illegal class were correctly classified ($FPR = 0$) (Tables 5 and 6).

5.2 Classification via Atanassov's Intuitionistic Fuzzy Sets

We solved the same problem but with additional possibilities giving by A-IFSs. In Sect. 3 it was shown how to convert relative frequencies to A-IFSs (and the meaning of all the parameters was discussed). So first we converted our training data set D_1 obtaining A-IFSs describing legal and illegal classes in

Table 7. Model of the data – intuitionistic fuzzy description in each of ten regions

	Interval									
	1	2	3	4	5	6	7	8	9	10
X=legal: Possibility:	0	0	0.2045	0.7273	1	1	0.7273	0.2045	0	0
X=legal: hesitation margin	0	0	0.1656	0.4643	0.5033	0.5033	0.4643	0.1656	0	0
X=legal:membership	0	0	0.2045	0.7273	1	1	0.7273	0.2045	0	0
X=illegal: Possibility:	1	1	0.9610	0.7370	0.5033	0.5033	0.7370	0.9610	1	1
X=illegal: hesitation margin	0	0	0.1656	0.4643	0.5033	0.5033	0.4643	0.1656	0	0
X=illegal:membership ↓	1	1	0.7955	0.2727	2.22E-16	2.22E-16	0.2727	0.7955	1	1
Y=legal: Possibility:	0	0.1818	0.6932	0.8523	1	1	0.8523	0.6932	0.1818	0
Y=legal: hesitation margin	0	0.1688	0.5828	0.6737	0.7305	0.7305	0.6737	0.5828	0.1688	0
Y=legal:membership	0	0.1818	0.6932	0.8523	1	1	0.8523	0.6932	0.1818	0
Y=illegal: Possibility:	1	0.9870	0.8896	0.8214	0.7305	0.7305	0.8214	0.8896	0.9870	1
Y=illegal: hesitation margin	0	0.1688	0.5828	0.6737	0.7305	0.7305	0.6737	0.5828	0.1688	0
Y=illegal:membership ↓	1	0.8182	0.3068	0.1477	1.11E-16	1.11E-16	0.1477	0.3068	0.8182	1

X and in Y – Table 7. We exploited the information about hesitation margins (making use of the fact that legal and illegal classes overlap). Taking into account that hesitation margins assign (the width of the) intervals where the unknown values of memberships lies, we applied in the model the following values:

- Maximal possible values of the memberships describing the legal class (see Table 7 – the values of memberships for the legal class both in Ω_X and Ω_Y are given in bolds).
- Minimal possible values of the memberships describing the illegal class (see Table 7 – the minimal possible values of the memberships for illegal class were obtained, both in Ω_X and Ω_Y, by subtracting the hesitation margins from the maximal possible values of the memberships for the illegal class – this operation is signed by: ↓ – Table 7).

This way in the training phase we formed Cartesian granule A-IFSs corresponding to the legal and illegal classes in such a way that the legal class should be seen as good as possible.

The results for tested data D_2 (the same rule (19) was used) are the following – the accuracy is equal to 94.4% – better result than those obtained when applying fuzzy set approach (88.9%). But the most interesting is the difference in separate classification of legal and illegal classes by both classifiers. General results are in Tables 4 and 8. The smaller class is better classified – 28 legal elements were correctly classified instead of 16 for fuzzy classifier. In effect TPR is bigger (0.875 instead of 0.5). Of course, in effect FPR is a little bigger (0.036 instead of 0) as the result of decreasing memberships for illegal class (4 incorrectly classified elements instead of 0 for fuzzy classifier). But now the smaller class is better classified, and the general accuracy is also better.

Table 8. The confusion matrix for intuitionistic fuzzy classifier

	Tested legal	Tested illegal
Actual legal	28	4
Actual illegal	108	4

Table 9. Intuitionistic Fuzzy classifier: tested results – confusion matrices for each of ten regions

Actual classes ↓	1 +	1 −	2 +	2 −	3 +	3 −	4 +	4 −	5 +	5 −	6 +	6 −	7 +	7 −	8 +	8 −	9 +	9 −	10 +	10 −
X: +	0	0	0	0	0	2	6	2	22	0	22	0	6	2	0	2	0	0	0	0
X: −	24	0	36	0	22	0	14	2	12	2	12	2	14	2	22	0	36	0	24	0
Y: +	0	0	2	0	6	0	8	0	12	4	12	4	8	0	6	0	2	0	0	0
Y: −	24	0	32	2	16	2	16	0	20	0	20	0	16	0	16	2	32	2	24	0

+ means: legal; − means: illegal

Table 10. Intuitionistic fuzzy classifier: errors for tested results in each of ten regions

	1	2	3	4	5	6	7	8	9	10
X: accuracy	1	1	0.92	0.83	0.94	0.94	0.83	0.92	1	1
X: TPR	0	0	0	0.75	1	1	0.75	0	0	0
X: FPR	0	0	0	0.13	0.14	0.14	0.13	0	0	0
Y: accuracy	1	0.94	0.92	1	0.89	0.89	1	0.92	0.94	1
Y: TPR	0	1	1	1	0.75	0.75	1	1	1	0
Y: FPR	0	0.059	0.11	0	0	0	0	0.11	0.059	0

It is also interesting to compare the results in separate intervals for both universes – Tables 5, 6 and 9, 10 for the fuzzy classifier and intuitionisticf fuzzy classifier respectively. The misclassified data is located for the most part around the regions of high curvature or high rates of change over both universes Ω_X and Ω_Y.

In intervals 4 and 7 only half of the points were classified correctly – Ω_X: $TPR = 0.5$ for fuzzy classifier, whereas for intuitionistic fuzzy classifier for the same intervals $TPR = 0.75$. Even better performance was obtained in intervals 5 and 6: $TPR = 0.55$ for fuzzy classifier, and $TPR = 1$ for intuitionistic fuzzy classifier (all points from the smaller legal class were properly classified). Only in intervals 3 and 8 $TPR = 0$ for both classifiers which were not able to see two points from the legal class (on the other side, accuracy of both classifiers in the intervals is the highest: 0.92).

For Ω_Y fuzzy classifier does not see legal class at all in intervals 2, 3 and 8, 9 – $TPR = 0$ whereas intuitionistic fuzzy classifier correctly classified all the points – $TPR = 1$. But for intervals 5 and 6 (for which we observed the best improvement by intuitionistic fuzzy classifier in Ω_X) we do not observe any changes – $TPR = 0.75$ and accuracy is equal to 0.89 for both classifiers.

As it has been already mentioned, we pay for the improved classification of the smaller legal class in the sense of increasing the values of FPR. But the changes are small and they have not influence at all on the general error/accuracy of the classifier. Opposite – the general accuracy of the intuitionistic fuzzy classifier is bigger.

But as the structure of the data considered in the ellipse example was specific, in the next section we examine the proposed methods using a benchmark data set.

6 Results for a Benchmark Data

We examined the performance of the discussed in Sect. 5 classifiers using a benchmark data, namely, *Wine* data set from the UCI ML Repository [24]. The classification was made on the basis of 13 attributes.

To illustrate the performance of the considered classifiers designed for imbalanced classes, we solved a two-class classification problem – class by class was treated as *legal* class (minority class) and the rest 2 classes as one (majority) *illegal* classes. In Table 11 there are listed the natural class distributions of the data sets expressed as the minority class percentage of the whole data set. For example, if Wine 1 (59 instances) is the minority class, the remaining two classes ($71 + 48 = 119$ instances) are treated as one majority class. Three separate experiments were performed – each one for classifying one type of wine. In each experiment (for a chosen class to be classified as *legal*) the database consisting of 178 instances was split into a training set and test set in such a way that the instances of a *legal* (minor) class, and *illegal* (major class consisting of the sum of the rest classes) were divided equally between D_1 and D_2 (every second instance was assigned to D_1; the remaining instances were assigned to D_2). Asymmetric triangular fuzzy partitioning was then defined for each attribute (Baldwin, and Karale, 2003), i.e., the training data set was divided so that each fuzzy partition had almost the same number of data (instances) associated with it. Next, for each case, the fuzzy models of data (for *legal* and *illegal* classes) were constructed (as described in Sect. 5.1 – separately for each attribute. In effect, for each experiment (each type of wine)

Table 11. Data set used in our experiment ([24])

Wine	Name	size	Minority class %
1	Wine 1	59	33
2	Wine 2	71	40
3	Wine 3	48	27

13 classifiers (for each of 13 attributes) were derived, and next – aggregated. The following aggregation was used

$$Agg_1 : Agg_1^{CLASS}(e) = \sum_{k=1}^{n} w_k \mu_{CLASS}^k(e) \tag{20}$$

where e – an examined instance from a database,
$w_k = \frac{n_k}{\sum\limits_{k=1}^{n} n_k}$ for $k = 1, \ldots, n$ is a set of weights for each attribute: n_k is the
number of correctly classified training data by k-th attribute.

Knowing the aggregation Agg_1, the classification of an examined instance was done by evaluating

$$D_1(e) = arg \max[Agg_1^{CLASS}(e), CLASS \in \{legal, illegal\}] \tag{21}$$

The described above procedure concerns a fuzzy classifier (*legal* and *illegal* data were given as fuzzy sets). The same procedure was repeated to construct an intuitionistic fuzzy classifier. The difference was that the data (given originally in the form of the frequency distributions) were converted (cf. Sect. 3) into A-IFSs. In effect each examined instance e was described due to the definition of A-IFSs) by a triplet: membership value to a *legal* (smaller) class, non-membership value to a *legal* class (equal to membership value to *illegal*-bigger class) and hesitation margin, i.e.

$$e : (\mu_e, \nu_e, \pi_e) \tag{22}$$

To enhance the possibility of a proper classification of the instances belonging to a smaller (*legal*) class, while training the intuitionistic fuzzy classifier, the values of the hesitation margins were divided so to "see" better the smaller class – each instance e (22) was expressed as

$$e : (\mu_e + \alpha\pi_e, \nu_e + (1 - \alpha)\pi_e) \tag{23}$$

where $\alpha \in (0.5, 1)$ is a parameter.

The results obtained are given in Table 12. Here we discuss the test results only. Wine 1 was recognized by the intuitionistic fuzzy classifier ($\alpha = 0.7$) with 97.75% accuracy, and TPR equal to 0.966 (better than for fuzzy classifier: 92.13%, and 0.793, respectively). More, the better behavior was not achieved at the price of FPR which in both cases is the same (0.017).

Wine 2 was also recognized with better *accuracy* (98.88%) by the intuitionistic fuzzy classifier ($\alpha = 0.7$) than for fuzzy classifier (94.38%). TPR for both classifiers is excellent (equals to 1), but the intuitionistic fuzzy classifier behaves worse in the sense of FPR which value increases from zero for fuzzy classifier to 0.015. It is the price paid for increasing the *accuracy* from 94.38 to 98.88%.

Wine 3 was recognized without any mistakes by both classifiers.

Table 12. Classification results obtained by a fuzzy classifier and intuitionistic fuzzy classifier

		Fuzzy classifier		Intuitionistic Fuzzy classifier	
		training	testing	training	testing
	Accuracy (%)	100	92.13	100	97.75
Wine 1	*TPR*	1	0.793	1	0.966
$\alpha = 0.7$	*FPR*	0	0.017	0	0.017
	Accuracy (%)	100	94.38	100	98.88
Wine 2	*TPR*	1	1	1	1
$\alpha = 0.7$	*FPR*	0	0	0	0.015
	Accuracy (%)	100	100	100	100
Wine 3	*TPR*	1	1	1	1
$\alpha = 0.5$	*FPR*	0	0	0	0

The results seem very promising. Each kind of wine is seen better (or, at least, not worse) by the intuitionistic classifier than by the fuzzy classifier. In other words, the intuitionistic fuzzy classifier better recognizes the smaller classes.

7 Conclusions

We proposed a simple intuitionistic fuzzy classifier for imbalanced and over-lapping data. Detailed analysis of the errors has shown that using intuitionistic fuzzy sets gives better results than the counterpart approach via fuzzy sets. Better performance of the intuitionistic fuzzy classifier concerns especially the recognition power of a smaller class. It is strongly connected with the fact that Atanassov's intuitionistic fuzzy sets, being a generalization of fuzzy sets, make use of more parameters (memberships, non-memberships, and hesitation margins) so the resulting models are more reliable.

In some cases we have noticed, besides the improvement of the better recognition power of the smaller class, also the improvement of the *accuracy* of the intuitionistic fuzzy classifier in comparison with a fuzzy classifier. We examine further benchmark data sets so to look for the connection (if any) between the the distributions of the data sets and the behavior of the proposed classifier.

References

1. Atanassov K. (1983), Intuitionistic Fuzzy Sets. VII ITKR Session. Sofia (Deposed in Central Sci. Techn. Library of Bulg. Acad. of Sci., 1697/84) (in Bulgarian).
2. Atanassov K. (1986) Intuitionistic fuzzy sets. Fuzzy Sets and Systems, 20, 87–96.

3. Atanassov K. (1999), Intuitionistic Fuzzy Sets: Theory and Applications. Springer, Berlin Heidelberg New York.
4. Baldwin J.F. (1991), Combining Evidences for Evidential Reasoning. International Journal of Intelligent Systems, 6, 569–616.
5. Baldwin J.F. (1992), The management of fuzzy and probabilistic uncertainties for knowledge based systems. In S.A. Shapiro (ed.), Encyclopaedia of AI, Wiley, New York (2nd edn.), 528–537.
6. Baldwin J.F., Martin T.P., Pilsworth B.W. (1995) FRIL – Fuzzy and Evidential Reasoning in Artificial Intelligence. Wiley, New York.
7. Baldwin J.F., Lawry J., Martin T.P. (1995), A Mass Assignment Theory of the Probability of Fuzzy Events. ITRC Report 229, University of Bristol, UK.
8. Baldwin J.F., Coyne M.R., Martin T.P. (1995), Intelligent reasoning using general knowledge to update specific information: a database approach. Journal of Intelligent Information Systems, 4, 281–304.
9. Baldwin J.F., Lawry J., Martin T.P. (1998), The application of generalized fuzzy rules to machine learning and automated knowledge discovery. International Journal of Uncertainty, Fuzzyness and Knowledge-Based Systems, 6(5), 459–487.
10. Dubois D. and Prade H. (1983) Unfair coins and necessity measures: towards a possibilistic interpretation of histograms. Fuzzy Sets and Systems 10, 15–20.
11. Dubois D. and Prade H. (1997) The three semantics of fuzzy sets. Fuzzy Sets and Systems, 90, 141–150.
12. Fawcett T. and Provost F. (1997) Adaptive fraud detection. Data Mining and Knowledge Discovery, 3(1), 291–316.
13. Japkowicz N. (2003) Class Imbalances: Are we Focusing on the Right Issue? ICML, Washington 2003.
14. Kubat M., Holte R., Matwin S. (1998) Machine learning for the detection of oil spills in satellite radar images. Machine Learning, 30, 195–215.
15. Lewis D. and Catlett J. (1994) Heterogeneous uncertainty sampling for supervised learning. Proc. 11th Conf. on Machine Learning, 148–156.
16. Mladenic D. and Grobelnik M. (1999) Feature selection for unbalanced class distribution and naive bayes. 16th Int. Conf. on Machine Learning, 258–267.
17. Provost F. (2001) Machine learning from imbalanced data sets (extended abstract). Available at citeseer.ist.psu.edu/387449.html.
18. Provost F., Fawcett T. and Kohavi R. (1998) The case against accuracy estimation for comparing classifiers. 5th Int. Conference on Machine Learning, San Francisco, Kaufman Morgan.
19. Szmidt E. and Baldwin J. (2003) New similarity measure for intuitionistic fuzzy set theory and mass assignment theory. Notes on IFSs, 9(3), 60–76.
20. Szmidt E. and Baldwin J. (2004) Entropy for intuitionistic fuzzy set theory and mass assignment theory. Notes on IFSs, 10(3), 15–28.
21. Szmidt E. and Baldwin J. (2005) Assigning the parameters for Intuitionistic Fuzzy Sets. Notes on IFSs, 11(6), 1–12.
22. Szmidt E. and Kacprzyk J. (2002). An intuitionistic fuzzy set base approach to intelligent data analysis (an application to medical diagnosis). In A. Abraham, L. Jain, J. Kacprzyk (Eds.): Recent Advances in Intelligent Paradigms and Applications. Springer, Berlin Heidelberg New York, 57–70.
23. Szmidt E. and Kacprzyk J. (2005). A new concept of similarity measure for intuitionistic fuzzy sets and its use in group decision making. In V. Torra,

Y. Narukawa, S. Miyamoto (Eds.): Modelling Decisions for AI. LNAI 3558, Springer, Berlin Heidelberg New York, 2005, 272–282.

24. UCI machine learning repository (in world wide web) http://www.ics.uci.edu/mlearn/MLRepository.html
25. Yager R.R. (1979) Level sets for membership evaluation of fuzzy subsets. Tech. Rep. RRY-79-14, Iona Colledge, New York.
26. Yamada K. (2001) Probability–possibility transformation based on evidence theory. Proc. IFSA–NAFIPS'2001, 70–75.
27. Zadeh L.A. (1965) Fuzzy sets. Information and Control, 8, 338–353.
28. Zadeh L.A. (1978) Fuzzy sets as the basis for a theory of possibility. Fuzzy Sets and Systems, 1, 3–28.

Representation of Value Imperfection with the Aid of Background Knowledge: H-IFS

Boyan Kolev[1], Panagiotis Chountas[2], Ermir Rogova[2], and Krassimir Atanassov[1]

[1] CLBME – Bulgarian Academy of Sciences, B1 105, Sofia-1113, Bulgaria
[2] HSCS – University of Westminster, London HA1 3TP, UK
chountpwmin.ac.uk, kratargo.bas.bg

Summary. An Integrated Data Management System is required for representing and managing indicative information from multiple sources describing the state of an enterprise. In such environments, information may be partially known because the related information from the real world corresponds to a set of possible values including the unknown. Here, we present a way to replace unknown values using background knowledge of data that is often available arising from a concept hierarchy, as integrity constraints, from database integration, or from knowledge possessed by domain experts. We present and examine the case of H-IFS to represent support contained in subsets of the domain as a candidate for replacing unknown values mostly referred in the literature as NULL values.

1 Introduction

Background knowledge of data is often available, arising from a concept hierarchy, as integrity constraints, from database integration, or from knowledge possessed by domain experts. Frequently integrated DBMSs contain incomplete data which we may represent by using H-IFS to declare support contained in subsets of the domain. These subsets may be represented in the database as partial values, which are derived from background knowledge using conceptual modelling to re-engineer the integrated DBMS. For example, we may know that the value of the attribute JOB-DESCRIPTION is unknown for the tuple relating to employee Natalie but also know from the attribute salary that Natalie receives an estimated salary in the range of €5 K $\sim$ Salary$_{25K}$. A logic program, using a declarative language can then derive the result that Natalie is a "Junior- Staff", which we input to the attribute JOB-DESCRIPTION of tuple Natalie in the re-engineered database. In such a manner we may use the knowledge base to replace much of the unknown in the integrated database environment.

Generalised relations have been proposed to provide ways of storing and retrieving data. Data may be imprecise, hence we are not certain about the

B. Kolev et al.: *Representation of Value Imperfection with the Aid of Background Knowledge: H-IFS*, Studies in Computational Intelligence (SCI) **109**, 473–492 (2008)
www.springerlink.com   © Springer-Verlag Berlin Heidelberg 2008

specific value of an attribute but only that it takes a value which is a member of a set of possible values. An extended relational model for assigning data to sets has been proposed by [1]. This approach may be used either to answer queries for decision making or for the extraction of patterns and knowledge discovery from relational databases. It is therefore important that appropriate functionality is provided for database systems to handle such information.

A model, which is based on partial values [2], has been proposed to handle imprecise data. Partial values may be thought of as a generalisation of null values where, rather than not knowing anything about a particular attribute value, we may identify the attribute as a set of possible values. A partial value is therefore a set such that exactly one of the values in the set is the true value.

We review the different types of NULL values and then we focus is on providing an integrated DBMS environment that will enable us to:

- Reconcile unknown information with the aid of background knowledge. We examine the appropriateness of the H-IFS as a form of Background knowledge for replacing unknown attribute values
- Utilise constraints as part of the integrated DBMS metadata to improve query execution
- Query imprecise information a part of integrated DBMS environment that may entail more than one sources of information

2 Review of NULL Values

A null value represents an *unknown* attribute value, is a value that is known to exist, but the actual value is unknown. The unknown value is assumed to be a valid attribute value, that is, some value in the domain of that attribute. This is a very common kind of ignorant information. For example, in an employee database, while everyone must have a surname, Alex's surname may be recorded as unknown. The unknown value indicates that Alex has a name, but we do not know her name. An unknown value has various names in the literature including *unknown null* [3], *missing null* [4], and *existential null* [5].

The meaning of a fact, F, with an unknown attribute value over an attribute domain of cardinality N is a multiset with N members; each member is a set containing an F instance with the unknown value being replaced by a different value from the attribute domain. For example, assume $f = \{\text{IBM}(\perp)\}$ where ($\perp$ represents an unknown value over a domain $\{20, 22\}$ with respect to IBM's share-price), then the meaning of f is

$$f = \{\{\text{IBM}(20)\}, \{\text{IBM}(22)\}\}$$

This corresponds to the notion that a fact with an unknown value is incomplete with respect to a fact where that unknown value is no longer unknown, but is now known to be a specific value (i.e. $f_1 = \{\{\text{IBM}(20)\}\}$).

Another generalization of an unknown fact is a *disjunctive* fact [6], so known as *indefinite* information [7]. A disjunction is a logical *or* applied to fact instances. Let F be an inclusive disjunctive fact with N disjuncts. The meaning of F is given by a multiset with N members; each member is a set containing one disjunct. For example, the share price of IBM may be £20 or £22. (i.e. "IBM (20), IBM (22)"). The disjunction is *exclusive* [8] or *inclusive* [9]. If it is an exclusive disjunction, one and only one disjunct is true. The meaning of an exclusive disjunctive fact is the same as that of an imprecise value. Let $f = \{\{IBM(20)\}, \{IBM(22)\}\}$ be an exclusive disjunctive then the meaning of f is $f = \{IBM(20)\} \vee \{IBM(22)\}$.

The meaning of an inclusive disjunctive fact is somewhat different than that of its exclusive complement, at least one alternative may be true. Let F be an inclusive disjunctive fact with N disjuncts. The meaning of F is given by a multiset with $2^N - 1$ members; each member is a unique subset of disjuncts. For example, assume, the inclusive disjunct $f = \{IBM(20)\{IBM(22)\}$ then the meaning of f, is $f = \{\{IBM(20)\}, \{IBM(22)\}, \{IBM(20), IBM(22)\}\}$, excluding the fact, $\{\{IBM(\perp)\}$. The empty ($\perp$) attribute represents the situation where a fact instance exists, but does not have a particular attribute-label value.

A *maybe value* is an attribute-label value which might or might not exist [10]. If it does exist, the value is known. A *maybe tuple or fact-instance* is similar to a maybe value, but the entire tuple might not be part of the relation. Maybe tuples are produced when one disjunct of an inclusive disjunctive fact-instance is found to be true.

A combination of inclusive disjunctive fact instance and a maybe fact instance can determine the semantics of *open information* or nulls [11]. The denotation of an open null is exact to inclusive disjunctive information with the addition of the empty set as a possible value. That is the attribute-lable value may not exist, could be exactly one value, or could be many values. For example, in the shares database an open value could be used to present IBM share prices. This value means that IBM share price possibly had a past record, (this could be the first appearance in the market); IBM share price may be one or many. The open value covers all this possibilities. A generalization of open information is *possible* information [12] (this differs from our use of the term "possible"). Possible information is an attribute value whose existence is undetermined, but if it does exist, it could be multiple values from a *subset* of the attribute domain.

A *no information* value is a combination of an open value and an unknown value [13]. The no information value restricts an open value to resemble an unknown value. A, no information value might not exist, but if it does, then it is a single value, which is unknown, rather than possibly many values. The meaning of a no information value is similar to that of an *unknown* value with the inclusion of the addition of the empty set as a possible value.

Unknown, partially known, open, no information, and maybe null values are different interpretations of a null value. There are other null value interpretations, but none of these is a kind of well cognisant information.

An *inapplicable* or *does not exist* null is a very common null value. An inapplicable null, appearing as an attribute value, means that an attribute does not have a value [14]. An inapplicable value neither contains nor represents any ignorance; it is known that the attribute value does not exist. Inapplicable values indicate that the schema (usually for reasons of efficiency or clarity) does not adequately model the data. The relation containing the inapplicable value can always be decomposed into an equivalent set of relations that do not contain it. Hence the presence of inapplicable values indicates inadequacies in the schema, but does not imply that information is being incompletely encoded.

Open nulls is the main representative of the possible-unweighted–unrestricted branch. Universal nulls may also be classified under this branch assuming the OWA semantics. Inclusive disjunctive information, possible information and maybe tuples or values are indicative representatives of the possible-unweighted-restricted school.

In [15] five different types of nulls are suggested. The labels and semantics of them are defined as follows. Let V be a function, which takes a label and returns a set of possible values that the label may have.

Intuitively, V(Ex-mar) = D says that the actual value of an existential marker can be any member of the domain D. Likewise, V(Ma-mar) = D$\cup\{\perp\}$ says that the actual value of a maybe marker can be either any member of D, or the symbol $\perp$, denoting a non-existent value. Similarly, V (Par-mar(V s)) = V_s says that the actual value of a partial null marker of the form pa mar (V_s) lies in the set V_s, a subset of the domain D.

An important issue is the use of $\perp$, which denotes that an attribute is inapplicable. However such an interpretation of the unknown information, is not consistent with the principles of conceptual modelling. Assuming the sample fact spouse, the individual, Tony, is a bachelor and hence, the wife field is inapplicable to him, $\perp$. Conceptually the issue can be resolved with the use of the subtypes (e.g. married, unmarried) as part of the entity class Person. A subtype is introduced only when there is at least one role recorded for that subtype. The conceptual treatment of null will permit us to reduce the table in Fig. 1 using only two types of null markers (Fig. 2).

Label (χ)	V(χ)
Ex-mar	D
Ma-mar	D $\cup \{\perp\}$
Pl-mar	$\{\perp\}$
Par-mar (V_s)	V_s
Pm-mar (V_s)	$V_s \cup \{\perp\}$

Fig. 1. Types of NULL and their semantics

Label ($_\chi$)	V($_\chi$)
V-mar (V)	{V}
P-mar (V_s)	{V_s}
Π-mar (D-V_s)	{D $-V_s$}

Fig. 2. The reduced set of NULL values

In the general case the algebraic issue under the use of subtypes is whether the population of the subtypes in relationship to the super type is:

- *Total and disjoint*: Populations are mutually exclusive and collectively exhaustive.
- *Non-total and disjoint*: Populations are mutually exclusive but not exhaustive.
- *Total and overlapping*: Common members between subtypes and collectively exhaustive, in relationship to super type.
- *Non-total and overlapping*: Common members between subtypes and not collectively exhaustive, in relationship to super type.

Conclusively it can be said that a null value is often semantically overloaded to mean either an unknown value or an inapplicable. For an extensive coverage of the issues related to the semantic overloading of null values somebody may further refer to [16].

3 NULL Values and Background Knowledge in DBMS

In a generalised relational database we consider an attribute A and a tuple t_i of a relation R in which n attribute value $t_i[A]$ may be a partial value. A partial value is formally defined as follows.

Definition 3.1. *A partial value is determined by a set of possible attribute values of tuple t of attribute A of which one and only one is the true value. We denote a partial value by $P = [a_1, \ldots, a_n]$ corresponding to a set of P possible values $\{a_1, \ldots, a_n\}$ of the same domain, in which exactly one of these values is the true value of. Here, P is the cardinality of $\{a_1, \ldots, a_n\}$ is a subset of the domain set $\{a_0, \ldots, a_{n+1}\}$ of attribute A of relation R, and $P \leq n+1$.*

Queries may require operations to be performed on partial values; this can result in a query being answered by means of bags, where the tuples have partial attribute values [17].

Example 3.1. We consider the attribute JOB_DESCRIPTION that has possible values 'Research Associate', 'Teaching Associate', 'Programmer', 'Junior Staff', and 'Senior Staff'. Then {'Junior Staff', 'Senior Staff'}, is an example of a partial value in terms of classical logic. In this case we know only that the individual is a staff but not whether he or she is a junior staff or a senior staff.

Definition 3.2. *An Intuitionistic Fuzzy partial value relation R, is a relation based on partial values for domains $D_1, D_2, \ldots, D_n$ of attributes $A_1, A_2, \ldots, A_n$ where $R \subseteq P_1 x P_2$ x x P_n and P_i is the set of all the partial values on power set of domain D_i. A pruned value of attribute A_i of the relation R corresponds then to a H-IFS which is a subset of the domain D_i. An example of a partial value relation is presented in Table 1 below:*

Let l be an element defined by a structured domain D_i. $U(e)$ is the set of higher level concepts, i.e. $U(e) = \{n | n \in D_i \wedge n$ is an ancestor of $l\}$, and $L(e)$ is the set of lower concepts $L(e) = \{n | n \in D_i \wedge n$ is a descendent of $l\}$. If l is a base concept then $L(e) = \emptyset$ and if l is a top level concept, then $U(e) = \emptyset$. Considering the H-IFS F = {Research Associate/<1.0, 0.0>, Programmer/<0.7, 0.2>, Teaching Associate/<0.4, 0.1>} as part of the Concept Employee in Fig. 3:

$$U(\text{Programmer}/<0.7, 0.2>) = \{\text{Technical Staff}/<0.7, 0.1>\}$$
$$L(\text{Programmer}/<0.7, 0.2>) = \emptyset$$

Rule-1: If $(|U(e)| > 1 \wedge L(e) = \emptyset)$, then it is simply declared that a child or base concept has many parents.

E.g. $|U(\text{Programmer}/<0.7, 0.2>)| = 3$, $L(\text{Programmer}/<0.7, 0.2>) = \emptyset$, Therefore a child or base concept acting as a selection predicate can claim any tuple (parent) containing elements found in $U(e)$, as its ancestor.

Table 1. Generalised Relation Staff with partial values in the form of hierarchical IFS

Name	JOB_DESCRIPTION	SALARY
Natalie	{Research Associate/<1.0,0>}	$\{ \sim \text{Salary}_{25K}\}$
Anna	{Programmer/<0.7,0.2>}	$\{ \sim \text{Salary}_{20K}\}$

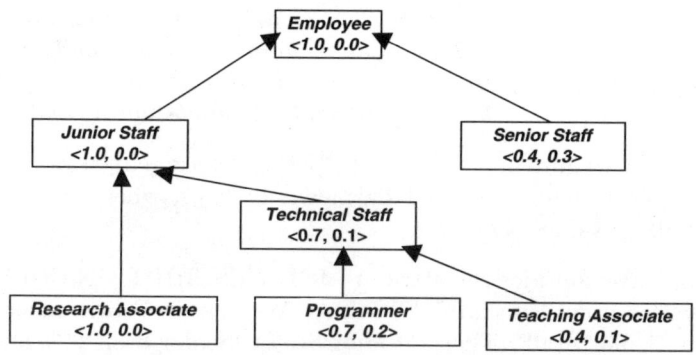

Fig. 3. Generalised H-IFS F – concept employee

Now let us consider the following case where l_1 = "Employee" and l_2 = "Programmer" then let B the function that defines the space between $l_1 \wedge l_2$. In this case $|B((l_1), (l_2))| = 2$, and > 1

$B((l_1), (l_2)) = U(L(l_1) \wedge (l_2))$ where l_1 is a high level concept, l_2 is a base concept are elements defined in a structured domain. If both arguments are high level concepts or low level concepts then $B((l_1), (l_2)) = \phi$.

Rule-2: If $B((l_1), (l_2)$ is defined and $|B((l_1), (l_2))| > 1$, then it is simply declared that multiple parents, high level concepts, are receiving a base concept as their own child. Therefore a parent or high level concept acting as a selection predicate can claim any tuple (child) containing elements found in $(L(l_1) \wedge (l_2))$, as its descendant, but with variants level of certainty.

Background knowledge may be specified as arising from a hierarchical Intuitionistic Fuzzy hierarchy, as integrity constraints, from the integration of conflicting databases, or from knowledge selected by domain experts. Using such information we offer to re-engineer the database by replacing missing, conflicting or unacceptable data by sets of the attribute domain.

Concept hierarchies have previously been used for attribute-induced knowledge discovery [18]. However the proposed use of background knowledge in this context is unique.

We assume that original attribute values may be given either as singleton sets, or subsets of the domain, or as concepts, which correspond to subsets of an attribute domain. In the last case the values may be defined in terms of a concept hierarchy. In addition there are rules describing the domain, and these may be formed in a number of ways: they may take the form of integrity constraints, where we have certain restrictions on domain values; functional dependencies and also rules specified by a domain expert.

An example of a concept hierarchy expressed with the aid of H-IFS F = {Research Associate/<1.0, 0.0>, Programmer/<0.7, 0.2>, Teaching Associate/<0.4, 0.1>} with values which are sets given in the generalised relation staff in Table 1. Here a value for the attribute JOB_DESCRIPTION may be a may be a concept from the concept hierarchy as defined in Fig. 3 (e.g. {Technical Staff/<0.7, 0.1>}). Then in terms of functional dependencies we may receive the following information Technical-Staff→~Salary$_{25K}$. To this extent in terms of any declarative query language it can be concluded that the salary in for a reaching associate or a Programmer must be in the range of Salary$_{25\,K}$. We can also use this knowledge to approximate the salary for all instances of Junior Staff in case where no further background knowledge after estimating firstly the $<\mu, v>$ degrees for the hierarchical concept Employee. Such a hierarchical concept like employee in Fig. 3 can be defined with the aid of Intuitionistic fuzzy set over a universe [19, 20] that has a hierarchical structure, named as H-IFS.

4 Definition of IFS and H-IFS

4.1 Principles of Intuitionistic Fuzzy Sets – Atanassov's Sets

Each element of an Intuitionistic fuzzy [21,22] set has degrees of membership or truth (μ) and non-membership or falsity (v), which do not sum up to 1.0 thus leaving a degree of hesitation margin (π).

As opposed to the classical definition of a fuzzy set given by $A' = \{<x, \mu_{A'}(x)>|x \varepsilon X\}$ where $\mu_A(x) \varepsilon [0,1]$ is the membership function of the fuzzy set A', an Intuitionistic fuzzy set A is given by

$$A = \{<x, \mu_A(x), v_A(x)>|x \varepsilon X\}$$

where: $\mu_A : X \to [0,1]$ and $v_A : X \to [0,1]$ such that $0 < \mu_A(x) + v_A(x) < 1$ and $\mu_A(x)\ v_A(x)\ \varepsilon[0,1]$ denote a degree of membership and a degree of non-membership of $x \varepsilon A$, respectively.

Obviously, each fuzzy set may be represented by the following Intuitionistic fuzzy set

$$A = \{<x, \mu'_A(x), (x), 1 - \mu'_A(x)>|x \varepsilon X\}$$

For each Intuitionistic fuzzy set in X, we will call $\pi_A(x) = 1 - \mu_A(x) - v_A(x)$ an Intuitionistic fuzzy index (or a hesitation margin) of $x \varepsilon A$ which expresses a lack of knowledge of whether x belongs to A or not. For each $x \varepsilon A 0 < \pi_A(x) < 1$.

Definition 4.1.1. *Let A and B be two fuzzy sets defined on a domain X. A is included in B (denoted $A \subseteq B$) if and only if their membership functions and non-membership functions satisfy the condition:* $(\forall \chi \in X)(\mu A(x) \leq \mu B(x)\ \&\ \nu A(x) \geq \nu B(x))$

Two scalar measures are classically used in classical fuzzy pattern matching to evaluate the compatibility between an ill- known datum and a flexible query, known as:

- A possibility degree of matching, $\Pi(Q;\ D)$
- A necessity degree of matching, $N(Q/D)$

Definition 4.1.2. *Let Q and D be two fuzzy sets defined on a domain X and representing, respectively, a flexible query and an ill-known datum:*

- *The possibility degree of matching between Q and D, denoted $\Pi(Q; D)$, is an "optimistic" degree of overlapping that measures the maximum compatibility between Q and D, and is defined by:*

$$\Pi(Q/D) = sup_x \in X\ min(<1 - \nu_Q(x), \nu_Q(x)>, <1 - \nu_D(x), \nu_D(x)>)$$

- *The necessity degree of matching between Q and D, denoted $N(Q; D)$, is a "pessimistic" degree of inclusion that estimates the extent to which it is certain that D is compatible with Q, and is defined by:*

$$N(Q/D) = inf_x \in X\ max(<\mu_Q(x), 1 - \mu_Q(x)>, <\mu_D(x), 1 - \mu_D(x)>)$$

4.2 H-IFS

The notion of hierarchical fuzzy set rose from our need to express fuzzy values in the case where these values are part of taxonomies as for food products or microorganisms for example.

The definition domains of the hierarchical fuzzy sets that we propose below are subsets of hierarchies composed of elements partially ordered by the "kind of" relation. An element l_i is more general than an element l_j (denoted $l_i \sim l_j$), if l_i is a predecessor of l_j in the partial order induced by the "kind of" relation of the hierarchy. An example of such a hierarchy is given in Fig. 1. A hierarchical fuzzy set is then defined as follows.

Definition 4.2.1. *A H-IFS is an Intuitionistic fuzzy set whose definition domain is a subset of the elements of a finite hierarchy partially ordered by the "kind of" $\leq$ relation.*

For example, the fuzzy set M defined as: $\{Milk <0.8, 0.1>, Whole-Milk <0.7, 0.1>, Condensed-Milk <0.4, 0.3>\}$ conforms to Definition-3. Their definition domains are subsets of the hierarchy given in Fig. 4.

We can note that no restriction has been imposed concerning the elements that compose the definition domain of a H-IFS. In particular, the user may associate a given $<\mu, \nu>$ with an element l_i and another degree $<\mu_1, \nu_1>$ with an element l_j more specific than l_i . $<\mu, \nu> \sim <\mu_1, \nu_1>$ represents a semantic of restriction for l_j compared to l_i, whereas $<\mu_1, \nu_1> \sim <\mu, \nu>$ represents a semantic of reinforcement for l_j compared to l_i. For example, if there is particular interest in condensed milk because the user studies the properties of low fat products, but also wants to retrieve complementary information about other kinds of milk, these preferences can be expressed using, for instance, the following Intuitionistic fuzzy set: $<1, 0>$/condensed milk $+ <0.5, 0.1>$/Milk.

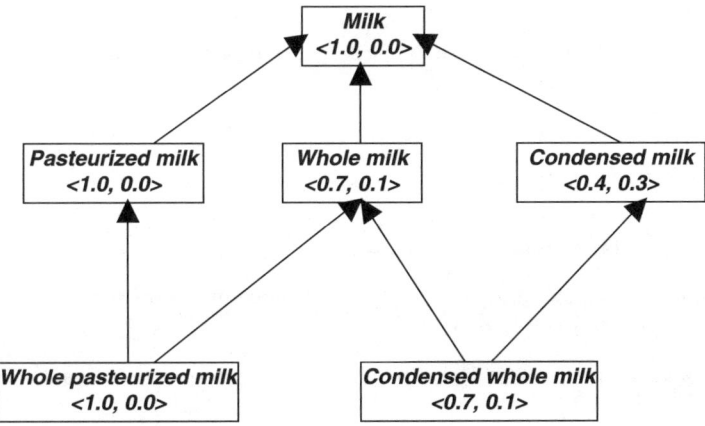

Fig. 4. Common closure of the H-IFSs Q and R

In this example, the element condensed milk has a greater degree than the more general element Milk, which corresponds to a semantic of reinforcement for condensed milk compared to Milk.

4.3 Closure of the H-IFS

We can make two observations concerning the use of H-IFSs:

- Let $<1, 0>$/condensed milk $+ <0.5, 0.1>$/Milk be an expression of liking in a query. We can note that this H-IFS implicitly gives information about elements of the hierarchy other than Condensed milk and Milk. One may also assume that any kind of condensed milk (i.e. whole condensed milk) interests the user with $<\mu, \nu> \Diamond <1, 0>$.
- Two different H-IFSs on the same hierarchy do not necessarily have the same definition domain, which means they cannot be compared using the classic comparison operations of Intuitionistic fuzzy set theory "see Sect. 4.1". For example, $<1, 0>$/condensed milk $+ <0.5, 0.1>$/Milk and 1/Milk $+ 0.2$/Pasteurised milk are defined on two different subsets of the hierarchy of "Fig. 1" and, thus, are not comparable.

These observations led to the introduction of the concept of closure of a Intuitionistic hierarchical fuzzy set, which is defined on the whole hierarchy. Intuitively, in the closure of a H-IFS, the "kind of, $\leq$" relation is taken into account by propagating the $<\mu, \nu>$ associated with an element to its sub-elements (more specific elements) in the hierarchy. For instance, in a query, if the user is interested in the element Milk, we consider that all kinds of Milk—Whole milk, Pasteurised milk, are also of interest. On the opposite, we consider that the super-elements (more general elements) of Milk in the hierarch are too broad to be relevant for the user's query.

Definition 4.3.1. *Let F be a H-IFS defined on a subset D of the elements of a hierarchy L. It degree is denoted as $<\mu, \nu>$. The closure of F, denoted $clos(F)$, is a H-IFS defined on the whole set of elements of L and its degree $<\mu, \nu>_{clos(F)}$ is defined as follows.*

For each element l of L, let $S_L = \{l_1, \ldots, l_n\}$ be the set of the smallest super-elements of in D:

- **If S_L is not empty,** $<\mu, \nu>_{clos(F)}(S_L) = <\max_{1 \leq i \leq n}(\mu(L_i))$, $\min_{1 \leq i \leq n}(\nu(L_i))>$**else,** $<\mu, \nu>_{clos(F)}(S_L) = <0, 0>$

In other words, the closure of a H-IFS F is built according to the following rules. For each element l_1 of L:

- If l_I belongs to F, then l_I keeps the same degree in the closure of F (case where $S_L = \{l_I\}$).
- If l_I has a unique smallest super-element l_1 in F, then the degree associated with l_I is propagated to L in the closure of F, $S_L = \{l_1\}$ with $l_1 > l_I$)

- If L has several smallest super-elements $\{l_1, \ldots, l_n\}$ in F, with different degrees, a choice has to be made concerning the degree that will be associated with l_I in the closure. The proposition put forward in Definition 4.3.1, consists of choosing the maximum degree of validity μ and minimum degree of non validity v associated with $\{l_1, \ldots, l_n\}$.
- All the other elements of L, i.e., those that are more general than, or not comparable with the elements of F, are considered as non-relevant. The degree $<0, 0>$ is associated with them.

Let us consider once more the H-IFS M defined as: {Milk $<0.8, 0.1>$, Whole-Milk $<0.7, 0.1>$, Condensed- Milk $<0.4, 0.3>$} which is presented in Fig. 1.

The case of whole condensed milk is different: The user has associated the degree $<0.8, 0.1>$ with Milk, but has given a restriction on the more specific element whole milk (degree $<0.7, 0.1>$). As whole condensed milk is a kind of whole milk it inherits the $<\mu, v>$ associated with whole milk, that is $<0.7, 0.1>$.

If the H-IFS expresses preferences in a query, the choice of the maximum allows us not to exclude any possible answers. In real cases, the lack of answers to a query generally makes this choice preferable, because it consists of enlarging the query rather than restricting it.

If the H-IFS represents an ill formulated concept, the choice of the maximum allows us to preserve all the possible values of the datum, but it also makes the datum less specific. This solution is chosen in order to homogenize the treatment of queries and data. In a way, it enlarges the query, answer.

4.4 Properties of H-IFS

In Sect. 4.3 we saw that each H-IFS has an associated closure that is defined on the whole hierarchy.

We focus on the fact that two different H-IFSs, defined on the same hierarchy, can have the same closure, as in the following example.

Example . The H-IFSs Q = {Milk $<1, 0>$, Whole-Milk $<0.7, 0.1>$, Pasteurised-milk $<1, 0>$, Condensed-Milk $<0.4, 0.3>$} and R = {Milk $<1, 0>$, Whole-milk$<0.7, 0.1>$, Pasteurised-milk $<1, 0>$, Whole-Pasteurised-milk $<1, 0>$, Condensed Milk $<0.4, 0.3>$} have the same closure, represented Fig. 4 below.

Such H-IFSs form equivalence classes with respect to their closures.

Definition 4.4.1. *Two H-IFSs Q and R, defined on the same hierarchy, are said to be equivalent $Q \equiv R$ if and only if they have the same closure.*

Property. Let Q and R be two equivalent Intuitionistic hierarchical fuzzy sets. If $l_I \in \mathrm{dom}(Q) \cap \mathrm{dom}(R)$, then $<\mu, v>(Q.l_I) = <\mu, v>(R.l_I)$

Proof. According to the definition of the closure of a H-IFS F, Definition 4.3.1, the closure of F preserves the degrees that are specified in F. As Q and R have the same closure (by definition of the equivalence), an element that belongs to Q and R necessarily has the same degree $<\mu, \nu>$ in both.

We can note that R contains the same element as Q with the same $<\mu, \nu>$, and also one more element Whole-Pasteurised-milk $<1, 0>$. The $<\mu, \nu>$ associated with this additional element is the same as in the closure of Q. Then it can be said that the element, Whole-Pasteurised-milk $<1, 0>$ is derivable in R through Q.

The same conclusions can be drawn in the case of condensed whole milk $<0.7, 0.1>$ □

Definition 4.3.2. *Let F be a hierarchical fuzzy set, with* $dom(F) = \{l_1, ..., l_n\}$, *and* F_{-k} *the H-IFS resulting from the restriction of F to the domain* $dom(F)\backslash\{l_k\}$. l_k *is deducible in F if*

$$<\mu, \nu>clos_{(F-k)}(l_k) = <\mu, \nu>clos_{(F)}(l_k)$$

As a first intuition, it can be said that removing a derivable element from a hierarchical fuzzy set allows one to eliminate redundant information. But, an element being derivable in F does not necessarily mean that removing it from F will have no consequence on the closure: removing k from F will not impact the degree associated with k itself in the closure, but it may impact the degrees of the sub-elements of k in the closure.

For instance, if the element Pasteurised milk is derivable in Q, according to Definition 4.3.2, removing Pasteurised-milk $<1, 0>$ from Q would not modify the degree of Pasteurised milk itself in the resulting closure, but it could modify the degree of its sub-element Whole-pasteurised-milk. Thus, Pasteurised-milk $<1, 0>$ can not be derived or removed. This remark leads us to the following definition of a minimal hierarchical fuzzy set.

Definition 4.3.3. *In a given equivalence class (that is, for a given closure C), a hierarchical fuzzy set is said to be minimal if its closure is C and if none of the elements of its domain is derivable.*

For instance the H-IFSs S_1 and S_2 are *minimal* (none of their elements is derivable). They cannot be reduced further.

$$S_1 = \text{Milk}<1, 0>$$

$S_2 = \{\text{Milk } <1, 0>, \text{Whole-Milk } <0.7, 0.1>, \text{Whole-Pasteurised-milk} <1, 0>, \text{Condensed-Milk } <0.4, 0.3>\}$

5 Replacing and Constraining Unknown Attribute Values

All descendents of an instance of a high-level concept are replaced with a *minimal H-IFS* has these descendents as members. A *null* value is regarded as a partial value with all base domain values as members. We refer to the resultant partial value, obtained as a result of this process, as a *primal* partial value. The replacement process is thus performed by the following procedure:

Procedure: replacement

Input: A concept table R consisting of partial values, or nulls
Output: A re-engineered partial value table U
Method: For each attribute value of R recursively replace the cell value by a *primal* partial value. For each cell of R replace, the primal partial value by a pruned prime-partial –value, until a minimal partial value is reached

---

If a particular member of a partial value violates the domain constraint (rule) then it is *pruned* from the *minimal H-IFS primal partial value*. This process is continued until all partial values have been pruned by the constraints as much as possible. We refer to the resultant partial value, obtained as a result of this process, as a *minimal* partial value.

In addition in an integrated DBMS environment it will be also useful not to query all sources, but only those that contain information relevant to our request. This is quite critical for achieving better query performance. For this reason we equip our Integrated architecture with a repository that contains various constraints (i.e. Intuitionistic Fuzzy Range Constraints, Intuitionistic Fuzzy Functional Dependencies, etc) that are related to the information sources that participate in the Integrated Architecture.

Range constraints: such as "The average income per person is estimated to be in the range of €50K". Considering a finite universe of discourse, say X whose cardinality is N. Let us suppose that $X = \{X_1, X_2, \ldots, X_n\}$ and the Intuitionistic fuzzy number $\sim a$ given by $\sim a = \{(x_i, \mu_i, v_i) : x_i \in X, I = 1, 2 \ldots N\}$ We can express the above constraint as follows $\sim Income50K$ $\{(49, .8, .1), (50, .9, .02)(51, .7, .15)\}$.

Classical data integrity constraints such as "All persons stored at a source have a unique identifier".

Functional Dependencies: for instance, a source relation S1 (Name, lives, income, Occupation) has a functional dependency Name $\rightarrow$ (Lives, $\sim$Income). These constraints are very useful to compute answers to queries.

There are several reasons we want to consider constraints separately from the query language. Describing constraints separately from the query language can allow us to do reasoning about the usefulness of a data source with respect to a valid user request.

Some of source constraints can be naturally represented as local constraints. Each *local constraint* is defined on one data source only. These constraints carry a rich set of semantics, which can be utilized in query processing. Any projected database instance of source, these conditions must be satisfied by the tuples in the database.

Definition 5.1. *Let $si, \ldots, sl$ be l sources in a data-integrated system. Let $P = \{pi, \ldots, pn\}$ be a set of global predicates, on which the contents of each* *source s are defined. A general global constraint is a condition that should be satisfied by any database instance of the global predicates P.*

General global constraints can be introduced during the design phase of such a data-integration system. That is, even if new sources join or existing ones leave the system, it is assumed that these constraints should be satisfied by any database instance of the global predicates. Given the global predicate *Income*, if a query asks for citizens with an average income above $\sim Income60K$, without checking the source contents and constraints, the integrated system can immediately know that the answer is empty.

To this extent we can interrogate the constraints repository to find out if a particular source contains relevant information with respect to particular request. We now consider the problem of aggregation for the partial value data model. In what follows we are concerned with symbolic attributes, which are typically described by counts and summarised by aggregated tables. The objective is to provide an aggregation operator which allows us to aggregate individual tuples to form summary tables.

6 Summary Tables and Aggregation

A summary table R, is represented in the form [23] of an Intuitionistic fuzzy relation (IFR).

Aggregation (A): An aggregation operator A is a function $A(G)$ where $G = \{<x, \mu_F(x), \nu_F(x)> | x \in X\}$ where $x = <att_1, \ldots, att_n>$ is an ordered tuple belonging to a given universe X, $\{att_1, \ldots, att_n\}$ is the set of attributes of the elements of X, $\mu_F(x)$ *and* $\nu_F(x)$ *are the degree of membership and non-membership of* x. The result is a bag of the type $\{<x', \mu_F(x'), \nu_F(x')> | x' \in X\}$. To this extent, the bag is a group of elements that can be duplicated and each one has a degree of μ and ν.
Input: $R_i = (l, F, H)$ and the function $A(G)$
Output: $R_o = (l_o, F_o, H_o)$ *where*

- l is a set of levels $l_1, \ldots, l_n$, that belong to a partial order $\leq O$
  To identify the level l as part of a hierarchy we use dl.

$$l_\perp : \text{base level} \quad l_\top : \text{top level}$$

for each pair of levels l_i and l_j we have the relation

$$\mu_{ij} : l_i \times l_j \to [0, 1] \quad \nu_{ij} : l_i \times l_j \to [0, 1] \quad 0 < \mu_{ij} + \nu_{ij} < 1$$

- F is a set of fact instances with schema $F = \{<x, \mu_F(x), \nu_F(x)> | x \in X\}$, where $x = <att_1, \ldots, att_n>$ is an ordered tuple belonging to a given universe X, $\mu_F(x)$ and $\nu_F(x)$ are the degree of membership and non-membership of x in the fact table F respectively.

- H is an object type history that corresponds to a structure (l, F, H') which allows us to trace back the evolution of a structure after performing a set of operators i.e. aggregation

The definition of the extended group operators allows us to define the extended group operators *Roll up* (Δ), and *Roll Down* (Ω).

Roll up (Δ): The result of applying Roll up over dimension d_i at level dl_r using the aggregation operator A over a relation $R_i = (l_i, F_i, H_i)$ is another relation $R_o = (l_o, F_o, H_o)$

Input: $R_i = (l_i, F_i, H_i)$
Output: $R_o = (l_o, F_o, H_o)$

An object of type history is a recursive structure $H = \begin{cases} \omega \text{ is the initial state} \\ \text{of the relation.} \\ (l, A, H') \text{ is the state} \\ \text{of the relation after} \\ \text{performing an} \\ \text{operation on it.} \end{cases}$

The structured history of the relation allows us to keep all the information when applying *Roll up* and get it all back when *Roll Down* is performed. To be able to apply the operation of *Roll Up* we need to make use of the IF_{SUM} aggregation operator.

Roll Down (Ω): This operator performs the opposite function of the *Roll Up* operator. It is used to roll down from the higher levels of the hierarchy with a greater degree of generalization, to the leaves with the greater degree of precision. The result of applying *Roll Down* over a relation $R_i = (l, F, H)$ having $H = (l', A', H')$ is another relation $R_o = (l', F', H')$.

Input: $R_i = (l, F, H)$
Output: $R_o = (l', F', H')$ where $F' \rightarrow$ set of fact instances defined by operator A.

To this extent, the *Roll Down* operative makes use of the recursive history structure previously created after performing the *Roll Up* operator.

The definition of aggregation operator points to the need of defining the IF extensions for traditional group operators [20], such as *SUM, AVG, MIN and MAX*. Based on the standard group operators, we provide their IF extensions and meaning.

IF_{SUM}: The IF$_{sum}$ aggregate, like its standard counterpart, is only defined for numeric domains. The relation R consists of tuples R_i *with* $1 \leq i \leq m$. The tuples R_i are assumed to take Intuitionistic Fuzzy values for the attribute att$_{n-1}$ for $i = 1$ *to* m we have $R_i[att_{n-1}] = \{<\mu_i(u_{ki}), \nu_i(u_{ki})>/u_{ki}|1 \leq k_i \leq n\}$. The IF$_{sum}$ of the attribute att_{n-1} of the relation R is defined by:

$$IF_{SUM}((att_{n-1})(R)) = \left\{ <u>/y| \left(\left(u = \min_{i=1}^{m} (\mu_i(u_{ki}), \nu_i(u_{ki})) \right. \right. \right.$$
$$\left. \left. \left. \wedge \left(y = \sum_{ki=k1}^{km} u_{ki} \right) \left(\forall_{k1,\ldots km} : 1 \leq k1, \ldots km \leq n \right) \right) \right) \right\}$$

Example:

$$IF_{SUM}((Amount)(ProdID))$$
$$= \{<.8, .1>/10\} + \{(<.4, .2>/11), (<.3, .2>/12)\}$$
$$+ \{(<.5, .3>/13), (<.5, .1>/12)\}$$
$$= \{(<.8 \wedge .4, .1 \wedge .2>/10 + 11), (<.8 \wedge .3, .1 \wedge .2>/10 + 12)\}$$
$$+ \{<.5, .3>/13, <.5, .1>/12\}$$
$$= \{(<.4, .2>/21), (<.3, .2>/22)\} + \{<.5, .3>/13, <.5, .1>/12\}$$
$$= \{(<.4 \wedge .5, .2 \wedge .3>/21 + 13), (<.4 \wedge .5, .2 \wedge .1>/21 + 12),$$
$$(<.3 \wedge .5, .2 \wedge .3>/22 + 13), (<.3 \wedge .5, .2 \wedge .1>/22 + 12)$$
$$= \{(<.4, .3>/34), (<.4, .2>/33), (<.3, .3>/35), (<.3, .2>/34)\}$$
$$= \{(<.3, .3>/34), (<.4, .2>/33), (<.3, .3>/35)\}$$

IF_{AVG}: The IF_{AVG} aggregate, like its standard counterpart, is only defined for numeric domains. This aggregate makes use of the IF_{SUM} that was discussed previously and the standard $COUNT$. The IF_{AVG} can be defined as:

$$IF_{AVG}((att_{n-1})(R)) = IF_{SUM}((att_{n-1})(R))/COUNT((att_{n-1})(R))$$

IF_{MAX}: The IF_{MAX} aggregate, like its standard counterpart, is only defined for numeric domains. The IF_{sum} of the attribute att_{n-1} of the relation R is defined by:

$$IF_{MAX}((att_{n-1})(R))$$
$$= \{<u>y|((u = \min_{i=1}^{m}(\mu_i(u_{ki}), \nu_i(u_{ki})) \wedge (y = \max_{i=1}^{m}(\mu_i(u_{ki}), \nu_i(u_{ki})))$$
$$(\forall_{k1,...km} : 1 \leq k1, ...km \leq n))\}$$

IF_{MIN}: The IF_{MIN} aggregate, like its standard counterpart, is only defined for numeric domains. Given a relation R defined on the schema X $(att_1, ..., att_n)$, let att_{n-1} defined on the domain $U = \{u_1, ..., u_n\}$. The relation R consists of tuples R_i with $1 \leq i \leq m$. Tuples R_i are assumed to take Intuitionistic Fuzzy values for the attribute att_{n-1} for $i = 1$ *to* m we have $R_i[att_{n-1}] = \{ < \mu_i(u_{ki}), \nu_i(u_{ki}) > /u_{ki}|1 \leq k_i \leq n\}$. The IF_{sum} of the attribute att_{n-1} of the relation R is defined by:

$$IF_{MIN}((att_{n-1})(R))$$
$$= \{<u>/y|((u = \min_{i=1}^{m}(\mu_i(u_{ki}), \nu_i(u_{ki})) \wedge (y = \min_{i=1}^{m}(\mu_i(u_{ki}), \nu_i(u_{ki})))$$
$$(\forall_{k1,...km} : 1 \leq k1, ...km \leq n))\}$$

We can observe that the IF_{MIN} is extended in the same manner as IF_{MAX} aggregate except for replacing the symbol *max* in the IF_{MAX} definition with *min*.

6.1 Summarisation Paths

The structure of any H-IFS can be described by a domain concept relation DCR = (Concept, Element), where each tuple describes a relation between elements of the domain on different levels.

The DCR can be used in calculating recursively [24] the different summarisation or selection paths as follows:

$$\text{PATH} \leftarrow \text{DCR} \underset{\{x=1...(n-2)|n>2\}}{\bowtie} \text{DCR}_x$$

If $n \leq 2$, then DCR becomes the Path table as it describes all summarisation and selection paths.

These are entries to a knowledge table that holds the metadata on parent-child relationships. An example is presented below:

Figure 5 shows how our Milk hierarchy knowledge table is kept. Paths are created by running a recursive query that reflects the 'PATH' algebraic statement. The hierarchical IFS used as example throughout this paper comprises of three levels, thus calling for the SQL-like query as below:

SELECT A.Concept as Grand-concept, b.concept, b.element
FROM DCR as A, DCR as B
WHERE A.child = B.parent;

This query will produce the following paths (Fig. 6):

DCR	
Concept	Element
Milk <1.0, 0.0>	Pasteurised Milk <1.0, 0.0>
Milk <1.0, 0.0>	Whole Milk <0.7, 0.1>
Milk <1.0, 0.0>	Condensed Milk <0.4, 0.3>
Pasteurised Milk <1.0, 0.0>	Whole Pasteurised Milk <1.0, 0.0>
Whole Milk <0.7, 0.1>	Whole Pasteurised Milk <1.0, 0.0>
Whole Milk <0.7, 0.1>	Whole Condensed Milk <0.7, 0.1>
Condensed Milk <0.4, 0.3>	Whole Condensed Milk <0.7, 0.1>

Fig. 5. Domain concept relation

Path			
Grand-concept	Concept	Element	Path Colour
Milk <1.0, 0.0>	Pasteurised Milk <1.0, 0.0>	Whole Pasteurised Milk <1.0, 0.0>	Red
Milk <1.0, 0.0>	Whole Milk <0.7, 0.1>	Whole Pasteurised Milk <1.0, 0.0>	Blue
Milk <1.0, 0.0>	Whole Milk <0.7, 0.1>	Whole Condensed Milk <0.7, 0.1>	Green
Milk <1.0, 0.0>	Condensed Milk <1.0, 0.0>	Whole Condensed Milk <0.7, 0.1>	Brown

Fig. 6. Path table

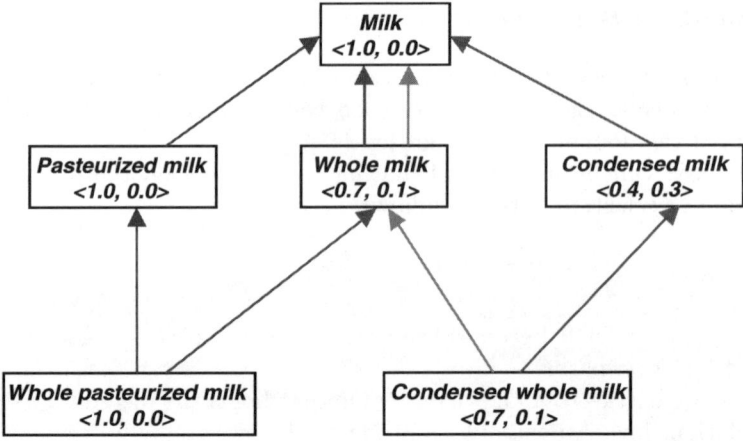

Fig. 7. Pictorial representation of paths

Figure 7 presents a pictorial view of the four distinct summarisation and selection paths.

These paths will be used in fuzzy queries to extract answers that could be either definite or possible. This will be realised with the aid of the predicate (θ).

A predicate (θ) involves a set of atomic predicates ($\theta_1, \ldots, \theta_n$) associated with the aid of logical operators p (i.e. $\wedge$, $\vee$, etc.). Consider a predicate θ that takes the value "Whole Milk", $\theta =$ "Whole Milk".

After utilizing the IFS hierarchy presented in Fig. 7, this predicate can be reconstructed as follows:

$$\theta = \theta_1 \vee \theta_2 \vee \ldots \vee \theta_n$$

In our example, $\theta_1 =$ "Whole Milk", $\theta_2 =$ "Whole Pasteurised Milk" and $\theta_n =$ "Condensed Whole Milk".

The reconstructed predicate $\theta =$ (Whole Milk $\vee$ Whole Pasteurised Milk $\vee$ Condensed Whole Milk) allows the query mechanism to not only definite answers, but also possible answers [25].

In terms a query retrieving data from a summary table, the output contains not only records that match the initial condition, but also those that satisfy the reconstructed predicate. Consider the case where no records satisfy the initial condition (Whole Milk). Traditional aggregation query would have returned no answer, however, based on our approach, the extended query would even in this case, return an answer, though only a possible one, with a specific belief and disbelief $<\mu, \nu>$. It will point to those records that satisfy the reconstructed predicateθ, more specifically, "Whole Pasteurised Milk and Condensed Whole Milk".

7 Conclusions

We provide a means of using background knowledge to re-engineer the data representation into a partial value representation with the aid of H-IFS and Intuitionistic Fuzzy relational representation.

The hierarchical links are defined by the "kind of, $\leq$" relation. The membership of an element in a H-IFS has consequences on the membership and non-membership of its sub elements in this set. The notion of H-IFS, that may be defined on a part of a hierarchy and the notion of closure of a H-IFS, that is explicitly defined on the whole hierarchy, using the links between the elements that compose the hierarchy.

H-IFSs that have the same closure define equivalence classes, called minimal H-IFS. Minimal fuzzy sets are used as a basis to define the generalization of a H-IFS fuzzy set. The proposed methodology aims at enlarging the user preferences expressed when defining a query, in order to obtain related and complementary answers.

We have discussed how domain knowledge presented in the form of background knowledge, such as integrity constraints, functional dependencies or details of the concept hierarchy, may be used to reduce the amount of missing data in the database.

We have presented a new multidimensional model that is able to operate over data with imprecision in the facts and the summarisation hierarchies. Classical models imposed a rigid structure that made the models present difficulties when merging information from different but still reconcilable sources.

This is likely to be a useful tool for decision support and knowledge discovery in, for example, data mediators, data warehouses, where the data are often subject to such imperfections. Furthermore we notice that our approach can be used for the representation of Intuitionistic fuzzy linguistic terms.

References

1. Bell, D., Guan, J., Lee, S.: Generalized union and project operations for pooling uncertain and imprecise information. DKE 18 (1996) 89–117
2. Chen, A., Tseng, F.: Evaluating aggregate operations over imprecise data. IEEE Transaction on Knowledge and Data Engineering, 8 (1996) 273–284
3. Zemankova, M., Kandel, A.: Implementing imprecision in Information systems. Information Science 37 (1985) 107–141
4. Dubois, D., Prade, H., Testamale, C.: Handling Incomplete or Uncertain Data and Vague Queries in Database Applications. Plenum Press, New York (1988)
5. Prade, H.: Annotated bibliography on fuzzy information processing. Readings on Fuzzy Sets in Intelligent Systems. Morgan Kaufmann Publishers Inc., San Francisco (1993)
6. Codd, E.: Extending the Data Base Relational Model to Capture More Meaning. ACM Transaction Database Systems, 4 (1979) 397–434
7. Goldstein, B.: Constraints on Null Values in Relational Databases. Proc. 7th Int. Conf. on VLDB, IEEE Press, Piscataway (1981) pp. 101–110

8. Biskup, J.: A Foundation of Codd's Relational Maybe-Operations. XP2 Workshop on Relational Database Theory (1981)
9. Liu, C., Sunderraman, R.: Indefinite and maybe information in relational databases. ACM Transaction Database System, 15(1) (1990) 1–39
10. Liu, K., Sunderraman, R.: On Representing Indefinite and Maybe Information in Relational Databases: A Generalization. ICDE. IEEE Computer Society (1990) 495–502
11. Ola, A.: Relational databases with exclusive disjunctions. Data Engineering (1992) 328–336
12. Homenda, W.: Databases with Alternative Information. IEEE Transaction on Knowledge and Data Engineering, 3(3) (1991) 384– 386.
13. Gessert, G.: Handling Missing Data by Using Stored Truth Values. SIGMOD Record, 20(1) (1991) 30–42
14. Zicari, R.: Closed World Databases Opened Through Null Values. Proc. 14th Int. Conf. on VLDB (1988) pp. 50–61
15. Zaniolo, C.: Database relations with null values. Journal of Computer Systems Science 28 (1984) 142–166.
16. Lipski, J.: On semantic issues connected with incomplete information databases. ACM Trans. Database System, 4(3) (1979) 262–296
17. Dhar, V., Tuzhilin, A.: Abstract-driven pattern discovery in databases. IEEE Transaction on Knowledge and Data Engineering 6 (1993) 926–938
18. Han, J., Fu, Y.: Attribute-oriented induction in data mining. Advances in Knowledge Discovery. AAAI Press/MIT Press, Cambridge, MA (1996) pp. 399–421
19. Rogova E., Chountas P., Atanassov, K.: Flexible Hierarchies and Fuzzy Knowledge-Based OLAP. FSKD 2007, IEEE CS pp. 7–11
20. Rogova E., Chountas P.: On imprecision intuitionistic fuzzy sets & OPLAP – The case for KNOLAP. IFSA 2007, to be published by LNAI Springer, Berlin Heidelberg New York pp. 11–20
21. Atanassov, K.: Intuitionistic Fuzzy Sets. Springer, Berlin Heidelberg New York (1999)
22. Atanassov, K.: Intuitionistic fuzzy sets, Fuzzy Sets and Systems 20 (1986) 87–96
23. Atanassov, K., Kolev, B., Chountas, P., Petrounias, I.: A mediation approach towards an intuitionistic fuzzy mediator. IPMU (2006) 2035–2063
24. Silberschatz, A., Korth, H., Sudarshan, S.: Database System Concepts. McGraw-Hill
25. Parsons, S.: Current approaches to handling imperfect information in data and knowledge bases. IEEE Transaction on Knowledge and Data Engineering 8(3) (1996) 353–372

Part VII

Tracking Systems

Tracking of Multiple Target Types with a Single Neural Extended Kalman Filter

Kathleen A. Kramer[1] and Stephen C. Stubberud[2]

[1] Department of Engineering, University of San Diego, 5998 Alcalá Park,
   San Diego, CA, USA
   kramer@sandiego.edu
[2] Department of Engineering, Rockwell Collins, 5998 Alcalá Park, San Diego
   CA, USA
   scstubberud@ieee.org

Summary. The neural extended Kalman filter is an adaptive state estimation routine that can be used in target-tracking systems to aid in the tracking through maneuvers. A neural network is trained using a Kalman filter training paradigm that is driven by the same residual as the state estimator and approximates the difference between the a priori model used in the prediction steps of the estimator and the actual target dynamics. An important benefit of the technique is its versatility because little if any a priori knowledge of the target dynamics is needed. This allows the neural extended Kalman filter to be used in a generic tracking system that will encounter various classes of targets. Here, the neural extended Kalman filter is applied simultaneously to three separate classes of targets each with different maneuver capabilities. The results show that the approach is well suited for use within a tracking system without prior knowledge of the targets' characteristics.

1 Introduction

In sensor data fusion, the concept of target tracking is the combination of multiple sensor reports from a variety of sensors and at different times to provide a filtered estimate of the target's dynamic state. A significant problem in this Level 1 data fusion problem [1, 2] is tracking a target through a maneuver. Estimation lagging or loss of filter smoothing in the state estimates often result. A wide variety of techniques to address the maneuver-tracking issue have been developed. The basic underlying concept of all of these methods, clearly stated in [3], is that the most important issue in target tracking is to model the target motion accurately. The result of accurate target motion models is significantly improved track estimates. The technique described in [4] uses a maneuver detection step that determines when a modified Kalman filter model should be employed. The model used in this modification is derived by an estimate of the severity of the maneuver. Other techniques that

K.A. Kramer and S.C. Stubberud: *Tracking of Multiple Target Types with a Single Neural Extended Kalman Filter*, Studies in Computational Intelligence (SCI) **109**, 495–512 (2008)

have been developed to track through a maneuver have been similar to that of [5]. Such techniques develop a parameter-based model that models particular maneuvers. These approaches provide an improved model, which more closely emulates potential motion of a target. One of the most widely used techniques to track through a maneuver is that of the interacting multiple model (IMM) approach [6] and [7]. (An enhanced version, the variable structure IMM (VS-IMM), reduces the computational overhead that can arise.) The basic premise of the IMM is to generate a subset of motion models over the span of the space of potential target motion models. The IMM approach requires a priori information about the capabilities of the target. While the different motion models in implemented systems have typically been simply a high process noise model and a low process noise model, current research often assumes that any maneuver the target would be able to perform is modeled to some degree.

Another concept is to continually adapt the model of the target motion or dynamics based on the current estimates. One such approach is a neural extended Kalman filter (NEKF) and is here applied to the tracking problem. The NEKF is an adaptive neural network technique that trains completely on-line. Unlike several other neural network techniques that change the process or measurement noise on-line [8], the NEKF learns a function that approximates the error between the a priori motion model and the actual target dynamics. The NEKF, originally developed for operations in a control loop [9] and [10], is comprised of a coupled Kalman filter. The two components are the standard tracking state estimator and a Kalman filter training paradigm similar to that first developed by Singhal and Wu [11] and applied to dynamic system identification in [12]. The NEKF is similar in concept to the Kalman filter parameter estimation techniques as in [13] and [14] where parameters of the dynamic model were modified using the Kalman filter as the estimator to improve the dynamic model. Unlike those earlier techniques which utilized a known functional structure, the NEKF uses the general function approximation property of a neural network. This allows the NEKF to be applied to a variety of targets without prior knowledge of capabilities and a large database to draw from as would be necessary in most standard IMM implementations. A single NEKF can be applied a wide variety of target classes without significant change in performance.

To demonstrate this generic property, the NEKF will be analyzed in its performance on three separate types of targets: a ground target that can stop and change directions quickly, an aircraft performing a series of racetracks in flight, and surface ship with longer requirements for velocity changes. The analysis shows that a single NEKF using the same fixed parameters can used to track these targets with similar accuracy performance results.

The development of the NEKF algorithm begins with a background discussion of maneuver tracking and the issues that the EKF can experience in this application. This is followed by an overview of the function approximation capabilities of a neural network. After this background information is presented, the actual NEKF algorithm is developed. Then, the example application of the NEKF to a generic tracking problem set starts with a description of the three target motions used in the analysis The performance results are provided and analyzed.

2 Neural Extended Kalman Filter Development

The NEKF is based on the standard extended Kalman filter (EKF) tracking algorithm and the use of a neural network that trains on-line. This development begins with the concept of maneuver target tracking. This is followed by a discussion of function approximation of a neural network. With this background information, the development of the NEKF algorithm is presented.

2.1 Maneuver Tracking

Target tracking is part of Level 1 sensor data fusion [1, 2]. The concept is to combine associated measurements (related to the same target). These measurements can come from one or more sensors and are reported at different times. These measurements all include uncertainty or random errors in their reports based on the quality of the sensors as well as environmental conditions and target properties. The estimator used in the target tracking system is usually designed to filter or smooth the measurements over time by incorporating the statistics of the measurement uncertainty and an estimate of the target motion. One of the standard estimation techniques used in target tracking is that of the extended Kalman filter.

The standard EKF equations are given as

$$\mathbf{K}_k = \mathbf{P}_{k|k-1} \frac{\partial \mathbf{h}\left(\hat{\mathbf{x}}_{k|k-1}\right)}{\partial \hat{\mathbf{x}}_{k|k-1}}^T \left(\frac{\partial \mathbf{h}\left(\hat{\mathbf{x}}_{k|k-1}\right)}{\partial \hat{\mathbf{x}}_{k|k-1}} \mathbf{P}_{k|k-1} \frac{\partial \mathbf{h}\left(\hat{\mathbf{x}}_{k|k-1}\right)}{\partial \hat{\mathbf{x}}_{k|k-1}} + \mathbf{R}_k \right)^{-1} \quad \text{(1a)}$$

$$\hat{\mathbf{x}}_{k|k} = \hat{\mathbf{x}}_{k|k-1} + \mathbf{K}_k \left(\mathbf{z}_k - \mathbf{h}\left(\hat{\mathbf{x}}_{k|k-1}\right) \right) \quad \text{(1b)}$$

$$\mathbf{P}_{k|k} = \left(\mathbf{I} - \mathbf{K}_k \frac{\partial \mathbf{h}\left(\hat{\mathbf{x}}_{k|k-1}\right)}{\partial \hat{\mathbf{x}}_{k|k-1}} \right) \mathbf{P}_{k|k-1} \quad \text{(1c)}$$

$$\hat{\mathbf{x}}_{k+1|k} = \mathbf{f}\left(\hat{\mathbf{x}}_{k|k}\right) \quad \text{(1d)}$$

$$\mathbf{P}_{k+1|k} = \left(\frac{\partial \mathbf{f}\left(\hat{\mathbf{x}}_{k|k}\right)}{\partial \hat{\mathbf{x}}_{k|k}} \right) \mathbf{P}_{k|k} \left(\frac{\partial \mathbf{f}\left(\hat{\mathbf{x}}_{k|k}\right)}{\partial \hat{\mathbf{x}}_{k|k}} \right)^T + \mathbf{Q}_k. \quad \text{(1e)}$$

Equations (1a)–(1c) are the update equations where $\mathbf{x}_k$ denotes the target states and $\mathbf{K}_k$ is the Kalman Gain. The output-coupling function, $\mathbf{h}(.)$, defines the mathematical relation between the states and the measurements. The variable $\mathbf{P}$ defines the state error covariance matrix, which provides a quality measure of the state estimate, while $\mathbf{R}$ is the covariance of the measurement noise. The subscript $k|k - 1$ indicates a prediction while the subscript $k|k$ indicates an update. The update equations use the weighted residual in (1c) to correct the predicted state estimate based on the provided measurements. In the prediction equations, (1d) and (1e), the matrix $\mathbf{Q}$ defines the process noise covariance or a quality measure of the target motion model or state-coupling function $\mathbf{f}(.)$. The matrices of partial derivatives in (1a), (1c), and (1e), are used to linearize the output-coupling and state-coupling functions and referred to as the Jacobians.

For standard tracking, the target state kinematics are described often as position and velocity. For a two-dimension problem, the state vector would be defined as

$$\mathbf{x}_k = [x_k \quad \dot{x}_k \quad y_k \quad \dot{y}_k]^T .\tag{2}$$

The quality of the target motion model has a significant impact on the capabilities of the EKF as an estimator. Unlike application in control systems [15, 16], with a tracking problem external inputs on the target are not often known. Thus, the developed models, especially for the target motion, do not include an external input term. Also, the motion model must be defined priori to operations and, therefore, is often defined to be a straight-line motion model:

$$\mathbf{x}_{k+1} = \mathbf{A}\mathbf{x}_k = \begin{bmatrix} 1 & dt & 0 & 0 \\ 0 & 1 & 0 & 0 \\ 0 & 0 & 1 & dt \\ 0 & 0 & 0 & 1 \end{bmatrix} \begin{bmatrix} x_k \\ \dot{x}_k \\ y_k \\ \dot{y}_k \end{bmatrix} .\tag{3}$$

If the target deviates from this model significantly, the EKF state estimate can lag the true trajectory. In Fig. 1, a maneuvering target trajectory and its associated measurements with noise are shown. Figure 2 shows the target-track estimates for the trajectory using the motion model of (3) in the EKF. As a result, the target track falls behind the trajectory. Such a problem can be compensated for if the process noise $\mathbf{Q}$ is inflated. This has the beneficial effect of reducing the impact of the prediction state estimates in the Kalman filter. However, the prediction equations smooth the estimates so that, while, as in Fig. 3, the tracking system follows the target through the maneuver, the resulting track is quite noisy.

To avoid this problem, two approaches have been tried. The first is to vary the process noise when a maneuver is detected and return it to its base setting once the maneuver is completed as in [6,17]. The second is to adapt the model to a maneuver as in [18].

Using this second type of approach, an adaptive Kalman filter is implemented that can change to a variety of maneuver models without a priori knowledge of the functional form of the target's dynamics. Also, unlike many other tracking techniques applied to maneuvering targets [18,19], the NEKF is a general approach that can be applied to each target. Thus, the technique can applied to various target types simultaneously with acceptable performance without the need to identify the target a priori.

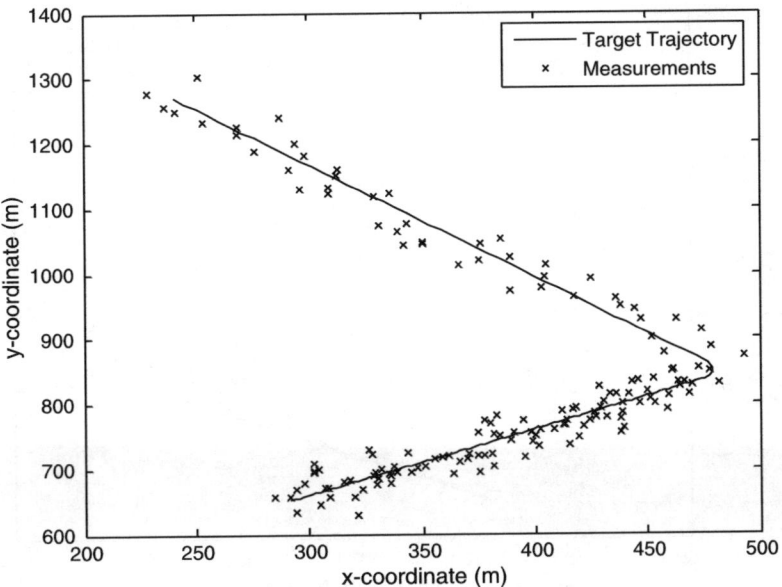

Fig. 1. Target and associated measurements with noise

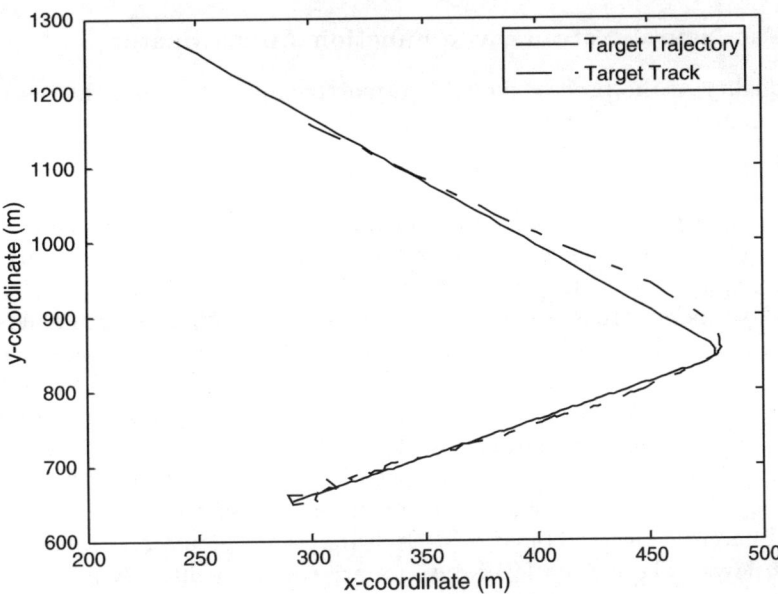

Fig. 2. The tracked target trajectory lags the true trajectory once the maneuver occurs

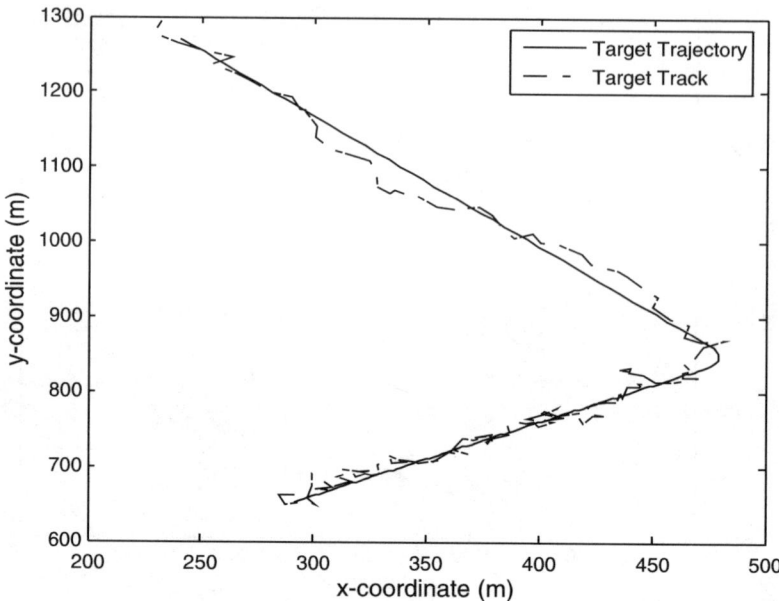

Fig. 3. The tracked target trajectory handles the maneuver but does not filter out the noise

2.2 The Neural Network as a Function Approximator

The quality of the process model is a matrix-valued function that estimates the dynamics of the target

$$\mathbf{f}_{true} = \mathbf{f}(\cdot) + \varepsilon. \tag{4}$$

The error in the model, ε, can be estimated arbitrarily closely using a function aproximation scheme that meets the criteria for the Stone–Weierstrauss Theorem [20]. One solution that fits this criterion is a neural network that uses multi-layer perceptrons (MLPs) as the squashing functions in each node of the hidden layer. (The MLP can also be replaced by a set of multi-dimensional Gaussians as described in [21] and [22]. These are also referred to as elliptical or radial basis functions.) A layer in a neural network refers to a set of operations performed in parallel. In this case, the neural network contains at least one hidden layer that maps a set of input values to an output values between ±1. Each layer contains one or more of the same function operations with different input combinations. Each function operation is a node in the hidden layer. The hidden layer outputs are combined linearly at the output layer.

Thus, for the MLP-based neural network in this effort, an improved function approximation

$$\mathbf{f}_{true} = \mathbf{f}(\cdot) + NN(\cdot) + \delta, \tag{5}$$

where $\delta < \varepsilon$, is generated.

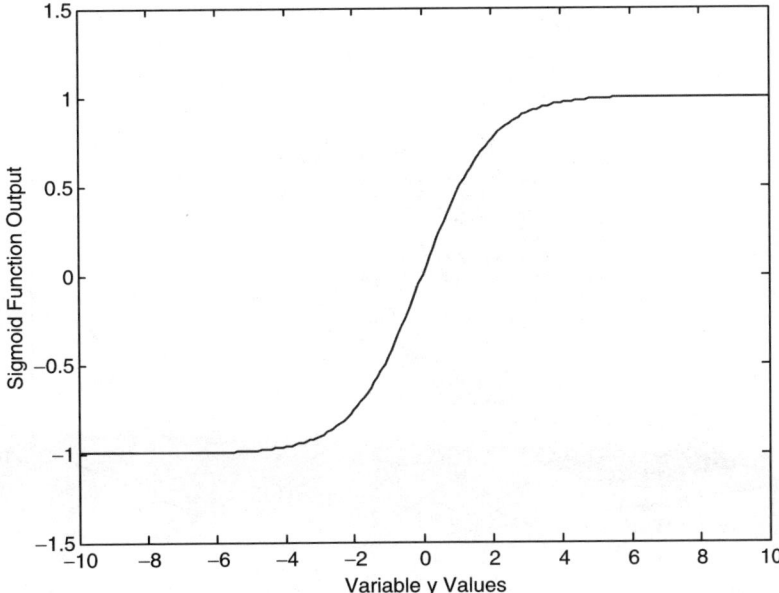

Fig. 4. The sigmoid squashing function used by the neural network in the NEKF

A wide variety of functions can be used in the hidden layer of the neural network. The function that is usually employed in the NEKF is

$$g(y) = \frac{1 - e^{-y}}{1 + e^{-y}}. \tag{6}$$

The function is shown in Fig. 4. This function is considered part of a class of function referred to as squashing functions due to the mapping of large magnitude values into values close to ± 1. In the neural network, the variable y is defined as

$$y = \sum_k w_k x_k, \tag{7}$$

where w is a weight and, for the tracking problem, the x's are the track states. The neural network combines several of these functions and the selection of the weights,

$$NN\left(\mathbf{x}, \mathbf{w}, \boldsymbol{\beta}\right) = \sum_j \beta_j g_j\left(\mathbf{x}, \mathbf{w}_j\right) = \sum_j \beta_j \frac{1 - e^{-\sum_k w_{jk} x_k}}{1 + e^{-\sum_k w_{jk} x_k}}. \tag{8}$$

Equation (8) creates a formulation referred to as a multi-layered perceptron (MLP).

The MLP results in wide variety of functions being developed. In Fig. 5, an output of a neural network of the form of (6), with two input variables,

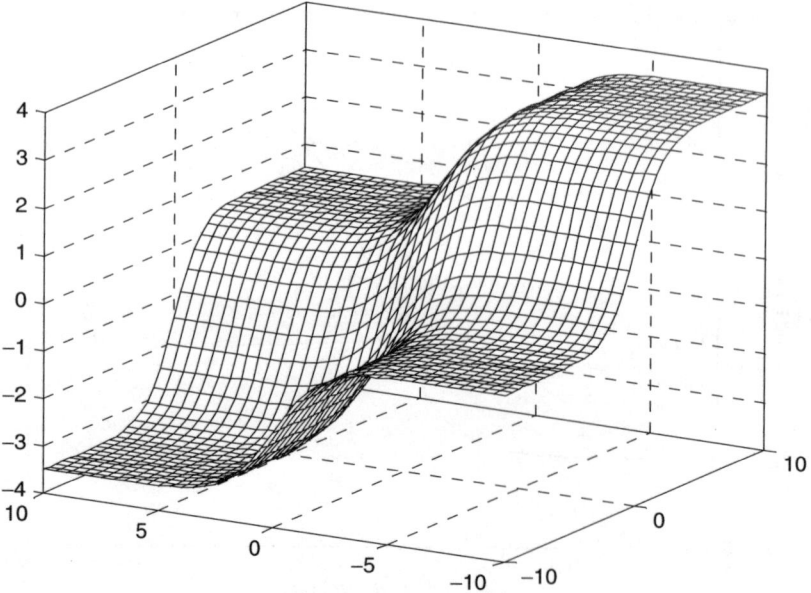

Fig. 5. The function of the MLP with only two output weights allows significant function approximation near the coordinate origin

the same two randomly-selected weights, w_{jk}, used in each squashing function and combined together as

$$-1.5g_1 + 2g_2, \tag{9}$$

is shown. By adjusting the input and output weights, other functions can be approximated. However, if the magnitudes of the input weights or the input variables become too large in any one direction, the function approximation capability of the MLP is reduced. When the input variables of the function shown become large, the function loses its complexity. It is noted that (7) is a sum which means that along the axis of opposite signs of the addends there exists a non-flat function. However, the greatest variation occurs in the region of small input variable values. The range of interest of the input variables can be scaled through the weights of (7). This indicates that the weights for each squashing function should scale the value of (7) based on the magnitude of the input variables used to track the maneuvering target.

Figure 5 also shows that a small number of nodes can form a capable function approximator. Greater complexity may require a larger number of nodes in the hidden layer of the MLP. However, if the selection of the number of these functions becomes too large, then other problems may arise. In addition to the increased computational complexity, a large number of hidden nodes can cause the NEKF to create "chirps" in the estimate.

2.3 The Neural Extended Kalman Filter Algorithm

The neural extended Kalman filter is a combination of the standard extended Kalman filter for state estimation and that of the neural network training algorithm. Using the same concept as in [14] for parameter estimation, the so-called parameter-free technique of the NEKF has a combined state vector of

$$
\bar{\mathbf{x}}_k = \begin{bmatrix} \mathbf{x}_k \\ \mathbf{w}_k \end{bmatrix} = \begin{bmatrix} \mathbf{x}_k \\ \omega_k \\ \beta_k \end{bmatrix}, \tag{10}
$$

where $\mathbf{x_k}$ denotes the target states and ω_k and β_k represent the input and output weights, respectively.

The coupling of the neural network training to the state estimation to form the NEKF is quite straightforward. In both applications of the EKF, as a tracker and as a training paradigm, the same residual is used to perform the same function: ensure that the output or estimated measurements of the system estimate make the residuals as small as possible.

With the NEKF-based tracking approach, the initial process model is the straight-line motion model, (3), with a neural network that is varied during operations, added to this model. This addition improves the target motion model, thus improving the state estimates of the target track. The continual updating of the neural network permits the use of a smaller neural network. This is a result of the fact that the NEKF only needs to approximate the local function for the maneuver and not the entire trajectory.

The implemented neural-augmented process model for the tracking system component of the NEKF becomes:

$$
\mathbf{f}(\mathbf{x}_k) + NN(\omega_k, \beta_k, \mathbf{x}_k) = \begin{bmatrix} 1 & dt & 0 & 0 \\ 0 & 1 & 0 & 0 \\ 0 & 0 & 1 & dt \\ 0 & 0 & 0 & 1 \end{bmatrix} \mathbf{x}_k + \begin{bmatrix} NN_1(\omega_k, \beta_{1k}, \mathbf{x}_k) \\ NN_2(\omega_k, \beta_{2k}, \mathbf{x}_k) \\ NN_3(\omega_k, \beta_{nk}, \mathbf{x}_k) \\ NN_4(\omega_k, \beta_{nk}, \mathbf{x}_k) \end{bmatrix}. \tag{11}
$$

Expanding (11) to incorporate training, the state-coupling function of the overall NEKF is then defined as:

$$
\bar{\mathbf{f}}(\bar{\mathbf{x}}_k) = \left[\begin{array}{cccc|c} 1 & dt & 0 & 0 & \\ 0 & 1 & 0 & 0 & \mathbf{0}_{4 \times num\_wts} \\ 0 & 0 & 1 & dt & \\ 0 & 0 & 0 & 1 & \\ \hline \mathbf{0}_{num\_wts \times 4} & & & & \mathbf{I} \end{array} \right] \bar{\mathbf{x}}_k + \begin{bmatrix} NN_1(\omega_k, \beta_{1k}, \mathbf{x}_k) \\ NN_2(\omega_k, \beta_{2k}, \mathbf{x}_k) \\ NN_3(\omega_k, \beta_{nk}, \mathbf{x}_k) \\ NN_4(\omega_k, \beta_{nk}, \mathbf{x}_k) \\ \mathbf{0}_{num\_of\_wts} \end{bmatrix}. \tag{12}
$$

$$
= \begin{bmatrix} \mathbf{F} & | & \mathbf{0} \\ \mathbf{0} & | & \mathbf{I} \end{bmatrix} \bar{\mathbf{x}}_k + NN(\omega_k, \beta_k, \mathbf{x}_k) = \bar{\mathbf{F}} \cdot \bar{\mathbf{x}}_k + NN(\omega_k, \beta_k, \mathbf{x}_k)
$$

The associated Jacobian would be:

$$\bar{\mathbf{F}} = \frac{\partial \bar{\mathbf{f}}\,(\bar{\mathbf{x}}_k)}{\partial \bar{\mathbf{x}}_k} = \left[\begin{array}{c|cc} \mathbf{F} + \frac{\partial NN(\omega_k, \beta_k, \mathbf{x}_k)}{\partial \mathbf{x}_k} & \frac{\partial NN(\omega_k, \ldots)}{\partial \omega} \quad k & \frac{\partial NN(\omega_k, \ldots)}{\partial \beta_k} \\ \hline \mathbf{0} & \mathbf{I} \end{array} \right]. \tag{13}$$

Equation (13) results in the state estimation and neural network training being coupled. Using these equations as the system dynamics, the equations of the NEKF are defined as

$$\mathbf{K}_k = \mathbf{P}_{k|k-1} \mathbf{H}_k^T \left(\mathbf{H}_k \mathbf{P}_{k|k-1} \mathbf{H}_k^T + \mathbf{R}_k \right)^{-1} \tag{14a}$$

$$\hat{\bar{\mathbf{x}}}_{k|k} = \begin{bmatrix} \hat{\mathbf{x}}_{k|k} \\ \hat{\mathbf{w}}_{k|k} \end{bmatrix} = \hat{\bar{\mathbf{x}}}_{k|k-1} + \mathbf{K}_k \left(\mathbf{z}_k - \mathbf{h}(\hat{\mathbf{x}}_{k|k-1}) \right) \tag{14b}$$

$$\mathbf{P}_{k|k} = \left(\mathbf{I} - \mathbf{K}_k \mathbf{H}_k \right) \mathbf{P}_{k|k-1} \tag{14c}$$

$$\hat{\bar{\mathbf{x}}}_{k+1|k} = \begin{bmatrix} \hat{\mathbf{x}}_{k+1|k} \\ \hat{\mathbf{w}}_{k+1|k} \end{bmatrix} = \begin{bmatrix} \mathbf{f}\left(\hat{\mathbf{x}}_{k|k}, \mathbf{u}_k \right) + NN\left(\hat{\mathbf{w}}_{k|k}, \hat{\mathbf{x}}_{k|k} \right) \\ \hat{\mathbf{w}}_{k|k} \end{bmatrix} \tag{14d}$$

$$\mathbf{P}_{k+1|k} = \left(\underline{\mathbf{F}} + \begin{bmatrix} \frac{\partial NN\left(\hat{\mathbf{w}}_{k|k}, \hat{\mathbf{x}}_{k|k} \right)}{\partial \hat{\bar{\mathbf{x}}}_{k|k}} \\ \mathbf{0} \end{bmatrix} \right) \mathbf{P}_{k|k} \left(\underline{\mathbf{F}} + \begin{bmatrix} \frac{\partial NN\left(\hat{\mathbf{w}}_{k|k}, \hat{\mathbf{x}}_{k|k} \right)}{\partial \hat{\bar{\mathbf{x}}}_{k|k}} \\ \mathbf{0} \end{bmatrix} \right)^T + \mathbf{Q}_k \tag{14e}$$

where $\hat{\bar{\mathbf{x}}}_{k|k}$ is the augmented state estimate vector of (10).

The measurement Jacobian is:

$$\mathbf{H} = \begin{bmatrix} \dfrac{\partial \mathbf{h}\left(\hat{\mathbf{x}}_{k|k-1} \right)}{\partial \hat{\mathbf{x}}_{k|k-1}} & \mathbf{0}_w \end{bmatrix}, \tag{15}$$

the same as the EKF Jacobian of (1a) and (1c) with the augmentation of zeros to handle the change in dimensionality.

3 Multi-Target Tracking and Analysis

A three target example is employed to demonstrate the ability of the NEKF to work in a generic tracking environment. The vehicle capabilities are not known a priori.

3.1 Target Motion Description

The test-case example was chosen to contain three separate target examples that could exist simultaneously in a battlespace. Such target types can be tracked simultaneously. Thus. to investigate and analyze the NEKF as a generalized tracking algorithm that can be applied to a wide variety of target

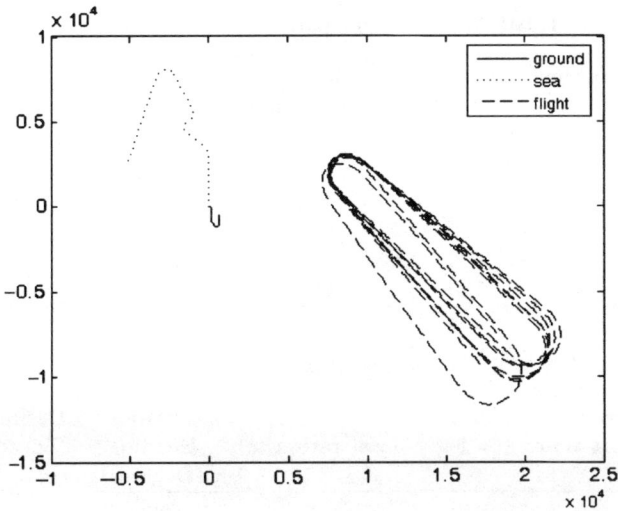

Fig. 6. Three targets (flight, ground, and sea) in the scenario

classes without a priori information, three different target motions were developed: an aircraft, a surface ship, and a ground vehicle. The two latter targets are generated by a simulator while the air target is from the truth data file of an F-4 aircraft in flight performing a series of racetrack-like maneuvers. Each target provided two thousand sample points at one second intervals. The one second intervals were chosen to use the same sample time of the real flight data. The targets' trajectories are shown in Fig. 6.

The aircraft data shows the flight path of a US Navy F-4 flying a series of racetrack-like patterns. As seen in Fig. 6, the racetracks are clearly non-overlapping and provide an additional challenge to the NEKF as the maneuvers, while similar, are different and require continual changes in the neural network weights. The reported positions are in absolute x–y coordinates with units of meters.

The simulated data was generated by a target simulator that defined various legs of the target trajectory. Each leg contained a desired heading in degrees where $0°$ is defined along the positive x-axis and a target speed. At the start of each new leg of the trajectory, a transition time is also provided. This time represents the time that the vehicle takes to go from the last heading to the desired heading at uniform changes. For example, a change from 10 to $20 \, \text{m s}^{-1}$ in 10 s would mean a constant $1 \, \text{m s}^{-2}$ acceleration would be applied to the target. If this transition time is longer than the time assigned to the leg of the trajectory, then the maneuver is not completed.

The first simulated target was generated to emulate a surface ship. Table 1 lists the parameters that were used to develop its trajectory. The second simulated target was designed to emulate a ground target. This included the

Table 1. Trajectory parameters For sea target

Leg start time (s)	Time on leg (s)	Length of acceleration (s)	Desired heading (Degrees)	Desired speed (knts)
0	300	0	90	10
300	280	120	145	15
580	120	90	30	15
700	600	120	120	10
1,300	700	180	250	20

Table 2. Trajectory parameters for ground target

Leg start time (s)	Time on leg (s)	Length Of acceleration (s)	Desired Heading (Degrees)	Desired Speed (m s^{-1})
0	200	0	360	1.0
200	500	30	270	1.5
700	450	450	360	1.5
1,150	150	10	360	0
1,300	700	1	90	1.0

target motion having a stop and an instantaneous turn. The parameters that described its motion are listed in Table 2.

For the NEKF tracking system, the measurements were assumed to be tagged such that the association was not considered an issue. The process noise for the track states in NEKF was defined as in (16):

$$\mathbf{Q} = 0.017^2 \begin{bmatrix} dt^3/3 & dt^2/2 & 0 & 0 \\ dt^2/2 & dt & 0 & 0 \\ 0 & 0 & dt^3/3 & dt^2/2 \\ 0 & 0 & dt^2/2 & dt \end{bmatrix} \qquad (16)$$

The process noise on the weights for neural network for each target was set to $10 \cdot \mathbf{I}$ for the input weights and to $10^{-4} \cdot \mathbf{I}$ for the output weights. The initial error covariance $\mathbf{P}$ was set to $1,000 \cdot \mathbf{I}$ for the track states and to $100 \cdot \mathbf{I}$ for the neural network weights. Each target state includes its kinematic information and their own weights for the NEKF based on their own maneuvers. The measurements were defined as x-y positions with independent measurement noise of ± 50 m on each coordinate.

3.2 Results and Analysis

The NEKF was compared to three Kalman filters. The linear Kalman filter could be used as a result of using the position measurement type. Each

Kalman filter used a straight-line motion model for the target dynamics but incorporated a different process noise. The process noise matrix defined in 16 provided the baseline used in the analysis. For the larger, second process noise, the baseline was multiplied by a factor of 100. The smaller, third process noise was set to that of the baseline was divided by a factor of 100. The initial covariance matrix $\mathbf{P}$ was set to the same value as the NEKF, 1,000·$\mathbf{I}$. The NEKF used the fixed motion model of (11) as the a priori motion model and the process noise defined in Sect. 3.1. The same filters were used to track each target.

Table 3 contains the average distance error throughout the target path and the associated standard deviation of the errors for each target resulting from the NEKF. Figures 7–9 show the Monte Carlo errors over each point of the trajectory for the air target, ground target, and sea target, respectively. Table 4 denotes the results for the Kalman filter where the process noise on the track states is the same as the NEKF. Tables 5 and 6 contain the error results for the Kalman filter trackers with 100 times and 0.01 times the NEKF track state process noise, respectively.

Table 3. NEKF error results

Target	Average error (m)	Standard deviation (m)
Air	415.5	290.6
Ground	23.18	4.633
Sea	22.5	3.36

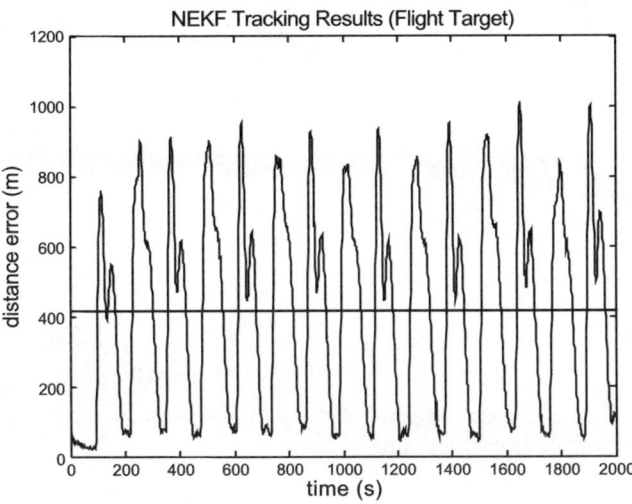

Fig. 7. NEKF tracking error results for flight target showing difficulty tracking through repeated sharp turns

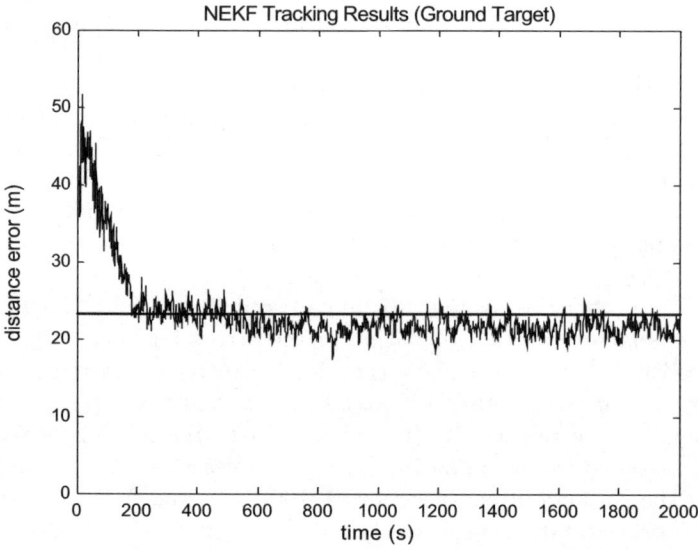

Fig. 8. NEKF tracking error results for ground target showing good tracking results, particularly following initial training period

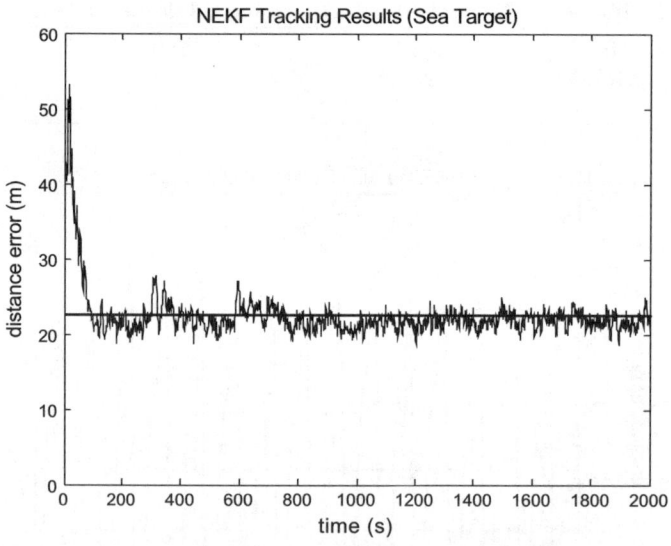

Fig. 9. NEKF tracking error results for sea target showing good tracking results, particularly following initial training period

Figures 10–12 show the Monte Carlo errors over each point of the trajectory for the equivalent process noise Kalman filter.

The results indicate that NEKF that is not tuned for a specific target type performs well in each case. The Kalman filter clearly is superior when tracking

Table 4. Kalman filter error results – baseline Q

Target	Average error	Standard deviation
Air	1,255.7	1,092.9
Ground	15.0	1.97
Sea	23.1	12.0

Table 5. Kalman filter error results – large Q

Target	Average error	Standard deviation
Air	332.0	646.1
Ground	24.9	1.61
Sea	26.5	1.911

Table 6. Kalman filter error results – small Q

Target	Average error	Standard deviation
Air	4,989.9	1,585.3
Ground	17.5	11.11
Sea	89.2	97.1

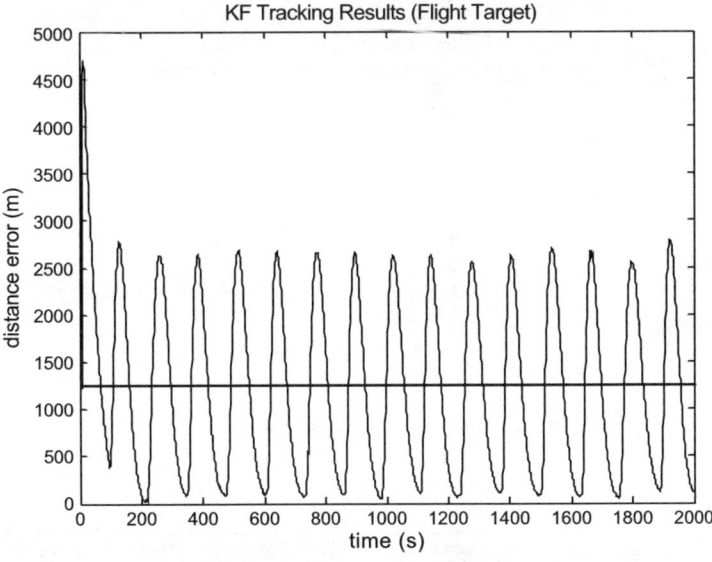

Fig. 10. Kalman filter tracking error results for flight target showing greater difficulty tracking through repeated sharp turns

the ground target over the entire trajectory. It has approximately a 50% improvement. In contrast, the Kalman filter performs slightly worse against the sea target on average. The standard deviation indicates that the performance

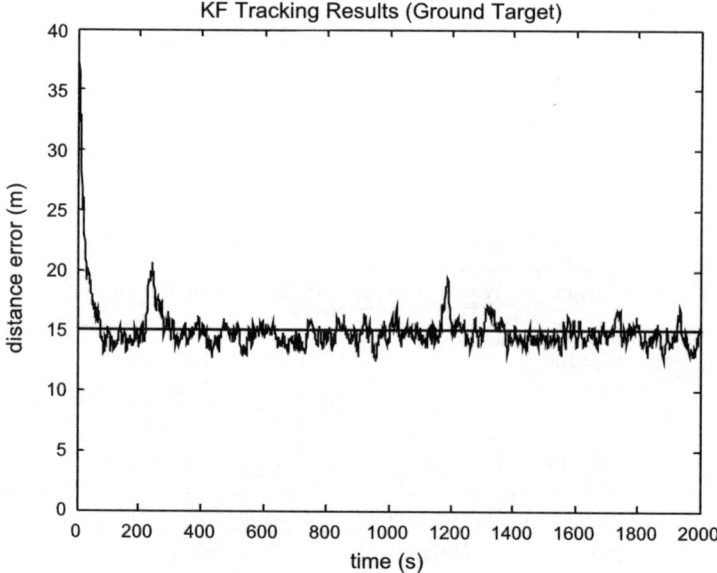

Fig. 11. Kalman filter tracking error results for ground target showing better tracking results than those of the NEKF

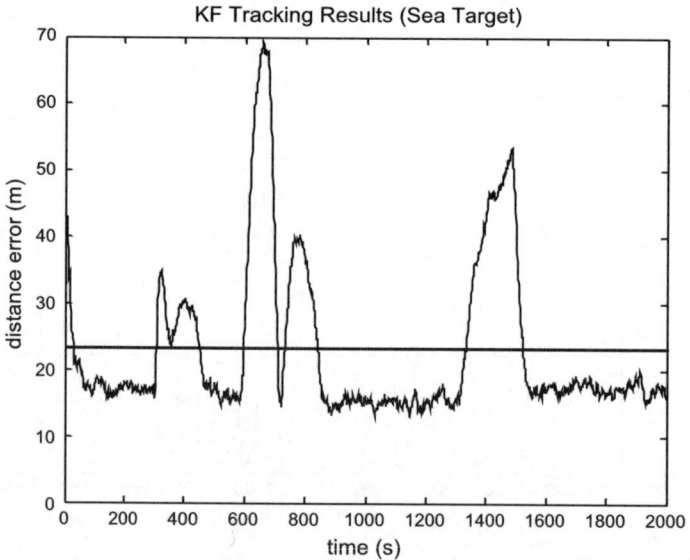

Fig. 12. Kalman tracking error results for sea target showing large variance in tracking error

is widely varying unlike the with NEKF. For the air target, performance differences are much more significant: the NEKF performs three times better. For the Kalman filter to perform better against the air target than the NEKF, it sacrifices its performance for the other two targets. The NEKF clearly shows that as a generalized tracking algorithm it can provide very accurate performance.

4 Conclusions

Results from using the NEKF to track multiple targets with different maneuver characteristics show that the NEKF was able to adapt and yield good tracking performance for a variety of targets. The NEKF is therefore an adaptive estimation routine that can be used as a generalized tracking algorithm. The results, presented here, show that, for generalized tracking where limited a priori information is available, the NEKF is an appropriate choice as a tracking engine because it has the ability to adapt on-line to improve the target motion model and hence tracking accuracy. The ability of the NEKF to learn on-line is a key quality for either a stand-alone tracker or an excellent model to be used in an IMM configuration.

Many modifications to the NEKF have been and can be incorporated to improve performance for specific problems. The use of the NEKF as a tracking system also has the advantage that it can learn a model of the current dynamics that can be used to better predict behavior of the target for problems such as target intercept and impact-point prediction.

References

1. Steinberg, A., Bowman, C., White, F.: Revisions to the JDL data fusion model. Proceedings of the SPIE Sensor Fusion: Architectures, Algorithms, and Applications III, 1999, 430–441
2. Llinas, J., Bowman, C., Rogova, G.L., Steinberg, A., Waltz, E., White, F.: Revisions and extensions to the JDL data fusion model II. Proceedings of Fusion 2004, Stockholm, Sweden, 2003, 1218–1230
3. Blackman, S., Popoli, R.: Design and analysis of modern tracking systems, Artech House, Norwood, MA, 1999
4. Best, R.A., Norton, J.P.: A new model and efficient tracker for a target with curvilinear motion. IEEE Transactions on Aerospace and Electronic Systems, Vol. 34, 1997, 1030–1037
5. Efe, M., Atherton, D.P.: Maneuvering target tracking with an adaptive Kalman filter. Proceedings of the 37th IEEE Conference on Decision and Control, 1998, 737–742
6. Kirubarajan, T., Bar-Shalom, Y.: Tracking evasive move-stop-move targets with a GMTI radar using a VS-IMM estimator. IEEE Transactions on Aerospace and Electronic Systems, Vol. 39(3), 2003, 1098–1103

7. Owen, M.W., Stubberud, A.R.: A neural extended Kalman filter multiple model tracker. Proceedings of OCEANS 2003 MTS/IEEE, 2003, 2111–2119

8. Jetto, L., Longhi, S., Venturini, G.: Development and experimental validation of an adaptive extended Kalman filter for the localization of mobile robots. IEEE Transactions on Robotics and Automation, Vol. 15(2), 1999, 219–229

9. Stubberud, S.C., Lobbia, R.N., Owen, M.: An adaptive extended Kalman filter using artificial neural networks. Proceedings of the 34th IEEE Conference on Decision and Control, 1995, 1852–1856

10. Stubberud, S.C., Lobbia, R.N., Owen, M: An adaptive extended Kalman filter using artificial neural networks. International Journal on Smart System Design, Vol. 1, 1998, 207–221

11. Singhal, S., Wu, L.: Training multilayer perceptrons with the extended Kalman algorithm. In: Touretsky, D.S. (ed.), Advances in Neural Processing Systems I, Morgan Kaufmann, Los Altos, CA (1989), 133–140

12. Stubberud, A.R., Wabgaonkar, H.M: Neural network techniques and applications to robust control. In: Leondes, C.T. (ed.), Advances in Control and Dynamic Systems (53), Academic, New York, 1992 427–522

13. Mack, G., Jain, V.: Speech parameter estimation by time-weighted-error Kalman filtering. IEEE Transactions on Acoustics, Speech, and Signal Processing, Vol. 31(5), 1983, 1300–1303

14. Iglehart, S., Leondes, C.: Estimation of a dispersion parameter in discrete Kalman filtering. IEEE Transactions on Automatic Control, Vol. 19 (3), 1974, 262–263

15. Rhodes, I.B.: A tutorial introduction to estimation and filtering. IEEE Transactions on Automatic Control, Vol. AC-16 (6), (1971), 688–706

16. Brown, R.G.: Introduction to random signal analysis and Kalman filtering, Wiley, New York, 1983

17. Owen, M.W., Stubberud, A.R.: NEKF IMM tracking algorithm. (ed.) Drummond, O.: Proceedings of SPIE: Signal and Data Processing of Small targets 2003, Vol. 5204, San Diego, California, 2003

18. Shea, P.J., Zadra, T., Klamer, D., Frangione, E., Brouillard, R.: Improved state estimation through use of roads in ground tracking. Proceedings of SPIE: Signal and Data Processing of Small Targets, Orlando, FL, 2000

19. Brouillard, R., Stubberud, S.: A parallel approach to GMTI tracking. Proceedings of the Military Sensing Symposia: 2003 National Symposium on Sensor Data Fusion, San Diego, CA, 2003

20. Hornik, K., Stinchcombe, M., White, H.: Multilayer feedforward networks are universal approximators. Neural Networks, Vol. 2, 1989, 35–36

21. Alspach, D., Sorenson, H.: Nonlinear Bayesian estimation using Gaussian sum approximations. IEEE Transactions on Automatic Control, Vol. 17(4), 1972, 439–448

22. Brotherton, T., Johnson, T., Chadderdon, G.: Classification and novelty detection using linear models and a class dependent neural network. Proceedings of World Congress for Computational Intelligence (IJCNN), Anchorage, Alaska, 1998, 876–879

Tracking Extended Moving Objects with a Mobile Robot

Andreas Kräußling

Research Establishment of Applied Sciences, Department of Communication, Information Processing and Ergonomics, Bad Neuenahrer Straße 20, D-53343 Wachtberg, Germany

Summary. Tracking moving objects is of central interest in mobile robotics. It is a prerequisite for providing a robot with cooperative behaviour. Most algorithms assume punctiform targets, which is not always suitable. In this work we expand the problem to extended objects and compare the algorithms that have been developed by our research group. These algorithms are capable of tracking extended objects. It is shown that there are great differences between tracking robots, where a certain shape can be assumed, and people.

1 Introduction and Related Work

Multirobot systems and service robots need cooperative capabilities to some extend. This holds for robots that are supposed to move in a certain formation or the interaction with people or robots. Apparently, there has to be knowledge about the objects we wish to interact with. More precisely, we want to know where these objects are. Thus, we need to track their positions.

Target tracking deals with the state estimation of one or more objects. This is a well studied problem in the field of aerial surveillance with radar devices [1] as well as in mobile robotics [2–5]. For their high accuracy and relatively low price, laser range scanners are a good choice in robotics [6, 7].

Due to the high resolution of these sensors, one target is usually the source of multiple returns of a laser scan. This conflicts with the assumption of punctiform targets, where each target is the origin of exactly one measurement. Therefore, one needs to be able to assign the obtained measurements to extended targets.

Tracking extended targets has similarities to tracking punctiform objects in clutter, which denotes false alarms nearby a target that do not originate from a target. We observe multiple readings which can be associated with the same target for punctiform objects in clutter as well as for extended targets. Hence, well known techniques for tracking punctual objects can be transferred to the

A. Kräußling: *Tracking Extended Moving Objects with a Mobile Robot*, Studies in Computational Intelligence (SCI) **109**, 513–530 (2008)

field of extended targets. These algorithms may apply the EM (Expectation Maximization) [8] or the Viterbi method [9]. In the past years our research group has established a number of these algorithms for tracking extended targets [10, 11].

This work mainly aims to compare several tracking algorithms, which were introduced by our research group. The focus is on the superiority of the methods presented in [12, 13] compared to other techniques. Due to limited space we were not able to motivate the use of these two algorithms in previous papers. Furthermore, we will discuss the question of what information we obtain about a targets position using these methods. We will mainly regard objects of circular shape, because we can treat such targets analytically. This already covers a wide range of interesting targets, such as many service robots. When tracking people, one has to deal with a variety of rapidly changing shapes. This problem will be outlined in Sect. 5.

This work is organized as follows. In Sect. 2 we introduce the model which we utilized for tracking extended targets and briefly summarize the characteristics of the used algorithms. Section 3 examines the problem of tracking circular objects. Then Sect. 4 presents experimental results. Furthermore, in Sect. 5 we discuss people tracking. In Sect. 6 we consider the problem of crossing targets. Finally, Sect. 7 summarizes the results and concludes with an outlook on future research.

2 Models and Methods

The dynamics of the object to be observed and the observation process itself are modeled by a hidden Gauss–Markov chain with the equations

$$x_k = Ax_{k-1} + w_{k-1} \tag{1}$$

and

$$z_k = Bx_k + v_k. \tag{2}$$

x_k is the objects state vector at time k, A is the state transition matrix, z_k is the observation vector at time k and B is the observation matrix. Furthermore, w_k and v_k are supposed to be uncorrelated zero mean white Gaussian noises with covariances Q and R.

Since the motion of a target in a plane has to be described a two dimensional kinematic model is used.

$$x_k = \left(\, x_{k1} \; x_{k2} \; \dot{x}_{k1} \; \dot{x}_{k2} \, \right)^\top \tag{3}$$

With x_{k1} and x_{k2} being the Cartesian coordinates of the target and $\dot{x}_{k1}$ and $\dot{x}_{k2}$ the corresponding velocities. z_k just contains the Cartesian coordinates of the target. In the usual applications of this model it is assumed that the target is punctiform and there is only one measurement from the target. Nevertheless, this model is also applicable for tracking extended targets. Then z_k will be a representative feature generated from all measurements of the target. For the coordinates the equation of a movement with constant velocity is holding, i.e. it is

$$x_{k,j} = x_{k-1,j} + \Delta T \dot{x}_{k-1,j}. \tag{4}$$

Thereby ΔT is the time interval between the measurements at time step $k-1$ and k. For the progression of the velocities the equation

$$\dot{x}_{k,j} = e^{-\Delta T/\Theta} \dot{x}_{k-1,j} + \Sigma \sqrt{1 - e^{-2\Delta T/\Theta}} u(k-1) \tag{5}$$

with zero mean white Gaussian noise $u(k)$ and $E[u(m)u(n)^{\top}] = \delta_{mn}$ is used. The choice of the first term in (5) ensures that the velocity declines exponentially, whereas the second term models the process noise and the acceleration. For the parameters Θ and Σ the values $\Theta = 20$ and $\Sigma = 60$ are good choices. The advantage of using this model is that the Kalman filter [14] can be applied for conducting the tracking process [1]. Therefore, the measurements are used in the update equation

$$x(k|k) = x(k|k-1) + K_k(z_k - y(k|k-1)) \tag{6}$$

with $x(k|k)$ being the estimate for the internal state x_k, $x(k|k-1)$ the prediction for the internal state, $y(k|k-1)$ the prediction for the measurement and K_k the Kalman gain. z_k is generated from all m_k measurements $\{z_{k,i}\}_{i=1}^{m_k}$ in the validation gate. Regarding punctiform targets with clutter there is at most one measurement from the target and the rest are false alarms. When dealing with extended objects in most cases all measurements in the validation gate are from one target. An exception is when the targets get close to another object, an obstacle or a wall. Only the point on the surface of the object, that generates the measurement, for instance by reflecting the laser beam, differs from measurement to measurement. The validation gate is implemented using the Kalman filter [1]. It selects the measurements from the target from all 360 measurements from the scan. All measurements with a distance from the prediction $y(k|k-1)$ beyond a certain bound are excluded.

In this work we analyze several ways to use the measurements in the gate. The first is to calculate z_k as an unweighted mean of these measurements, i.e. it is

$$z_k = \frac{\sum_{i=1}^{m_k} z_{k,i}}{m_k}. \tag{7}$$

The corresponding tracking algorithm is called Kalman filter algorithm (KFA) and has been introduced in [13]. The second way is to calculate a weighted mean as

$$z_k = \sum_{i=1}^{m_k} \alpha_i z_{k,i}. \tag{8}$$

For the weights α_i there are predominantly two possible choices. The first is in analogy to the weights of the Probabilistic Data Association Filter (PDAF)

[15] and the second is in analogy to the Probabilistic Multi Hypothesis Tracker (PMHT) [8]. In the following we only refer to the PMHT algorithm, because the results are similar. Here it is

$$\alpha_i = \frac{\exp\left(-\frac{1}{2}\left[\nu_i(k)\right]^\top \left[R_{eff}\right]^{-1} \nu_i(k)\right)}{\sum_{j=1}^{m_k} \exp\left(-\frac{1}{2}\left[\nu_j(k)\right]^\top \left[R_{eff}\right]^{-1} \nu_j(k)\right)}. \tag{9}$$

$\nu_i(k)$ is the innovation with

$$\nu_i(k) = z_{k,i} - y(k|k-1) \tag{10}$$

and

$$R_{eff} = R + E \tag{11}$$

with the measurement covariance R and an additional positive definite matrix

$$E = \begin{pmatrix} 780 & 0 \\ 0 & 780 \end{pmatrix}, \tag{12}$$

which models the extendedness of the targets. This algorithm is called Weighted Mean Algorithm (WMA). A further improvement might be the following approach: the weights can be viewed as parameters. Thus, the EM (Expectation–Maximization) algorithm [16] can be applied using the Kalman smoother [17] in order to further improve the estimates. The corresponding algorithm is called EM based algorithm (EMBA) and has been introduced in [10].

Another method is to calculate a separate estimate $x(k|k)_i$ for each measurement $z_{k,i}$. For the calculation of the estimates $x(k|k)_i$ in the update equation the measurement $z_{k,i}$ and the predictions $x(k|k-1)_j$ and $y(k|k-1)_j$ from the predecessor j are used. For the determination of the predecessor there are two possibilities: firstly, the length $d_{k,j,i}$ of the path ending in $z_{k,i}$ through the measurements $z_{k-1,j}$ $(j = 1, \ldots, m_{k-1})$ can be minimized. The lengths are defined recursively. They are

$$d_{k,j,i} = d_{k-1,j} + a_{k,j,i} \tag{13}$$

with

$$a_{k,j,i} = \frac{1}{2}\nu_{k,j,i}^\top [S_k]^{-1} \nu_{k,j,i} + \ln\left(\sqrt{|2\pi S_k|}\right) \tag{14}$$

and the innovation

$$\nu_{k,j,i} = z_{k,i} - y(k|k-1)_j. \tag{15}$$

S_k is the innovations covariance derived from the Kalman filter. Once the predecessor $z_{k-1,j(k-1,i)}$ has been found the estimate $x(k|k)_i$ can be calculated and the length $d_{k,i}$ of the path ending in $z_{k,i}$ is defined as $d_{k,j(k-1,i),i}$. A similar procedure is used for a punctiform object in clutter [9]. When regarding extended targets in most cases all measurements in the validation gate are from the target. Therefore, it is not meaningful to consider the lengths $d_{k-1,j}$ of the paths ending in the possible predecessors $z_{k-1,j}$ when determining the predecessor. A better choice for the predecessor is the one for which the Mahalanobis distance [18]

$$\nu_{k,j,i}^{\top}[S_k]^{-1}\nu_{k,j,i} \tag{16}$$

is kept to a minimum. This procedure is similar to a nearest neighbour algorithm [1].

When applying the Viterbi algorithm, the application of the validation gate is performed in the following way: at first for every measurement $z_{k-1,j}$ the gate is applied to all measurements at time k. This results in the sets $Z_{k,j}$ of measurements which have passed the particular gate for the measurement $z_{k-1,j}$ successfully. The set of all measurements $z_{k,i}$, which are associated with the target, is then just the union of these sets. The corresponding algorithms can deal with multimodal distributions to some extend, which is a major improvement when dealing with crossing targets.

There are two different ways to deal with the estimates delivered by the Viterbi algorithm. The first is to choose one of these estimates as an estimate of the position of a target. It can be chosen between the firstly delivered estimate for instance or the one with the shortest corresponding path. The corresponding algorithm is called Viterbi based algorithm (VBA) and has been introduced in [11]. The second is to calculate an unweighted mean of all estimates and use this mean as an estimate for the position of the target. The corresponding algorithm is called Viterbi average algorithm (VAA), in case the predecessors are defined minimizing the length of the path or modified Viterbi average algorithm (MVAA), if the predecessors are defined minimizing the Mahalanobis distance.

3 Tracking a Circular Object

In mobile robotics there are mainly two classes of objects to track – other robots or people. Since a lot of service robots are of circular shape and these objects can be treated analytically we will concentrate on this class of objects in this section. We start with the following conjecture: the algorithms that use a mean (i.e. KFA, WMA, EMBA, VAA or MVAA) estimate a point in the interior of the object, that is the mean of the points on the surface of the object, which are in the view of the observer. This mean, that we will call balance point, will be calculated as follows. To simplify the problem, it is assumed, that the centre of the observed circular object is at the origin of the planar coordinate system and the centre of the observer lies on the x-axis. The radius of the observed object is r and the distance from its centre to the centre of the observer is denoted by d. The coordinates of the mean S

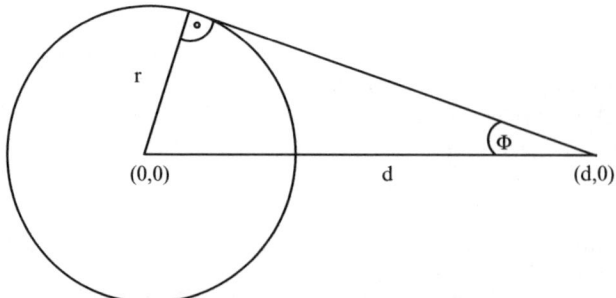

Fig. 1. Derivation of the angle ϕ

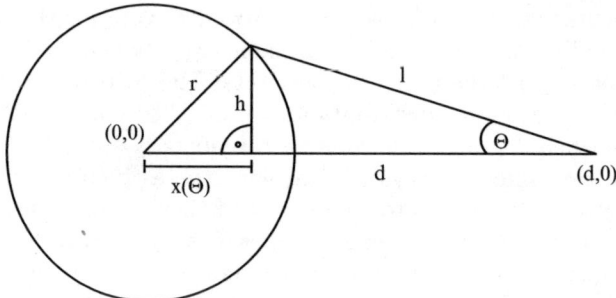

Fig. 2. Derivation of the distance $x(\theta)$

in the interior are denoted by (x, y). Because of the problem's symmetry it immediately follows that $y = 0$. Moreover,

$$x = \frac{1}{\phi} \int_0^\phi x(\theta) d\theta. \tag{17}$$

For the definition of the angle ϕ and the definition of the distance $x(\theta)$ we refer to Figs. 1 and 2. The latter is calculated from the known values d and r and the angle θ as follows: By the proposition of Pythagoras

$$r^2 = h^2 + x^2(\theta) \tag{18}$$

and

$$(d - x(\theta))^2 + h^2 = l^2. \tag{19}$$

Furthermore

$$\sin \theta = \frac{h}{l} . \tag{20}$$

From these three equations the term

$$x(\theta) = d \sin^2 \theta + \cos \theta \sqrt{r^2 - d^2 \sin^2 \theta} \tag{21}$$

for calculating $x(\theta)$ can be derived. Together with (17) this results in

$$x = \frac{1}{\phi} \int_0^\phi \left(d \sin^2 \theta + \cos \theta \sqrt{r^2 - d^2 \sin^2 \theta} \right) d\theta. \tag{22}$$

According to Fig. 1 the expression

$$\sin \phi = \frac{r}{d} \tag{23}$$

can be used for the derivation of the angle ϕ. The antiderivative of the first term in the integral can be found in textbooks, e.g. [19]. Therefrom it is

$$\int \sin^2 \theta \, d\theta = \frac{1}{2} \theta - \frac{1}{2} \sin \theta \cos \theta = \frac{1}{2} \theta - \frac{1}{2} \sin \theta \sqrt{1 - \sin^2 \theta}. \tag{24}$$

The antiderivative of the second term can be found by integration by substitution. We use $u = d \sin \theta$. Therefrom it is

$$\int \cos \theta \sqrt{r^2 - d^2 \sin^2 \theta} \, d\theta = \frac{1}{d} \int \sqrt{r^2 - u^2} \, du. \tag{25}$$

The antiderivative of $\int \sqrt{r^2 - u^2} \, du$ is [19]

$$\frac{1}{2} \left(u \sqrt{r^2 - u^2} + r^2 \arcsin \frac{u}{r} \right). \tag{26}$$

The combination of these results after some algebraic manipulations finally delivers

$$x = \frac{d}{2} + \frac{r}{2d \arcsin \frac{r}{d}} \left(\frac{\pi r}{2} - \sqrt{d^2 - r^2} \right). \tag{27}$$

4 Experimental Results

Table 1 shows the results for different values of d used in our simulations. They are in the range of the typical distances between the laser and the object, which occur in the field of mobile robotics. For the radius r of the object we have set $r = 27$ cm, which is in the range of the dimension of a typical mobile

Table 1. Angle ϕ and distance x

d (cm)	ϕ (rad)	ϕ (deg)	x (cm)
100	0.2734	15.6647	23.3965
200	0.1354	7.7578	22.3607
400	0.0676	3.8732	21.7837
600	0.0450	2.5783	21.6103
800	0.0338	1.9366	21.4803

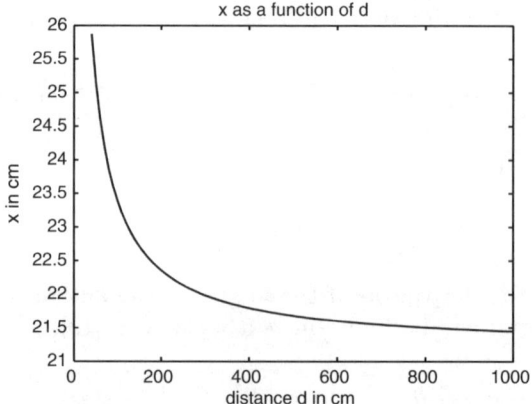

Fig. 3. x As a function of the distance d

robot. Table 1 and Fig. 3 show the effect of the radius d on the balance point. With growing radius d the observable area of the object increases and hence the balance point S moves closer to the centre of the object.

In the following we consider the movement of the circular object around the laser range scanner on a circle with radius R. To evaluate the performance of the algorithms solving this problem simulated data has been used, because we needed to know the true position of the target very accurately. This is hard to achieve using data from a real experiment and has already been mentioned by other authors [20]. Since we considered a movement on a circle, the process noise only originates from the centripetal force, that keeps the object on the circle. The simulations have been carried out for the values of the radius R introduced in Table 1. The values for the standard deviation σ of the measurement noise have been 0, 1, 3, 5, 7.5 and 10 cm. These values are in the typical range of the errors of the commercial laser distance sensors, which are commonly used in mobile robotics [21]. For each pair of R and σ 20 replications have been carried out. For each time step k the Euclidian distance of the tracking algorithms' output from the balance point has been calculated and therefrom the average of these distances over the whole run has been calculated. Finally, from these averages the average over the 20 cycles had been calculated. The results with unit 1 cm are presented in the Tables 2–6. The corresponding standard deviations calculated from the 20 cycles are small. They reach from about 0.01 cm for the smaller standard deviations of the measurement noise to about 0.1 cm for the larger variants.

It is apparent, that the outputs of all algorithms, except of the VAA, produce a good estimate for the balance point S. Thereby, it should be pointed out, that the KFA performs best. Only for the radius $R = 100$ cm the EMBA performs slightly better. When comparing the WMA and the EMBA it turns out, that the WMA performs better for some combinations of radius R and measurement noise σ. This result might be of interest not only for tracking

Table 2. Average distance from S for the KFA

R (cm)	100	200	400	600	800
$\sigma = 0$	0.9689	0.3654	0.2154	0.5597	0.2478
$\sigma = 1$	0.9698	0.3654	0.2202	0.5609	0.2575
$\sigma = 3$	0.9710	0.3805	0.2962	0.5744	0.3487
$\sigma = 5$	0.9853	0.4266	0.4133	0.6531	0.5392
$\sigma = 7.5$	1.0497	0.5617	0.5673	0.7845	0.7811
$\sigma = 10$	1.1054	0.6317	0.7409	0.9626	1.0119

Table 3. Average distance from S for the WMA

R (cm)	100	200	400	600	800
$\sigma = 0$	1.6247	1.0523	0.9999	1.2498	1.0860
$\sigma = 1$	1.6223	1.0499	1.0010	1.2496	1.0884
$\sigma = 3$	1.6062	1.0350	1.0124	1.2512	1.0753
$\sigma = 5$	1.5714	1.0164	1.0069	1.2625	1.1530
$\sigma = 7.5$	1.5592	1.0326	1.0661	1.2796	1.2661
$\sigma = 10$	1.5239	1.0125	1.1030	1.3799	1.4187

Table 4. Average distance from S for the EMBA

R (cm)	100	200	400	600	800
$\sigma = 0$	0.3324	0.5599	0.7941	1.0399	1.0744
$\sigma = 1$	0.3381	0.5704	0.8052	1.0555	1.0758
$\sigma = 3$	0.3882	0.6157	0.9085	1.1524	1.1583
$\sigma = 5$	0.4569	0.7386	1.0644	1.3136	1.4272
$\sigma = 7.5$	0.6274	0.9567	1.3267	1.5937	1.8012
$\sigma = 10$	0.7809	1.1097	1.6078	1.9553	2.2215

Table 5. Average distance from S for the VAA

R (cm)	100	200	400	600	800
$\sigma = 0$	6.9678	3.0463	3.0767	6.5293	6.4645
$\sigma = 1$	11.5044	4.5481	1.7062	5.8976	6.5338
$\sigma = 3$	12.2924	5.0634	2.4768	5.2141	7.1276
$\sigma = 5$	12.4933	6.6898	3.6525	5.2175	6.7693
$\sigma = 7.5$	14.2701	7.6690	4.6502	5.4802	6.7554
$\sigma = 10$	14.5553	8.3581	5.7584	6.2627	7.1146

applications, because it is widely assumed that the estimates can be bettered by conducting the iterations of the EM algorithm.

Comparing the results for the VAA and the MVAA it is apparent, that it is much better to choose the predecessor with respect to the Mahalanobis distance. Since the balance point is estimated very accurately by most of the algorithms, an estimate for the centre of the circular object can be derived

Table 6. Average distance from S for the MVAA

R (cm)	100	200	400	600	800
$\sigma = 0$	2.1260	1.2587	1.7045	2.3523	2.6191
$\sigma = 1$	1.8316	1.1886	1.6577	2.4278	2.5511
$\sigma = 3$	1.5376	1.1199	1.6894	2.6091	2.7851
$\sigma = 5$	1.6535	1.3429	1.9600	2.8546	3.1441
$\sigma = 7.5$	1.9116	1.7742	2.4516	3.3260	3.6828
$\sigma = 10$	2.2212	2.1866	3.0103	3.9349	4.2878

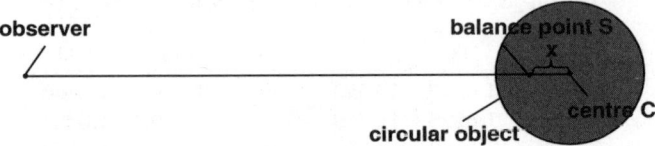

Fig. 4. Determination of the centre C

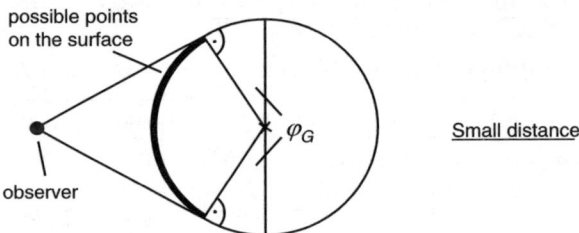

Fig. 5. Possible points, small distance between observer and object

directly. This is due to the fact, that the Euclidian distance x of the balance point to the centre of the object can be calculated depending on R as above. Furthermore, the observer, the balance point S and the centre C of the object are lying on a straight line as indicated in Fig. 4. Thus, KFA, WMA, EMBA and MVAA deliver good information about the position of the object.

The VBA algorithm calculates one position estimate for every single range reading that originates from one target. So each estimate corresponds to one point on the surface of this object. The algorithm then chooses one of these estimates without having the knowledge to which point the estimation is corresponding. Therefore, we have a great uncertainty about the estimated position. This is illustrated in Figs. 5–9. In Figs. 5 and 6 the points on the surface of the object, which are in the view of the observer, are reproduced. The figures show that for a greater distance of the object to the observer there is a larger amount of points in the view of the observer, as already has been mentioned. Thus, for a greater distance there is a greater uncertainty about which point on the surface is estimated. The angle φ_G indicated in the figures is the same

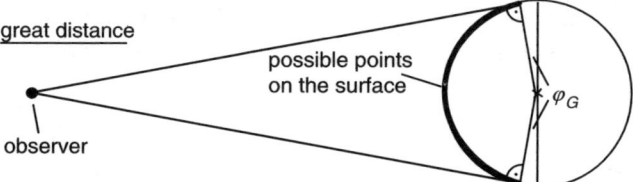

Fig. 6. Possible points, large distance between observer and object

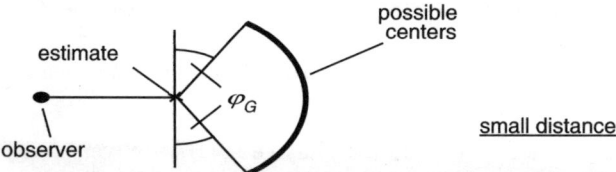

Fig. 7. Possible centers, small distance between observer and object

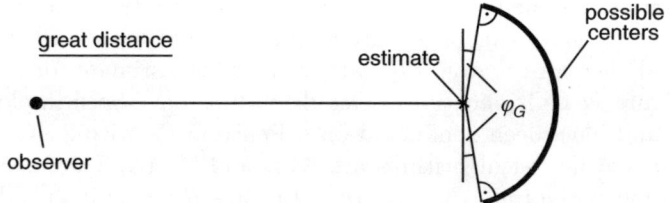

Fig. 8. Possible centers, large distance between observer and object

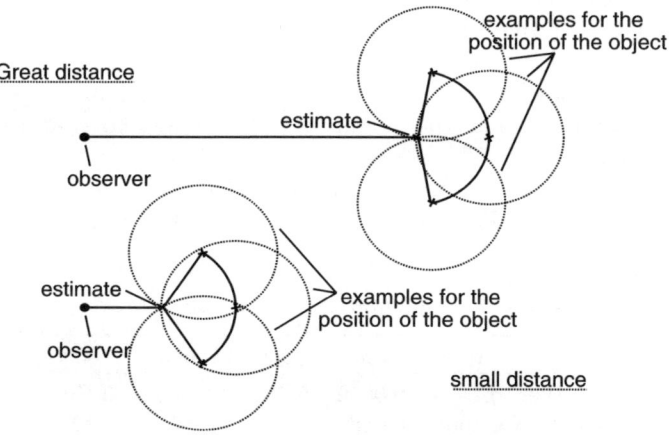

Fig. 9. Examples for the position of the object

as the angle ϕ introduced in Fig. 1. Figures 7 and 8 show the possible positions of the centre of the object. Again, there is a greater uncertainty for a greater distance. Finally, Fig. 9 presents some examples for possible positions of the object.

From Figs. 7 and 8 the following statement can be concluded: there is a great uncertainty in the estimate of the centre of the object when applying VBA. For large distances this uncertainty is in the range of the diameter of the object. There are two further problems that complicate the situation:

– First, the point on the surface that is hit by the laser beam, changes from time step to time step.
– Second, there is an additional error in form of the measurement noise, which corrupts the data.

Recapitulating it can be concluded that the VBA delivers only sparse information about the true position of the target.

Now it is referred to a second criterion for the comparison of the algorithms, the computational complexity. As a measure for this property the need of computing time for the fulfilment of the calculations for one time step is used. The reason for this procedure is, that some of the algorithms are very complex and therefore it would be very difficult to estimate for instance the number of matrix multiplications. The algorithms have been implemented in MATLAB and have been conducted on a Pentium IV with 2.8 GHz. The by far simplest and fastest algorithms are KFA and WMA. They needed about 20 ms per time step for all combinations of radius R and standard deviation σ of the measurement noise. The results for the EMBA and the VBA are given in Tables 7 and 8. The results for VAA and MVAA are similar to those of the VBA. The tables show that the computation time varies from about 70 ms to about 200 ms for the EMBA and from about 80 ms to about 1.5 s for the VBA. Of course the time for the EMBA depends on the number of iterations and thus on the choice of the stop criterion. In our implementation the algorithm stops iterating when the estimates of two consecutive iterations differ less than one centimeter. The values of the VBA depend on the radius R. This is due to the fact that the number of measurements from the target highly depends on the radius. The complexity of the VBA strongly depends on the

Table 7. Computation time for the EMBA

R (cm)	100	200	400	600	800
$\sigma = 0$	0.1635	0.0859	0.0696	0.0832	0.1015
$\sigma = 1$	0.1662	0.0881	0.0701	0.0828	0.1026
$\sigma = 3$	0.1660	0.0870	0.0729	0.0900	0.1152
$\sigma = 5$	0.1646	0.0868	0.0784	0.1024	0.1359
$\sigma = 7.5$	0.1627	0.0897	0.0891	0.1201	0.1643
$\sigma = 10$	0.1627	0.0955	0.1008	0.1412	0.1959

Table 8. Computation time for the VBA

R (cm)	100	200	400	600	800
$\sigma = 0$	1.4500	0.4079	0.1673	0.1023	0.0765
$\sigma = 1$	1.4615	0.4122	0.1712	0.1043	0.0779
$\sigma = 3$	1.4581	0.4199	0.1686	0.1026	0.0776
$\sigma = 5$	1.4739	0.4144	0.1648	0.1048	0.0783
$\sigma = 7.5$	1.4712	0.4124	0.1682	0.1028	0.0778
$\sigma = 10$	1.4814	0.4231	0.1652	0.1044	0.0783

number of measurements from the object. For example the predecessor, which has to be determined for every new measurement, has to be chosen among all the measurements from the last time step.

5 People Tracking

People tracking is a more difficult task. One cannot just assume a certain shape, but has to deal with arbitrary, possibly shifting shapes, that can change rapidly from scan to scan. As opposed to circular objects, where one just takes the center of the target as the targets position, it is even hard to define which point to use as an estimate. The problem becomes even more complex when one of the legs is occluded by the other.

It is possible to assume the mean of all points on the scanned surface to be the center. Then the KFA algorithm would produce a very good estimate of the position. If there are more scans from one leg than from the other, the estimated position shifts to the leg for which we have more returns.

To solve the problem more accurately, one could make sure that there is a good coverage of the surface of a target by multiple sensors. For a known circular shape we compute an estimate of the targets position for every single sensor using KFA. Then we produce an unweighted mean of the results as an estimate of the object position.

When tracking people, one should use all measurements from all sensors as an input for KFA. However, one factor should be considered thereby. It might happen that the same part of the object's surface is covered by more than one sensor. Then the corresponding measurements should be weighted less depending on the number of those sensors. Otherwise the estimate delivered by the KFA might shift to this part of the object.

An important prerequisite for this approach is that the coordinate systems of the sensors almost coincide. Otherwise, we would not gain any advantage by fusing the data, because KFA estimates the position of an object with an accuracy of less than one centimeter as shown in the previous section. Right now, our equipment only allows an accuracy of approximately ten centimeters for the coordinate systems.

6 Tracking Two Crossing Objects

Crossing targets is a central issue in tracking. When two targets cross, they might not be distinguishable until they split again. Most existing algorithms tend to loose one of the objects after crossing occurred. Thus, crossing targets is a good benchmark for target tracking algorithms.

Figures 10 and 11 illustrate this problem using data from the EMBA algorithm. There the targets are represented by ellipses, with the centre being the estimated position. The ellipses themselves approximate the shapes of the targets.

In Fig. 10, which is created from simulated data, there is always a minimal distance between the targets, even during the crossing and the algorithm separates the targets after the crossing. When the two objects cross each

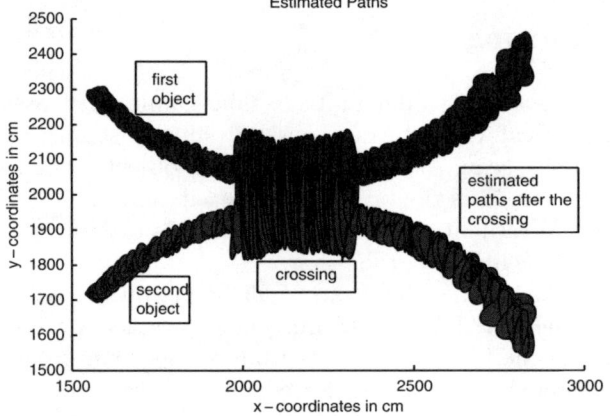

Fig. 10. Crossing of two objects, EMBA, simulated data

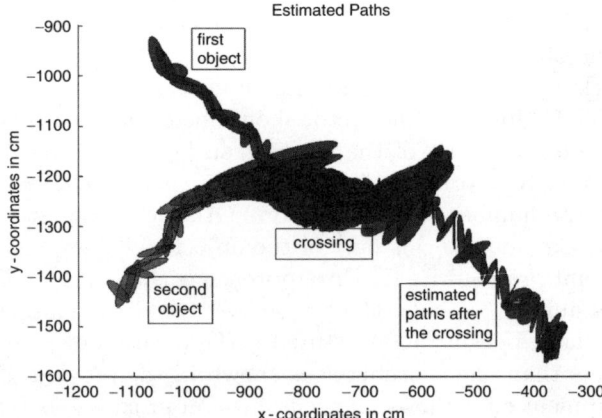

Fig. 11. Crossing of two objects, EMBA, real data

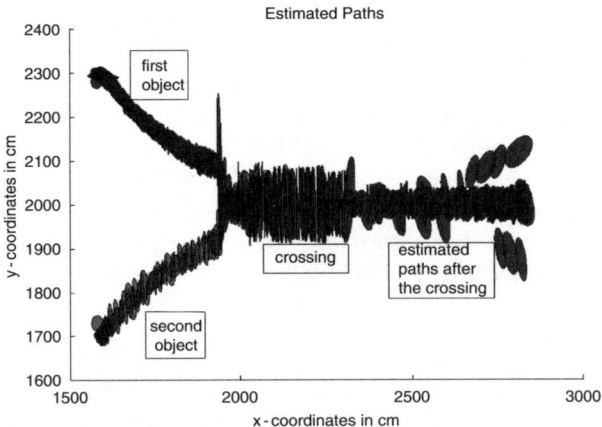

Fig. 12. Crossing of two objects, MVAA, simulated data

others path, we cannot always assign the readings correctly and hence, the ellipses grow larger. The results look alike for WMA. VBA and KFA are not able to solve this situation.

In Fig. 11 real data from an experiment with two persons walking around was used. Here the situation is more complex, because of the close proximity of the two persons. Furthermore, one of them is walking in the shadow of the other for some time. Both EMBA and WMA fail to keep track of both targets.

Figure 12 illustrates the behaviour of MVAA using simulated data. The results for VAA are similar. Starting with two separately tracked objects, this algorithm fails to assign the readings after the crossing for either object. Instead, the readings from both objects are used to calculate the position of a non existent object using the mean of the estimates. A similar behaviour is known in the case of punctiform targets in clutter using the PDAF [1]. The reason for this behaviour is the fact that we use the Viterbi algorithm to calculate a validation gate for each measurement and the MVAA algorithm just takes a mean of all estimated positions. Thus, we obtain a position estimate that is between the real positions of the targets. When we obtain more readings from one object than from the other, the estimate moves closer to that object, what frequently occurs in the figure.

The fact that the VBA algorithm still processes the readings of both targets after the crossing can be utilized. This is the basis for a method which is able to track the crossing targets stable. After the crossing occurred VBA still tracks both objects using two clusters of estimates. Since both objects are assigned to the same cluster, only the assignment of the clusters to the targets is wrong. Thus, we developed a method to correctly assign the clusters to the objects. It uses the results of the VBA algorithm and geometrical considerations, which characterize the problem. The derived algorithm is called Cluster Sorting Algorithm (CSA) and was introduced in [12].

The major disadvantage of CSA is that it is using VBA for the whole tracking process, even if no crossing occurred. This leads to higher computational demands and lower accuracy compared to KFA as outlined in Sect. 4. Hence, we use KFA as long as there is no crossing and switch to CSA if we detect crossing. Such a switching algorithm has been introduced in [13]. Both algorithms, CSA and the switching method, solve the problem of crossing targets. Furthermore, CSA can be strongly accelerated using an additional filter step to make the switching algorithm capable for real time applications also during a crossing when it uses CSA [22].

7 Conclusions and Future Work

In this work several algorithms for tracking extended targets have been discussed. The problem was outlined for known circular shapes as for people tracking, where no particular shape can be assumed. It was shown that the algorithms, which use a mean of the measurements, produce a good state estimation for fixed shape, circular objects. Here the KFA showed to be superior to other methods in terms of the quality of the estimate and computational complexity.

According to Table 2 KFA is able to estimate the position of the balance point of an object with circular shape with a very high accuracy in the range between 2 mm and 1.1 cm. Furthermore, due to 27 and Fig. 4, there is exact knowledge about the position of the centre of the object, once the position of the balance point is known. Thus, a mobile robot with circular shape can be tracked very accurately using KFA and laser range scanners as sensors. For this reason the problem of localization can be solved by our method with an accuracy and computational complexity, which is superior to the formerly known methods like [23–25]. Thereby both, the local and the global localization problem can be solved. Local localization or tracking means the consecutive position estimation of a mobile robot, when the initial position is known. Global localization means the determination of the initial position from scratch. As an application example of the method one might think of the surveillance of an art museum when it is closed. Then one has to deal with wide, empty spaces, which can well be covered by some laser scanners embedded into the walls. The mobile robots, which need to have a circular shape in the plane, that is observed by the laser scanners, can then be localized with a very high accuracy when they move from room to room to monitor them. The costs for the laser scanners and the mobile robots are negligible compared to the adversity of the theft of a famous painting. A paper with further details is under review [26].

Most of the methods presented here do not solve the problem of crossing targets. Only EMBA and WMA are able to keep track of both targets when using simulated data, but they fail in real experiments. For punctual objects in the field of aerial surveillance this problem is well known and several solutions exist [8,27,28]. However, when tracking extended targets in the field of mobile robotics, there are major disagreements [29]. Hence, these techniques will fail when being applied for tracking extended targets in mobile robotics.

The problem of extended crossing targets is well known in the field of mobile robotics [2,30]. There it has been explicated that this problem is particularly difficult to solve. Nevertheless, our research group developed two algorithms that are able to keep track of the targets [12,13]. The first is called Cluster Sorting Algorithm (CSA) and uses VBA with an additional clustering routine [12]. The second is a switching algorithm that utilizes CSA only during crossing and KFA else [13]. Therefore, these two algorithms are predominant to the other represented methods. Since CSA and the switching algorithm only solve the problem in the case of just two targets, finally an improved switching algorithm has been developed, that solves the problem of an arbitrary number of crossing extended targets in the context of mobile robotics [29].

References

1. Bar-Shalom, Y., Fortmann, T.: Tracking and Data Association. Academic Press, New York (1988)
2. Prassler, E., Scholz, J., Elfes, E.: Tracking people in a railway station during rush-hour. Computer Vision Systems **1542** (1999) 162–179. Springer Lecture Notes
3. Schulz, D., Burgard, W., Fox, D., Cremers, A. B.: Tracking multiple moving objects with a mobile robot. IEEE Computer Society Conference on Computer Vision and Pattern Recognition (CVPR 2001)
4. Fod, A., Howard, A., Mataric, M. J.: Laser-based people tracking. IEEE International Conference on Robotics and Automation (ICRA 2002) 3024–3029
5. Fuerstenberg, K. C., Linzmeier, D. T., Dietmayer, K. C. J.: Pedestrian recognition and tracking of vehicles using a vehicle based multilayer laserscanner. Intelligent Vehicles Symposium **1** (2002) 31–35
6. Thrun, S.: Learning metric-topological maps for indoor mobile robot navigation. Artificial Intelligence **99(1)** (1998) 21–71
7. Thrun, S., Fox, D., Burgard, W.: Markov localization for mobile robots in dynamic environments. Journal of Artificial Intelligence Research **11** (1999) 391–427
8. Streit, R. L., Luginbuhl, T. E.: Maximum Likelihood Method for Multi-Hypothesis Tracking. Signal and Data Processing of Small Targets **2335** (1994)
9. Quach, T., Farrooq, M.: Maximum Likelihood Track Formation with the Viterbi Algorithm. 33rd Conference on Decision and Control (1994)
10. Stannus, W., Koch, W., Kräußling, A.: On robot-borne extended object tracking using the EM algorithm. 5th Symposium on Intelligent Autonomous Vehicles, Lisbon, Portugal (IAV 2004)
11. Kräußling, A., Schneider, F. E., Wildermuth, D.: Tracking expanded objects using the Viterbi algorithm. IEEE Conference on Intelligent Systems, Varna, Bulgaria (IS 2004)

12. Kräußling, A., Schneider, F. E., Wildermuth, D.: Tracking of extended crossing objects using the Viterbi algorithm. 1st International Conference on Informatics in Control, Automation and Robotics, Setubal, Portugal (ICINCO 2004)
13. Kräußling, A., Schneider, F. E., Wildermuth, D.: A switching algorithm for tracking extended targets. 2nd International Conference on Informatics in Control, Automation and Robotics, Barcelona, Spain (ICINCO 2005), also to be published in the Springer book of best papers of ICINCO 2005
14. Kalman, R. E.: A new approach to linear filtering and prediction problems. Transaction of the ASME Journal of Basic Engineering **82** (1960) 34–45
15. Bar-Shalom, Y., Tse, E.: Tracking in a cluttered environment with probabilistic data association. Automatica **11** (1975) 451–460
16. Dempster, A. P., Laird, N., Rubin, D. B.: Maximum likelihood from incomplete likelihood via the EM algorithm. Journal of the Royal Statistical Society B **39** (1977) 1–38
17. Shumway, R. H., Stoffer, D. S.: Time series analysis and its applications. Springer, Berlin Heidelberg New York (2000)
18. Mahalanobis, P. C.: On the generalized distance in statistics. Proceedings of the National Institute of Science **12** (1936) 49–55
19. Bronstein, I. N., Semendjajew, K. A.: Taschenbuch der Mathematik. Verlag Harri Deutsch Thun und Frankfurt/Main (1987)
20. Zhao, H., Shibasaki, R.: A novel system for tracking pedestrians using multiple single-row laser-range scanners. IEEE Transactions on Systems Man and Cybernetics – Part A: Systems and Humans **35(2)** (2005) 283–291
21. Technical information sheet lms/lms 200/lms 211/ lms 29. **Sick Inc.** 6900 West 110th Street Minneapolis MN 55438
22. Kräußling, A.: A fast algorithm for tracking multiple interacting objects. Submitted to the 3rd European Conference on Mobile Robots (ECMR 2007)
23. Burgard, W., Fox, D., Hennig, D., Schmidt, T.: Estimating the absolute position of a mobile robot using position probability grids. Proc. of AAAI (1996)
24. Fox, D., Burgard, W., Thrun, S., Cremers, A., B.: Position estimation for mobile robots in dynamic environments. Proc. of AAAI (1998)
25. Fox, D., Burgard, W., Dellaert, F., Thrun, S.: Monte Carlo localization: effizient position estimation for mobile robots. Proc. of AAAI (1999)
26. Kräußling, A.: A novel approach to the mobile robot localization problem using tracking methods. Submitted to the 13th IASTED International Conference on Robotics and Applications (RA 2007)
27. Reid, D. B.: An algorithm for tracking multiple targets. IEEE Transaction on Automatic Control **24** (1979) 843–854
28. Fortmann, T. E., Bar-Shalom, Y., Scheffe M.: Sonar tracking of multiple targets using joint probabilistic data association. IEEE Journal of Oceanic Engineering **OE-8(3)** (1983)
29. Kräußling, A.: Tracking multiple objects using the Viterbi algorithm. 3rd International Conference on Informatics in Control, Automation and Robotics, Setubal, Portugal, (ICINCO 2006), also to be published in the Springer book of best papers of ICINCO 2006
30. Schumitch, B., Thrun, S., Bradski, B., Olukotun, K.: The information – form data association filter. 2005 Conference on Neural Information Processing Systems (NIPS) MIT Press (2006)

A Bayesian Solution to Robustly Track Multiple Objects from Visual Data

M. Marrón, J.C. García, M.A. Sotelo, D. Pizarro, I. Bravo, and J.L. Martín

Department of Electronics, University of Alcala, Edificio Politécnico, Campus Universitario, Alcalá de Henares, 28871 (Madrid), Spain
marta@depeca.uah.es, jcarlos@depeca.uah.es, sotelo@depeca.uah.es, pizarro@depeca.uah.es, ibravo@depeca.uah.es, jlmartin@depeca.uah.es

Summary. Different solutions have been proposed for multiple objects tracking based on probabilistic algorithms. In this chapter, the authors propose the use of a single particle filter to track a variable number of objects in a complex environment. Estimator robustness and adaptability are both increased by the use of a clustering algorithm. Measurements used in the tracking process are extracted from a stereo-vision system, and thus, the 3D position of the tracked objects is obtained at each time step. As a proof of concept, real results are obtained in a long sequence with a mobile robot moving in a cluttered scene.

1 Introduction

Probabilistic algorithms in their different implementations (Multi-Hypothesis Techniques – MHT – [1], Particle Filters – PF – [2,3] and their diversifications [4,5]) have fully shown their reliability in estimation tasks. Nowadays these methods are widely applied to solve positioning problems in robot autonomous navigation [6,7].

The idea of tracking multiple objects appeared with the first autonomous navigation robot to overcome the obstacle avoidance problem, and soon probabilistic algorithms, such as PFs [8,9] and Kalman Filters (KFs) [10,11], were applied to achieve this aim. The objective is, in any case, to calculate the posterior probability $(p(\vec{x}_t|\vec{y}_{1:t})^1)$ of the state vector $\vec{x}_t$, that informs about the position of the objects to track, in the recursive two steps standard estimation process (prediction-correction), in which, at least, some of the involved variables are stochastic, and by means of the Bayes rule.

To solve the multiplicity problem, the use of an expansion of the state vector $(\vec{\chi}_t = \{\vec{x}_t^1, \vec{x}_t^2, \dots, \vec{x}_t^k\})$ that includes the model of all objects to track was the first solution proposed in [12].

---

[1] Definition of all variables is included in Table 1.

M. Marrón et al.: *A Bayesian Solution to Robustly Track Multiple Objects from Visual Data*, Studies in Computational Intelligence (SCI) **109**, 531–547 (2008)
www.springerlink.com

Table 1. Variables definition

Variables	Definition
$\vec{x}_t$	State vector. In the tracking application this vector contains the 3D position and the 2D ground speed in Cartesian coordinates
$\vec{x}_{t|t-1}$	State vector prediction
$\vec{y}_t$	Measurements vector. In the tracking application this vector contains the 3D position in Cartesian coordinates
$f(\vec{x}_t, \vec{u}_t, \vec{o}_t)$	Transition model. $\vec{u}_t$ is the input vector and $\vec{o}_t$ is the noise vector related with the states
$p(\vec{x}_t|\vec{x}_{t-1})$	Transition model in the model Markovian definition
$p(\vec{y}_{1:t})$	Measurements distribution
$h(\vec{x}_t, \vec{r}_t)$	Observation model. $\vec{r}_t$ is the noise vector related with the measurements
$p(\vec{y}_t|\vec{x}_t)$	Observation model in the model Markovian definition. This density informs about measurements likelihood
$p(\vec{x}_t|\vec{y}_{1:t})$	Belief or posterior distribution. Result of the state vector probabilistic estimation
$p(\vec{x}_t|\vec{y}_{1:t-1})$	Prior distribution. Probabilistic prediction of the state vector
$S_t = \left\{ \vec{x}_t^{(i)}, \tilde{w}_t^{(i)} \right\}_{i=1}^n$	Particle set. Discrete representation of the belief used in the PF. Defined by n normal weighed $\tilde{w}_t^{(1:n)}$ evaluations of the state vector $\vec{x}_t^{(1:n)}$
$S_{t|t-1}$	Prediction of the particle set
n	Total number of particles
$n_{m,t} = \gamma_t \cdot n$	Number of particles to be inserted at the re-initialization step
$w(\vec{x}_{0:t}) \equiv \vec{w}_t = \left\{ w_t^{(i)} \right\}_{i=1}^n$	Importance sampling function. Continuous representation of the weights array $\vec{w}_t$
$q(\vec{x}_t|\vec{x}_{0:t-1}\vec{y}_{1:t})$	Best approximation to the belief
m	Number of measurements in the set
$Y_t = \{\vec{y}\}_{i=1}^m$	Measurements set
k	Number of clusters
$G_{1:k,t} \equiv \{\vec{g}_{1:k,t}, L_{1:k,t}\}$	Clusters set. Each cluster is defined by its centroide $\vec{g}_{1:k,t}$ in the clustering characteristics space, and its member set $L_{1:k,t}$
$\{d_{i,j}\}_{i=1,j=1}^{m,k}$	Distance defined in the clustering characteristic space between the centroides $\vec{g}_{1:k,t}$ and the data set $Y_t = \{\vec{y}\}_{i=1}^m$ in the tracking application

The computational load of the resultant estimator does not allow achieving a real time execution of the algorithm for more than four or five objects [13].

Another solution for the multiple objects tracker is to use a standard estimator to track each object but, apart from the inefficiency of the final algorithm [14], it cannot deal easily with a dynamic number of objects [15].

In any case, in order to achieve a robust multi-tracking system, it is necessary to include an association algorithm to correctly insert the information included in the observation model to the estimation process. Most of the association solutions are based on the Probabilistic Data Association (PDA) theory [16], such as the Joint Probabilistic Particle Filter (JPDAF) like in [17] or in [18]. Again, the problem related to these techniques is the execution time.

In this context the authors propose in [19] another solution to the multi-tracking problem based on a PF. In this case, the multi-modality of the filter is exploited to perform the estimation task for various models with a single PF, and a clustering algorithm is used as association process in the multi-modal estimation, whose deterministic behavior is also exploited in order to increase the multi-tracker robustness.

The algorithm obtained is called Extended Particle Filter with Clustering Process (XPFCP). This solution has been tested in complex indoor environments with sonar [19] and vision data [20] with good results.

The choice of vision sensors to implement the observation system of the tracking application guarantees a rich amount of information from the objects in the world. For this reason, the final development described here is based on visual information.

In this chapter, a general revision of the global tracking system is included and a complete analysis of the results obtained with the multi-tracking proposal is exposed.

2 System Description

The complete obstacle detection and tracking system proposed is described in Fig. 1. The objective is to design a tracker that detects and predicts the movement and position of dynamic and static objects in complex environments, so two main constraints are taken into account in the development:

- *Indoor environment is unknown*, because no map information is available, and complex, because hard dynamic and crowded situations are frequent.
- *A real time application in a modular organization has to be achieved*, in order to attach it to any robotic autonomous navigator.

As it can be noticed in Fig. 1, three main processes are included in the global tracking system:

1. *A stereovision system* is used to obtain 3D position information from the elements in the environment.
2. *The extracted 3D position data is then classified in two types*: measurements related with the objects to track; and information from the environmental structure that can be used in a partial-reconstruction process.

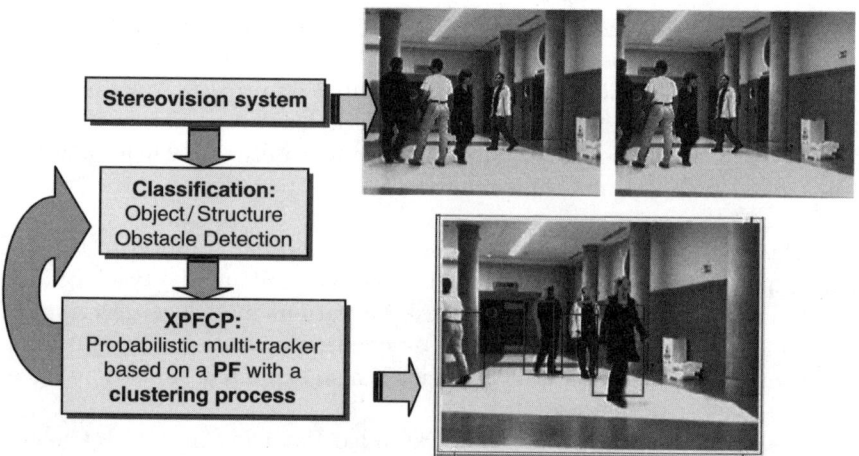

Fig. 1. Block diagram of the global tracking system

3. A probabilistic algorithm, the XPFCP, with two main components:
   - *An extended PF* is used to implement the multi-modal tracker. Using this kind of algorithm it is possible to estimate a variable number of probabilistic non-linear and non-Gaussian models with a single density function.
   - *A clustering algorithm* is inserted in the PF to develop the association process task and to increase the robustness and adaptability of the multi-modal estimator.

Descriptions of each one of the modules presented are completed in the following sections.

3 The Estimation Model

The main objective of XPFCP is to estimate the movement of objects around an autonomous navigation platform. In order to develop the tracking process a position estimation model has to be defined.

State vector encoding the objects position and speed in Cartesian coordinates at time t is represented by $\vec{x}_t$.

From a probabilistic point of view, this state vector can be expressed by a density function $p(\vec{x}_t|\vec{y}_{1:t})$, also called belief.

The evolution of this belief $p(\vec{x}_t|\vec{y}_{1:t-1})$ is defined by a Markov Process, with transition kernel $p(\vec{x}_t|\vec{x}_{t-1})$, as follows:

$$p(\vec{x}_t|\vec{y}_{1:t-1}) = \int p(\vec{x}_t|\vec{x}_{t-1}) \cdot p(\vec{x}_{t-1}|\vec{y}_{1:t-1}) \cdot \partial\vec{x} \tag{1}$$

The transition kernel is derived from a simple motion model, which can be expressed as follows:

$$p(\vec{x}_t|\vec{y}_{1:t-1}) \equiv \vec{x}_{t|t-1} = f(\vec{x}_{t-1}, \vec{o}_{t-1}),$$

$$\vec{x}_{t|t-1} = \begin{bmatrix} 1 & 0 & 0 & t_s & 0 \\ 0 & 1 & 0 & 0 & t_s \\ 0 & 0 & 1 & 0 & 0 \\ 0 & 0 & 0 & 1 & 0 \\ 0 & 0 & 0 & 0 & 1 \end{bmatrix} \cdot \vec{x}_{t-1} + \vec{o}_{t-1} \tag{2}$$

On the other hand, the measurements vector $\vec{y}_t$ contains the 3D position information sensed by the vision system (see Table 1).

The probabilistic relation between this vector $\vec{y}_t$ and the state one $\vec{x}_t$ is given by the likelihood $p(\vec{y}_t|\vec{x}_t)$, that defines the observation model from a stochastic approach.

The observation model, that describes the deterministic relation expressed by the likelihood, is defined as follows:

$$p(\vec{y}_t|\vec{x}_t) \equiv \vec{y}_t = h(\vec{x}_t, \vec{r}_t),$$

$$\vec{y}_t = \begin{bmatrix} 1 & 0 & 0 & 0 & 0 \\ 0 & 1 & 0 & 0 & 0 \\ 0 & 0 & 1 & 0 & 0 \end{bmatrix} \cdot \vec{x}_t + \vec{r}_t \tag{3}$$

Both observation and motion models are used to estimate the state vector over time. As commented in the introduction section, different algorithms can be used in order to achieve this functionality. Our contribution in this point is to use a single PF to obtain a multi-modal distribution $p(\vec{x}_t|\vec{y}_{1:t})$ that describes the estimated stochastic position of every object being tracked at each sample time t.

4 The Stereovision Classifier

Most of tracking systems developed in last years for autonomous navigation and surveillance applications are based on visual information; this is due to the diverse and vast amount of information included in a visual view of the environment.

Developing obstacle tracking tasks for robot's navigation requires 3D information about the objects position in the robot moving environment.

As shown in Fig. 1, position information obtained with the stereovision system is related both with the environment and the objects to track. Therefore it is needed a classification algorithm in order to organize measurements coming from the vision system in two groups or classes:

- *Objects class.* Formed by points that inform about position of objects. These conform the data set that is input in the multiple objects tracker as the measurement vector $\vec{y}_t$.

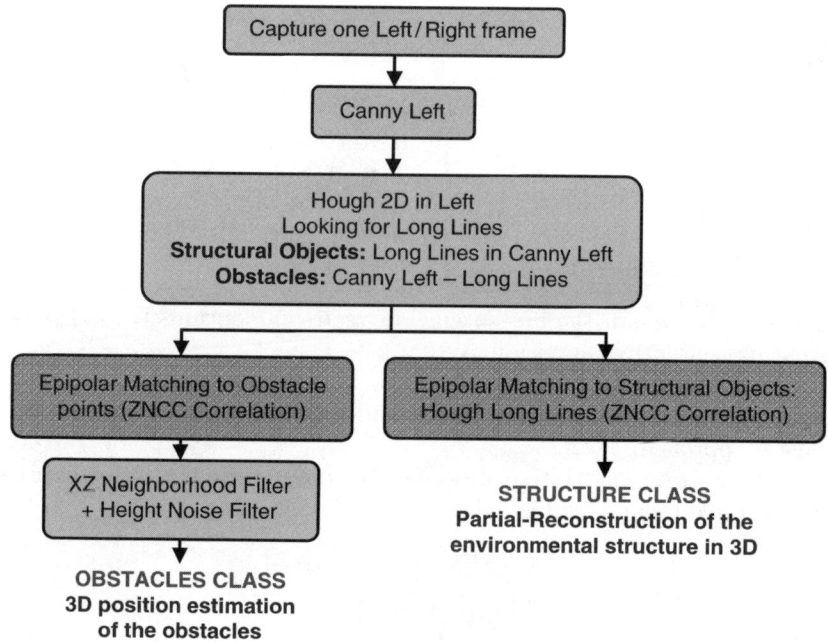

Fig. 2. Block diagram of the stereovision classifier and object detector

- *Structure class.* Formed by points related to elements in environmental structure (such as floor and walls). This data set can be used to implement a partial reconstruction of the environment in which the tracked objects and the robot itself are moving.

Figure 2 shows the proposal to classify the measurements extracted with the stereovision system. The detection and classification process is deeply described by the authors in [21], but a slight revision of its functionality is included in the following paragraphs:

1. The stereovision system proposed is formed by two synchronized black and white digital cameras statically mounted to acquire left and right images.
2. As the amount of information in each image is too big, a canny filter is applied to one of the pair of frames.
3. The classification process is performed to the edge pixels that appear in the canny image. Environmental structures edges have the common characteristic of forming long lines in the canny image. Due to this fact, the Hough transform has been chosen as the best method to define the pixels from the canny image that should be part of the structure class. The rest of points in the canny image are assigned to the objects class.
4. Two correlation processes are used in order to find the matching point of each member in both classes. 3D position information is obtained with a

matching process applied to the pixels of each pair of frames, using the epipolar geometry that relates the relative position of the cameras.

With the described functionality, the classification process here proposed behaves as an obstacle detection module.

Also, this pre-processing algorithm selects wisely the most interesting data points from the big set of measurements that is extracted from the environment. This fact is especially important in order to achieve the real time specification pursuit. In fact, a processing rate of 15–33 fps has been achieved in different tests run with this classifier.

Some results of the classification process described are included in the results section of this chapter.

5 The Estimation Algorithm

A particle filter (PF) is used as a multi-modal tracker to estimate position and speed of objects in the environment, from the measurement array obtained in the classification process.

PF is a particularization of the Bayesian estimator in which the densities related to the posterior estimation (also called belief) is discretized. A detailed description of the PF mathematical base can be found in [2] and in [5].

As the state vector is not discretized, like it is in most of Bayes filter implementations, the PF is more accurate in its estimation than the KF or estimators based on a grid (MonteCarlo estimators). Moreover, due to the same reason, the computational load of this Bayes filter form is lower in this than in other implementations, and thus more adequate to implement real time estimation.

Finally, PFs include an interesting characteristic for multi-tracking applications: the ability of representing multiple estimation hypotheses with a single algorithm, through the multi-modality of the belief. This facility is not available in the optimal implementation of the Bayes estimator, the KF.

For all these reasons, the PF has been thought as the most appropriated algorithm to develop a multi-tracking system.

5.1 The XPFCP

Most of the solutions to the tracking problem, based on a PF, do not use the multi-modal character of the filter in order to implement the multiple objects position estimation task. The main reason of this fact is that the association process needed to allow the multi-modality of the estimator is very expensive in execution time (this is the case of the solutions based on the JPDAF) or lacks of robustness (as it is the case in the solution presented in [22]).

The XPFCP here presented is a multi-modal estimator based on a single PF that can be used with a variable number of models, thanks to a clustering process that is used as association process in the estimation loop. The functionality of the XPFCP is presented in the following paragraphs.

The main loop of a standard Bootstrap PF [12] based on the SIR algorithm [13] starts at time t with a set $S_{t-1} = \left\{ \vec{x}_{t-1}^{(i)}, \tilde{w}_{t-1}^{(i)} \right\}_{i=1}^{n}$ of n random particles representing the posterior distribution of the state vector estimated $p(\vec{x}_{t-1}|\vec{y}_{1:t-1})$ at the previous step. The rest of the process is developed in three steps, as follows:

1. *Prediction step.* The particles are propagated by the motion model $p(\vec{x}_t|\vec{x}_{t-1})$ to obtain a new set $S_{t|t-1} = \left\{ \vec{x}_{t|t-1}^{(i)}, \tilde{w}_{t-1}^{(i)} \right\}_{i=1}^{n}$ that represents the prior distribution of the state vector at time t, $p(\vec{x}_t|\vec{y}_{1:t-1})$.

2. *Correction step.* The weight of each particle $\vec{w}_t = \left\{ w_t^{(i)} \right\}_{i=1}^{n} \equiv w(\vec{x}_{0:t})$ is then obtained comparing the measurements vector $\vec{y}_t$ and its predicted value based on the prior estimation $h(\vec{x}_{t|t-1})$. In the Bootstrap version of the filter, these weights are obtained directly from the likelihood function $p(\vec{y}_t|\vec{x}_t)$, as follows:

$$w(\vec{x}_{0:t}) = w(\vec{x}_{0:t-1}) \cdot \frac{p(\vec{y}_t|\vec{x}_t) \cdot p(\vec{x}_t|\vec{x}_{t-1})}{q(\vec{x}_t|\vec{x}_{0:t-1}, \vec{y}_{1:t})} \tag{4}$$

$$\xrightarrow[q(\vec{x}_t|\vec{x}_{0:t-1}, \vec{y}_{1:t}) \propto p(\vec{x}_t|\vec{x}_{t-1})]{} w(\vec{x}_{0:t}) = w(\vec{x}_{0:t-1}) \cdot p(\vec{y}_t|\vec{x}_t)$$

3. *Selection step.* Using the weights vector $\vec{w}_t = \left\{ w_t^{(i)} \right\}_{i=1}^{n}$, and applying a re-sampling scheme, a new set $S_t = \left\{ \vec{x}_t^{(i)}, \tilde{w}_t^{(i)} \right\}_{i=1}^{n}$ is obtained with the most probable particles, which will represent the new belief $p(\vec{x}_t|\vec{y}_{1:t})$.

The standard PF can be used to robustly estimate the position of any kind of a single object defined through its motion model, but it cannot be directly used to estimate the position of appearing objects because there is not a process to assign particles to the new estimations.

In order to adapt the standard PF to be used to track a variable number of elements, some modifications must be included in the basic algorithm. In [22] an adaptation of the standard PF for the multi-tracking task is proposed. The algorithm described there was nevertheless not finally used because it is not robust enough.

The extension of the PF proposed by the authors in [20] includes a clustering algorithm to improve the behavior of the first extended PF, giving as a result the XPFCP process, shown in Fig. 3.

The clustering algorithm, whose functionality is presented in next section, organizes the vector of measurements in clusters that represent all objects in the scene. These clusters are then wisely used in the multi-modal estimator. Two innovations are included in the standard PF to achieve the multi-modal behavior:

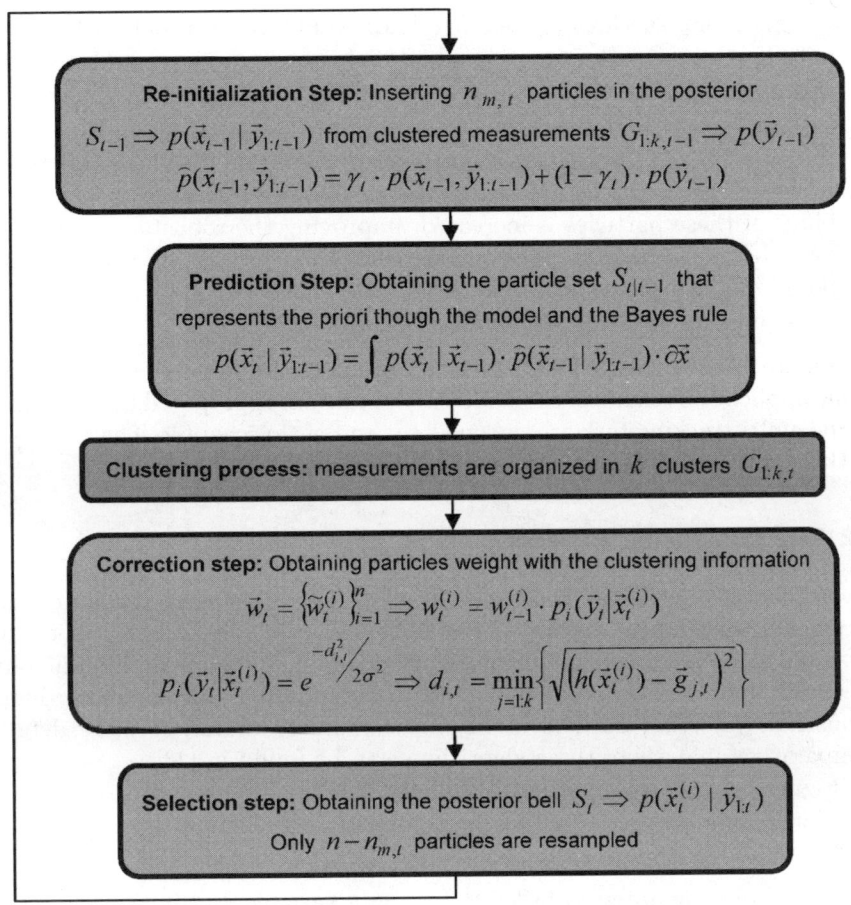

Fig. 3. Description of the XPFCP functionality

- *With a new re-initialization step.* $n_{m,t}$ from the n total number of particles that form the belief $p(\vec{x}_t|\vec{y}_{1:t})$ in the PF are directly inserted from the measurements vector $\vec{y}_t$ in this step previous to the prediction one. With this modification, measurements related to newly appearing objects in the scene have a representation in the priori distribution $p(\vec{x}_{t-1}|\vec{y}_{1:t-1})$. To improve the robustness of the estimator, the inserted particles are not selected randomly from the array of measurements $\vec{y}_{t-1}$ but from the k clusters $G_{1:k,t-1}$. Choosing measurements from every cluster ensures a probable representation of all objects in the scene, and therefore, an increased robustness of the multi-tracker. Thanks to this re-initialization step the belief dynamically adapts itself to represent the position hypothesis of the different objects in the scene.
- *At the Correction step.* This step is also modified from the standard PF. On one hand, only $n - n_{m,t}$ samples of the particle set have to be extracted

in this step, as the $n_{m,t}$ resting ones would be inserted with the re-initialization. On the other hand, the clustering process is also used in this step, because the importance sampling function $p_i(\vec{y}_t|\vec{x}_t^{(i)})$ used to calculate each particle weight $w_t^{(i)}$ is obtained from the similarity between the particle and the k cluster centroides $\vec{g}_{1:k,t}$. Using the cluster centroides to weight the particles related to the newly appeared objects, the probability of these particles is increased, improving the robustness of the new hypotheses estimation. Without the clustering process, the solution proposed in [22] rejects these hypotheses, and thus, the multi-modality of the PF cannot be robustly exploited.

Figure 3 shows the XFPCP functionality, described in previous paragraphs. Some application results of the multi-modal estimator proposed by the authors to the multi-tracking task are shown at the end of this chapter. The robustness of this contribution is demonstrated there.

5.2 The Clustering Process

Two different algorithms have been developed for clustering the set of measurements: an adapted version of the K-Means for a variable number of clusters; and a modified version of the Subtractive fuzzy clustering. Its reliability is similar, but the proposal based on the standard K-Means shows higher robustness rejecting outliers in the measurements vector. A more detailed comparative analysis of these algorithms can be found in [23].

Figure 4 shows the functionality of the proposed version of the K-Means. Two main modifications to the standard functionality can be found in the proposal:

1. It has been adapted in order to handle a variable and initially unknown number k of clusters $G_{1:k}$, by defining a threshold $distM$ in the distance $d_{i,1:k}$ used in the clustering process.
2. A cluster centroides' prediction process is included at the beginning of the algorithm in order to minimize its execution time. Whit this information, the process starts looking for centroides near their predicted values $\vec{g}_{0,1:k,t} = \vec{g}_{1:k,t|t-1}$.

A validation process is also added to the clustering algorithm in order to increase the robustness of the global algorithm to spurious measurements. This process is useful when noisy measurements or outliers produce a cluster creation or deletion erroneously. The validation algorithm functionality is the following:

- *When a new cluster is created,* it is converted into a candidate that will not be used in the XPFCP until it is possible to follow its dynamics.
- *The same procedure is used to erase a cluster* when it is not confirmed with new measurements for a specific number of times.

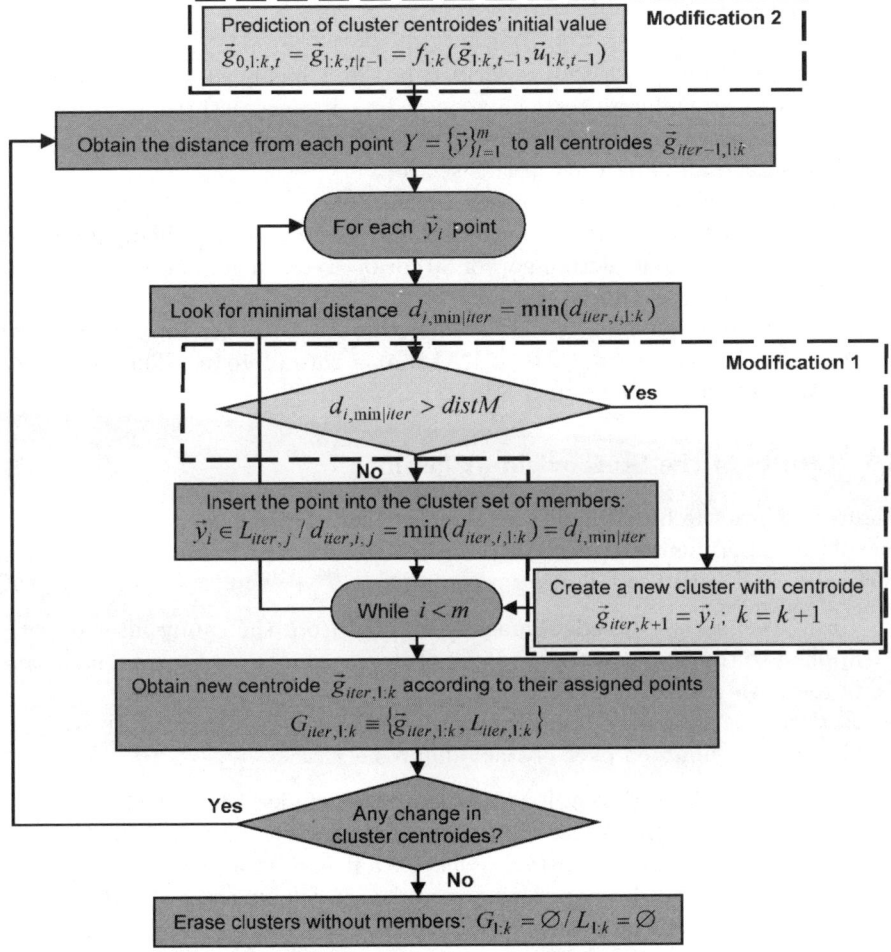

Fig. 4. Description of the K-Means clustering algorithm

The validation process is based in two parameters, which are calculated for each cluster:

- *Distance between the estimated and the resulting clusters centroide.* The centroides estimation process, already commented, is also used in the validation process. The estimated value of the centroides $\vec{g}_{1:k,t|t-1}$ is compared with its final value at the end of the clustering process $\vec{g}_{1:k,t}$, in order to obtain a confidence value for the corresponding cluster validation.
- *Cluster likelihood.* A cluster probability value is calculated as a function of number of members in each cluster $L_{1:k}$.

The effectiveness of the clustering proposal is demonstrated in the following section, with different results.

6 Results

The global tracking algorithm described in Fig. 1 has been implemented in a mobile platform. Different tests have been done in unstructured and unknown indoor environments. Some of the most interesting results extracted from these tests are shown and analyzed in this section.

The stereovision system is formed by two black and white digital cameras synchronized with a Firewire connection and located on the robot in a static mounting arrangement, with a gap of 30 cm between them, and at a height of around 1.5 m from the floor.

The classification and tracking algorithms run in an Intel Dual Core processor at 1.8 GHz with 1 GB of RAM, at a rate of 10 fps. The mean execution time of the application is 80 ms.

6.1 Results of the Stereovision Classifier

Figure 5 shows the functionality of the classifier. Three sequential instants of one of the experiments are described in the figure by a pair of images organized vertically, and with the following meaning:

- *Upper row* shows the edge images obtained from the canny filter directly applied to the acquired frame. Both obstacles and environmental structure borders are mixed in those images.
- *Bottom row* shows the final frames in which points assigned to the objects class are highlighted over obstacle figures.

From the results shown in Fig. 5, it can be concluded that the classification objective has been achieved. Only points related to the obstacles in the scene have been classified in the obstacles class. As it can be noticed, the analyzed experiment has been developed in a complex and unstructured indoor environment, where five static and dynamic objects are present and cross their

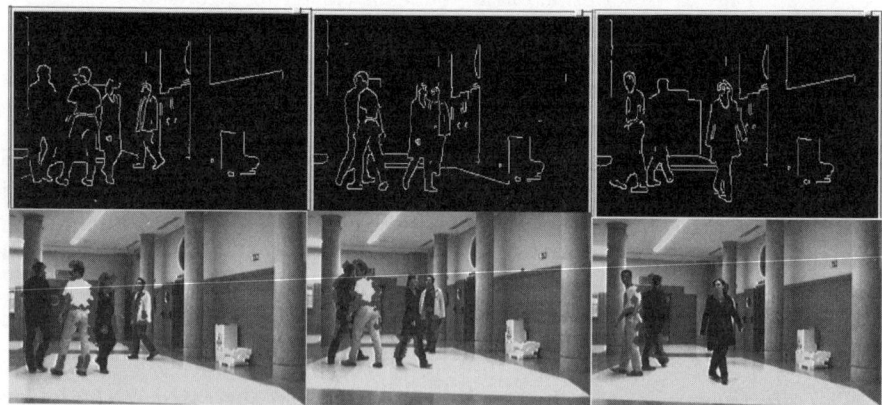

Fig. 5. Results of the classification algorithm in a real situation

paths generating partial and global occlusions. In any case, the proposed classification algorithm is able to extract 3D position points from every object in the scene.

The set of 3D position points assigned to the objects class can now be used in the multi-tracking task.

Nevertheless, the number of objects present in each final frame in Fig. 5 cannot be easily extracted from the highlighted set of points.

Furthermore, it can be noticed that the set of points are not equally distributed among all objects in the environment, and hence, the tracking algorithm should be able to manage object hypotheses with very different likelihood.

6.2 Results of the Estimation Algorithm

Figure 6 displays the functionality of the XPFCP in one of the tested situations. Three sequential instants of the estimation process are represented by a pair of images.

- *Upper row* displays the initial frames with highlighted dots representing the measurement vector contents obtained from the classification process, and rectangles representing the K-Means output.
- *Lower row* shows the same frame with highlighted dots representing each of the obstacle position hypotheses that the set of particles define at the XPFCP output. This final set of particles has also been clustered using the same K-Means proposal in order to obtain a deterministic output for the multi-tracker. Rectangles in this lower frame represent the clustered particles.

Fig. 6. Results of the multi-tracking algorithm XPFCP in a real situation

Table 2. Rate of different types of errors obtained with the XPFCP at the output of the K-Means, and at the end of the multi-tracking task in a 1,054 frames experiment complex situations with five and six objects

	K-Means (% frames with error)	XPFCP (% frames with error)
Missing	10.9	2.9
Duplicates	6.1	0
2 as 1	3.9	0
Total	20.9	2.9

Comparing the upper and lower image in Fig. 6, it can be noticed that the tracker based on the XPFCP can solve tracking errors such as object duplications generated in the input clustering process. An example of an object duplication error generated by the K-Means and successfully solved by the XPFCP can be seen in the third vertical pair of images (on the right side, in the last sequential instant) of Fig. 6.

Table 2 shows a comparison between the errors at the output of the clustering process and at the end of the global XPFCP estimator. In order to obtain these results an experiment of 1,054 frames of complex situations similar to the ones presented Figs. 5 and 6 has been run. The results displayed in Table 2 demonstrate the reliability and robustness of the tracker facing up to occlusions and other errors.

Figure 7 displays the tracking results extracted from the XPFCP output in another real time experiment. In this case 9 sequential instants of the experiment are shown, and each image represents one of them, from (a) to (i). The meaning of every frame is the same as in the lower row in Fig. 6.

The results displayed in Fig. 7 show that the tracker estimates correctly each obstacle position in the dynamic and unstructured indoor environment.

7 Conclusions

In this chapter the authors describe the functionality of a global tracking system based on vision sensors to be used by the navigation or obstacle avoidance module in an autonomous robot.

In order to achieve this objective, a specific classification algorithm for stereovision data has been developed. This process is used to separate visual position information related with obstacles from the one related with the environment.

An algorithm, called XPFCP, is used to estimate obstacles' movement and position in an unstructured environment. It has been designed as the kernel of the multi-tracking process. The XPFCP is based on a probabilistic multimodal filter, a PF, and is completed with a clustering process based on a standard K-Means.

Fig. 7. Sequential images of a real time experiment with stereovision data

Results of the different processes involved in the global tracking system have been presented, demonstrating the successful behaviour of the different contributions. The main conclusions of these proposals are:

- The proposed tracking system has shown high reliability in complex situations where a variable number of static and dynamic obstacles are constantly crossing, and no preliminary knowledge of the environment is available.
- It has been demonstrated that the estimation of a variable number of systems can be achieved with a single algorithm, the XPFCP, and without imposing model restrictions.
- The use of a clustering process as association algorithm makes possible a robust multi-modal estimation with a single PF, and without the computational complexity of some other association proposals such as the PDAF.
- Thanks to the simplicity of its functional components (a PF and a modified K-Means) the XPFCP accomplishes the real time specification pursuit.
- Though vision sensors are used in the tracking process presented in the chapter, some other e XPFCP designed can easily handle data coming up from different kinds of sensors. This fact makes the tracker proposed more flexible, modular, and thus, easy to use in different robotic applications than other solutions proposed in the related literature.

Acknowledgments

This work has been financed by the Spanish administration (CICYT: DPI2005-07980-C03-02).

References

1. D.B. Reid, An algorithm for tracking multiple targets, IEEE Transactions on Automatic Control, vol. 24, no 6, pp. 843–854, December 1979
2. M.S. Arulampalam, S. Maskell, N. Gordon, T. Clapp, A tutorial on particle filters for online nonlinear non-gaussian bayesian tracking, IEEE Transactions on Signal Processing, vol. 50, no 2, pp. 174–188, February 2002
3. N.J. Gordon, D.J Salmond, A.F.M. Smith, Novel approach to nonlinear/non-gaussian bayesian state estimation, IEE Proceedings Part F, vol. 140, no 2, pp. 107–113, April 1993
4. A. Doucet, J.F.G. de Freitas, N.J. Gordon, Sequential montecarlo methods in practice. Springer, New York, ISBN: 0-387-95146-6, 2000
5. R. Van der Merwe, A. Doucet, N. de Freitas, E. Wan, The unscented particle filter, Advances in Neural Information Processing Systems, NIPS13, November 2001
6. S. Thrun, Probabilistic algorithms in robotics, Artificial Intelligence Magazine, Winter 2000
7. D. Fox, W. Burgard, F. Dellaert, S. Thrun, Montecarlo localization. Efficient position estimation for mobile robots, Proceedings of the Sixteenth National Conference on Artificial Intelligence (AAAI99), pp. 343–349, Orlando, July 1999
8. M. Isard, A. Blake, Condensation: Conditional density propagation for visual tracking, International Journal of Computer Vision, vol. 29, no 1, pp. 5–28, 1998
9. K. Okuma, A. Taleghani, N. De Freitas, J.J. Little, D.G. Lowe, A boosted particle filter: multi-target detection and tracking, Proceedings of the Eighth European Conference on Computer Vision (ECCV04), Lecture Notes in Computer Science, ISBN: 3-540-21984-6, vol. 3021, Part I, pp. 28–39 Prague, May 2004
10. T. Schmitt, M. Beetz, R. Hanek, S. Buck, Watch their moves applying probabilistic multiple object tracking to autonomous robot soccer, Proceedings of the Eighteenth National Conference on Artificial Intelligence (AAAI02), ISBN: 0-262-51129-0, pp. 599–604, Edmonton, July 2002
11. K.C. Fuerstenberg, K.C.J. Dietmayer, V. Willhoeft, Pedestrian recognition in urban traffic using a vehicle based multilayer laserscanner, Proceedings of the IEEE Intelligent Vehicles Symposium (IV02), vol. 4, no 80, Versailles, June 2002
12. J. MacCormick, A, Blake, A probabilistic exclusion principle for tracking multiple objects, Proceedings of the Seventh IEEE International Conference on Computer Vision (ICCV99), vol. 1, pp. 572–578, Corfu, September 1999
13. H. Tao, H.S. Sawhney, R. Kumar, A sampling algorithm for tracking multiple objects, Proceedings of the International Workshop on Vision Algorithms at (ICCV99), Lecture Notes in Computer Science, ISBN: 3-540-67973-1, vol. 1883, pp. 53–68, Corfu, September 1999

14. J. Vermaak, A. Doucet, P. Perez, Maintaining multimodality through mixture tracking, Proceedings of the Ninth IEEE International Conference on Computer Vision (ICCV03), vol. 2, pp. 1110–1116, Nice, June 2003
15. P. Pérez, C. Hue, J. Vermaak, M. Gangnet, Color-based probabilistic tracking, Proceedings of the Seventh European Conference on Computer Vision (ECCV02), Lecture Notes in Computer Science, ISBN: 3-540-43745-2, vol. 2350, Part I, pp. 661–675, Copenhagen, May 2002
16. Y. Bar-Shalom, T. Fortmann, Tracking and data association (Mathematics in Science and Engineering, V.182), Academic Press, New York, ISBN: 0120797607, January 1988
17. C. Rasmussen, G.D. Hager, Probabilistic data association methods for tracking complex visual objects, IEEE Transactions on Pattern Analysis and Machine Intelligence, vol. 23, no 6, pp. 560–576, June 2001
18. D. Schulz, W. Burgard, D. Fox, A.B. Cremers, People tracking with mobile robots using sample-based joint probabilistic data association filters, International Journal of Robotics Research, vol. 22, no 2, pp. 99–116, February 2003
19. M. Marron, M.A. Sotelo, J.C. García, Design and applications of an extended particle filter with a pre-clustering process -XPFCP-, Proceedings of the International IEEE Conference Mechatronics and Robotics 2004 (MECHROB04), ISBN: 3-938153-50-X, vol. 2/4, pp. 187–191, Aachen, September 2004
20. M. Marrón, J.C. García, M.A. Sotelo, D. Fernandez, D. Pizarro. XPFCP: An extended particle filter for tracking multiple and dynamic objects in complex environments, Proceedings of the IEEE International Symposium on Industrial Electronics 2005 (ISIE05), ISBN: 0-7803-8738-4, vol. I–IV, pp. 1587–1593, Dubrovnik, June 2005
21. M. Marrón, M.A. Sotelo, J.C. García, D. Fernández, I. Parra. 3D-visual detection of multiple objects and environmental structure in complex and dynamic indoor environments, Proceedings of the Thirty Second Annual Conference of the IEEE Industrial Electronics Society (IECON06), ISBN: 1-4244-0136-4, pp. 3373–3378, Paris, November 2006
22. E.B. Koller-Meier, F. Ade, Tracking multiple objects using a condensation algorithm, Journal of Robotics and Autonomous Systems, vol. 34, pp. 93–105, February 2001
23. M. Marrón, M.A. Sotelo, J.C. García, J. Brodfelt. Comparing improved versions of 'K-Means' and 'Subtractive' clustering in a tracking applications, Proceedings of the Eleventh International Workshop on Computer Aided Systems Theory, Extended Abstracts (EUROCAST07), ISBN: 978-84-690-3603-7, pp. 252–255, Las Palmas de Gran Canaria, February 2007

Printing: Krips bv, Meppel, The Netherlands
Binding: Stürtz, Würzburg, Germany